国家出版基金项目
NATIONAL PUBLICATION FOUNDATION

中国长翅目昆虫
原色图鉴

（中英文对照）

A Color Atlas of the Chinese Mecoptera

(Chinese and English)

王吉申　花保祯 / 著
Jishen Wang & Baozhen Hua

河南科学技术出版社
·郑州·

内容提要

本图鉴是作者在多项国家自然科学基金（30070101，30370179，30670255，30970386，31172125，31301898，31672341）研究工作的基础上，对中国长翅目昆虫研究进展的系统总结。共收录产自中国的长翅目昆虫 3 科 10 属 170 种；系统编制了本图鉴收录的中国长翅目昆虫分科、分属和分种检索表；概述部分介绍了长翅目昆虫的分类地位、研究简史、分类系统、地理分布、形态特征、生物学和生态学特性等；每种昆虫介绍了中名、拉丁学名、测量数据、形态特征、发生期、地理分布等，并提供有成虫生态照片或标本照片，雄性外生殖器图片以及地理分布图。本图鉴是世界上第一部长翅目昆虫分类研究的重要著作，全图鉴中英文对照，可作为昆虫学工作者及爱好者的重要参考资料。

图书在版编目（CIP）数据

中国长翅目昆虫原色图鉴 / 王吉申，花保祯著 . — 郑州：河南科学技术出版社，2018.12

ISBN 978-7-5349-9376-3

Ⅰ . ①中… Ⅱ . ①王… ②花… Ⅲ . ①长翅目 – 昆虫 – 中国 – 图集 Ⅳ . ① Q969.39-64

中国版本图书馆 CIP 数据核字 (2018) 第 221089 号

出版发行：河南科学技术出版社
地址：郑州市经五路 66 号　邮编：450002
电话：（0371）65737028　65788613
网址：www.hnstp.cn
总策划人：周本庆
策划编辑：陈淑芹
责任编辑：李义坤　曲　先
责任校对：司丽艳
整体设计：张　伟
责任印制：张艳芳
地图审图号：GS（2018）5192 号
地图编制：湖南地图出版社
印　　刷：河南瑞之光印刷股份有限公司
经　　销：全国新华书店
幅面尺寸：889 mm × 1194 mm　1/16　印张：23　字数：762 千字
版　　次：2018 年 12 月第 1 版　2018 年 12 月第 1 次印刷
定　　价：498.00 元

序言

谈到昆虫，我们最先想到的往往是蝴蝶、蜻蜓、蜜蜂、蚊蝇、蟑螂等耳熟能详的类群。这些常见的昆虫与人类日常生活最为密切，在影视文化作品中上镜率也最高。但少有人知，在已发现的100余万种现生昆虫中，有一类奇特的成员，它们的成虫通常具有延长的喙；一些种类由于雄性的生殖节膨大并向上翘起，形如蝎尾，因此俗称“蝎蛉 scorpionflies”；其外部特征和生物学特性变化多端；它们在昆虫纲 Insecta 的演化及系统发育关系研究中具有十分独特的地位。这类昆虫被昆虫学家归为长翅目 Mecoptera。由于缺少专门的资料，对广大昆虫学研究者和爱好者来说，长翅目 Mecoptera 昆虫的鉴定一直以来都较为困难。

现生的长翅目 Mecoptera 是昆虫纲 Insecta 中的一个孑遗小目，全世界已命名的现生种仅超过700种，归属于9个科近40个属。相较之下，已报道的化石类群却约有620种，归属于39个科、210个属；大量新分类单元（科、属、种）还在不断被发现，其中3/4以上的科在现今已绝灭。长翅目 Mecoptera 昆虫有着漫长的起源和演化历史，它们的化石记录最早出现于早二叠纪，距今约2.7亿年。中生代（距今2.52亿至6 500万年）是长翅目 Mecoptera 昆虫承前启后、发展演化的重要时期。长翅目 Mecoptera 昆虫在这一时期达到了其多样性的顶峰，在中生代全变态昆虫化石中占据着很大比例。现生长翅目 Mecoptera 的9个科在侏罗–白垩纪均已出现。许多灭绝的科（如支脉蚊蝎蛉科 Cimbrophlebiidae、直脉蝎蛉科 Orthophlebiidae 和蝶蝎蛉科 Choristopsychidae）不仅丰富了人们对长翅目 Mecoptera 多样性的认识，而且为研究长翅目 Mecoptera 的起源与演化提供了宝贵的实证资料。

长翅目 Mecoptera 昆虫的口器、翅、外生殖器等在不同的类群中差异很大，对应着它们丰富多样的生物学特性。将古老绝灭的长翅目 Mecoptera 昆虫与现生类群对比，其生物学特性变化之大，往往超出我们的想象！例如阿纽蝎蛉科 Aneuretopsychidae、中生蝎蛉科 Mesopsychidae 以及长喙蝎蛉科 Pseudopolycentropodidae 昆虫具有特殊的细长虹吸式口器，可以取食裸子植物的传粉滴，是古老的传粉者，与当时的植物之间存在着复杂的协同演化关系；支脉蚊蝎蛉科 Cimbrophlebiidae 的银杏侏罗支脉蚊蝎蛉 *Juracimbrophlebia ginkgofolia* 与银杏植物之间存在着复杂的拟态和互利共生关系，这种关系在地质历史上共同走过了1亿多年；缺脉蝎蛉科 Holcorpidae 一些种类的雄性腹部和生殖器十分夸张地延长，这种特殊结构不仅被用于雄性之间的武力争斗，更是一种对雌性的性展示。长翅目 Mecoptera 昆虫还有大量未知的形态结构、生物学特性以及演化规律在等待着人们发现和研究。

由西北农林科技大学花保祯教授领导的长翅目 Mecoptera 科研团队，长期以来致力于长翅目 Mecoptera 昆虫的形态学、行为学、分类学和系统演化研究，多年来取得了一系列令人瞩目的成果，并在国内外经典学术刊物上发表了一批高水平论文，使得西北农林科技大学成为了国际上现生长翅目 Mecoptera 昆虫的研究中心之一。尤为难得的是，花保祯教授在科学研究的同时，还从事大量的教学工作，培养了一大批优秀的年轻昆虫学者。他开设的科技论文写作、普通昆虫学、昆虫形态学、昆虫分类学等课程深受学生的喜爱。他还因独特的教学理念和严谨的工作态度，获得了“陕西省教学名师”的称号。

《中国长翅目昆虫原色图鉴》是一本难得的昆虫分类图鉴，它用直观精美的画面和科学准确的语言，将长翅目 Mecoptera 各方面的知识呈现出来。这也是一本融科学性、艺术性、趣味性于一体的精彩原创著作，适合广大昆虫学者及爱好者阅读，也可作为第一线农林工作者的实用参考资料。我很荣幸地将这本图鉴推荐给广大读者。同时，我也期待着花保祯教授所领导的长翅目 Mecoptera 科研团队更多优秀著作的诞生。

首都师范大学生命科学学院教授
国际古昆虫学会副主席 任东

2018 年 9 月

Foreword

Speaking of insects, some common groups immediately appear in our minds: butterflies, dragonflies, bees, flies, and cockroaches. These insects are closely related to the daily life of humans, and frequently occur in the screen and other cultural works. Few people know, however, that among the one million species of extant insects, there are a group of peculiar members with elongated rostrum in their adults; some of them have enlarged and recurved male genitalia resembling the tail of scorpions, hence the common name "scorpionflies"; their external morphology and biological characteristics are highly varied; and they play a significant role in the study of the evolution and phylogeny of Insecta. This group of insects have been assigned in the order Mecoptera by entomologists. Because of the lack of specific literatures, the identification of Mecoptera is relatively difficult in a long period of time for the general entomological researchers and hobbyists.

The extant Mecoptera, a small relic order in the class Insecta, consist of over 700 species in nine families and ca. 40 genera. By contrast, the fossil Mecoptera include ca. 620 species in 39 families and 210 genera; a great number of new taxa (families, genera and species) have unceasingly been found. Among them, more than three-fourths of families were extinct. Mecopterans have a long history of origin and evolution, with the earliest fossil records in the Early Permian (ca.270 mya). The Mesozoic era (252–65 mya) was an important period which links the past with the present days for the evolving Mecoptera. The mecopterans reached their peak of diversity in this era, and held a large proportion in the Mesozoic fossil of Holometabola. Nine extant families in Mecoptera have been found in the Jurassic-Cretaceous period. Many extinct families (*e.g.*, Cimbrophlebiidae, Orthophlebiidae, and Choristopsychidae) not only enrich our knowledge of the diversity of Mecoptera, but also provide precious solid evidence for the study of the origin and evolution of this order.

The Mecoptera have their mouthparts, wings and genitalia highly varied among different groups, corresponding to their diverse biological characteristics. Frequently, by comparing ancient, extinct mecopterans with extant ones, we are so amazed at their highly varied biological characteristics. For instance, members of the families Aneuretopsychidae, Mesopsychidae and Pseudopolycentropodidae were ancient pollinators sharing a complex co-evolution with plants, and their special slender siphoning mouthparts was able to feed the pollination drops of the gymnosperms; *Juracimbrophlebia ginkgofolia*, belonging to the family Cimbrophlebiidae, have a complex mimicry and symbiotic relationship with the plant *Ginkgo*, the relationship between them lasted over a hundred billion years; some male Holcorpidae have exaggeratedly elongated abdomen and genitalia, special structures not only utilized to forcedly fight with other males, but also an sexual display to the females. There are enormous number of unknown structures, biological characteristics and evolutionary laws await our further discovery and research.

For the past many years, the research team of Mecoptera, Northwest A&F University, leading by professor Baozhen Hua, dedicated to the study of morphology, behaviors, taxonomy and phylogeny of Mecoptera, and made a lot of remarkable achievements. They published some excellent works in many Chinese and abroad journals, making Northwest A&F University one of the global research center of Mecoptera. Meanwhile, professor Bao-Zhen Hua also dedicated to teaching works and trained a group of young entomologists. His college courses are popular among students, such as scientific writing, general entomology, insect morphology, and insect taxonomy are popular among students. A title of "Prominent Educator of Shaanxi" was granted for his unique teaching idea and rigorous working attitude.

A Color Atlas of the Chinese Mecoptera is a valuable and unmatched atlas, in which all the aspects of knowledge of Mecoptera are shown by distinct elegant pictures and accurate scientific texts. It is a great original work that gathered scientific, artistic, and enjoyable qualities, suitable for the general entomological researchers and hobbyists, and also a practical reference for the forefront workers in agriculture and forestry. I am honored to recommend this atlas to the readers. Meanwhile, I also expect more excellent works from the research team of Mecoptera leading by professor Baozhen Hua.

Professor of the College of Life Sciences, Capital University
Vice-President of the International Palaeoentomological Society
Dong Ren

September 2018

前言

长翅目 Mecoptera 是全变态类昆虫中的一个小目，全世界已知 700 多种，包括蝎蛉、蚊蝎蛉等。长翅目昆虫对生态环境要求严格，是重要的生态指示昆虫。它们常常表现出复杂的求偶和交配行为，是研究昆虫行为学的绝佳材料。长翅目昆虫是全变态类昆虫中较古老的成员，在昆虫纲的系统演化研究中占据重要地位。

中国是世界上长翅目昆虫最丰富的国家，已知有分布的长翅目昆虫包括 3 科 10 属 300 多种。然而，目前国内长翅目昆虫鉴定方面的资料十分匮乏。依托西北农林科技大学昆虫博物馆丰富的标本收藏和近几年来野外考察的积累，我们历时 3 年编写了《中国长翅目昆虫原色图鉴》一书。

本图鉴收录了 170 种产自中国的长翅目昆虫。仅凭生态照片，多数长翅目昆虫难以鉴定到种，因此我们也选用了大量针插或酒精浸泡标本的照片。书中的大多数物种，我们都提供了成虫和雄性外生殖器特征图。另外，为了方便了解分布情况，我们还为每个种配上了地理分布图。

本图鉴中用到的标本，主要保存于西北农林科技大学（NWAFU）昆虫博物馆，也有一些借阅或交换自澳大利亚国家昆虫标本馆（ANIC）、中国农业大学（CAU）、中国科学院动物研究所（IZAS）、南开大学（NKU）、华南农业大学 (SCAU)、中山大学（SYSU）、天津自然博物馆（TJNHM）以及堪萨斯大学自然历史博物馆（KUNHM）等单位。除注明拍摄者的外，所有图片皆为第一作者拍摄。

本图鉴的编写工作，离开了许多朋友和同事的无私帮助是难以完成的。对长翅目昆虫物种的鉴定，首先得益于西北农林科技大学已故周尧教授几十年来对中国长翅目昆虫标本、资料的积累以及开创性的研究；也与西北农林科技大学长翅目昆虫研究团队的研究生们，尤其是黄蓬英、谭江丽、陈静、蔡立君、马娜、钟问、姜碌、高琼华、王萌、胡桂林、苗颖、谢莎、彭韬、侯小燕、刘书宇、颜刚、徐颢、李雪、付强、张俊霞、高超、杜薇、高凯等的工作分不开。感谢董志巍、高凯文、何力、黄贵强、黄思遥、姜日新、蒋世翔、蒋卓衡、李泽建、林业杰、吕林、邱见玥、邱鹭、王洪建、王勇、吴超、许浩、杨峤志、余锟、詹程辉、张杰、赵明智、郑昱辰、周超、周德尧、周文一（中国台湾）等对珍贵标本的采集，使它们能够呈现在读者面前。感谢首都师范大学任东教授、华南农业大学王敏教授、中南林业科技大学魏美才教授捐赠标本。感谢中国科学院动物研究所陈军和刘虹为我们安排检视和借阅标本。作者历年的野外考察，得到了胡彦卿、薛爽、花远、Flinks A. Bakric（印度尼西亚）和 Sepni Juhansah（印度尼西亚）等的大力支持。感谢 Dalton S. Amorim

（巴西）、Cristiano Lopes-Andrade（巴西）、铃木智也（日本）、Jason G. H. Londt（南非）热情地提供文献资料。感谢柴义君、姜碌、雷波、林义祥（中国台湾）、徐晗、张巍巍、钟冋提供摄影作品。还要特别感谢河南科学技术出版社周本庆编审，在他的鼓励和督促下，我们才开始了本图鉴的编写；同时得益于他优秀的编辑工作，本图鉴才得以面世。我们的工作得到了国家自然科学基金项目（30070101，30370179，30670255，30970386，31172125，31301898，31672341）和2018年度国家出版基金资助项目的支持。

由于我们水平有限，书中遗漏和不妥之处，敬请专家学者和广大读者批评指正。

王吉申　花保祯

2018年9月于杨凌

Preface

Mecoptera is an small order of holometabolous insects, and consists of over 700 known species worldwide, including scorpionflies and hangingflies, *etc*. They are important indicators of ecological environments, and are often ideal materials for behavioral studies due to their interesting courtship and mating behaviors. They are ancient members in Holometabola, and play a significant role in the evolutionary research of Insecta.

China is rich in Mecoptera all over the world. Almost 300 species in 10 genera of three families have been recorded in China hitherto. However, literature regarding the identification of the Chinese Mecoptera is very scarce. Hence, we spent three years to compile this color atlas based on the abundant collections from the Entomological Museum, Northwest A&F University, and our intensive field investigations in recent years, to introduce the common species of the Chinese Mecoptera.

In this atlas, we list and illustrate 170 species of the Chinese Mecoptera concisely. It is often difficult to identify to the species-level based exclusively on live photographs without careful examinations under a microscope. Therefore we also provide lots of pictures of pinned or alcohol-preserved specimens. For most species, we present pictures of the adults and the male genitalia. A geographical distribution map is also provided for each species.

The examined specimens are mainly deposited in the Entomological Museum, Northwest A&F University (NWAFU). Others were loaned or interchanged with the Australian National Insect Collections (ANIC), the Entomological Museum, China Agricultural University (CAU), the Institute of Zoology, Chinese Academy of Sciences (IZAS), Nankai University (NKU), South China Agricultural University (SCAU), Sun Yat Sen University (SYSU), Tianjin Natural History Museum (TJNHM) and the Natural History Museum, University of Kansas (KUNHM). All the photographs in this atlas were taken by the first author except those specified.

This work cannot be finished without the generous helps from numerous friends and colleagues during our writing and editing courses. For the identifications of species, we mostly benefit from the late Professor Io Chou, who had collected specimens and literature, and initiated the studies of the Chinese fauna of Mecoptera for decades. We also profited from the postgraduate students of the research team of Mecoptera, Northwest A&F University, including Pengying Huang, Jiangli Tan, Jing Chen, Lijun Cai, Na Ma, Wen

Zhong, Lu Jiang, Qionghua Gao, Meng Wang, Guilin Hu, Ying Miao, Sha Xie, Tao Peng, Xiaoyan Hou, Shuyu Liu, Gang Yan, Hao Xu, Xue Li, Qiang Fu, Junxia Zhang, Chao Gao, Wei Du and Kai Gao. Our sincere thanks also go to Zhiwei Dong, Kaiwen Gao, Li He, Guiqiang Huang, Siyao Huang, Rixin Jiang, Shixiang Jiang, Zhuoheng Jiang, Zejian Li, Yejie Lin, Lin Lyu, Jianyue Qiu, Lu Qiu, Hongjian Wang, Yong Wang, Chao Wu, Hao Xu, Qiaozhi Yang, Kun Yu, Chenghui Zhan, Jie Zhang, Mingzhi Zhao, Yuchen Zheng, Chao Zhou, Deyao Zhou, and Wen-i Chou (Taiwan). They collected many valuable specimens and made it feasible to show those species to readers.We thank Professor Dong Ren from the Capital Normal University, Professor Min Wang from the South China Agricultural University, and Professor Meicai Wei from the Central South University of Forestry and Technology for donating specimens. We also appreciate Jun Chen and Hong Liu of the Institute of Zoology, Chinese Academy of Sciences (Beijing) for arranging the access to examine and loan of the specimens. Our recent field works were greatly supported by Yanqing Hu, Shuang Xue, Yuan Hua, Flinks A. Bakrie (Indonesia) and Sepni Juhansah (Indonesia). We would also express our thanks to Dalton S. Amorim (Brasil), Cristiano Lopes-Andrade (Brasil), Tomoya Suzuki (Japan), and Jason G.H. Londt (South Africa) for fervidly sending precious literatures. We also thank Yijun Chai, Lu Jiang, Bo Lei, Yechan Lin (Taiwan), Han Xu, Weiwei Zhang and Wen Zhong for providing photographs. A special thank is due to Mr. Benqing Zhou from the Henan Science and Technology Press. Under his encouragement and agitation, this atlas began to compile. The publication of the current form is also due to his excellent editorial work. Our works were financially supported by the National Natural Science Foundation of China (30070101, 30370179, 30670255, 30970386, 31172125, 31301898, 31672341) and the 2018 National Publication Foundation.

Due to our limited knowledge, there might be some mistakes and flaws in this atlas. We expect criticism and correction from experts and readers.

Jishen Wang & Baozhen Hua

at Yangling

September 2018

目录
Contents

第一部分　概述
Part Ⅰ Introduction

第二部分　蚊蝎蛉科
Part Ⅱ　Bittacidae

第三部分 蝎蛉科

Part Ⅲ Panorpidae

第四部分　拟蝎蛉科

Part Ⅳ Panorpodidae

第一部分 概述

Part Ⅰ Introduction

图 1-a　新蝎蛉 ***Neopanorpa* sp.** 雌成虫侧面观

Fig. 1-a　Female adult of ***Neopanorpa* sp.**, lateral view

一、分类地位 Systematic Status

长翅目 Mecoptera 一词来源于古希腊语 μέκο（= meco，长）和 πτερόν（= pteron，翅），指的是这一类昆虫的翅通常比较狭长。

长翅目 Mecoptera 是全变态类昆虫中的一个小目，全世界目前已超过 700 种（Bicha, 2018），包括 9 个现生科（Penny & Byers, 1979），是全变态类昆虫中最古老的类群之一（Kristensen, 1981; Dunford & Somma, 2008）。

大量化石记录表明，长翅目 Mecoptera 昆虫起源于早二叠纪（Lower Permian，距今约 2.7 亿年），并且在中生代（Mesozoic，距今 2.5 亿年至 6 500 万年）达到繁盛的顶峰，占据了各种各样的生态位（Grimaldi & Engel, 2005）。有记录的化石物种将近 400 种，分属于 34 科 87 属（Dunford & Somma, 2008; Byers, 2009）（图 1-1, 图 1-2）。在现今的传粉昆虫（如双翅目 Diptera 的蝇类、膜翅目 Hymenoptera 的蜂类）出现之前，一些具有长喙的长翅目 Mecoptera 昆虫很可能是最早的传粉者，与当时的裸子植物有着密切的协同演化关系（Ren *et al*., 2009）。已灭绝的支脉蚊蝎蛉科 Cimbrophlebiidae 昆虫也有拟态银杏叶片的例子（Wang *et al*., 2012）。一些长翅目 Mecoptera 昆虫的后翅甚至退化，与双翅目 Diptera 昆虫的平衡棒极其相似（Grimaldi *et al*., 2005）。然而，随着气候变化和几次大灭绝事件的发生（Erwin, 1990; Benton, 1995），长翅目 Mecoptera 昆虫大多未能幸免，因而成为了一个孑遗类群。

图 1-1　小蝎蛉科 Nannochoristidae 昆虫化石（任东赠）

Fig. 1-1　Fossil Nannochoristidae (donated by Dong Ren)

1

2

图 1-2　琥珀中的古长翅目 Mecoptera 昆虫（张巍巍摄）

Fig. 1-2　Ancient mecopterans in amber (photo by Weiwei Zhang)

1. 双翅长喙蝎蛉 *Parapolycentropus* sp.　2. 缅甸美蝎蛉 *Burmomerope eureka*

长翅目 Mecoptera 昆虫在昆虫纲的演化研究中占据着重要的地位。双翅目 Diptera 昆虫、鳞翅目 Lepidoptera 昆虫和毛翅目 Trichoptera 昆虫有可能起源于与长翅目 Mecoptera 昆虫十分近缘的祖先(Grimaldi & Engel, 2005)。长翅目 Mecoptera 昆虫、蚤目 Siphonaptera 昆虫和双翅目 Diptera 昆虫雄性生殖系统中存在精子泵构造，因而被归入具泵总目 Antliophora 之中（Hennig, 1969）。具泵总目 Antliophora 的单系性也得到了分子数据的支持（Whiting *et al.*, 1997; Misof *et al.*, 2014）。

长翅目 Mecoptera 昆虫在全变态类昆虫中的地位长期以来一直存在争议（Dunford & Somma, 2008; Friedrich *et al.*, 2013）。Hennig（1969）基于形态特征，认为长翅目 Mecoptera 与双翅目 Diptera 互为姊妹群，但 Kristensen（1999）则认为长翅目 Mecoptera 更接近于蚤目 Siphonaptera。随后，一项基于多基因联合数据的系统发育研究（Whiting, 2002）指出，蚤目 Siphonaptera 昆虫与长翅目 Mecoptera 昆虫具有较近的亲缘关系，是长翅目 Mecoptera 雪蝎蛉科 Boreidae 的姊妹群。蚤目 Siphonaptera 与长翅目 Mecoptera 的雪蝎蛉具有两个共有衍征：一是体内都具有节肢动物弹性蛋白（resilin）；二是卵巢管为无滋式。因而长翅目 Mecoptera 昆虫被认为是并系群（paraphyletic group）。然而，这一观点受到了许多质疑。Dallai *et al.*（2003）对精子超微结构的研究，发现了雪蝎蛉科 Boreidae 与蚤目 Siphonaptera 的明显差异，认为长翅目 Mecoptera 昆虫是单系群。Misof *et al.*（2014）基于 1 478 个蛋白质编码基因和氨基酸序列的昆虫纲系统发育研究，也支持长翅目 Mecoptera 的单系性，具泵总目 Antliophora 内部的系统发育关系应为：双翅目 Diptera +(长翅目 Mecoptera + 蚤目 Siphonaptera)（图 1-3）。

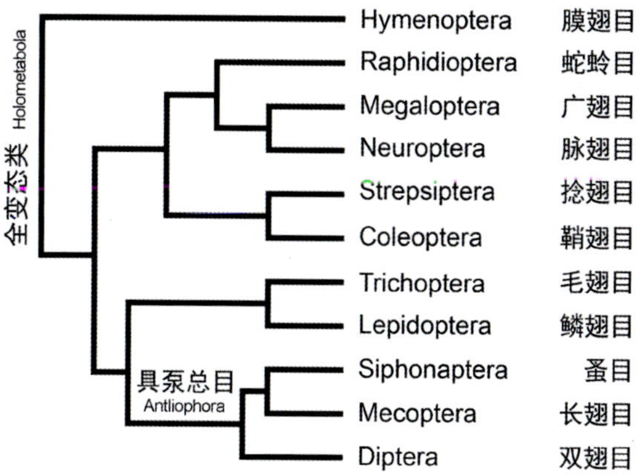

图 1-3　全变态类昆虫系统发育关系（仿 Misof *et al.*, 2014）

Fig. 1-3　Phylogenetic relationships among Holometabola (after Misof *et al.*, 2014)

The term Mecoptera is derived from the ancient Greek μέκο (=meco, long) and πτερόν (=pteron, wing), refering to the elongated wings of these insects.

Mecoptera is a small order of holometabolous insects,

and comprise approximately 700 species (Bicha, 2018) in nine extant families (Penny & Byers, 1979). They are one of the most ancient groups of holometabolous insects (Kristensen, 1981; Dunford & Somma, 2008).

According to fossil records, mecopterans probably originated in the Lower Permian (ca. 270 mya) and reached their peak of diversity in Mesozoic (250–65 mya), occupied various ecological niches (Grimaldi & Engel, 2005) (Fig. 1-1, Fig. 1-2). Nearly 400 fossil species in 87 genera and 34 families were recorded (Dunford & Somma, 2008; Byers, 2009). Some long proboscis-bearing mecopterans may have been the earliest pollinators and shared a close co-evolution with the gymnosperm seed plants, long before the present-day pollinating insects, such as flies in Diptera and bees in Hymenoptera (Ren *et al.*, 2009). In addition, an extinct hangingfly belonged to the extinct Cimbrophlebiidae shows a case of leaf-mimicry to ginkgo leaves (Wang *et al.*, 2012). Some mecopterans have their hindwings reduced and resemble the halteres of Diptera (Grimaldi *et al.*, 2005). With the occurrence of drastic climatic changes and several mass extinctions (Erwin, 1990; Benton, 1995), many mecopterans failed to survive and the rest of them become a relict group.

The species of Mecoptera play a significant role in the study of the evolution of Insecta. Diptera, Lepidoptera and Trichoptera may originate from ancestors that are closely related to Mecoptera (Grimaldi & Engel, 2005). Mecoptera, Siphonaptera and Diptera are grouped into the Antliophora, a superorder proposed for having a sperm pump in their male genitalia (Hennig, 1969). The monophyly of Antliophora was supported by molecular evidence (Whiting *et al.*, 1997; Misof *et al.*, 2014).

The systematic position of Mecoptera in Holometabola is open to debate for considerably a long time (Dunford & Somma, 2008; Friedrich *et al.*, 2013). Previously, Hennig (1969) regarded Mecoptera as a sister group to Diptera, though Kristensen (1999) considered the order to be more related to Siphonaptera. Subsequently, phylogenetic research based on concatenated data of multi-genes(Whiting, 2002) points to the close relationship between Mecoptera and Siphonaptera, and Siphonaptera is sister to the snow fleas Boreidae. There are two synapomorphies shared by Siphonaptera and snow fleas: 1) the possession of resilin (an elastomeric protein found in many arthropods) within their bodies; and 2) the panoistic ovariole (ovarioles that have no specialized nurse cells) in their female reproductive systems. Therefore, Mecoptera were regarded as a paraphyletic group. This standpoint of view, however, was in a lot of doubts. After investigating the ultrastructure of sperm, Dallai *et al.* (2003) discovered a significant difference between Boreidae and Siphonaptera, and reconfirmed the monophyly of Mecoptera. Misof *et al.* (2014) also supported the monophyly of Mecoptera based on their phylogenetic analysis of Insecta from 1,478 protein-coding genes and amino acid sequences. Therefore, the phylogenetic relationship within Antliophora should be as follows: Diptera + (Mecoptera + Siphonaptera) (Fig. 1-3) .

二、研究简史 Historical Review

早在古罗马时期，著名的博物学家老普林尼（Pliny the Elder，公元 23—79 年）所著《自然史》（*Naturalis Historia*）一书描述蝎子的章节中，就曾经提到“它们其中的一些长有翅膀”。之后，古生物学家乔治•居维叶（Georges Cuvier）指出，老普林尼所描述的动物可能就是长翅目 Mecoptera 的蝎蛉（Bostock & Riley, 1855）。

林奈（Carolus Linnaeus，公元 1707—1778 年）1758 年出版的《自然系统》第 10 版（*Systema Naturae*, 10th edition）（图 1-4）中记载了 2 种蝎蛉，普通蝎蛉 *Panorpa communis* 和德国蝎蛉 *Panorpa germanica*，它们当时被归入脉翅目 Neuroptera 中。同时被描述的，还有后来被归入脉翅目 Neuroptera 旌蛉科 Nemopteridae 的 *Panorpa coa*，其学名后变更为 *Nemoptera coa* (Linnaeus, 1758)。林奈对蝎蛉属 *Panorpa* 进行描述是用拉丁语，“*Rostrum corneum, Cylindricum: palpis … Antennae thorace longiores. Cauda maris chelata*”，意为“喙角状，须（当指下颚须和下唇须）圆柱形；触角长于胸部；尾部呈爪状”。

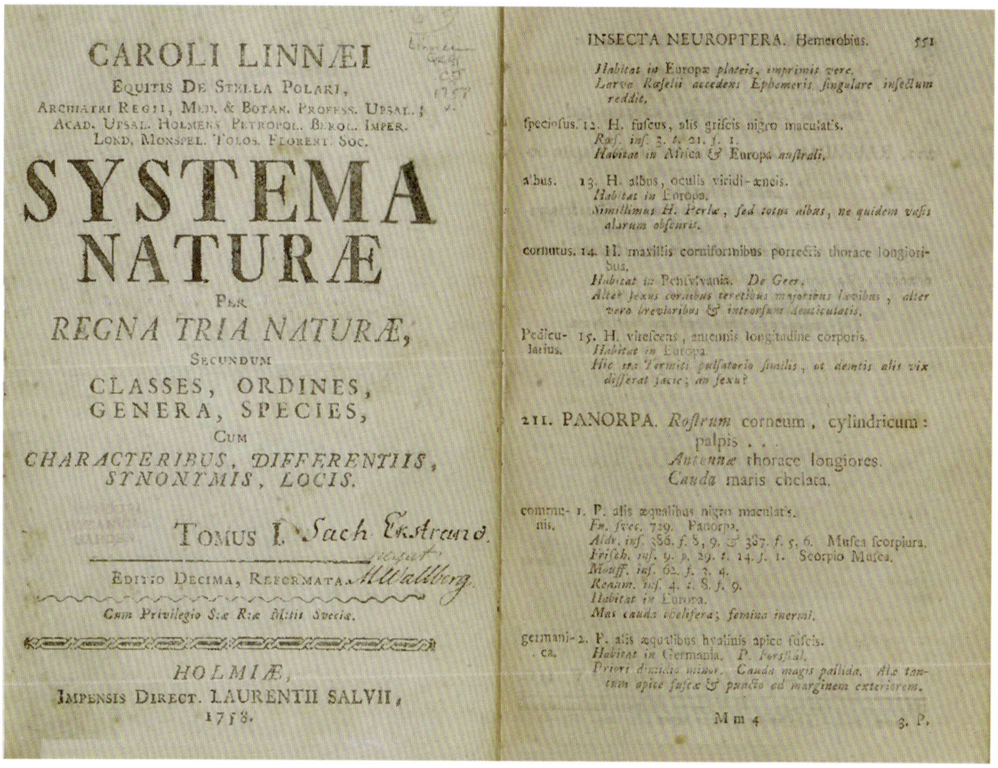
CAROLI LINNÆI
EQUITIS DE STELLA POLARI,
ARCHIATRI REGII, MED. & BOTAN. PROFESS. UPSAL.; ACAD. UPSAL. HOLMENS. PETROPOL. BEROL. IMPER. LOND. MONSPEL. TOLOS. FLORENT. SOC.
SYSTEMA NATURÆ
PER
REGNA TRIA NATURÆ,
SECUNDUM
CLASSES, ORDINES, GENERA, SPECIES,
CUM
CHARACTERIBUS, DIFFERENTIIS, SYNONYMIS, LOCIS.
TOMUS I.
EDITIO DECIMA, REFORMATA.
Cum Privilegio S:æ R:æ M:tis Sveciæ.
HOLMIÆ,
IMPENSIS DIRECT. LAURENTII SALVII,
1758.

INSECTA NEUROPTERA. Hemerobius. 551

Habitat in Europæ plateis, imprimis vere.
Larva Roeselii accedens Ephemeris singulare insectum reddit.

speciosus. 12. H. fuscus, alis griseis nigro maculatis.
Rœs. inf. 3. t. 21. f. 1.
Habitat in Africa & Europa australi.

albus. 13. H. albus, oculis viridi-æneis.
Habitat in Europa.
Simillimus H. Perlæ, sed totus albus, ne quidem venis alarum obscuris.

cornutus. 14. H. maxillis corniformibus porrectis thorace longioribus.
Habitat in Pensylvania. De Geer.
Alter sexus cornibus teretibus majoribus lævibus, alter vero brevioribus & introrsum denticulatis.

Pedicularius. 15. H. virescens, antennis longitudine corporis.
Habitat in Europa.
Hic ita Termiti pulsatorio similis, ut demtis alis vix differat facie; an sexu?

211. PANORPA. *Rostrum* corneum, cylindricum: palpis . . .
Antennæ thorace longiores.
Cauda maris chelata.

communis. 1. P. alis æqualibus nigro maculatis.
Fn. svec. 729. Panorpa.
Aldr. inf. 386. f. 8, 9. & 387. f. 5, 6. Musca scorpiura.
Frisch. inf. 9. p. 29. t. 14. f. 1. Scorpio Musca.
Mouff. inf. 62. f. 3. 4.
Reaum. inf. 4. t. 8. f. 9.
Habitat in Europa.
Mas cauda chelifera; femina inermi.

germanica. 2. P. alis æqualibus hyalinis apice fuscis.
Habitat in Germania. P. Forsskål.
Priori dimidio minor. Cauda magis pallida. Alæ tantum apice fuscæ & puncto ad marginem exteriorem.

Mm 4　　3. P.

图 1-4 《自然系统》第 10 版封面及蝎蛉属 ***Panorpa*** 物种描述

Fig. 1-4 Cover and descriptions of ***Panorpa*** in ***Systema Naturae***, 10th edition

早期的长翅目 Mecoptera 昆虫研究学者有 MacLachlan（1875, 1894）、Navás（1908, 1911, 1936）、Lieftink（1933, 1936）等。Esben-Petersen（1921）出版的 *Mecoptera*: *Monographic Revision* 是最早的、针对全世界长翅目 Mecoptera 昆虫分类的奠基之作，其中收录了长翅目 Mecoptera 昆虫 5 科 90 余种。

Issiki（1927, 1929a, 1929b, 1931, 1933）和 Miyaké（1908, 1913）都记述过产自日本和周边地区的长

翅目 Mecoptera 昆虫；Tjeder（1936, 1951, 1973）记述了产自中国、东南亚等地的长翅目 Mecoptera 昆虫；Carpenter 发表了一系列有关中国（1938, 1945, 1948）和北美洲（1931, 1935, 1936）等地长翅目 Mecoptera 昆虫的论文；Riek（1954, 1973）则进行了澳洲长翅目 Mecoptera 昆虫区系的研究。Byers 对东南亚（1965, 1966, 1999）、中国（1970）、北美洲（1958, 1962, 1973, 1993, 1996）、南美洲（1989）、非洲（1975）等地的长翅目 Mecoptera 昆虫都进行了一些研究。郑凤瀛发表了一系列中国的长翅目 Mecoptera 昆虫（Cheng, 1949, 1952, 1957a）研究论文，并将中国的长翅目 Mecoptera 昆虫进行了阶段性总结（1957b）。Rust & Byers（1976）、Chau & Byers（1978）和 Cheung *et al.*（2006）分别记述了印度、印度尼西亚和加拿大等地的长翅目 Mecoptera 昆虫。Penny & Byers（1979）发表了全世界长翅目 Mecoptera 昆虫名录。Bicha 记述了一些北美洲（Bicha, 1983, 2006, 2007, 2015a）和东南亚（Bicha, 2010, 2011, 2015b; Bicha *et al.*, 2017）地区的长翅目 Mecoptera 昆虫。

周尧等（Chou *et al.*, 1981, 1987, 1988）的研究主要针对陕西、湖南两省。花保祯（1997, 1998, 1999, 2004）、花保祯和周尧（1997, 1998a, 1998b, 1999）、花保祯等（2001）陆续记述了中国吉林、辽宁、河南、福建、海南、西藏、四川等地的长翅目 Mecoptera 昆虫。谭江丽和花保祯（2008a, 2008b, 2008c, 2009a, 2009b）、蔡立君等（2008）、马娜和花保祯（2012）、钟问和花保祯（2013）、高超等（2016）等陆续发表了中国长翅目 Mecoptera 昆虫的一些属及种。周文豹（2000, 2001, 2003, 2005, 2006）也发表了一系列关于分布在浙江、贵州等地中国的长翅目 Mecoptera 昆虫研究报告。

长翅目 Mecoptera 昆虫的形态学研究也吸引着诸多学者的兴趣。Ferris & Rees（1939）针对蝎蛉科 Panorpidae 昆虫进行了全面的解剖学研究。Mickoleit（1976）对整个长翅目 Mecoptera 昆虫的雌性生殖系统构造做了总结。Kaltenbach（1978）对整个长翅目 Mecoptera 昆虫的形态学研究进行了详细的综述。长翅目 Mecoptera 昆虫生物学方面的研究者，有 Miyaké（1912）、Yie（1951）、Byers（1963）、Gassner（1963）等人。过去的研究由于受到技术条件的限制，仅局限于整体形态和毛序特征等。蔡立君和花保祯（2009）对长翅目 Mecoptera 昆虫秦岭蝎蛉 *Panorpa qinlingensis* 幼期各虫态，利用电子显微镜进行了形态学研究。姜碌和花保祯对中国的长翅目 Mecoptera 昆虫幼虫进行了系统性的形态学研究（Jiang & Hua, 2013, 2014, 2015）。

长翅目 Mecoptera 昆虫中的蝎蛉科 Panorpidae、蚊蝎蛉科 Bittacidae 等，由于具有特殊的"献礼行为"（nuptial feeding）而受到关注（Zhong & Hua, 2013b; Zhong *et al.*, 2015a, 2015b）。此外，有关系统发育（Ma *et al.*, 2012；Hu *et al.*, 2015）、视觉器官构造（Chen & Hua, 2016）、胚胎发育（Du *et al.*, 2009；Yue & Hua, 2010；Kou & Hua, 2016）以及细胞分类学（Miao *et al.*, 2017）等方面的研究也不断涌现。

As early as the Ancient Rome, Pliny the Elder (AD 23–79), a famous natural historian, had mentioned that "some of which really have wings" in the chapter of scorpions in his *Naturalis Historia* (The Nature History). Later, the paleontologist George Cuvier claimed that the creatures Pliny described might be scorpionflies in the order Mecoptera (Bostock & Riley, 1855).

Carolus Linnaeus (AD 1707–1778) recorded two species of scorpionflies, *Panorpa communis* and *Panorpa germanica*, which were assigned to Neuroptera in his famous *Systema Naturae* (10th edition) published in 1758 (Fig. 1-4) . Meanwhile, another species, *Panorpa coa* was described along with them, but this species is actually a real neuropteran, and its name has been modified into *Nemoptera coa* (Linnaeus, 1758), which belongs to Nemopteridae, the spoonwings or thread-winged-antlions. Linnaeus's original descriptions of the genus *Panorpa* was in Latin: "*Rostrum corneum, Cylindrium: palpis... Antennae thorace longiores. Cauda maris chelata*", meaning "rostrum horn-like; palpi (including maxillary and labial palpus) cylindrical; antennae longer than thorax; and tail claw-like".

The early researchers of Mecoptera include MacLachlan (1875, 1894), Navás (1908, 1911, 1936), and Lieftink (1933, 1936), *etc*. Esben-Petersen's (1921) *Mecoptera*: *Monographic Revision* was the earliest fundamental work at the global scale. Over 90 species in five families Mecoptera were included in this monograph.

Issiki (1927, 1929a, 1929b, 1931, 1933) and Miyaké (1908, 1913) recorded and described the mecopterans from Japan and adjoining areas. Tjeder (1936, 1951, 1973) recorded some mecopterans from China and Southeast Asia. Carpenter published a series of papers dealing with the Chinese (1938, 1945, 1948) and North American (1931, 1935, 1936) faunas, respectively. Riek (1954, 1973) studied the Australian fauna. Byers's work included the Mecoptera faunas of Southeast Asia (1965, 1966, 1999), China (1970), North America (1958, 1962, 1973, 1993, 1996), South America (1989) and Africa (1975). Cheng published a series of papers (1949, 1952, 1957a), and made a revision on the Chinese fauna (Cheng, 1957b). Rust and Byers (1976), Chau and Byers (1978) and Cheung *et al.* (2006) recorded the Indian, Indonesian and Canadian faunas. Penny and Byers (1979) published a check-list of the extant Mecoptera species of the world. Bicha recorded some mecopterans from North America (1983, 2006, 2007, 2015a) and Southeast Asia (Bicha, 2010, 2011, 2015b; Bicha *et al.*, 2017).

Chou *et al.* (1981, 1987, 1988) investigated the mecopterous fauna in Shaanxi and Hunan Provinces. Afterwards, Hua (1997, 1998, 1999, 2004), Hua and Chou (1997, 1998a, 1998b, 1999) and Hua *et al.* (2001) recorded a series of mecopterans from Jilin, Liaoning, Henan, Fujian, Hainan, Sichuan and Tibet. Tan and Hua (2008a, 2008b, 2008c, 2009a, 2009b), Cai *et al.* (2008), Ma and Hua (2012), Zhong and Hua (2013), Gao (2016) established and described several Chinese genera and species. Zhou (2000, 2001, 2003, 2005, 2006) also published a series of research articles regarding the Chinese mecopterans.

The morphological studies also draw great interests from scholars. Ferris & Rees (1939) conducted a comprehensive anatomical study of Panorpidae. Mickoleit (1976) summarized the female reproductive systems of Mecoptera. Kaltenbach (1978) reviewed the morphological studies of the whole order. Biological researchers of Mecoptera include Miyaké (1912), Yie (1951), Byers (1963), and Gassner (1963), *etc.* Constrained by poor technical conditions, these studies mainly focused on the external morphology and larval chaetotaxy. The scorpionfly *Panorpa qinlingensis* was described and illustrated in detail for its immature stages by Cai & Hua (2009) using electron microscopy. A series of systematic work of larval morphology were conducted by Jiang & Hua (2013, 2014, 2015).

Some groups of Mecoptera, especially Panorpidae and Bittacidae, exhibit an interesting "nuptial feeding" behavior, which draws great attentions from entomologists (Zhong & Hua, 2013b; Zhong *et al.*, 2015a, 2015 b). In addition, studies regarding other fields are also constantly emerged in recent years, including phylogeny (Ma *et al.*, 2012; Hu *et al.*, 2015), structure of visual organs (Chen & Hua, 2016), embryonic development (Du *et al.*, 2009; Yue & Hua, 2010; Kou & Hua, 2016) and cytotaxonomy (Miao *et al.*, 2017).

三、分类系统 Classification System

现生的长翅目 Mecoptera 昆虫共分为 9 个科，分别为小蝎蛉科 Nannochoristidae、蚊蝎蛉科 Bittacidae、雪蝎蛉科 Boreidae、美蝎蛉科（又称仙女蝎蛉科）Meropeidae、原蝎蛉科 Eomeropidae、无翅蝎蛉科 Apteropanorpidae、异翅蝎蛉科 Choristidae、拟蝎蛉科 Panorpodidae 和蝎蛉科 Panorpidae（图 1-5），常见的有 6 个科（图 1-6）。

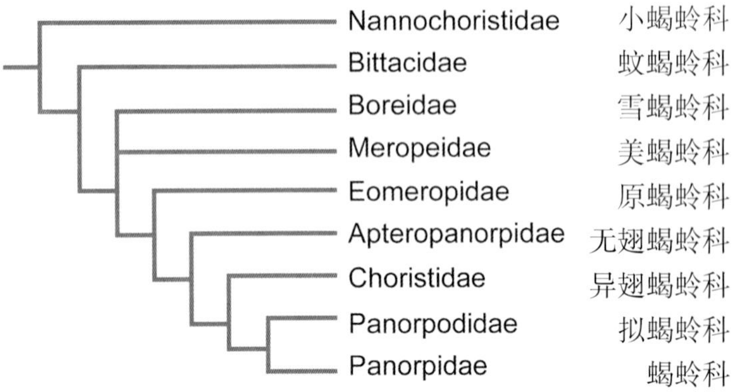

图 1-5 长翅目 **Mecoptera** 各科系统发育关系（仿 Willmann, 1987）

Fig. 1-5 Phylogenetic relationships among families in Mecoptera（after Willmann, 1987）

Tillyard（1917）根据翅的宽窄，将原来的长翅目 Mecoptera 分为原长翅目 Protomecoptera 和真长翅目 Eumecoptera，这二者后来被认为是长翅目 Mecoptera 的两个亚目。周尧（1963）提出，蚊蝎蛉科 Bittacidae 由于翅尤其狭长，足的跗节仅具 1 枚爪，生殖器构造特殊且不能行走，趋向于在茂密的灌木丛和草丛中生活，应为其建立束长翅亚目 Zygomecoptera。Kaltenbach（1978）按翅的有无、长度和生殖器特征，将长翅目 Mecoptera 划分为 3 亚目：原长翅亚目 Protomecoptera（包括美蝎蛉科 Meropeidae 和原蝎蛉科 Eomeropidae）、真长翅亚目 Eumecoptera（包括异翅蝎蛉科 Choristidae、小蝎蛉科 Nannochoristidae、拟蝎蛉科 Panorpodidae、蝎蛉科 Panorpidae、无翅蝎蛉科 Apteropanorpidae 及蚊蝎蛉科 Bittacidae）和新长翅亚目 Neomecoptera（仅包括雪蝎蛉科 Boreidae）。Mickoleit（1978）提出，蚊蝎蛉科 Bittacidae 应位于长翅目 Mecoptera 系统树的基部。Willmann（1987, 1989）把长翅目 Mecoptera 划分为 2 亚目，即小蝎蛉亚目 Nannomecoptera（仅有小蝎蛉科 Nannochoristidae ）和具塞亚目 Pistillifera（包括其他 8 科），其中后者又分为捕捉足次目 Raptipedia（仅有蚊蝎蛉科 Bittacidae）和后生殖孔次目 Opisthogonopora。Hünefeld & Beutel（2005）指出，精子泵的功能和相关结构以及位置在蚤目 Siphonaptera、具塞亚目 Pistillifera 和双翅目 Diptera 中很不相同，长翅目 Mecoptera 真正的基部分支应该是小蝎蛉科 Nannochoristidae，而不是同样不具有精子泵的雪蝎蛉科 Boreidae。

由于演化历程的漫长，长翅目 Mecoptera 各科之间表现出各自迥异的形态特征。比如，美蝎蛉科 Meropeidae 和原蝎蛉科 Eomeropidae 昆虫的翅都较宽大，且具有发达的网状翅脉，但蝎蛉科 Panorpidae、小蝎蛉科 Nannochoristidae、拟蝎蛉科 Panorpodidae 和蚊蝎蛉科 Bittacidae 昆虫的翅则较为狭长，且横脉稀少（Dunford & Somma, 2008）；雪蝎蛉科 Boreidae 雄虫的翅特化为抱握钩，而雌虫的翅退化成鳞片状；无翅蝎蛉科 Apteropanorpidae 昆虫的翅则完全消失。拟蝎蛉科 Panorpodidae 和蚊蝎蛉科 Bittacidae 昆虫也有短翅或无翅的种。此外，它们在喙、复眼、腹部及幼虫形态等方面差异也很大。

图 1-6 长翅目 **Mecoptera** 中常见的 6 个科

1. 小蝎蛉科 Nannochoristidae 2. 美蝎蛉科 Meropeidae 3. 蝎蛉科 Panorpidae 4. 雪蝎蛉科 Boreidae 5. 蚊蝎蛉科 Bittacidae 6. 拟蝎蛉科 Panorpodidae

Fig. 1-6 Six common families in Mecoptera

1. Nannochoristidae 2. Meropeidae 3. Panorpidae 4. Boreidae 5. Bittacidae 6. Panorpodidae

The extant Mecoptera consist of nine families, *i.e.* Nannochoristidae, Bittacidae, Boreidae, Meropeidae, Eomeropidae, Apteropanorpidae, Choristidae, Panorpodidae and Panorpidae (Fig. 1-5)，Among them, six families can commonly be found (Fig. 1-6).

Based on the width of wings, Tillyard (1917) proposed two orders, Protomecoptera and Eumecoptera, which were later regarded as two suborders in Mecoptera. Chou (1963) claimed that Bittacidae should be assigned to the suborder Zygomecoptera for the following reasons: relatively narrower wings; a single tarsal claw; specialized male genitalia; and unable to walk but can only hang themselves in dense shrubs and grasses. Based on the presence/absence and length of wings and genital characters, Kaltenbach (1978) proposed a three-suborder system: Protomecoptera, including Eomeropidae and Meropeidae; Eumecoptera, including Apteropanorpidae, Bittacidae, Choristidae, Nannochoristidae, Panorpodidae and Panorpidae; and Neomecoptera, including only Boreidae. Mickoleit (1978) noted that Bittacidae should be a basal lineage on the phylogenetic tree of Mecoptera. Willmann (1987, 1989) divided Mecoptera into two suborders: Nannomecoptera (only including Nannochoristidae) and Pistillifera (including the remaining eight families). Pistillifera is further split into two infraorders, Raptipedia (including only Bittacidae) and Opisthogonopora (including the remaining seven families). Hünefeld and Beutel (2005) pointed out that the function and related structures and position of the sperm pumps differed greatly among Siphonaptera, Pistillifera and Diptera. Therefore, the real basal lineage in Mecoptera should be Nannochoristidae instead of Boreidae, which lack a sperm pump.

After such a long evolutionary course, various families of Mecoptera have exhibited greatly distinctive morphological characters. For instance, Meropeidae and Eomeropidae have dense reticulate venations on their broad wings, while other families have much simpler venations on their much narrower wings (Dunford & Somma, 2008). In addition, the male wings of Boreidae are specialized into clasping hooks, but their female wings are greatly reduced and scale-like. The males and females lack wings in Apteropanorpidae. Panorpodidae and Bittacidae have some short-winged or apterous species. Moreover, these insects show wide distinctions not only in the wings, but also in the rostrum, compound eyes, abdomen and larval morphology.

四、地理分布 Geographical Distribution

（一）蚊蝎蛉科 Bittacidae

蚊蝎蛉科 Bittacidae 昆虫广泛分布于除南极洲之外的各大动物地理区（图 1-7）。蚊蝎蛉属 *Bittacus* 是分布最广的属。其他各属大多局限于某些较狭窄的区域内，如分布于智利的安娜蚊蝎蛉属 *Anabittacus*，分布于南非的异蚊蝎蛉属 *Anomalobittacus*，分布于北美洲的无翅蚊蝎蛉属 *Apterobittacus*，分布于南美洲的一色蚊蝎蛉属 *Issikiella*，分布于中国南部的双尾蚊蝎蛉属 *Bicaubittacus* 和地蚊蝎蛉属 *Terrobittacus*，以及分布于澳大利亚的曲蚊蝎蛉属 *Harpobittacus* 等。

Penny（1975）对蚊蝎蛉科 Bittacidae 昆虫分布及演化的解释是，蚊蝎蛉起源于冈瓦纳古陆（Gondwana），约在白垩纪末期（约 6 000 万年前）逐渐分化出许多独立的属；南美洲的蚊蝎蛉扩散至北美洲后进一步演化，并经过白令陆桥向欧亚大陆迁移。然而，欧亚大陆地区丰富的种、属多样性，以及越来越多古老化石的发现，证明蚊蝎蛉很可能最早发生于劳亚古陆（Laurasia）；超级古陆解体后，蚊蝎蛉逐渐扩散，形成了现在的分布格局（任东，1993）。李艳丽和任东（2009）在探讨蚊蝎蛉的分布及演化时，认为欧亚大陆很可能是蚊蝎蛉科 Bittacidae 的起源地，亚洲是其分化和扩散的中心，蚊蝎蛉在始新世（Eocene）从这里扩散到其他地区。

中国是目前世界上蚊蝎蛉报道种类最多的国家，绝大多数地区都有蚊蝎蛉科 Bittacidae 昆虫的分布记录（图 1-8），只有几个省份至今尚无蚊蝎蛉分布的记录。

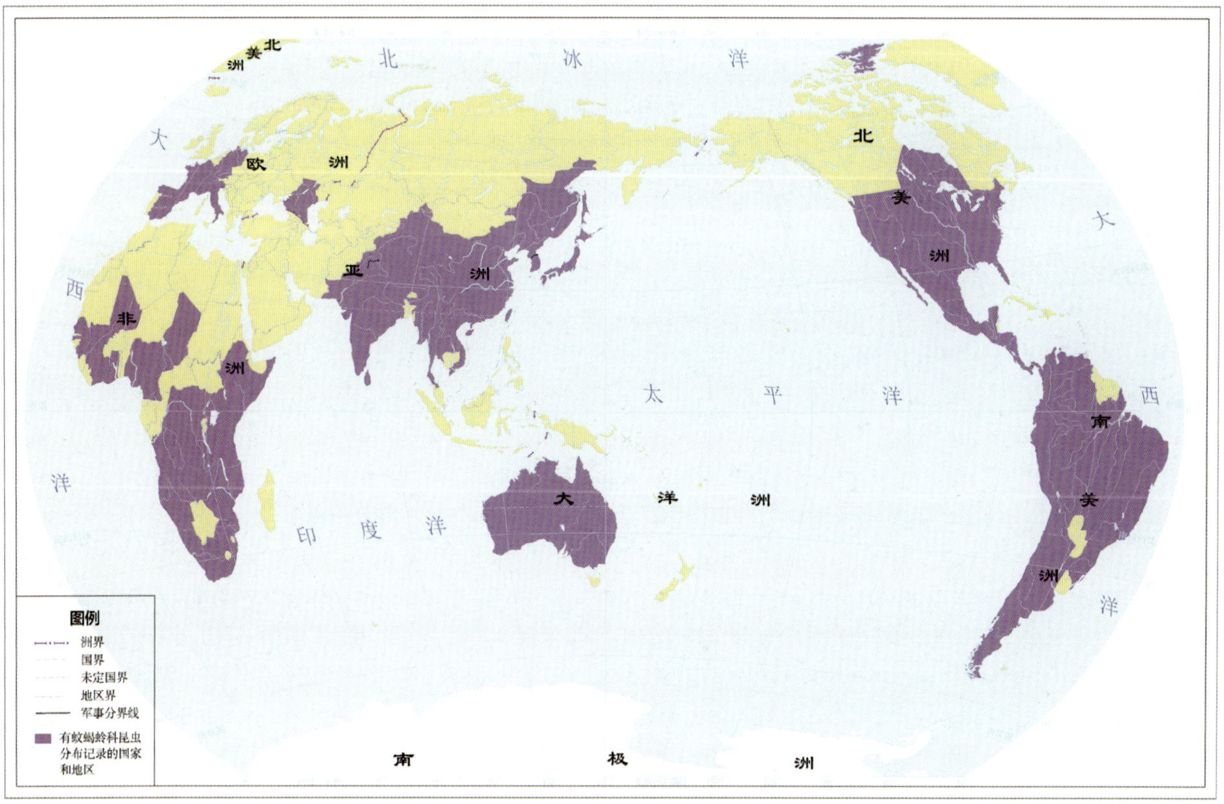

图 1-7　有蚊蝎蛉科 **Bittacidae** 昆虫分布记录的国家和地区

Fig. 1-7　Countries and regions with distribution records of Bittacidae

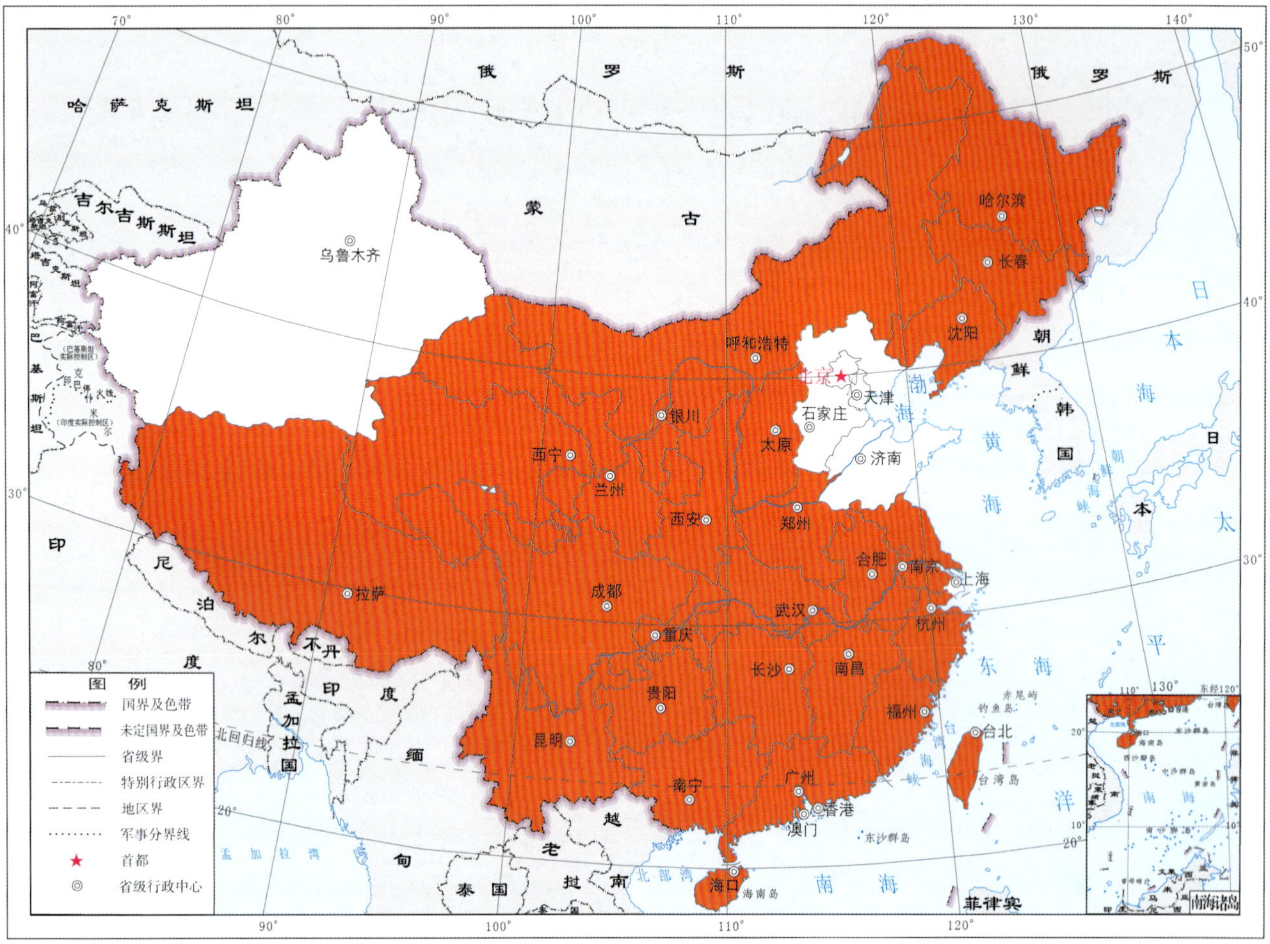

图 1-8 中国有蚊蝎蛉科 **Bittacidae** 昆虫分布记录的省、市、自治区（红色为有分布，下同）

Fig. 1-8 Provinces of China with distribution records of Bittacidae

(red indicates distribution; the same otherwise)

The species of Bittacidae are distributed in all geographical regions of the world except Antarctica (Fig. 1-7) . The genus *Bittacus* is the most widespread, while other genera are mostly constrained in narrow ranges. For instance, *Anabittacus* is only distributed in Chile, *Anomalobittacus* in South Africa, *Apterobittacus* in North America, *Issikiella* in South America, *Bicaubittacus* and *Terrobittacus* in south China, and *Harpobittacus* in Australia.

For the distributional and evolutionary explanations, Penny (1975) claimed that the family Bittacidae probably originated in Gondwana, and was split into several distinct genera around late Cretaceous (approximately 60 mya). The hangingflies spread across from South to North America, and then migrated into the Eurasian continent. The high diversity at both specific and generic levels in Eurasia, however, indicates that these insects are more likely originated from Laurasia instead of Gondwana. After the ancient supercontinent Pangaea broken apart, the hangingflies spread and formed the distributional pattern as we know today (Ren, 1993). Li & Ren (2009) discussed the distribution and evolution of Bittacidae and claimed that Eurasia was likely the area of origin of these insects, and Asia was the differentiating and dispersal center where hangingflies started their migration to other regions in Eocene.

China has the maximum number of recorded species of hangingflies all over the world. Most of its territories have distribution records of Bittacidae (Fig. 1-8) . Only several provinces lack records of hangingflies.

（二）蝎蛉科 Panorpidae

蝎蛉科 Panorpidae 昆虫分布于全北区（the Holarctic Region）和东洋区（the Oriental Region），包括欧洲、亚洲和北美洲的大部分地区（图 1-9）。Byers（1988）推测，北美洲的蝎蛉科 Panorpidae 昆虫是在始新世（Eocene）至渐新世（Oligocene）时，通过当时的白令陆桥从亚洲迁移而来。东亚（中国、日本等国）是蝎蛉科 Panorpidae 昆虫多样性最高的地区，很有可能是蝎蛉科 Panorpidae 昆虫的起源与扩散中心。

中国是蝎蛉科 Panorpidae 物种最多的国家，许多地区都有蝎蛉的分布记录。仅少数几个省份目前尚无蝎蛉的分布记录，包括山东、河北、新疆等（图 1-10）。

The species of Panorpidae are distributed in the Holarctic and Oriental Regions, including most areas of Eurasia, Asia and North America (Fig. 1-9). Byers (1988) presumed that the Panorpidae in North America might be migrated from Asia through Beringia during Eocene to Oligocene. East Asia bears the highest diversity of Panorpidae than other regions, suggesting that it is the center of origin and dispersal of Panorpidae.

China has the maximum number of recorded species of Panorpidae in the world. Most provinces have distribution records of Panorpidae. Only a few provinces lack records of scorpionflies, including Shandong, Hebei and Xinjiang *et al.*(Fig. 1-10).

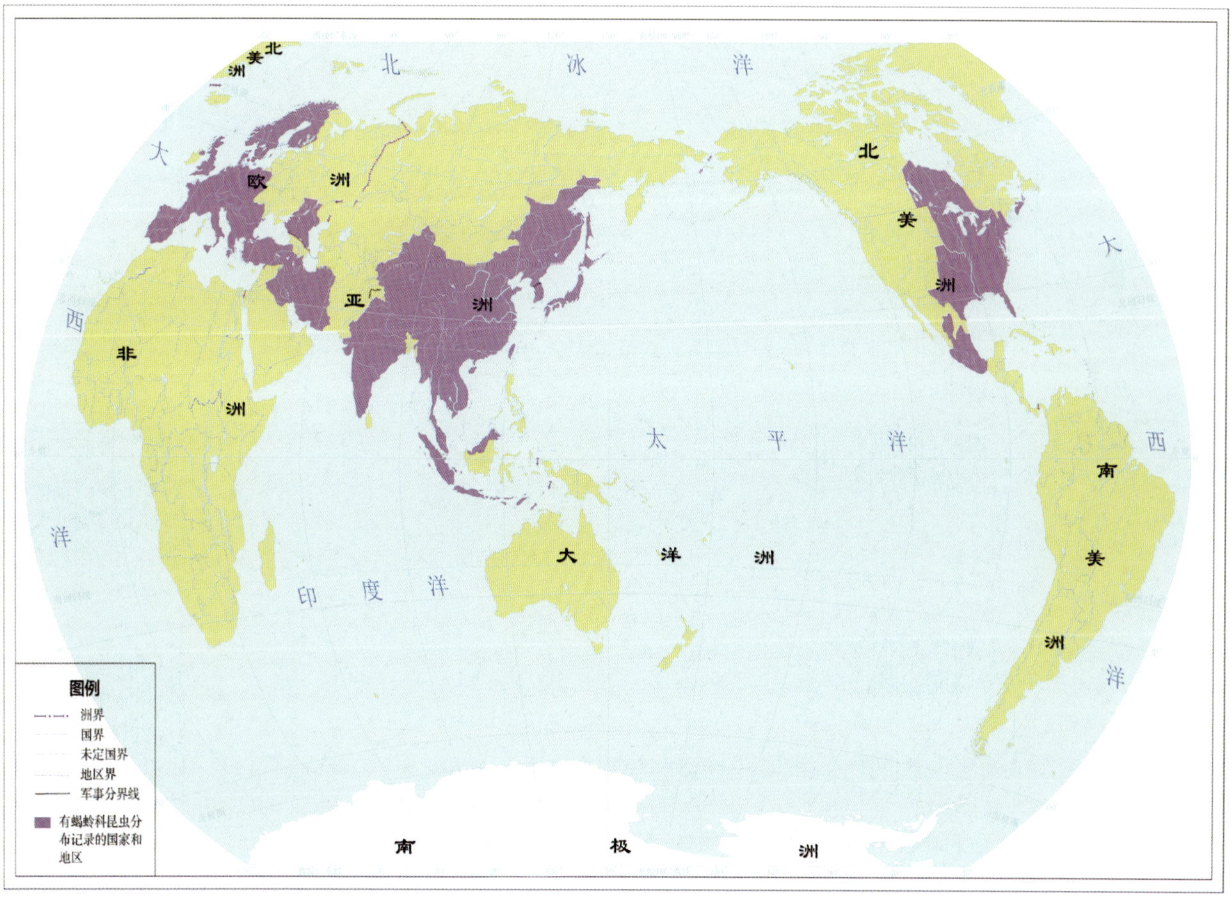

图 1-9 有蝎蛉科 **Panorpidae** 昆虫分布记录的国家和地区

Fig. 1-9 Countries and regions with distribution records of Panorpidae

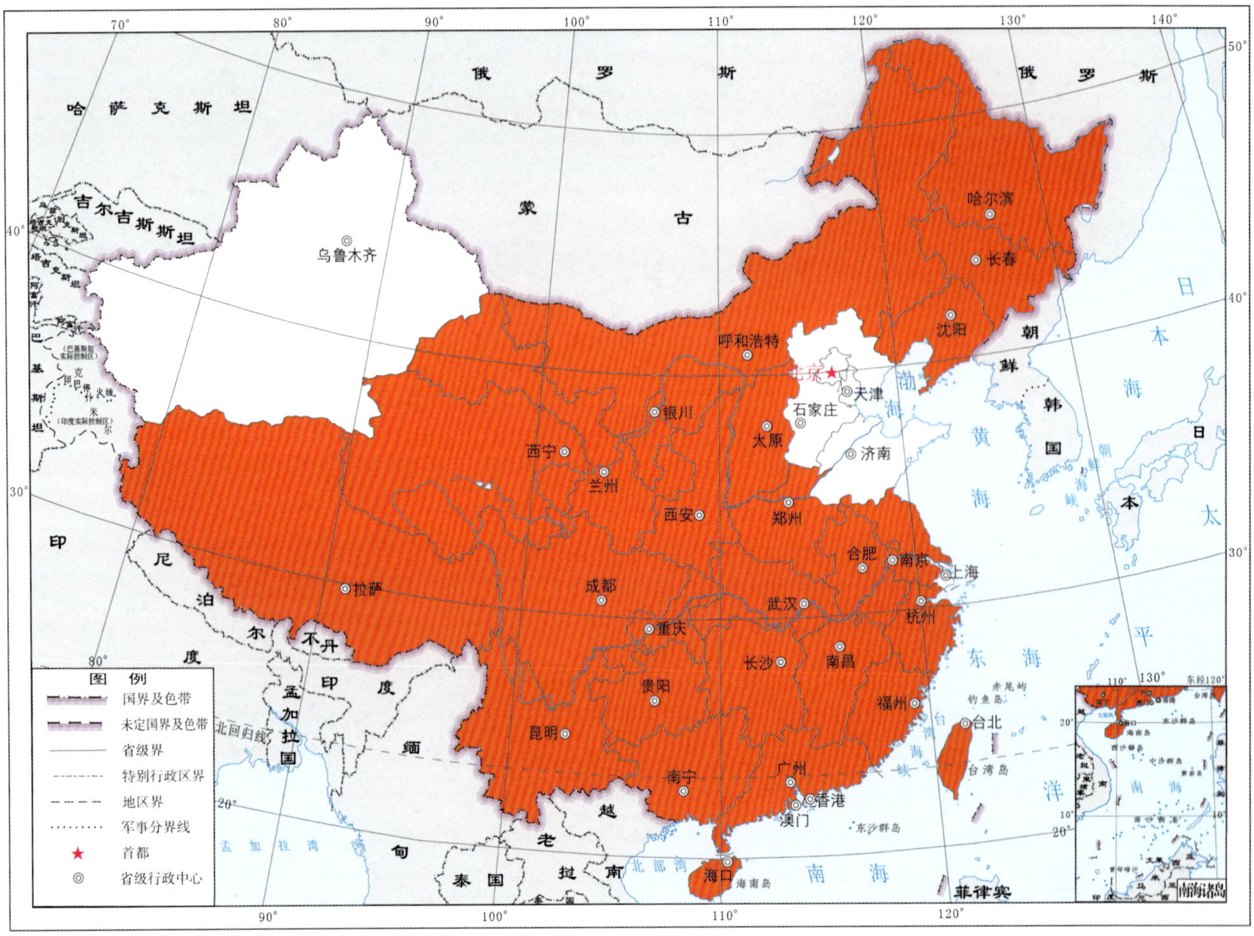

图 1-10 中国有蝎蛉科 Panorpidae 昆虫分布记录的省、市、自治区

Fig. 1-10 Provinces of China with distribution records of Panorpidae

（三）拟蝎蛉科 Panorpodidae

拟蝎蛉科 Panorpodidae 昆虫分布于北太平洋周边地区，如美国、朝鲜、韩国、日本、中国等（图 1-11）。在欧洲、南美洲等地都有拟蝎蛉科 Panorpodidae 昆虫化石的发现，说明拟蝎蛉科 Panorpodidae 昆虫在始新世可能广泛分布于欧亚大陆和南、北美洲（Hu & Hua, 2016），后来随着气候变化，许多物种灭绝，只有一些孑遗物种留存了下来。

中国的拟蝎蛉科 Panorpodidae 昆虫仅发现于吉林长白山、辽宁宽甸和重庆北碚（图 1-12）。

The species of Panorpodidae are distributed in the north Pacific Rim, including U.S.A., the Korean peninsula, Japan and China (Fig. 1-11). Fossil records have been found in Europe and South America, indicating that this family likely had a larger distribution range in Eurasia and North & South America during the Eocene (Hu & Hua, 2016). After a series of climate changes and mass extinctions, only a few of them survived to date.

In China, only the Changbai Mountains in Jilin, Kuandian in Liaoning, Beibei in Chongqing have distribution records of the short-faced scorpionflies (Fig. 1-12).

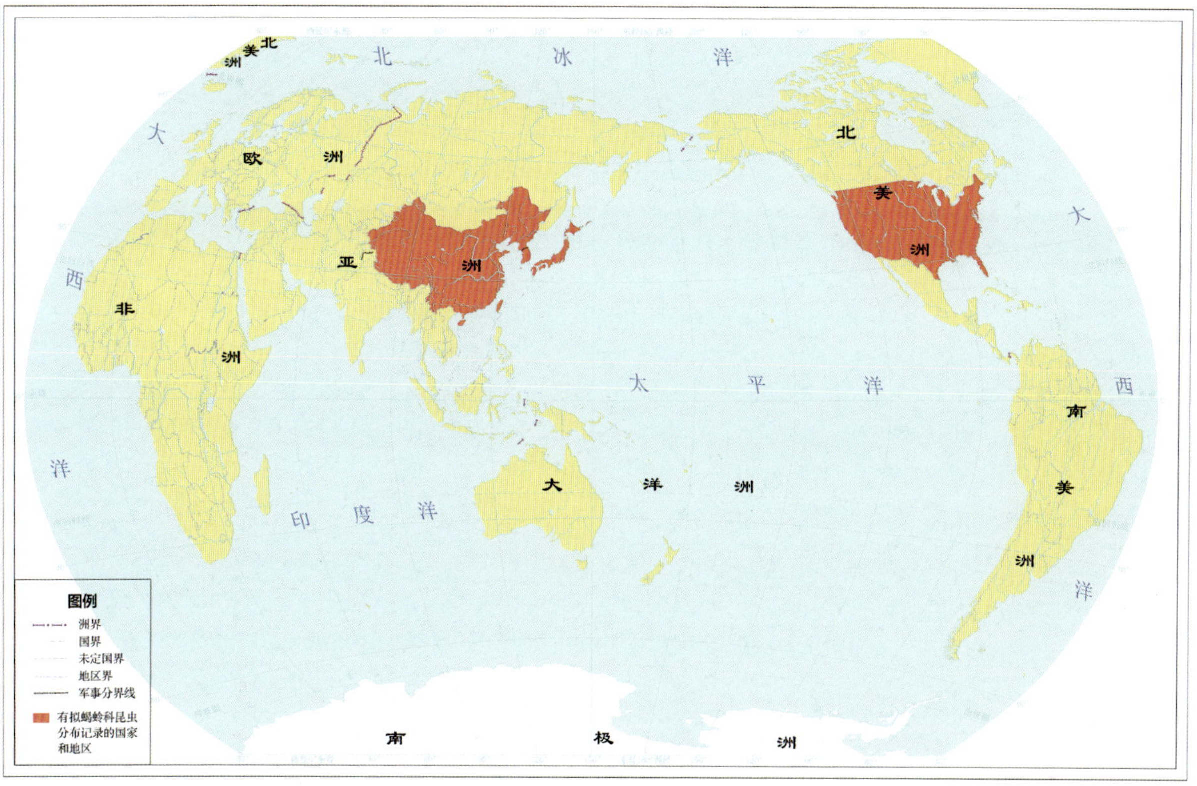

图 1-11 有拟蝎蛉科 Panorpodidae 昆虫分布记录的国家和地区

Fig. 1-11 Countries and regions with distribution records of Panorpodidae

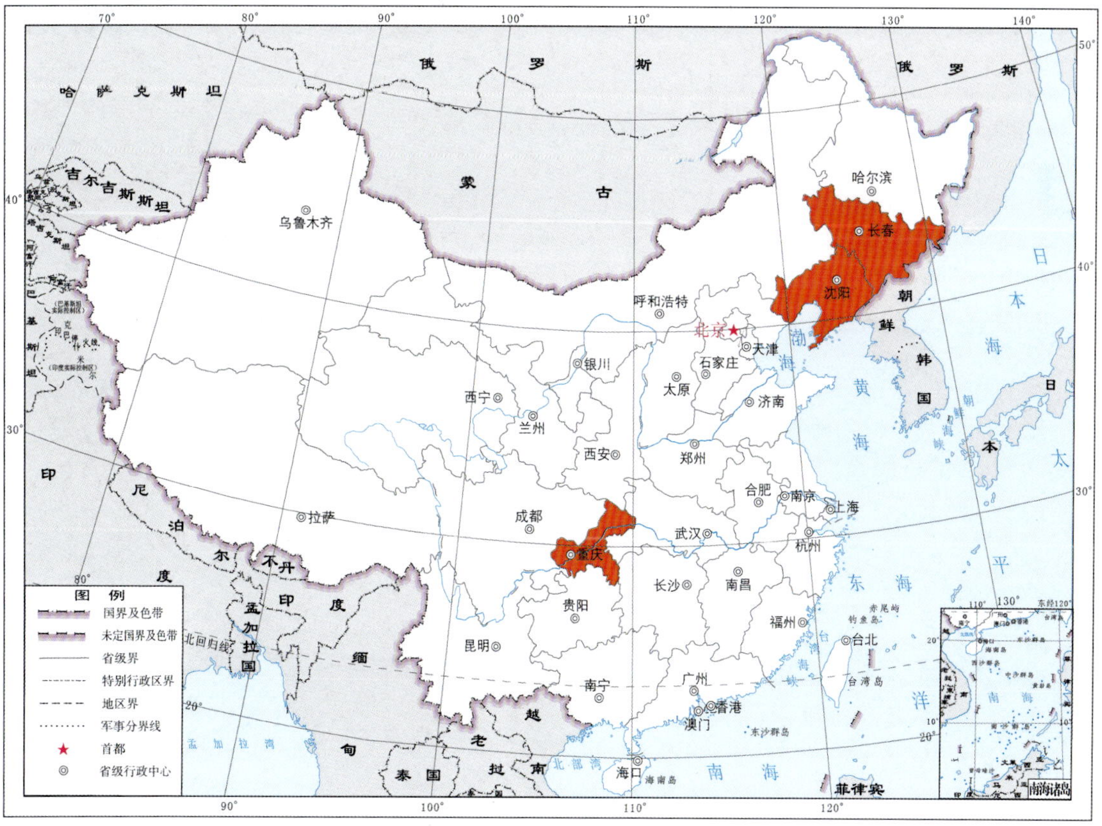

图 1-12 中国有拟蝎蛉科 Panorpodidae 昆虫分布记录的省、市、自治区

Fig. 1-12 Provinces of China with distribution records of Panorpodidae

五、形态特征 Morphology

（一）长翅目 Mecoptera 昆虫的主要识别特征

○头部向下不同程度地延长，呈喙状，咀嚼式口器位于喙的末端。
○翅大多狭长，翅脉接近原始脉序；前、后翅的大小、形状和脉序相似。
○雄性腹部生殖节不同程度膨大，有的呈蝎尾状反曲。
○尾须一般短小；雌虫无产卵器。
长翅目 Mecoptera 昆虫形态特征见图 1-13。

Main Diagnostic Characters of Mecoptera

● Head more or less elongated ventrad into a rostrum, with the chewing mouthparts on the distal end of the rostrum.

● Wings normally narrow with venations close to primitive pattern; forewings and hindwings similar in size, shape and venations.

● Male genital segments enlarged to some extent, and some of them curved upward like the stinger of scorpions.

● Cerci usually short and small; females lacking an ovipositor.

The morphological character of Mecoptera are shown in Fig. 1-13.

图 1-13　尼尔森新蝎蛉 *Neopanorpa nielseni* 雄成虫背面观

Fig. 1-13　Male adult of *Neopanorpa nielseni*, dorsal view

（二）蚊蝎蛉科 Bittacidae

1. 成虫

蚊蝎蛉科昆虫英文通称“hangingflies”，字面意思为“挂蛉”，因其成虫具有 3 对捕捉足，并常在不飞行时，用前、中足可抓握的跗节将虫体悬挂于植物叶片或枝条的边缘而得名（图 1-14）。它们与双翅目的大蚊极其相似，但不同之处在于 2 对大小相似的翅，而大蚊的后翅极度退化成了 1 对平衡棒。因此，它们的中文俗称为“蚊蝎蛉”。在中国台湾，它们则被称作“拟大蚊”。

（1）头部：蚊蝎蛉头部的分类特征，主要有颜色、单眼三角区（ocellar triangle）的斑纹、背单眼（dorsal ocellus）的大小、复眼及触角的形态等。头部的颜色一般为黄褐色至深褐色、红褐色，有的头顶和喙颜色不一。3 枚背单眼（dorsal ocellus）构成单眼三角区，其周围的斑纹一般颜色较深，有时向周围延伸。有的种在背单眼之间具有显著的单眼鬃。复眼（compound eye）大多长椭球形，两复眼的大小、形状及其间距常常是重要的分类特征。触角（antenna）柄节、梗节较短而粗壮；鞭节细丝状，有的被有密毛（图 1-15）。

图 1-14 纹翅蚊蝎蛉 ***Bittacus strigatus*** 雄成虫侧面观

Fig. 1-14 Male adult of ***Bittacus strigatus***, lateral view

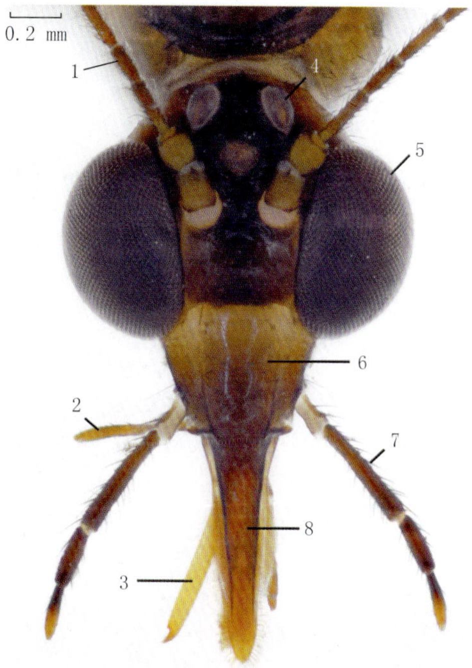

图 1-15 狭地蚊蝎蛉 ***Terrobittacus angustus*** 头部正面观

1. 触角 2. 下唇须 3. 上颚 4. 背单眼

5. 复眼 6. 唇基 7. 下颚须 8. 上唇

Fig. 1-15 Frontal view of the head of ***Terrobittacus angustus***

1. antenna 2.labial palp 3. mandible 4. dorsal ocellus

5. compound eye 6. clypeus 7. maxillary palp 8. labrum

（2）胸部：胸部强壮，背部稍隆拱。前胸（prothorax）背板小而窄，前缘有时具数根黑色刚毛。中胸背板（mesonotum）和后胸背板（metanotum）颜色多样：有的均为黄褐色或黑褐色，有的前后颜色不一，有的具有明显的背中带。侧板一般黄褐色或乳黄色（图 1-16）。

蚊蝎蛉的 3 对足均十分细长。足的颜色较均一，为黄褐色或红褐色，有的在股节和胫节（tibia）两端颜色加深。基跗节（basitarsus）的长短、第 4 跗分节（4th tarsomere）两侧黑刺的数量等常用于分类鉴定。第 5 跗分节（5th tarsomere）末端仅具 1 枚爪（claw）。第 5 跗分节能够折回，置于膨大的第 4 跗分节之上，此两节的内侧均有粗大的刺，形成可收放的抉持构造（图 1-17），能够用于悬挂身体或捕捉猎物。

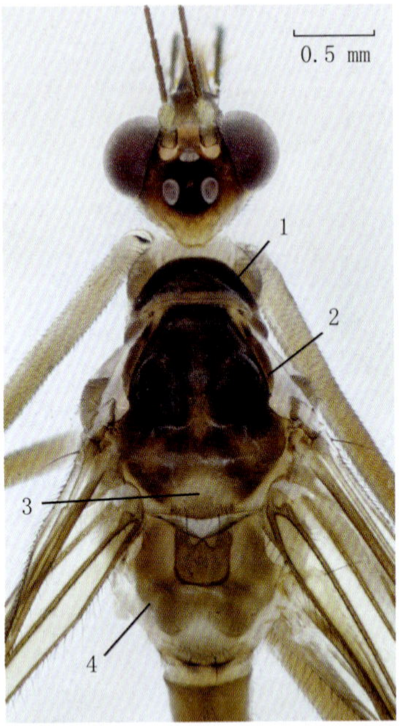

图 1-16 钩地蚊蝎蛉 ***Terrobittacus rostratus*** 胸部背面观

1. 前胸背板 2. 中胸背板 3. 中胸小盾片 4. 后胸背板

Fig. 1-16 Thorax of ***Terrobittacus rostratus***, dorsal view

1. pronotum 2. mesonotum 3. mesoscutellum 4. metanotum

蚊蝎蛉科 Bittacidae 属级单元划分的依据，主要集中在翅面特征上。在描述翅的特征时，采用的缩写和术语如下：A，臀脉；Av，臀横脉；C，前缘脉；CuA，前肘脉；CuP，后肘脉；CuV，肘横脉；FM，中脉第一分叉点；FM_{3+4}，M_3 和 M_4 脉分叉点；FRs，径分脉分叉点；h，肩横脉；M，中脉；nygmata，角斑（翅面上的角质圆斑）；OM，中脉起源点；ORs，径分脉起源点；Pcv，痣下横脉；pterostigma，翅痣；Rs，径分脉；Sc，亚前缘脉；Scv，亚前缘横脉；thyridium，明斑（位于 FM 处的一个半透明斑块）（图 1-18）。

常用的特征有翅的颜色、形状，Sc 末端位置，Pcv 的数目，Rs 在端部是否愈合，1A 的长度或末端所在位置，后翅 1A 与 CuP 愈合情况等。不同属的翅，形状变化较大，有的基部呈柄状，有的基部宽阔；有的端部略呈钩状，有的则较圆钝。翅面的斑纹也是重要的鉴定特征。

（3）腹部：雄性第 9 节背板特化为抱握构造，称为上生殖瓣（epandrial lobes），其形状是重要的鉴别特征。生殖肢并不行使抱握的功能，因为其生殖刺突（gonostylus）十分短小。阳茎着生于愈合的生殖肢基节（gonocoxite）形成的凹窝中，在大多数中国的物

图 1-17 中华蚊蝎蛉 ***Bittacus sinensis*** 后足跗节侧面观

1. 胫节距 2. 第 4 跗分节 3. 第 5 跗分节 4. 爪 5. 第 3 跗分节 6. 第 2 跗分节 7. 第 1 跗分节 8. 胫节

Fig. 1-17 Hind tarsus of ***Bittacus sinensis***, lateral view

1. tibial spurs 2. 4th tarsomere 3. 5th tarsomere 4. claw 5. 3rd tarsomere 6. 2nd tarsomere 7. 1st tarsomere 8. tibia

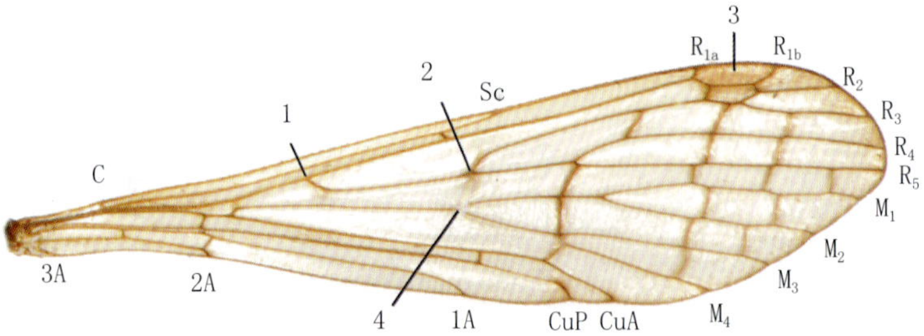

图 1-18 扁蚊蝎蛉 ***Bittacus planus*** 前翅及翅脉命名

1. 径分脉起源点 2. 径分脉分叉点 3. 翅痣 4. 中脉第一分叉点

Fig. 1-18 Forewing and venation of ***Bittacus planus***

1. ORs 2. FRs 3. pterostigma 4. FM

种中，阳茎末端极其延长，呈细丝状环状卷曲，称为阳茎丝（penisfilum）；阳茎基部常具叶状的阳茎叶（phallomeres）。第 10 节背板（tergum X）常不同程度退化。第 11 节的背板和腹板特化延长，称为载肛突（proctiger），分为上、下两瓣，其中上瓣从上生殖瓣之间穿出，其末端常膨大、钩状或具毛簇；下瓣一般短钩状，向下弯曲（图 1-19）。

雌性的分类特征主要是下生殖板（subgenital plate）的形态。下生殖板形似两瓣愈合而来，两瓣之间有时具膜质区域；其侧缘向背侧延伸出 1 对细长的突起；端半部常具粗壮的刚毛。第 10 节背板向腹面延伸与否也是重要的特征（图 1-20）。

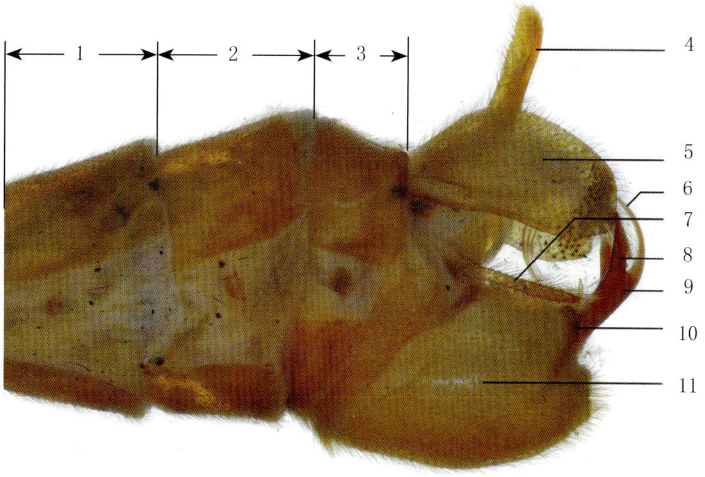

图 1-19 扁蚊蝎蛉 ***Bittacus planus*** 腹部与雄性外生殖器侧面观

1. 第 6 腹节 2. 第 7 腹节 3. 第 8 腹节 4. 载肛突上瓣 5. 上生殖瓣 6. 阳茎丝 7. 尾须 8. 阳茎叶 9. 阳茎 10. 生殖刺突 11. 生殖肢基节

Fig. 1-19 Male genitalia of ***Bittacus planus***, lateral view

1. 6th abdominal segment 2.7th abdominal segment 3. 8th abdominal segment 4. upper branch of proctiger 5. epandrial lobe 6. penisfilum 7. cercus 8. phallomeres 9. aedeagus 10. gonostylus 11. gonocoxite

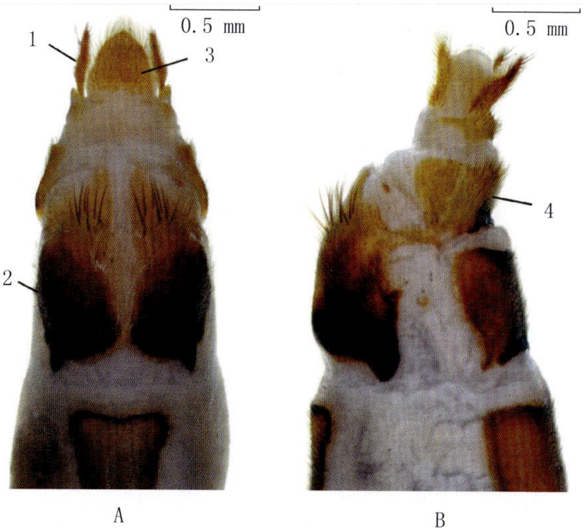

图 1-20 缅甸双尾蚊蝎蛉 ***Bicaubittacus burmanus*** 雌性腹部末端

A. 腹面观 B. 侧面观

1. 尾须 2. 下生殖板 3. 肛下板 4. 第 10 背板

Fig. 1-20 Female terminalia of ***Bicaubittacus burmanus***

A.ventral view B.lateral view

1. cercus 2. subgenital plate 3. subanal plate 4. tergum X

2. 卵

蚊蝎蛉的卵一般圆球形。蚊蝎蛉属 *Bittacus* 昆虫的卵表面一般较光滑，而地蚊蝎蛉属 *Terrobittacus* 昆虫一些种的卵表面具纬线一样的纹饰（图 1-21）。少数种，如环带蚊蝎蛉 *Bittacus cirratus* 等的卵则接近立方体形（图 1-22）。

图 1-21　蚊蝎蛉 Bittacidae spp. 的圆形卵

1. 扁蚊蝎蛉 *Bittacus planus*　2. 缠绕地蚊蝎蛉 *Terrobittacus implicatus*

Fig. 1-21　Spherical eggs of Bittacidae spp.

1. *Bittacus planus*　2. *Terrobittacus implicatus*

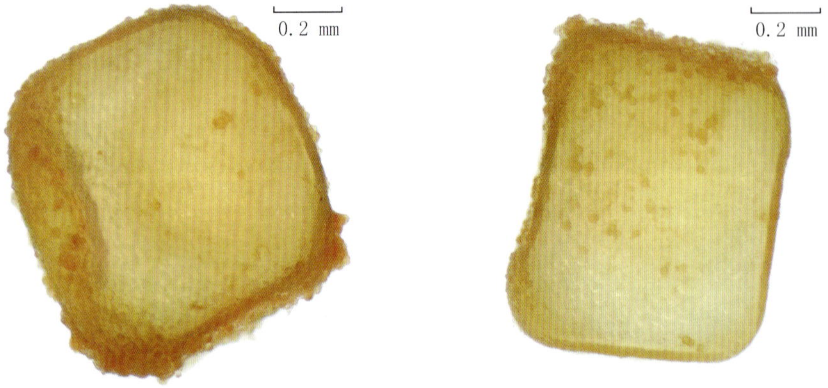

图 1-22　环带蚊蝎蛉 ***Bittacus cirratus*** 的近立方体形卵

Fig. 1-22　Cuboidal eggs of ***Bittacus cirratus***

3. 幼虫

幼虫蠋式（eruciform）。头部下口式，具 1 对复眼（compound eyes）及 1 枚中单眼（median ocellus）；

图 1-23　扁蚊蝎蛉 ***Bittacus planus*** 幼虫侧面观

Fig. 1-23　Larva of ***Bittacus planus***, lateral view

各体节具肉质“枝”状突起，末端着生棒状刚毛；前 8 节腹节各具 1 对腹足（prolegs），尾节具可伸缩的吸盘（protrusible sucker）（图 1-23，图 1-24）。大多地表生活，取食节肢动物尸体或其他有机物，常在体背黏附泥土及各种碎屑作为伪装。老熟幼虫建造土室，并在其中化蛹。

4. 蛹

蛹为强颚离蛹（decticous exarate pupa），头部、腹部能做有限的运动。外观与成虫相似，但喙、翅和腹部都较短，触角、足和翅疏松地贴于体表（图 1-25）。

图 1-24 正在蜕皮的缠绕地蚊蝎蛉 ***Terrobittacus implicatus*** 幼虫侧面观

Fig. 1-24 Molting larva of ***Terrobittacus implicatus***, lateral view

图 1-25 缠绕地蚊蝎蛉 ***Terrobittacus implicatus*** 雌蛹

1. 腹面观 2. 侧面观 3. 背面观

Fig. 1-25 Female pupa of ***Terrobittacus implicatus***

1. Ventral view 2. Lateral view 3. Dorsal view

1. Adult

The Bittacidae are commonly called hangingflies, because the adults have three pairs of raptorial legs, and usually use their prehensile tarsi of front and middle legs to hang themselves from the edge of leaves or twigs of plants at repose (Fig. 1-14). They are very similar to the crane flies in Diptera, but can be differentiated from the latter by their two pairs of wings that are similar in size, while the hindwings of crane flies are greatly reduced into a pair of halteres. Therefore, they are commonly called “mosquito scorpionflies” literally in Chinese. In Taiwan,China, they are commonly called “false crane flies” .

(1) Head: The diagnostic characters in the head include the color, the stripes on ocellar triangle, the size of dorsal ocelli, and the morphology of compound eyes and antennae. The head is usually yellowish, dark, or reddish brown. In some species, the dorsum and rostrum are different in color. The ocellar triangle is generally surrounded by a deep pattern. Some species bear distinct ocellar bristles between the ocelli. The compound eyes are mostly long ellipsoidal. The size, shape of compound eyes and the distance between them are often important traits of identification. The scape and pedicel of antenna are short and thick, while the flagellum is filiform and sometimes ciliated (Fig. 1-15).

(2) Thorax: Strong and slightly humped. The pronotum is narrow, and often bears a few black setae along its anterior margin. The meso- and metanotum are variously colored and sometimes are uniformly tawny or dark brown. In some species the notum is different in color between the anterior and posterior portions. Some species possess a distinct notal stripe on the notum. Pleura are usually yellowish brown or cream-colored (Fig. 1-16).

Hangingflies have three pairs of slender raptorial legs. The legs are usually uniformly yellowish or reddish brown. In some species the two ends of femur and tibia are darkened. The length of basitarsus and the number of black spines on the two sides of the fourth tarsomere are frequently used in taxonomy. The fifth tarsomere with a single claw can fold back and secure with the inner margin of the enlarged fourth tarsomere. The inner margins of these two tarsomeres bear strong spines that allow them to form a retractable clasp (Fig. 1-17). This structure allows hangingflies to hang themselves or catch preys.

The generic characters of hangingflies are mainly concentrated on the wings. When describing the wings, we mainly use the following abbreviations and terms: A, anal vein; Av, anal crossvein; C, costa; CuA, anterior Cubitus; CuP, posterior Cubitus; CuV, cubital crossvein; FM, fork of Media; FM_{3+4}, fork of M_3 and M_4; FRs, fork of radial sector; h, humeral crossvein; M, Media; nygmata (cicular and chitinized spots on wing membrane); OM, origin of Media; ORs, origin of radial sector; Pcv, pterostigmal crossvein; pterostigma; Rs, radial sector; Sc, subcosta; Scv, subcostal crossvein; thyridium (a translucent spot on FM) (Fig. 1-18).

The frequently used characters of wings include the color, shape, ending position of Sc, the number of Pcv, coalesce condition of Rs, the length of 1A and its ending position, the coalesce condition between 1A and CuP in hindwings, *etc*. The shapes are varied greatly among different genera. For instance, the wing base is broad or stalked, and the apex is acute and slightly hook-like or rounded. The bands and spots are also important traits in taxonomy.

(3) Abdomen: The tergum Ⅸ (or epandrium) in males is modified into a pair of claspers, which are normally called "epandrial lobes". The shape of these lobes are significant characters in taxonomy. The gonopods in Bittacidae are not used as grasping organs, since the gonostylus is greatly reduced. The male aedeagus is located in the cavity formed by joined gonocoxites. In most Chinese species, the apex of aedeagus is greatly elongated into a coiled filament that is called the "penisfilum". In addition, there is often a pair of aedeagal lobes on the two basal sides of aedeagus. The tergum Ⅹ is reduced to various extents. The tergum Ⅺ and sternum Ⅺ, called proctigers, are modified and elongated, and are regarded as upper and lower branches. The upper branch of proctiger often extends beyond the two epandrial lobes, with its apex enlarged, hook-like or tufted in some species. The lower branch is usually shorter, hook-like and curving ventrad (Fig. 1-19).

The main diagnostic character in females lies in the morphology of the subgenital plate. The subgenital plate looks like fused by two valves, and sometimes with a membranous area medially. A pair of acute processes each project from the lateral margins of the subgenital plate distally. A number of thick setae are on the apical half.

Another diagnostic character is the tenth tergum that extends ventrad (Fig. 1-20).

2. Egg

Mostly spherical. The chorion is smooth in *Bittacus*, but is covered with latitude-like ornaments in some *Terrobittacus* species (Fig. 1-21). However , the eggs are nearly cuboidal in a few species, such as *Bittacus cirratus* (Fig. 1-22).

3. Larva

Eruciform. The head is hypognathous, with a pair of compound eyes and a single median ocellus. The body is furnished with rows of thick protuberances terminated in clavate setae on each somite. The first eight abdominal segments each bear a pair of ventral prolegs. The anal segment is terminated in a protrusible sucker (Fig. 1-23, Fig. 1-24). The larvae mostly live on the ground and feed on dead arthropods or other kinds of organic materials. The larvae camouflage by adhering soils and other debris on their backs. Once matured, the larvae build a chamber in earth and prepare to pupate.

4. Pupa

The pupae of hangingflies are decticous and exarate, with their heads and abdomens capable of making limited movements. The pupae resemble adults but with undeveloped rostrum, abdomen and wings. Their antennae, legs and wings are loosely attached to the body (Fig. 1-25).

（三）蝎蛉科 Panorpidae

蝎蛉科 Panorpidae 昆虫通称为蝎蛉(scorpionflies)，因雄虫腹部末端常膨大并上举，形似蝎子的尾巴(图 1-26) 而得名，中文俗名又称为举尾虫（ tail-lifting insects ）。

1. 成虫

（1）头部：蝎蛉的头部向下延伸成喙（rostrum），唇基（clypeus）极度延长，上唇（labrum）较短小；

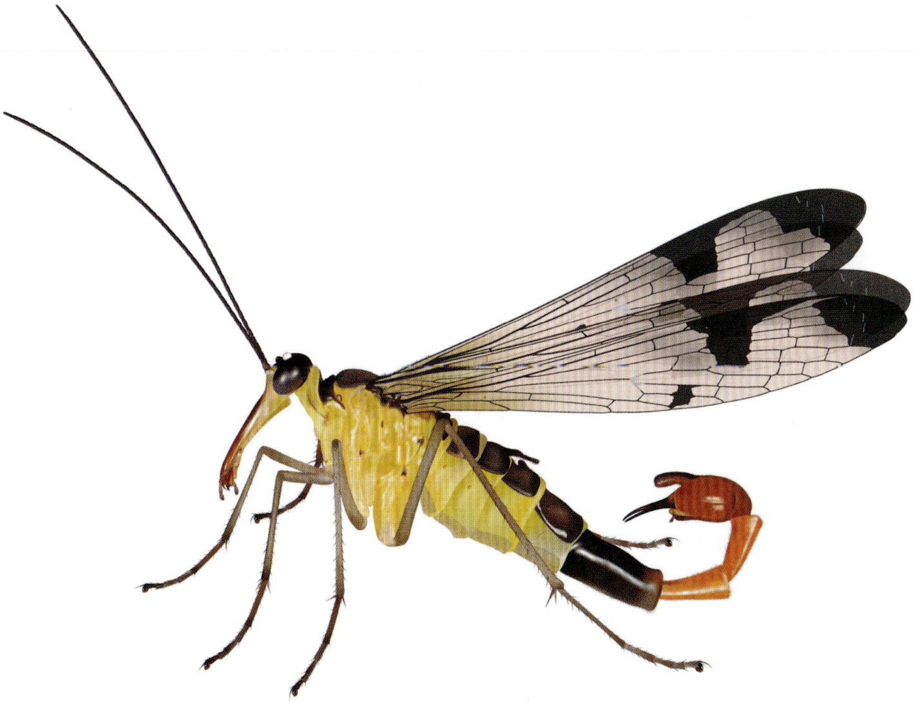

图 1-26 尼尔森新蝎蛉 *Neopanorpa nielseni* 雄虫侧面观

Fig. 1-26 Male adult of *Neopanorpa nielseni*, lateral view

咀嚼式口器位于喙的末端。一般地，长腹蝎蛉属 *Leptopanorpa* 昆虫和新蝎蛉属 *Neopanorpa* 昆虫的喙相对其他属昆虫要细长很多。喙一般黄色、黄褐色或黑褐色，有时两侧具明显的深色纵带。复眼（compound eyes）长椭球形。头顶大多黄色、黄褐色或黑褐色，有时具有各种斑纹。单眼三角区（ocellar triangle）常隆起且颜色加深，有时周围具各式斑纹。触角柄节和梗节短小且较粗，鞭节丝状、细长（图 1-27）。

（2）胸部：前胸背板（pronotum）略呈马鞍状，前缘常具 1 列粗刚毛。中、后胸背板（mesonotum and metanotum）颜色多样，大多为黄褐色或黑褐色，中央有时有浅色或深色的背中带（图 1-28）。足细长，胫节具 2 根端距（apical spurs），跗节末端具 2 枚爪（claws），爪内缘具梳状齿。

A  B

图 1-27 太白单角蝎蛉 *Cerapanorpa obtusa* 头部

A. 正面观 B. 侧面观

1. 复眼 2. 背单眼 3. 颊 4. 唇基 5. 下颚须 6. 上唇

Fig. 1-27 Head of *Cerapanorpa obtusa*

A. Frontal view B. Lateral view

1. compound eyes 2. dorsal ocellus 3. gena 4. clypeus 5. maxillary palp 6. labrum

翅 2 对，膜质，透明或半透明，大多无色，有时带有黄色或烟褐色。翅面常具一些斑纹，其形状和排列常被作为鉴定依据。许多种的翅面斑纹有规律地排列，典型的前翅斑纹由内向外分别为基斑（basal spot）、基带（basal band）、缘斑（marginal spot）、痣带（pterostigmal band）和端带（apical band）；其中痣带的后半部常常分为基支（basal branch）和端支（apical branch）(图 1-29)。蝎蛉科 Panorpidae 昆虫翅脉的命名系统与蚊蝎蛉科 Bittacidae 昆虫基本一致。M_4 基部强烈弯曲，是蝎蛉科 Panorpidae 的典型特征。1A 终点是否超过 Rs 起源点（ORs）是重要的属级分类特征。

（3）腹部：雄性腹部第 3 节背板后缘中央常有 1 个向后延伸的突起，称为背中突（notal organ），与第 4 节背板的后背突（postnotal organ）形成挟持构造（图 1-30），在交配时能夹住雌性翅的前缘。其有无、形状、发达程度等也是重要的分类依据。

雄性腹部第 6 ~ 8 节背、腹板愈合成筒形。有时第 6 节背板末端具 1 或 2 个指状突起，称为臀角（anal horns），也是重要的分类依据。第 7、8 节短于或长于第 6 节；有的基部不缢缩，整体呈筒形；有的基部稍微缢缩，整体呈倒锥形；有的基部强烈缢缩成为柄状。在长腹蝎蛉属 *Leptopanorpa* 昆虫中，雄性的腹部（尤其是第 6 ~ 8 节）极度延长（图 1-31）。

腹部末端膨大，称为生殖球（genital bulb），如蝎尾状向体背反曲（图 1-32）。第 9 节腹板又称为下生殖板（hypandrium），其末端常分裂为两瓣状构造，称为下瓣（hypovalves）；第 9 节背板又称为上生殖板

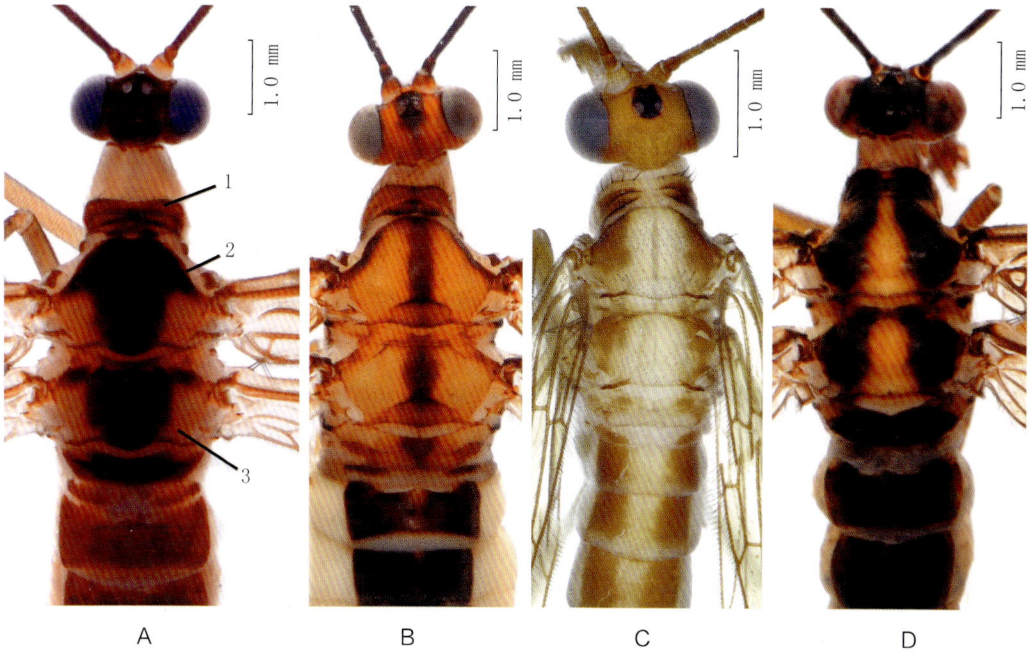

图 1-28 蝎蛉胸部背面观

A. 丽新蝎蛉 *Neopanorpa pulchra* B. 网翅新蝎蛉 *Neopanorpa caveata*
C. 昆明蝎蛉 *Panorpa kunmingensis* D. 楚南蝎蛉 *Panorpa sonani*
1. 前胸背板 2. 中胸背板 3. 后胸背板

Fig. 1-28 Thoraces of Panorpidae spp., dorsal view

A. *Neopanorpa pulchra* B. *Neopanorpa caveata* C. *Panorpa kunmingensis* D. *Panorpa sonani*
1. pronotum 2. mesonotum 3. metanotum

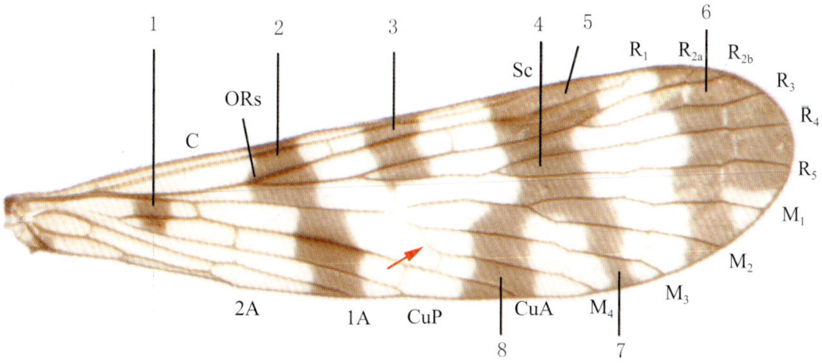

图 1-29 杨氏蝎蛉 ***Panorpa yangi*** 前翅脉序、脉斑（红色箭头示弯曲的 M_4 基部）
1. 基斑 2. 基带 3. 缘斑 4. 痣带 5. 翅痣 6. 端带 7. 端支 8. 基支

Fig. 1-29 Venation and markings of forewing of ***Panorpa yangi*** (arrow points to the curved base of M_4)
1. basal spot 2. basal band 3. marginal spot 4. pterostigma band 5. pterostigma
6. apical band 7. apical branch 8. basal branch

（epandrium），末端有时平截，有时有缺刻并在两侧形成 1 对细长多毛的突起。第 9 节的生殖肢（gonopod）发达，分为 1 对生殖肢基节（gonocoxites）和 1 对生殖刺突（gonostyli, 又称生殖肢端节）。生殖肢基节腹面端部内缘常有 1 簇长刚毛，有时内缘中部或端部形成一些向内的突起；左右基节基半部愈合，形成外生殖器的主体；生殖刺突钳状，内缘常具 1 个近三角形的中齿（median tooth）和 1 个形状多变的亚基齿（subbasal tooth），在交配时发挥抱握作用，能挟持雌虫的腹部末端。阳茎（aedeagus）位于生殖肢基节愈合形成的凹窝中，为左右对称的瓣状构造，分为 1 对腹瓣（ventral valves）和 1 对背瓣（dorsal valves）。在阳茎两侧着生有 1 对形状不一的骨化构造，称为阳基侧突（parameres），其形状是重要的分类特征。在

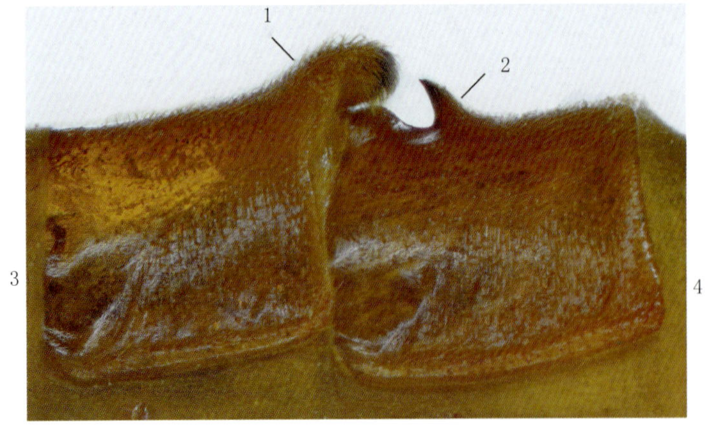

图 1-30　刘氏蝎蛉 ***Panorpa liui*** 雄性腹部第 3 和 4 节背板侧面观（示背中突及后背突）

1. 背中突 2. 后背突 3. 腹部第 3 节背板 4. 腹部第 4 节背板

Fig. 1-30　Male terga III and IV of ***Panorpa liui***, lateral view (showing the notal and postnotal organs)

1. notal organ 2. postnotal organ 3. tergum Ⅲ 4. tergum Ⅳ

图 1-31　蝎蛉科 Panorpidae 雄性腹部侧面观（比例尺：1.0 mm）

1. 刘氏蝎蛉 *Panorpa liui* 2. 长白山蝎蛉 *Panorpa changbaishana* 3. 昆明蝎蛉 *Panorpa kunmingensis* 4. 反曲蝎蛉 *Panorpa reflexa* 5. 长瓣叉蝎蛉 *Furcatopanorpa longihypovalva* 6. 任氏单角蝎蛉 *Cerapanorpa reni* 7. 大双角蝎蛉 *Dicerapanorpa magna* 8. 染翅华蝎蛉 *Sinopanorpa tincta* 9. 河南新蝎蛉 *Neopanorpa longiprocessa* 10. 尼尔森新蝎蛉 *Neopanorpa nielseni*

Fig. 1-31　Abdomens of male Panorpidae, lateral view (Scale bars：1.0 mm)

1. *Panorpa liui* 2. *Panorpa changbaishana* 3. *Panorpa kunmingensis* 4. *Panorpa reflexa* 5. *Furcatopanorpa longihypovalva* 6. *Cerapanorpa reni* 7. *Dicerapanorpa magna* 8. *Sinopanorpa tincta* 9. *Neopanorpa longiprocessa* 10. *Neopanorpa nielseni*

阳茎基部的腹面，有 1 对叶状的性信息素腺（sex pheromone glands），在求偶时能膨大翻出。第 10、11 节短小，通常被第 9 节背板盖住。第 10 节上着生 1 对不分节的尾须（cerci）。第 11 节末端为肛门（anus）（图 1-33）。

雌性腹部第 9 节腹面具 1 个片状构造，其末端钝圆、平截或具 V 形缺刻，形似两瓣愈合而成，称为下板（subgenital plate）。在下生殖板上方的生殖腔内有 1 个高度骨化的结构，称作生殖板（genital plate）或中板（medigynium），有些作者称为内骨（internal skeleton），其形态是雌性比较重要的分类依据。生殖板常由 1 个片状的主板（main plate）和中央的中轴（axis）构成；其中主板后方两侧常具 1 对突起，称为后臂（posterior arms）；中轴的后端愈合，腹面具 1 块膨大的花饰区，交配孔位于其末端；前端分为平行或分离的两支，超出主板或完全藏于主板之内；受精囊导管从中轴的中央穿过，通向交配孔。有的种生

图 1-32 雅各布森长腹蝎蛉 *Leptopanorpa jacobsoni* 雄成虫侧面观（摄于印度尼西亚爪哇岛温加岚火山）

Fig. 1-32 Male adult of ***Leptopanorpa jacobsoni***, lateral view（Photo taken from Mount Ungaran, Java island, indonesia）

图 1-33 滴蝎蛉 *Panorpa guttata* 雄性外生殖器

A. 腹面观 B. 右侧面观 C. 背面观

1. 生殖刺突 2. 阳茎背瓣 3. 阳基侧突 4. 下瓣 5. 生殖肢基节 6. 第 9 节腹板基柄 7. 尾须 8. 上生殖板

Fig. 1-33 Male genitalia of ***Panorpa guttata***

A. Ventral view B.Right-lateral view C. Dorsal view

1. gonostylus 2. dorsal valve of aedeagus 3. paramere 4. hypovalve 5. gonocoxite 6. basal stalk of hypandrium 7. cercus 8. epandrium

图 1-34　蝎蛉科 Panorpidae 雌性下生殖板和生殖板（比例尺：0.2 mm）

A. B. 南五台单角蝎蛉 *Cerapanorpa nanwutaina*　C. D. 长瓣叉蝎蛉 *Furcatopanorpa longihypovalva*　E. F. 福建蝎蛉 *Panorpa fukiensis*　G. H. 染翅华蝎蛉 *Sinopanorpa tincta*　I. J. 大双角蝎蛉 *Dicerapanorpa magna*　K. L. 丽新蝎蛉 *Neopanorpa pulchra*　M. N. 刘氏蝎蛉 *Panorpa liui*

1. 后臂　2. 主板　3. 基片　4. 中轴

Fig. 1-34　Subgenital and genital plates of female Panorpidae（Scale bars: 0.2 mm）

A. B. *Cerapanorpa nanwutaina*　C. D. *Furcatopanorpa longihypovalva*　E. F. *Panorpa fukiensis*　G. H. *Sinopanorpa tincta*　I. J. *Dicerapanorpa magna*　K. L. *Neopanorpa pulchra*　M. N. *Panorpa liui*

1. posterior arm　2. main plate　3. basal plates　4. axis

殖板两侧还附有 1 或 2 对基片（basal plates）（图 1-34）。

2. 卵

卵一般为椭球形，卵壳表面具有网状、多边形的几丁质纹饰（图 1-35）。一般由雌虫将腹部插入松软的浅层土壤后产下（图 1-36，图 1-37）。

图 1-35　单角蝎蛉 ***Cerapanorpa* sp.** 卵

Fig. 1-35　An egg of ***Cerapanorpa*** sp.

图 1-36　正在产卵的雌性单角蝎蛉 ***Cerapanorpa* sp.** 成虫侧面观

Fig. 1-36　Eggs-laying female adult of ***Cerapanorpa*** sp., lateral view

图 1-37　正在产卵的雌性大双角蝎蛉 ***Dicerapanorpa magna*** 侧面观

Fig. 1-37　Eggs-laying female adult of ***Dicerapanorpa magna***, lateral view

3. 幼虫

幼虫为蠋式（eruciform），体圆柱形，被稀疏的短毛（图 1-38，图 1-39）；头部具 1 对复眼（compound eyes），无侧单眼（stemmata）（图 1-40）；胸部具 3 对胸足；腹部前 8 节每节各具 1 对腹足（prolegs），尾节具可伸缩的吸盘（protrusible sucker）；生活于土壤表面、腐殖质层或土壤中，以各种有机质碎屑为食，但主要取食死亡的昆虫和其他节肢动物（Byers & Thornhill，1983）。老熟幼虫建造土室，在其中化蛹。

4. 蛹

蛹为强颚离蛹（decticous exarate pupa），外形与成虫十分相似，但喙、翅和腹部较短；雄蛹可见膨大并反曲的外生殖器（图 1-41）。

图 1-38　刘氏蝎蛉 ***Panorpa liui*** 幼虫侧面观

Fig. 1-38　Larva of ***Panorpa liui***, lateral view

图 1-39　尼尔森新蝎蛉 ***Neopanorpa nielseni*** 幼虫侧面观

Fig. 1-39　Larva of ***Neopanorpa nielseni***, lateral view

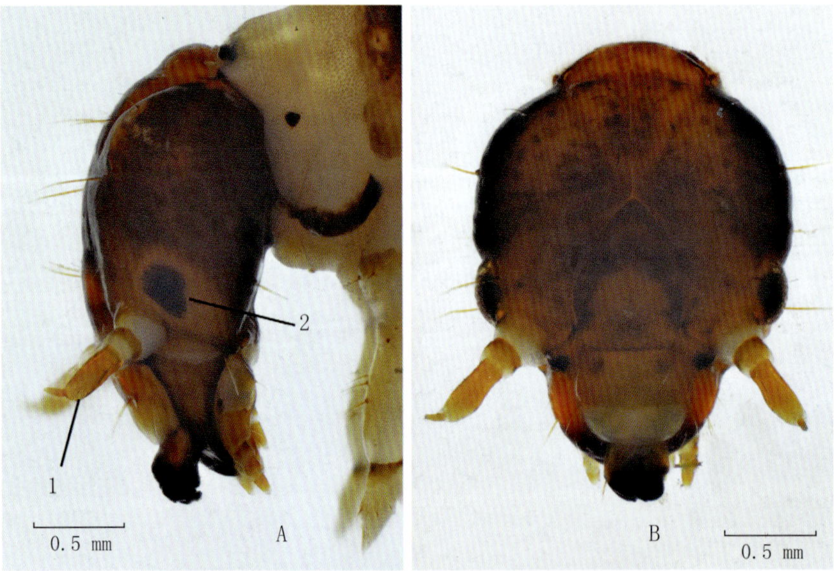

图 1-40　大双角蝎蛉 ***Dicerapanorpa magna*** 幼虫头部

A. 侧面观　B. 正面观　1. 触角　2. 复眼

Fig. 1-40　Head of the larva of ***Dicerapanorpa magna***

A. Lateral view　B. Frontal view　1. Antenna　2. Compound eyes

The insects of Panorpidae are commonly called scorpionflies, because the male genitalia are enlarged and recurved backward, resembling the stinger of scorpions (Fig. 1-26). They are also called "tail-lifting insects" literally in Chinese.

1. Adult

(1) Head: Prolonged ventrad into a rostrum, with a greatly elongated clypeus and a smaller labrum. The

5 6

图 1-41 蝎蛉蛹侧面观（比例尺：1.0 mm）

1. 3. 5. 雄 2. 4. 6. 雌 1. 2. 长蝎蛉 *Panorpa macrostyla* 3. 4. 昆明蝎蛉 *Panorpa kunmingensis* 5. 6. 刘氏蝎蛉 *Panorpa liui*

Fig. 1-41 Pupae of Panorpidae, lateral view (Scale bars : 1.0 mm)

1. 3. 5. males 2. 4. 6. females 1. 2. *Panorpa macrostyla* 3. 4. *Panorpa kunmingensis* 5. 6. *Panorpa liui*

chewing mouthparts are located at the end of the rostrum. In generally, the rostrum is relatively slender in *Leptopanorpa* and *Neopanorpa* than in other genera. The rostrums are mostly yellow, yellowish or dark brown, with a pair of dark longitudinal stripes laterally in some species. The compound eyes are mostly long ellipsoidal. The vertex is usually yellow, yellowish or dark brown, with various patterns. The ocellar triangle is usually protuberated and darkened in color. Various patterns enclose the ocellar triangle in some species. The antennae have short and thick scape and pedicel, but with slender and filiform flagellum (Fig. 1-27).

(2) Thorax: The pronotum is somewhat saddle-shaped and often bears a row of thick setae on its anterior margin. The mesonotum and metanotum are variously colored from yellowish to dark brown, with a paler or darkened mesal stripe in some species (Fig. 1-28). The legs are slender, with a pair of apical spurs on tibia. The pretarsus has a pair of claws, which bear comb-like teeth on their inner margins.

Two pairs of membranous wings are hyaline or translucent, mostly colorless or tinged with yellow or smoky brown. Varied markings on wings are important traits in taxonomy. In most species, the wing markings are arranged in a series of regular patterns, and are named as follows from the base to the apex: basal spot, basal band, marginal spot, pterostigmal band, and apical band(Fig. 1-29). The posterior half of the pterostigmal band is often split into a basal and an apical branch. The nomenclatural system of venation is almost the same with that in Bittacidae. The curved base of M_4 is a typical trait of Panorpidae. It is significant whether 1A ends beyond the level of ORs or not in the assignment of genera.

(3) Abdomen: In male scorpionflies, a protuberance on the posterior margin of tergum Ⅲ is called the notal organ, and is used as a clamp device with the postnotal organ on tergum Ⅸ. This device is used by the males to clamp the wings of the female during copulation (Fig. 1-30). The presence or absence, shape and development degree of the notal organ are very important diagnostic characters.

The terga and sterna of male abdominal segments Ⅵ (A6) – Ⅷ (A8) are fused altogether without membranous pleura. In some species, the tergum Ⅵ has one or two finger-like horns along the posterior margin, called "anal horns", important characters in taxonomy. A7 and A8 are shorter or longer than A6. Some of these two segments are cylindrical, some are constricted basally and inverted cone-shaped, and the others are intensively constricted and stalk-like basally. The abdomen (especially A6–A8) is greatly elongated in the males of *Leptopanorpa* (Fig. 1-31).

The terminal segments are enlarged and called the "genital bulb", which turns backward and resembles the stinger of scorpions(Fig. 1-32). Sternum Ⅸ (or hypandrium) is often split into a pair of hypovalves. Tergum Ⅸ (or epandrium) has its terminal margin truncated or emarginate and forming a pairs of slender setose lobes laterally. The gonopods on A9 are greatly developed and formed by a pair of fused gonocoxites and a pair of gonostyli. In general, the gonocoxites each bear a tuft of long setae on their ventral apex. In some species, the inner margin of the gonocoxite has a triangular protuberance. The paired gonocoxites are fused at basal half to form the main body of the genital bulb. Usually, the pincer-like gonostyli each have a subtriangular median tooth, and a variously shaped subbasal process on the inner margin. The gonostyli serve to grasp the female's terminalia during copulation. The aedeagus is composed of two pairs of phallomeres, bilaterally symmetrical, and is located in the cavity formed by the fused gonocoxites, with a pair of ventral valves and a pair of dorsal valves. Parameres are a pair of sclerotized structures located besides the aedeagus, and are variously shaped and are significant in taxonomy. A pair of sex pheromone glands is present ventral to the base of aedeagus, and can be expanded during the courtship. A10 and A11 are greatly reduced and usually covered by the epandrium. A pair of unsegmented cerci is located on the two sides of tergum Ⅹ. The anus is on the apex of A11 (Fig. 1-33) .

In female's A9, there is a plate-like structure ventrally, called the subgenital plate. Subgenital plate is located below tergumm Ⅸ , superficially fused by two separate plates, and rounded, truncated or V-shaped emargination terminally. The genital plate (or medigynium, or internal skeleton) is inside the genital chamber above the subgenital plate. The morphology of the genital plate is significant in the diagnose of species based on the females. The genital plate is made up of a main plate and a median axis. A pair of posterior arms are located on the posterior portion of the main plate. The copulatory pore lies in the terminal end of the enlarged and decorated area of axis. The anterior half of the axis is split into two parallel or divergent apodemes, with their anterior apexes concealed in or extending beyond the main plate. The spermathecal duct lies in the axis and opening at the copulatory pore. In some species, the main plate is accompanied by one or two pairs of basal plates laterally (Fig. 1-34).

2. Egg

The eggs of scorpionflies are usually long ellipsoidal, and furnished with sclerotized, polygonal networks on the surface of the chorion (Fig. 1-35). The eggs are usually laid in shallow layers of soft soil, which is penetrated by the female's terminal abdomen (Fig. 1-36, Fig. 1-37).

3. Larva

Eruciform, with sparse short setae on the cylindrical trunk (Fig. 1-38, Fig. 1-39) . The head bears a pair of prominent compound eyes instead of stemmata on the lateral side (Fig. 1-40) . They possess three pairs of thoracic legs and eight pairs of ventral prolegs. A protrusible sucker is on the last abdominal segment. The larvae live on the surface of soil, or in the layer of organic residue or in the soil, and feed on various organic debris, but they mainly feed on dead insects and other arthropods (Byers & Thornhill, 1983). Once matured, the larvae builds a chamber in the soil and prepares to pupate.

4. Pupa

Decticous exarate, and resemble fully grown adults but with much shortened rostrum, wings and abdomen. The male pupa has enlarged and recurved male genitalia (Fig. 1-41).

（四）拟蝎蛉科 Panorpodidae

拟蝎蛉科 Panorpodidae 昆虫通称为拟蝎蛉（short-faced scorpionflies），与蝎蛉科 Panorpidae 昆虫相似，但身体许多部分较短（图 1-42）。

图 1-42 宽甸拟蝎蛉 ***Panorpodes kuandianensis*** 雄成虫侧面观

Fig. 1-42 Male adult of ***Panorpodes kuandianensis***, lateral view

1. 成虫

（1）头部：拟蝎蛉的头部与蝎蛉科 Panorpidae 昆虫相似，但喙较短，从正面可见明显的颊齿，上唇端部具 V 形缺刻。单眼三角区有 4 根明显的单眼鬃（图 1-43）。

（2）胸部：前胸背板前缘一般有 3 对粗刚毛，后缘有 2 对。跗节爪简单，内缘光滑无齿。前、后翅 M_4 基部平直，不像蝎蛉科 Panorpidae 昆虫那样弯曲（图 1-44）。

（3）腹部：雄虫腹部第 6 ~ 8 节与之前的节几乎等长，并不像蝎蛉科 Panorpidae 那样延长。第 7、8 节背、腹板分离，而不像蝎蛉科 Panorpidae 昆虫那样完全愈合呈筒状。

2. 幼虫

拟蝎蛉科 Panorpodidae 幼虫蠋式，体表具短刚毛；头部下口式，具 1 对粗壮的触角，没有任何明显的视觉器官；3 对胸足粗壮；8 个腹足位于腹中线，仅稍隆起，尾节具 1 可伸缩的吸盘（图 1-45）。

The Panorpodidae are commonly called short-faced scorpionflies. They are very similar to the true scorpionflies in Panorpidae, but have shorter body parts (Fig. 1-42).

1. Adult

(1) Head: The head is similar to that of Panorpidae, but has a rather shorter rostrum. In addition, a pair of cheek teeth is distinct and can be observed in frontal view. The labrum has a deep V-shaped emargination distally. Four distinct ocellar bristles are present in the ocellar triangle (Fig. 1-43).

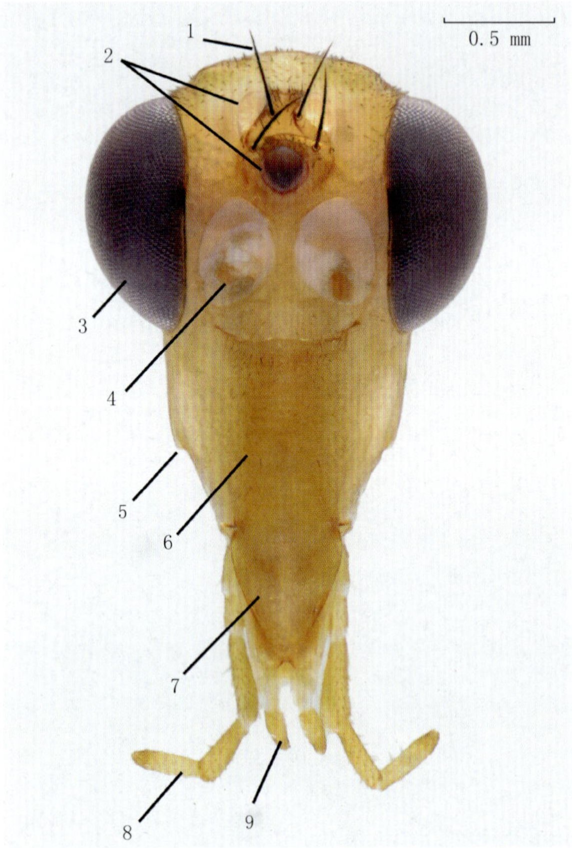

图 1-43 宽甸拟蝎蛉 ***Panorpodes kuandianensis*** 头部正面观（触角已去除）

1. 单眼鬃 2. 背单眼 3. 复眼 4. 触角窝 5. 颊刺 6. 唇基 7. 上唇 8. 下颚须 9. 下唇须

Fig. 1-43 Head of ***Panorpodes kuandianensis*** with antennae removed, frontal view

1. ocellar bristle 2. dorsal ocellus 3. compound eyes 4. antennal socket 5. genal spine 6. clypeus 7. labrum 8. maxillary palp 9. labial palp

(2) Thorax: In general, three pairs of thick setae are on the anterior margin of pronotum, while two pairs are on the posterior margin. A pair of claws is simple and lacks comb-like teeth on their inner margin. The vein M_4 in both forewings and hindwings are straight and not curved as in Panorpidae (Fig. 1-44).

(3) Abdomen: A6–A8 of males are nearly as long as the proceeding segments, not elongated as in Panorpidae. In addition, the A7 and A8 are membranous in the pleural region, with separated tergum and sternum, different from those fully-fused cylindrical segments in Panorpidae.

2. Larva

The larvae are eruciform with a hypognathous head and sparse short setae on the trunk. They have one pair

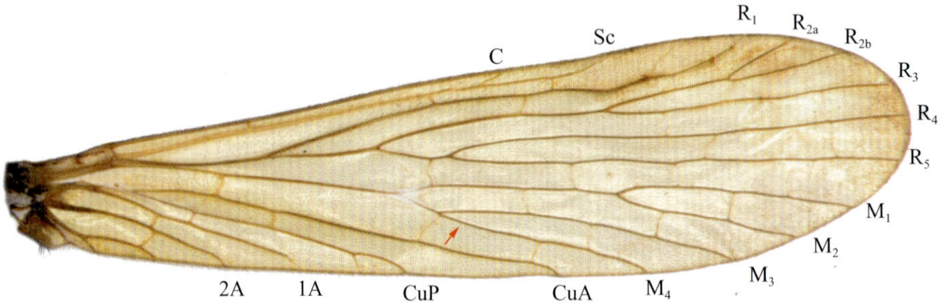

图 1-44 宽甸拟蝎蛉 ***Panorpodes kuandianensis*** 前翅（红色箭头示平直的 M_4 基部）

Fig. 1-44 Forewing of ***Panorpodes kuandianensis*** (arrow points to the straight base of M_4)

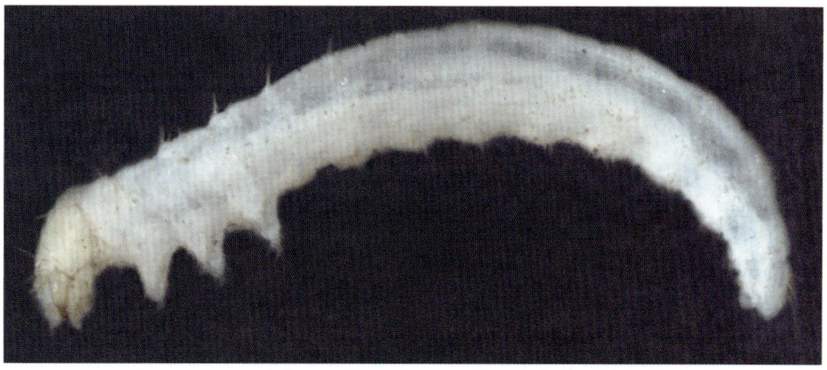

图 1-45 宽甸拟蝎蛉 ***Panorpodes kuandianensis*** 幼虫侧面观（姜碌摄）

Fig. 1-45 Larva of ***Panorpodes kuandianensis,*** lateral view (photo by Lu Jiang)

of stout antennae but lack any distinct visual organs. They have three pairs of stout thoracic legs and eight slightly protruded prolegs on the midventral line. A protrusible sucker is on the terminal segment (Fig. 1-45).

六、生物学和生态学 Biology and Ecology

图 1-46　正在清理触角的刘氏蝎蛉 ***Panorpa liui*** 雄虫

Fig. 1-46　A male of ***Panorpa liui*** cleans his antennae

（一）生物学和栖息环境 Biology and Habitats

适宜长翅目 Mecoptera 昆虫生活的环境一般为湿润的森林或灌丛（图 1-47，图 1-48，图 1-49）。

Environments suitable for Mecoptera often include moist forest or shrubs (Fig. 1-47, Fig. 1-48, Fig. 1-49).

图 1-47 亚热带常绿阔叶林下茂密的草本、灌木植物丛（云南永德棠梨山）
Fig. 1-47 Dense grasses and shrubs under a subtropical evergreen broad-leaved forest（Mount Tangli, Yongde, Yunnan）

图 1-48 林场道路两侧茂密的植物（吉林长白山）
Fig. 1-48 Dense plantations aside a road of a forest farm（Changbai Mountains, Jilin）

图 1-49　山区农家周围茂密的森林（陕西天台山国家森林公园）

Fig. 1-49　Dense forests around a farmyard of a mountainous region (Tiantaishan National Forest Park,Shaanxi)

图 1-50　挂在植物叶片上的钩地蚊蝎蛉 ***Terrobittacus rostratus*** 雌成虫侧面观

Fig. 1-50　Female adult of ***Terrobittacus rostratus*** hanging on a leaf, lateral view

1. 蚊蝎蛉科 Bittacidae

蚊蝎蛉一般一年发生 1 代，以卵或幼虫滞育越冬（Byers & Thornhill, 1983）。成虫大多于翌年 5 ~ 8 月羽化。

蚊蝎蛉一般生活于湿润的环境中，尤以海拔 1 000 m 以上的冷凉林区较为常见。在有溪水流动的山沟、林下开阔的平地，常常长有茂盛的草本和灌木植物，蚊蝎蛉就在这些植物丛中活动。由于足的特化，它们不能站立或行走，只能用前足和中足将自己挂在物体上（图 1-50）。蚊蝎蛉大多对湿度要求十分严格，在被带离高湿度的原始环境之后，很容易失去活力而死亡。

1. Bittacidae

The Bittacidae are mostly univoltine insects. They overwinter in diapause as eggs or larvae (Byers & Thornhill, 1983). The adults usually emerge from May to August next year.

Hangingflies mainly live in habitats with high moisture. They are frequently found in the cool forests at the elevation higher than 1 000 m. The suitable habitats that hangingflies prefer include dense herbs and shrubs aside a stream, or an open field under the woods. Owing to their specialized prehensile legs, they are unable to walk like other mecopterans, but can only hang themselves on other objects with fore and sometimes also middle legs (Fig. 1-50) . They are quite strict for the humidity of their habitats. Once left the high-moisture habitats, they might soon be moribund and perish.

2. 蝎蛉科 Panorpidae

蝎蛉科 Panorpidae 昆虫一年发生 1 代或 2 代，一般以老熟幼虫越冬。在中国北方，成虫于每年 4 月下旬开始出现。在南方的一些省份，如台湾、广东和海南，全年都有蝎蛉的发生，但不同物种的出现月份不同。

相对而言，蝎蛉科 Panorpidae 昆虫对湿度的要求并不如蚊蝎蛉科 Bittacidae 昆虫那样严格，因此分布范围时常更广。蝎蛉较喜冷凉的温度，在炎热的季节或地区一般没有成虫出现。在中国北方，蝎蛉大多发生在夏初至秋末，海拔 1 000 m 以上、气候凉爽的山区。而纬度较高的东北，平原上也有蝎蛉分布。在中国南方，蝎蛉可以分布于低至 200 m 的丘陵，高到超过 3 000 m 的高山地区。蝎蛉大多活动于草本、灌木的叶片上（图 1-51）。在阳光强烈、降水或风较大时，常躲在植物叶片下方而不易被发现。在被惊扰时，也会以类似“假死”的方式掉落到地面以躲避危险。

2. Panorpidae

The Panorpidae are univoltine or bivoltine insects. They usually overwinter in diapause as grown larvae. In the northern China, the adults begin their emergence in late April. In the southern provinces such as Taiwan, Guangdong and Hainan, however, scorpionflies are active during the whole year, with different species emerging in different months.

Relatively, scorpionflies have a wider range of humidity requirements than hangingflies, therefore they are often more widespread in distribution. They prefer cool temperature, and are usually absent in some hot seasons or areas. In the northern China, scorpionflies emerge from early summer to late autumn, and inhabit cool mountains of an elevation over 1 000 m. In northeastern China, however, they can live well in plains with high latitudes. In the southern China, scorpionflies have their distributional ranges as low as 200 m in hills, or as high as above 3 000 m in mountains. Scorpionflies are usually active on the leaves of grasses and shrubs. They often hide below the leaves under direct sunlight, rains or strong winds, and are hard to be find (Fig. 1-51). A "feign death" strategy is utilized to avoid danger by falling down to the ground when they encounter disturbance.

图 1-51　停歇于植物叶片上的长瓣叉蝎蛉 *Furcatopanorpa longihypovalva* 雄成虫侧面观

Fig. 1-51　Male adult of *Furcatopanorpa longihypovalva* resting on a leaf, lateral view

3. 拟蝎蛉科 Panorpodidae

拟蝎蛉科 Panorpodidae 生物学、栖息环境与蝎蛉科 Panorpidae 类似。

3. Panorpodidae

The biology and habitats of Panorpodidae are similar to those of Panorpidae.

（二）食性 Feeding Habits

1. 蚊蝎蛉科 Bittacidae

蚊蝎蛉为捕食性昆虫（图 1-52），捕食方式大致分为以下 3 种：一是用前足将身体悬挂于植物上，伸展后足，伏击撞到的小虫；二是悬挂时用后足在周围主动扫动捕猎；三是在飞行中捕捉猎物（Byers & Thornhill, 1983）。少数情况下，蚊蝎蛉也会从其他个体那里抢夺或盗取猎物，甚至同类相残。抓到猎物后，蚊蝎蛉立即将猎物送到口器旁，用上颚刺破后者的体壁，将喙插入取食；取食完毕后，将剩余的残渣丢弃（谭江丽和花保祯，2006）。

蚊蝎蛉的幼虫为腐食性，取食土壤中各类有机物残渣，主要为死亡的节肢动物尸体等（图 1-53）。

1. Bittacidae

Hangingflies are predatory insects (Fig. 1-52). They mainly adopt three predatory strategies: (1) hang themselves on plants using forelegs, and ambush the contacted small insects using their hindlegs; (2) proactively catch preys with sweeping hindlegs while hanging; (3) catch preys in flight (Byers & Thornhill, 1983). In some rare scenes, they rob or steal prey from other individuals, even attack their own kind. Once a prey is caught,

图 1-52　纹翅蚊蝎蛉 ***Bittacus strigatus*** 捕猎大蚊

Fig. 1-52　***Bittacus strigatus*** catching a crane fly

图 1-53　用家蝇围蛹饲喂扁蚊蝎蛉 ***Bittacus planus*** 幼虫

Fig. 1-53　Rearing larvae of ***Bittacus planus*** with house fly puparia

the hangingfly immediately serves it to its mouthparts and pierces the body wall of the victim with mandibles.

The hangingfly may insert its rostrum into the victim's body and enjoys a meal. The food residues will then be discarded (Tan and Hua, 2006).

The larvae of hangingflies are saprophagous. They feed on various kinds of organic debris in the soil, especially dead arthropods (Fig. 1-53).

2. 蝎蛉科 Panorpidae

蝎蛉科 Panorpidae 昆虫成虫为腐食性，主要取食死亡或濒死的节肢动物（图 1-54），偶尔也能取食花粉、花蜜或水果（Palmer, 2010）。少数情况下，蝎蛉也取食死亡的脊椎动物（图 1-55），甚至人类尸体，因而可以被用于法医学研究（Lindgren *et al*., 2015）。蝎蛉一般靠嗅觉发现食物。食物不充足时会出现几头成虫同时取食同一食物的现象。有时甚至强壮的个体会攻击并取食弱小或病残的个体。取食时，蝎蛉先用上颚刺破猎物外骨骼，之后将喙伸进猎物内部取食，一般仅取食猎物的软组织而留下坚硬的体壁。

图 1-54 刘氏蝎蛉 ***Panorpa liui*** 取食死亡的叶甲幼虫（左：雄 右：雌）

Fig. 1-54 Adults of ***Panorpa liui*** feed on a dead leaf beetle larva (Left: male right: female)

蝎蛉还有一种独特的取食行为，即盗食蜘蛛网中的昆虫尸体。为了防止被蜘蛛捕食，它们甚至能吐出具有消化功能的棕色液体来溶解缠住自己的蛛丝（Thornhill, 1975）。

蝎蛉科 Panorpidae 昆虫幼虫为腐食性，通常取食节肢动物的尸体或其他有机物（Jiang & Hua, 2013, 2015）（图 1-56）。

2. Panorpidae

The adults are saprophagous, mainly feeding on dead or dying arthropods (Fig. 1-54), rarely on pollens, nectar or fruits (Palmer, 2010). They are also recorded to feed on dead vertebrates (Fig. 1-55), even human corpse in a weird case. Therefore, they can be valuable in forensic research (Lindgren *et al*., 2015). They often forage by means of smelling. In a shortage of food, a troop of two to a dozen individuals may gather and feed on a single prey. The stronger individuals will attack and try to prey on the smaller, weaker or crippled ones in some situations. They often tear the prey's exoskeleton apart with their mandibles, and then insert their rostrum

图 1-55 单角蝎蛉 ***Cerapanorpa* sp.** 雌虫取食死亡的蛙

Fig. 1-55 Female adult of ***Cerapanorpa*** sp. feeds on a dead frog

图 1-56 用剪碎的黄粉虫饲养蝎蛉幼虫

Fig. 1-56 Feeding scorpionfly larvae with snipped mealworm larvae

inside the prey's body in order to feed. They usually consume the soft tissue and fluids, but discard the tough exoskeletons.

A weird but unique behavior of scorpionflies is that they stealthily eat the host's prey as kleptoparasites of web-building spiders. To avoid being attacked by the spider, scorpionflies can regurgitate a brown enteric fluid to dissolve the web that entangled them (Thornhill, 1975).

The larvae are saprophagous and feed generally on dead arthropods or other organic debris (Jiang & Hua, 2013, 2015) (Fig. 1-56).

3. 拟蝎蛉科 Panorpodidae

拟蝎蛉科 Panorpodidae 昆虫成虫仅被观察到取食植物性食物，如叶表皮和叶肉组织（Byers, 1997）。

拟蝎蛉科 Panorpodidae 昆虫幼虫的食性至今未知（Byers, 1997; Jiang & Hua, 2014）。

3. Panorpodidae

The adults of Panorpodidae have merely been observed to feed on vegetable materials, such as epidermis and mesophyll of leaves (Byers, 1997).

Their larval feeding habits remain a mystery to date (Byers, 1997; Jiang & Hua, 2014).

（三）求偶和交配行为 Courtship and Mating Behavior

1. 蚊蝎蛉科 Bittacidae

图 1-57　扁蚊蝎蛉 ***Bittacus planus*** 正在交配，同时共同取食猎物（姜碌摄）

Fig. 1-57　Adults of ***Bittacus planus*** in a face-to-face copulation while eating a prey together（photo by Lu Jiang）

蚊蝎蛉科 Bittacidae 昆虫在求偶时有着独特的“献礼”行为。雄虫捕捉到猎物后，会进行短暂取食。如果猎物符合需要，它将用后足继续抓住猎物，在植物间做短距离飞行或挂在叶片上，翻出腹部的性信息素腺，释放性信息素以吸引雌虫（Thornhill, 1976; Kaltenbach, 1978）（图 1-57）。这种以猎物作为彩礼的求偶行为，在其他昆虫中比较少见。有些蚊蝎蛉，如分布于日本的马斯特里蚊蝎蛉 *Bittacus mastrillii* 的雄虫，也能向雌虫提供唾腺分泌物作为彩礼（Iwasaki, 1996）。

蚊蝎蛉两性之间采用面对面的方式进行交配，交配方式迥异于长翅目 Mecoptera 其他科昆虫，甚至在整个昆虫纲中都是比较罕见的。扁蚊蝎蛉 *Bittacus planus* 交配时，雄虫面对雌虫，将腹部扭转大约 180° 后，用其铗状的上生殖瓣夹住雌虫的腹部末端，以便它们的生殖器相互接触（Gao & Hua, 2013）。在交配期，猎物会被雌、雄虫从两侧共同取食，或由雌虫单独取食，或由雄虫单独取食。

1. Bittacidae

The Bittacidae exhibit a unique courtship behavior called "nuptial feeding" . When a male catches a prey, he first feeds on it briefly. If the prey is proper, he hovers over vegetation or hang himself on a leaf while carrying the prey with his hind legs. Later, he everts his abdominal sex pheromone glands, and disperses a "sexual perfume" to attract the females (Thornhill, 1976; Kaltenbach, 1978) (Fig. 1-57) . It is uncommon to use a prey as a nuptial gift in other insects. However, the male of *Bittacus mastrillii* is capable of providing a salivary mass as a nuptial gift to the female (Iwasaki, 1996).

Hangingflies are unique in the members of Mecoptera for taking a face-to-face hanging copulation. This is rare even in the whole Insecta. During copulation, the male of *Bittacus planus* twists his abdomen approximately 180° to accommodate the face-to-face mating position. In addition, his paired epandrial lobes are used to grasp the female′s abdomen, providing an assistance for the genital contact between the mating pair (Gao & Hua, 2013). During the mating course, the prey might be fed by the male and the female from both sides, or only fed by a single individual.

2. 蝎蛉科 Panorpidae

蝎蛉的求偶与交配行为复杂且多样化，其中献礼行为最为突出，吸引了许多学者的研究兴趣。蝎蛉雄虫有 3 种交配策略：一是提供昆虫或节肢动物尸体；二是提供唾腺分泌物，一般将唾腺分泌团附着于植物叶片上供雌虫取食；三是不提供任何“彩礼”而强行交配。雄虫腹部的背中突（notal organ）、后背突（postnotal organ）、臀角（anal horn）、生殖肢（gonopods）等结构发挥着固定雌虫身体的某一部分，以维持交配姿势的作用。

典型的蝎蛉交配姿势一般为 V 形。河南新蝎蛉 *Neopanorpa longiprocessa* 雄虫腹部第 3 节的背中突极其长，在交配过程中能够有力地挟持雌虫的翅，不向雌虫提供任何“彩礼”（Zhong *et al.*, 2013b）。刘氏蝎蛉 *Panorpa liui* 雄虫也不向雌虫提供唾液分泌物，而是将昆虫或节肢动物尸体作为“彩礼”献给雌虫，以获取交配机会（图 1-58）；同时，也会用钳状的背中突夹持雌虫翅的前缘（Ma & Hua, 2011）。大双角蝎蛉 *Dicerapanorpa magna* 雄虫使用腹部第 6 节的 1 对臀角以及柄状的第 7 节基部来挟持住雌虫的腹部末端，以维持交配体位；同时，利用背中突挟持住雌虫翅的前缘，并向雌虫提供唾腺分泌物作为“彩礼”（Zhong *et al.,* 2015a）（图 1-59）。长瓣叉蝎蛉 *Furcatopanorpa longihypovalva* 雄虫腹部第 3 节无背中突，第 6 节也无臀角；在交配时，雄虫通过嘴对嘴的方式，向雌虫持续不断地提供唾腺分泌物。因此，它们在交配时仅头部和生殖器相接触，为 O 形的非典型姿势（Zhong *et al.*, 2015b）。

2. Panorpidae

The courtship and copulation of scorpionflies are complicated and diverse. The most appealing behavior, nuptial feeding, has attracted research interests from many scholars. Three mating strategies are usually adopted by the male scorpionflies: (1) providing a dead arthropod; (2) providing a salivary mass, which is often attached

图 1-58 刘氏蝎蛉 ***Panorpa liui*** 成虫正在取食与交配（红色箭头示蛾类尸体）

Fig. 1-58 Feeding and copulating adults of ***Panorpa liui*** (arrow points to the dead moth)

图 1-59 大双角蝎蛉 ***Dicerapanorpa magna*** 成虫正在交配（红色箭头示唾腺分泌物，钟问摄）

Fig. 1-59 Copulating adults of ***Dicerapanorpa magna*** (arrow points to the salivary mass, photo by Wen Zhong)

to leaves of plants for the female to feed; (3) raping without providing any gifts. The notal organ, postnotal organ, anal horn, and gonopods serve to grasp the corresponding body parts of the female, and maintain the mating posture.

The typical mating posture of scorpionflies is in a V-shape. The male of *Neopanorpa longiprocessa* possess an extremely elongated notal organ with which it is capable of holding the female′ s wings effectively, and provide nothing to his mate (Zhong *et al.*, 2013b). Similarly, the male of *Panorpa liui* also uses his clamp-like notal organ to hold the anterior margin of the female′s wings (Fig. 1-58), but provides a dead arthropod as a nuptial gift instead of a salivary mass to earn the mating opportunity (Ma & Hua, 2011). The two finger-like anal horns on the A6, and the stalked base of A7 of male *Dicerapanorpa magna* serve to hold the female′ s terminalia and maintain the mating posture. Meanwhile, he also uses his notal organ to hold the anterior margin of the female′s wings, and provides a salivary mass as a nuptial gift (Zhong *et al.*, 2015a) (Fig. 1-59). Peculiarly, the male of *Furcatopanorpa longihypovalva* lacks a notal organ on A3 and anal horns on A6, and provides liquid salivary secretion through a mouth-to-mouth mode to the female. Therefore, they copulate in an atypical O-shape, with only their mouthparts and genitalia connected to each other (Zhong *et al.*, 2015b).

3. 拟蝎蛉科 Panorpodidae

拟蝎蛉的求偶和交配模式与蝎蛉科 Panorpidae 昆虫类似。宽甸拟蝎蛉 *Panorpodes kuandianensis* 的雄虫向雌虫提供唾腺分泌物作为“彩礼”，以 V 形姿势开始交配，之后再将雌虫腹部末端旋转 180°，并切换到尾对尾的交配姿势（Tong *et al.*, 2017）。

3. Panorpodidae

A similar courtship and mating pattern to Panorpidae was observed in Panorpodidae. The male of *Panorpodes kuandianensis* offers a salivary mass as nuptial gift to the female and starts a V-shaped copulation, then twists the terminal segments of the female by 180° and changes to an end-to-end position (Tong *et al.*, 2017).

七、中国长翅目 Mecoptera 昆虫名录[1]
Checklist of the Chinese Mecoptera

蚊蝎蛉科 Bittacidae

（一）双尾蚊蝎蛉属 *Bicaubittacus* Tan & Hua, 2009

1. 具肢双尾蚊蝎蛉 *Bicaubittacus appendiculatus* (Esben-Petersen, 1927)
2. 缅甸双尾蚊蝎蛉 *Bicaubittacus burmanus* (Tjeder, 1973)
3. 长突双尾蚊蝎蛉 *Bicaubittacus longiprocessus* (Huang & Hua, 2005)
4. 勐养双尾蚊蝎蛉 *Bicaubittacus mengyangicus* Tan & Hua, 2009
5. 杨氏双尾蚊蝎蛉 *Bicaubittacus yangi* Tan & Hua, 2009

（二）蚊蝎蛉属 *Bittacus* Latreille, 1805

6. 双叉蚊蝎蛉 *Bittacus bifurcatus* Hua & Tanin Li in Hua *et al*., 2008
7. 卡本特蚊蝎蛉 *Bittacus carpenteri* Cheng, 1957
8. 周氏蚊蝎蛉 *Bittacus choui* Hua & Tan, 2007
9. 中条蚊蝎蛉 *Bittacus chujoi* Issiki & Cheng, 1947
10. 环带蚊蝎蛉 *Bittacus cirratus* Tjeder, 1956
    异名：*Bittacus zoensis* Cheng, 1957, synonymized by Byers, 1970
11. 短颚蚊蝎蛉 *Bittacus contumax* Tjeder, 1956
12. 朝鲜蚊蝎蛉 *Bittacus coreanus* Issiki, 1929
13. 大沙河蚊蝎蛉 *Bittacus dashahensis* Zhou & Zhou, 2005
14. 吊罗山蚊蝎蛉 *Bittacus diaoluoshanus* Chen & Hua, 2011
15. 镰形蚊蝎蛉 *Bittacus falcatus* Zhou & Zhou, 2007
16. 淡黄蚊蝎蛉 *Bittacus flavidus* Huang & Hua, 2005
17. 蓬莱蚊蝎蛉 *Bittacus formosanus* Issiki, 1927
18. 嘉理斯蚊蝎蛉 *Bittacus gressitti* Cheng, 1957
19. 海南蚊蝎蛉 *Bittacus hainanicus* Tan & Hua, 2008
20. 九连山蚊蝎蛉 *Bittacus jiulianshanus* Chen, Tan & Hua, 2013
21. 宽阔水蚊蝎蛉 *Bittacus kuankuoshuensis* Zhou & Zhou, 2012
22. 李氏蚊蝎蛉 *Bittacus lii* Zhou, 2003
23. 长角蚊蝎蛉 *Bittacus longantennatus* Chen, Tan & Hua, 2013
24. 斑蚊蝎蛉 *Bittacus maculatus* Issiki, 1927
25. 莫纳斯蚊蝎蛉 *Bittacus monastyrskiyi* Bicha, 2007
26. 暗蚊蝎蛉 *Bittacus obscurus* Huang & Hua, 2005
27. 璧尔蚊蝎蛉 *Bittacus pieli* Navás, 1935
28. 扁蚊蝎蛉 *Bittacus planus* Cheng, 1949
29. 净翅蚊蝎蛉 *Bittacus puripennis* Cai & Hua, 2006

注：1. 根据 Wang & Hua（2017）修订。，截止 2018 年 10 月。

30. 具毛蚊蝎蛉 *Bittacus setigerus* Chen, Tan & Hua, 2013
31. 中华蚊蝎蛉 *Bittacus sinensis* Walker, 1853
    异名：*Bittacus quaternipunctatus* Enderlein, 1910, synonymized by Esben-Petersen, 1921
    异名：*Bittacus strategus* Navás, 1913, synonymized by Esben-Petersen, 1921
32. 中国蚊蝎蛉 *Bittacus sinicus* Issiki, 1931
33. 楚南蚊蝎蛉 *Bittacus sonani* Issiki, 1929
34. 多斑蚊蝎蛉 *Bittacus stigmosus* Zhou & Zhou, 2005
35. 条纹蚊蝎蛉 *Bittacus striatus* Issiki, 1927
36. 纹翅蚊蝎蛉 *Bittacus strigatus* Hua & Chou, 1998
37. 天目山蚊蝎蛉 *Bittacus tienmushana* Cheng, 1957
38. 四边蚊蝎蛉 *Bittacus trapezoideus* Huang & Hua, 2005
39. 三角蚊蝎蛉 *Bittacus triangularis* Issiki, 1929
40. 乌苏里蚊蝎蛉 *Bittacus ussuriensis* Plutenko, 1985
41. 旗蚊蝎蛉 *Bittacus vexilliferus* Byers, 1970
42. 吴氏蚊蝎蛉 *Bittacus wui* Zhou, 2001
43. 浙江蚊蝎蛉 *Bittacus zhejiangicus* Tan & Hua, 2008

（三）地蚊蝎蛉属 *Terrobittacus* Tan & Hua, 2009

44. 狭地蚊蝎蛉 *Terrobittacus angustus* Du & Hua, 2017
45. 具刺地蚊蝎蛉 *Terrobittacus echinatus* (Hua & Huang in Hua *et al.*, 2008)
46. 缠绕地蚊蝎蛉 *Terrobittacus implicatus* (Huang & Hua in Cai *et al.*, 2006)
47. 长毛地蚊蝎蛉 *Terrobittacus longisetus* Tan & Hua, 2009
48. 刀地蚊蝎蛉 *Terrobittacus xiphicus* Tan & Hua, 2009
49. 钩地蚊蝎蛉 *Terrobittacus rostratus* Du & Hua, 2017

蝎蛉科 Panorpidae

（四）单角蝎蛉属 *Cerapanorpa* Gao, Ma & Hua, 2016

50. 尖翅单角蝎蛉 *Cerapanorpa acutipennis* (Hua, 1998)
51. 二角单角蝎蛉 *Cerapanorpa bicornifera* (Chou & Wang in Chou *et al.*, 1981)
52. 短角单角蝎蛉 *Cerapanorpa brevicornis* (Hua & Li in Li *et al.*, 2007)
53. 波利斯单角蝎蛉 *Cerapanorpa bonis* (Cheng, 1949)
54. 拜尔斯单角蝎蛉 *Cerapanorpa byersi* (Hua & Huang in Li *et al.*, 2007)
55. 中部单角蝎蛉 *Cerapanorpa centralis* (Tjeder, 1936) (Sichuan)
56. 拟华山单角蝎蛉 *Cerapanorpa dubia* (Chou & Wang in Chou *et al.*, 1981)
57. 华山单角蝎蛉 *Cerapanorpa emarginata* (Cheng, 1949)
58. 伏牛山单角蝎蛉 *Cerapanorpa funiushana* (Hua & Chou, 1997)
59. 埔里单角蝎蛉 *Cerapanorpa horiensis* (Issiki, 1929)
60. 六盘山单角蝎蛉 *Cerapanorpa liupanshana* Gao, Ma & Hua, 2016
61. 南五台单角蝎蛉 *Cerapanorpa nanwutaina* (Chou in Chou *et al.*, 1981)
62. 太白单角蝎蛉 *Cerapanorpa obtusa* (Cheng, 1949)
    异名：*Panorpa leei* Cheng, 1949, synonymized by Ma *et al.*, 2014

异名: *Panorpa alticola* Zhou, 2000, synonymized by Gao *et al.*, 2016

63. 壮突单角蝎蛉 *Cerapanorpa protrudens* Gao, Ma & Hua, 2016
64. 任氏单角蝎蛉 *Cerapanorpa reni* (Chou in Chou *et al.*, 1981)
65. 弯曲单角蝎蛉 *Cerapanorpa sinuata* Gao, Ma & Hua, 2016
66. 王屋山单角蝎蛉 *Cerapanorpa wangwushana* (Huang, Hua & Shen, 2004)

（五）双角蝎蛉属 *Dicerapanorpa* Zhong & Hua, 2013

67. 白云山双角蝎蛉 *Dicerapanorpa baiyunshana* Zhong & Hua, 2013
68. 双角蝎蛉 *Dicerapanorpa diceras* (MacLachlan, 1894)
    异名: *Panorpa grahami* Carpenter, 1938, synonymized by Carpenter, 1948
69. 基明斯双角蝎蛉 *Dicerapanorpa kimminsi* (Carpenter, 1948)
70. 大双角蝎蛉 *Dicerapanorpa magna* (Chou in Chou *et al.*, 1981)
71. 神农双角蝎蛉 *Dicerapanorpa shennongensis* Zhong & Hua, 2013
72. 西康双角蝎蛉 *Dicerapanorpa stotzneri* (Esben-Petersen, 1934)
73. 谢德双角蝎蛉 *Dicerapanorpa tjederi* (Carpenter, 1938)
74. 三支双角蝎蛉 *Dicerapanorpa triclada* (Qian & Zhou, 2001)

（六）叉蝎蛉属 *Furcatopanorpa* Ma & Hua, 2011

75. 长瓣叉蝎蛉 *Furcatopanorpa longihypovalva* (Hua & Cai, 2009)

（七）新蝎蛉属 *Neopanorpa* van der Weele, 1909

76. 暗新蝎蛉 *Neopanorpa abstrusa* Zhou & Wu in Zhou *et al.*, 1993a
77. 锚形新蝎蛉 *Neopanorpa anchoroides* Zhou, 2003
78. 尖翅新蝎蛉 *Neopanorpa apicata* Navás, 1922
79. 耳状新蝎蛉 *Neopanorpa auriculata* Zhou, 2005
80. 班克斯新蝎蛉 *Neopanorpa banksi* Carpenter, 1938
81. 短瓣新蝎蛉 *Neopanorpa brevivalvae* Chou & Wang in Chou *et al.*, 1988
82. 布里斯新蝎蛉 *Neopanorpa brisi* (Navás, 1930)
    异名: *Neopanorpa dimidiata* Navás, 1930, synonymized by Wang & Hua, 2018
    异名: *Neopanorpa pilosa* Carpenter, 1945, synonymized by Byers, 1970
83. 苍山新蝎蛉 *Neopanorpa cangshanensis* Zhou, 2006
84. 广东新蝎蛉 *Neopanorpa cantonensis* Cheng, 1957
85. 卡本特新蝎蛉 *Neopanorpa carpenteri* Cheng, 1957
86. 卡氏新蝎蛉 *Neopanorpa cavaleriei* (Navás, 1908)
87. 网翅新蝎蛉 *Neopanorpa caveata* Cheng, 1957
88. 修复新蝎蛉 *Neopanorpa chaohsiufui*, Wang & Hua, 2017
    = *Neopanorpa chaoi* Zhou, 2003, preoccupied by *N. chaoi* Cheng, 1957
89. 赵氏新蝎蛉 *Neopanorpa chaoi* Cheng, 1957
    同名: *Neopanorpa chaoi* Zhou, 2003
90. 爪新蝎蛉 *Neopanorpa chelata* Carpenter, 1938
91. 契氏新蝎蛉 *Neopanorpa chillcotti* Byers, 1971

92. 周氏新蝎蛉 *Neopanorpa choui* Cheng, 1949
93. 圆突新蝎蛉 *Neopanorpa circularis* Ju & Zhou, 2003
94. 显斑新蝎蛉 *Neopanorpa clara* Chou & Wang in Chou *et al.*, 1988
95. 明翅新蝎蛉 *Neopanorpa claripennis* Carpenter, 1938
96. 曲瓣新蝎蛉 *Neopanorpa curva* Zhou, 2000
97. 尖齿新蝎蛉 *Neopanorpa cuspidata* Byers, 1965
98. 点苍新蝎蛉 *Neopanorpa diancangshanensis* Wang & Hua, 2018
99. 差异新蝎蛉 *Neopanorpa dispa*r Issiki & Cheng, 1947
100. 疑似小新蝎蛉 *Neopanorpa dubis* Chou & Wang in Chou *et al.*, 1988
101. 方祥新蝎蛉 *Neopanorpa fangxianga* Zhou & Zhou, 2007
102. 台湾新蝎蛉 *Neopanorpa formosana* (Navás, 1911)
103. 蓬莱新蝎蛉 *Neopanorpa formosensis* Navás, 1930
104. 叉状新蝎蛉 *Neopanorpa furcata* Zhou, 2005
105. 梯状新蝎蛉 *Neopanorpa gradana* Cheng, 1952
106. 古蔺新蝎蛉 *Neopanorpa gulinensis* Zhou & Zhou, 2005
107. 海南新蝎蛉 *Neopanorpa hainanica* Hua & Chou, 1998
108. 哈曼德新蝎蛉 *Neopanorpa harmandi* (Navás, 1908)
异名：*Neopanorpa harmandiconjuncta* Navás, 1930, synonymized by Byers, 1965
109. 何氏新蝎蛉 *Neopanorpa hei* Zhou & Fan, 1998
110. 郝氏新蝎蛉 *Neopanorpa heii* Cheng, 1949
111. 华氏新蝎蛉 *Neopanorpa hualizhongi* Hua & Chou, 1998
112. 黄山新蝎蛉 *Neopanorpa huangshana* Cheng, 1957
113. 湖南新蝎蛉 *Neopanorpa hunanensis* Hua, 2002
= *Neopanorpa magna* Chou & Wang in Chou *et al.*, 1988, preoccupied by *N. magna* Issiki, 1927
114. 胡氏新蝎蛉 *Neopanorpa hushengchangi* Hua & Chou, 1999
115. 鸡公山新蝎蛉 *Neopanorpa jigongshanensis* Hua, 1998
116. 九龙新蝎蛉 *Neopanorpa jiulongensis* Zhou in Zhou *et al.*, 1993b
117. K 纹新蝎蛉 *Neopanorpa k-maculata* Cheng, 1952
118. 光泽新蝎蛉 *Neopanorpa kwangtsehi* Cheng, 1957
119. 小孔新蝎蛉 *Neopanorpa lacunaris* Navás, 1930
120. 宽翅新蝎蛉 *Neopanorpa latipennis* Cheng, 1949
121. 雷公山新蝎蛉 *Neopanorpa leigongshana* Zhou & Zhou, 2007
122. 利川新蝎蛉 *Neopanorpa lichuanensis* Cheng, 1957
123. 李氏新蝎蛉 *Neopanorpa lifashengi* Hua & Chou, 1999
124. 蔺江新蝎蛉 *Neopanorpa linjiangensis* Zhou, 2005
125. 黎坪新蝎蛉 *Neopanorpa lipingensis* Cai & Hua, 2009
126. 长杆新蝎蛉 *Neopanorpa longistipitata* Wang & Hua, 2018
127. 河南新蝎蛉 *Neopanorpa longiprocessa* Hua & Chou, 1997
128. 路氏新蝎蛉 *Neopanorpa lui* Chou & Ran in Chou *et al.*, 1981
129. 龙斗山新蝎蛉 *Neopanorpa lungtausana* Cheng, 1957
130. 马氏新蝎蛉 *Neopanorpa maai* Cheng, 1957
131. 大新蝎蛉 *Neopanorpa magna* Issiki, 1927
同名：*Neopanorpa magna* Chou & Wang in Chou *et al.*, 1988
132. 后藤新蝎蛉 *Neopanorpa makii* Issiki, 1927

133. 莽山新蝎蛉 *Neopanorpa mangshanensis* Chou & Wang in Chou *et al.*, 1988
134. 茂兰新蝎蛉 *Neopanorpa maolanensis* Zhou & Bao, 2002
135. 巨突新蝎蛉 *Neopanorpa magnatititana* Wang & Hua, 2018
136. 麻阳新蝎蛉 *Neopanorpa mayangensis* Zhou & Zhou, 2010
137. 勐海新蝎蛉 *Neopanorpa menghaiensis* Zhou & Wu in Zhou *et al.*, 1993a
138. 小新蝎蛉 *Neopanorpa minuta* Chou & Wang in Chou *et al.*, 1988
139. 莫干山新蝎蛉 *Neopanorpa moganshanensis* Zhou & Wu in Zhou *et al.*, 1993a
140. 莫山新蝎蛉 *Neopanorpa mokansana* Cheng, 1957
141. 山地新蝎蛉 *Neopanorpa montana* Zhou in Zhou *et al.*, 1993b
142. 异新蝎蛉 *Neopanorpa mutabilis* Cheng, 1957
143. 尼尔森新蝎蛉 *Neopanorpa nielseni* Byers, 1965
144. 黑色新蝎蛉 *Neopanorpa nigritis* Carpenter, 1938
145. 眼新蝎蛉 *Neopanorpa ophthalmica* (Navás, 1911)
146. 卵翅新蝎蛉 *Neopanorpa ovata* Cheng, 1957
147. 微新蝎蛉 *Neopanorpa parva* Carpenter, 1945
148. 淡瓣新蝎蛉 *Neopanorpa pallivalva* Zhou, 2003
149. 垂齿新蝎蛉 *Neopanorpa pendula* Qian & Zhou, 2001
150. 璧尔新蝎蛉 *Neopanorpa pielina* Navás, 1936
151. 丽新蝎蛉 *Neopanorpa pulchra* Carpenter, 1945
152. 净翅新蝎蛉 *Neopanorpa puripennis* Chou & Wang in Chou *et al.*, 1988
153. 方痣新蝎蛉 *Neopanorpa quadristigma* Wang & Hua, 2018
154. 网纹新蝎蛉 *Neopanorpa retina* Chou & Li in Chou *et al.*, 1988
155. 绍德新蝎蛉 *Neopanorpa sauteri* (Esben-Petersen, 1912)
156. 半圆新蝎蛉 *Neopanorpa semiorbiculata* Wang & Hua, 2018
157. 具毛新蝎蛉 *Neopanorpa setigera* Wang & Hua, 2018
158. 申氏新蝎蛉 *Neopanorpa sheni* Hua & Chou, 1997
159. 泰国新蝎蛉 *Neopanorpa siamensis* Byers, 1965
160. 相似新蝎蛉 *Neopanorpa simulans* Byers, 2008
    = *Neopanorpa similis* Byers, 2002, preoccupied by *N. similis* Byers, 1999
161. 刀瓣新蝎蛉 *Neopanorpa spatulata* Byers, 1965
162. 陶氏新蝎蛉 *Neopanorpa taoi* Cheng, 1949 (Sichuan)
163. 腾冲新蝎蛉 *Neopanorpa tengchongensis* Zhou & Wu in Zhou *et al.*, 1993a
164. 狭瓣新蝎蛉 *Neopanorpa tenuis* Zhou, 2000
165. 西藏新蝎蛉 *Neopanorpa tibetensis* Hua & Chou, 1999
166. 天目山新蝎蛉 *Neopanorpa tienmushana* Cheng, 1957
167. 天平山新蝎蛉 *Neopanorpa tienpingshana* Chou & Wang in Chou *et al.*, 1988
168. 染翅新蝎蛉 *Neopanorpa tincta* Wang & Hua, 2018
169. 透翅新蝎蛉 *Neopanorpa translucida* Cheng, 1957
170. 三角新蝎蛉 *Neopanorpa triangulata* Wang & Hua, 2018
171. 钩曲新蝎蛉 *Neopanorpa uncata* Zhou, 2000
172. 钩齿新蝎蛉 *Neopanorpa uncinella* Qian & Zhou, 2001
173. 粗脉新蝎蛉 *Neopanorpa validipennis* Cheng, 1949
174. 变新蝎蛉 *Neopanorpa varia* Cheng, 1949
175. 旺草镇新蝎蛉 *Neopanorpa wangcaensis* Zhou & Zhou, 2012

176. 习水新蝎蛉 *Neopanorpa xishuiensis* Zhou, 2005
177. 盈江新蝎蛉 *Neopanorpa yingjiangensis* Zhou, 2000
178. 杨氏新蝎蛉 *Neopanorpa youngi* Byers, 1994

（八）蝎蛉属 *Panorpa* Linnaeus, 1758

179. 刺叶蝎蛉 *Panorpa acanthophylla* Zhou, 2006
180. 赤坂蝎蛉 *Panorpa akasakai* Issiki, 1929
181. 翼蝎蛉 *Panorpa alata* Zhou & Zhou, 2005
182. 弯杆蝎蛉 *Panorpa anfracta* Ju & Zhou, 2003
183. 狭带蝎蛉 *Panorpa angustistriata* Issiki, 1929
184. 安仁蝎蛉 *Panorpa anrenensis* Chou & Wang in Chou *et al.*, 1987
185. 端黑蝎蛉 *Panorpa apiconebulosa* Issiki, 1929
186. 金身蝎蛉 *Panorpa aurea* Cheng, 1957
187. 宝华山蝎蛉 *Panorpa baohwashana* Cheng, 1957
188. 巴山蝎蛉 *Panorpa bashanicola* Hua, Tao & Hua, 2018
189. 二支蝎蛉 *Panorpa biclada* Zhang & Hua, 2012
190. 双带蝎蛉 *Panorpa bifasciata* Chou & Wang in Chou *et al.*, 1981
191. 双纹蝎蛉 *Panorpa bistriata* Issiki, 1929
192. 短突蝎蛉 *Panorpa brevititilana* Issiki, 1929
193. 布农蝎蛉 *Panorpa bunun* Issiki, 1929
194. 卡本特蝎蛉 *Panorpa carpenteri* Cheng, 1957
195. 长白山蝎蛉 *Panorpa changbaishana* Hua in Hou & Hua, 2008
    = *Panorpa choui* Hua, 1998, preoccupied by *Panorpa choui* Zhou, 1993
196. 郑氏蝎蛉 *Panorpa chengi* Chou in Chou *et al.*, 1981
197. 陈氏蝎蛉 *Panorpa cheni* Cheng, 1957
198. 周氏蝎蛉 *Panorpa choui* Zhou & Wu in Zhou *et al.*, 1993a
    同名：*Panorpa choui* Hua, 1998
199. 支蝎蛉 *Panorpa cladocerca* Navás, 1935
200. 同色蝎蛉 *Panorpa concolor* Esben-Petersen, 1915
201. 库曼蝎蛉 *Panorpa coomani* Cheng, 1957
202. 弯曲蝎蛉 *Panorpa curva* Carpenter, 1938
203. 曲杆蝎蛉 *Panorpa curvata* Zhou, 2006
204. 大沙河蝎蛉 *Panorpa dashahensis* Zhou & Zhou, 2005
205. 大卫蝎蛉 *Panorpa davidi* Navás, 1908
206. 斑蝎蛉 *Panorpa deceptor* Esben-Petersen, 1913
207. 淡色蝎蛉 *Panorpa decolorata* Chou & Wang in Chou *et al.*, 1981
208. 难蝎蛉 *Panorpa difficilis* Carpenter, 1938
209. 江崎蝎蛉 *Panorpa esakii* Issiki, 1929
210. 峨眉山蝎蛉 *Panorpa emeishana* Hua, Sun & Li, 2001
211. 伪蝎蛉 *Panorpa falsa* Issiki & Cheng, 1947
212. 女儿蝎蛉 *Panorpa filina* Chou & Wang in Chou *et al.*, 1987
213. 黄体蝎蛉 *Panorpa flavicorporis* Cheng, 1957
214. 黄翅蝎蛉 *Panorpa flavipennis* Carpenter, 1938

215. 实蝎蛉 *Panorpa fructa* Cheng, 1949
216. 福建蝎蛉 *Panorpa fukiensis* Tjeder, 1950
217. 淡黄蝎蛉 *Panorpa fulvastra* Chou in Chou *et al.*, 1981
218. 叉形蝎蛉 *Panorpa furcata* Zhou & Zhou, 2007
219. 格雷艾姆蝎蛉 *Panorpa grahamana* Cheng, 1957
220. 嘉理斯蝎蛉 *Panorpa gressitti* Byers, 1970
221. 桂东蝎蛉 *Panorpa guidongensis* Chou & Li in Chou *et al.*, 1987
222. 滴蝎蛉 *Panorpa guttata* Navás, 1908
223. 钩蝎蛉 *Panorpa hamata* Issiki & Cheng, 1947
224. 织纹蝎蛉 *Panorpa implicata* Cheng, 1957
225. 岛生蝎蛉 *Panorpa insularis* Hua & Chou, 1998
226. 一色蝎蛉 *Panorpa issikiana* Byers, 1970
227. 台湾一色蝎蛉 *Panorpa issikii* Penny & Byers, 1979
  = *Panorpa acuta* Issiki & Cheng, 1947, preoccupied by *P. acuta* Carpenter, 1931
228. 吉林蝎蛉 *Panorpa jilinensis* Zhou, 2000
229. 金川蝎蛉 *Panorpa jinchuana* Hua, Sun & Li, 2001
230. 凯氏蝎蛉 *Panorpa kellogi* Cheng, 1957
231. 尤氏蝎蛉 *Panorpa kiautai* Zhou & Wu in Zhou *et al.*, 1993a
232. 克拉帕利希蝎蛉 *Panorpa klapperichi* Tjeder, 1950
233. 昆明蝎蛉 *Panorpa kunmingensis* Fu & Hua, 2009
234. 拉氏蝎蛉 *Panorpa lachlani* Navás, 1930
235. 廖氏蝎蛉 *Panorpa liaoi* Zhou & Zhou, 2007
236. 林田山蝎蛉 *Panorpa lintienshana* Cheng, 1952
237. 刘氏蝎蛉 *Panorpa liui* Hua, 1997
238. 长侧蝎蛉 *Panorpa longiramina* Issiki & Cheng, 1947
239. 长突蝎蛉 *Panorpa longititilana* Issiki, 1929
240. 黄蝎蛉 *Panorpa lutea* Carpenter, 1945
241. 长蝎蛉 *Panorpa macrostyla* Hua, 1998
242. 莽山蝎蛉 *Panorpa mangshanensis* Chou & Wang in Chou *et al.*, 1987
243. 莫干山蝎蛉 *Panorpa mokansana* Cheng, 1957
244. 新刺蝎蛉 *Panorpa neospinosa* Chou & Wang in Chou *et al.*, 1981
245. 能高蝎蛉 *Panorpa nokoensis* Issiki, 1929
246. 裸茎蝎蛉 *Panorpa nudiramus* Byers, 2002
247. 斜蝎蛉 *Panorpa obliqua* Carpenter, 1945
248. 斜带蝎蛉 *Panorpa obliquifascia* Chou & Wang in Chou *et al.*, 1987
249. 赭尾蝎蛉 *Panorpa ochraceocauda* Issiki, 1927
250. 淡纹蝎蛉 *Panorpa pallidimaculata* Issiki, 1929
251. 平支蝎蛉 *Panorpa parallela* Wang & Hua, 2016
252. 栉蝎蛉 *Panorpa pectinata* Issiki, 1929
253. 彼得蝎蛉 *Panorpa peterseana* Issiki, 1927
254. 璧尔蝎蛉 *Panorpa pieli* Cheng, 1957
255. 平江蝎蛉 *Panorpa pingjiangensis* Chou & Wang in Chou *et al.*, 1987
256. 微蝎蛉 *Panorpa pusilla* Cheng, 1949
257. 黔蝎蛉 *Panorpa qiana* Zhou & Zhou, 2010

258. 秦岭蝎蛉 *Panorpa qinlingensis* Chou & Ran in Chou *et al*., 1981
259. 四带蝎蛉 *Panorpa quadrifasciata* Chou & Wang in Chou *et al*., 1987
260. 郡大山蝎蛉 *Panorpa rantaisanensis* Issiki, 1929
261. 反曲蝎蛉 *Panorpa reflexa* Wang & Hua, 2016
262. 半带蝎蛉 *Panorpa semifasciata* Cheng, 1949
263. 六刺蝎蛉 *Panorpa sexspinosa* Cheng, 1949
    六刺蝎蛉终南亚种 *Panorpa sexspinosa zhongnanensis* Chou & Ran in Chou *et al*., 1981
264. 六带蝎蛉 *Panorpa sextaenia* Zhou & Bao, 2002
265. 山阳蝎蛉 *Panorpa shanyangensis* Chou & Wang in Chou *et al*., 1981
266. 柴田蝎蛉 *Panorpa shibatai* Issiki, 1929
267. 楚南蝎蛉 *Panorpa sonani* Issiki, 1929
268. 宋氏蝎蛉 *Panorpa songes* Zhou & Zhou, 2005
269. 痣带蝎蛉 *Panorpa stigmalis* Navás, 1908
270. 斑点蝎蛉 *Panorpa stigmosa* Zhou, 2006
271. 淡珀蝎蛉 *Panorpa subambra* Chou & Tong in Chou *et al*., 1987
272. 亚曙蝎蛉 *Panorpa subaurea* Chou & Li, 1987
273. 太平山蝎蛉 *Panorpa taiheisanensis* Issiki, 1929
274. 台湾蝎蛉 *Panorpa taiwanensis* Issiki, 1929
275. 盖蝎蛉 *Panorpa tecta* Byers, 2002
276. 四段蝎蛉 *Panorpa tetrazonia* Navás, 1935
277. 三线蝎蛉 *Panorpa trifasciata* Cheng, 1957
278. 三带蝎蛉 *Panorpa tritaenia* Chou & Wang in Chou *et al*., 1987
279. 模蝎蛉 *Panorpa typicoides* Cheng, 1949
280. 王氏蝎蛉 *Panorpa waongkehzengi* Navás, 1935
281. 莱特蝎蛉 *Panorpa wrightae* Cheng, 1957
282. 杨氏蝎蛉 *Panorpa yangi* Chou in Chou *et al*., 1981
283. 易氏蝎蛉 *Panorpa yiei* Issiki & Cheng, 1947

（九）华蝎蛉属 *Sinopanorpa* Cai & Hua in Cai *et al*., 2008

284. 指形华蝎蛉 *Sinopanorpa digitiformis* Huang & Hua in Cai *et al*., 2008
285. 南宫山华蝎蛉 *Sinopanorpa nangongshana* Cai & Hua in Cai *et al*., 2008
286. 染翅华蝎蛉 *Sinopanorpa tincta* (Navás, 1931)
    异名：*Panorpa statura* Cheng, 1949, synonymized by Hua & Peng, 2005

拟蝎蛉科 Panorpodidae

（一〇）拟蝎蛉属 *Panorpodes* MacLachlan, 1875

287. 短柄拟蝎蛉 *Panorpodes brachypodus* Tan & Hua, 2008
288. 短拟蝎蛉 *Panorpodes brevicaudatus* (Hua, 1998)
289. 宽甸拟蝎蛉 *Panorpodes kuandianensis* Zhong, Zhang & Hua, 2011

第二部分 蚊蝎蛉科

Part Ⅱ Bittacidae

图 2-a 蚊蝎蛉 ***Bittacus* sp.** 雌成虫侧面观（雷波摄）

Fig. 2-a Female adult of ***Bittacus* sp.**, lateral view (photo by Bo Lei)

蚊蝎蛉六足纤细，捕捉式，跗节仅具 1 爪；只能悬挂在植物枝条或叶片边缘，不能在物体表面行走（图 2-a）。

蚊蝎蛉科 Bittacidae 昆虫全世界共 200 余种，中国已记录 40 余种，本图鉴收录 3 属 31 种。

Hangingflies have six slender raptorial legs, each bearing a single tarsal claw. They can only hang themselves on twigs or edges of leaves of plants, but are unable to walk on a surface (Fig.2-a).

More than 200 species have been described worldwide. Over 40 species have been recorded in China and 31 of them are included in this atlas.

中国蚊蝎蛉科 Bittacidae 分属检索表

1. 生殖肢基节端部具 1 对弯曲的突起（图 2-b-1）………………………… 双尾蚊蝎蛉属 ***Bicaubittacus***
   生殖肢基节端部无上述突起 ………………………………………………………… 2
2. 上生殖瓣短于生殖肢基节之半（图 2-b-2）……………………………… 地蚊蝎蛉属 ***Terrobittacus***
   上生殖瓣长于生殖肢基节之半（图 2-b-3）……………………………………… 蚊蝎蛉属 ***Bittacus***

Key to genera of Bittacidae in Chinese

1. Gonocoxites with a pair of curved prolongations apically (Fig. 2-b-1) ……………………… ***Bicaubittacus***
   Gonocoxites lacking such prolongations ……………………………………………………… 2
2. Epandrial lobes shorter than half gonocoxites (Fig. 2-b-2) ……………………………… ***Terrobittacus***
   Epandrial lobes longer than half gonocoxites (Fig. 2-b-3) ………………………………… ***Bittacus***

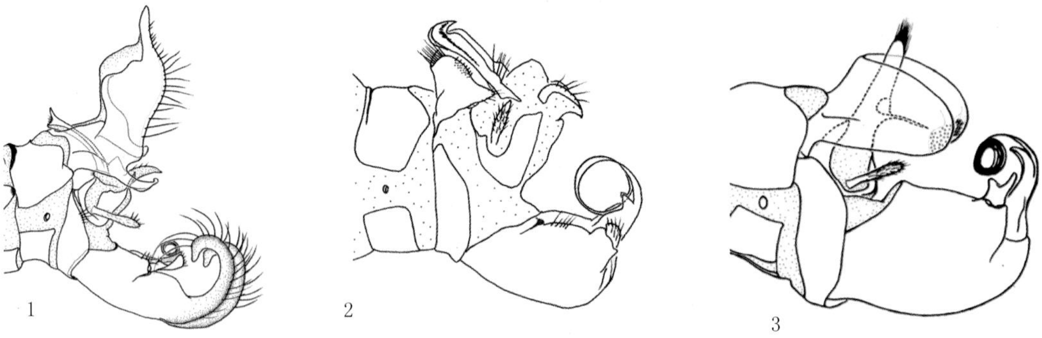

图 2-b 蚊蝎蛉科 Bittacidae 昆虫雄性外生殖器侧面观

1. 双尾蚊蝎蛉属 *Bicaubittacus* 2. 蚊蝎蛉属 *Bittacus* 3. 地蚊蝎蛉属 *Terrobittacus*

Fig. 2-b Male genitalia of Bittacidae, lateral view

1. *Bicaubittacus* 2. *Bittacus* 3. *Terrobittacus*

一、双尾蚊蝎蛉属 *Bicaubittacus* Tan & Hua, 2009

双尾蚊蝎蛉属 *Bicaubittacus* 昆虫主要识别特征为雄性外生殖器的生殖肢基节末端具有 1 对弯曲的长突起（图 2-c）。

双尾蚊蝎蛉属 *Bicaubittacus* 昆虫全世界共 5 种，中国全部有分布，本图鉴收录 4 种。

This genus can be recognized by the male gonocoxites with a pair of long curved distal prolongations (Fig. 2-c). Five species endemic to China are known, and four of them are included in this atlas.

图 2-c 缅甸双尾蚊蝎蛉 ***Bicaubittacus burmanus*** 雄成虫侧面观（云南永德棠梨山）

Fig. 2-c Male adult of ***Bicaubittacus burmanus***, lateral view (Mount Tangli, Yongde, Yunnan)

本图鉴双尾蚊蝎蛉属 *Bicaubittacus* 分种检索表

（4 种，雄性）

1. 上生殖瓣末端具 2 根延长的尾突 ……………………………… 2. 缅甸双尾蚊蝎蛉 *Bicaubittacus burmanus*
   上生殖瓣末端仅具 1 根延长的尾突 ……………………………… 2
2. 上生殖瓣末端近尾突处均匀变窄；尾突末端不分叉 ……………………………… 3
   上生殖瓣末端近尾突处平截，后缘有 2 ~ 3 个尖齿；尾突末端稍分为 2 叉 ……………………………… 4. 杨氏双尾蚊蝎蛉 *Bicaubittacus yangi*
3. 上生殖瓣腹缘近基部具 1 个向下的指形突起……………………1. 具肢双尾蚊蝎蛉 *Bicaubittacus appendiculatus*
   上生殖瓣无上述突起……………………………………3. 长突双尾蚊蝎蛉 *Bicaubittacus longiprocessus*

Key to species of *Bicaubittacus* in this atlas

(Four species, males)

1. Epandrial lobe with two elongated caudal prongs ……………………………… 2. *Bicaubittacus burmanus*
   Epandrial lobe with only one elongated caudal prong ……………………………… 2
2. Epandrial lobe evenly tapering towards unfurcated caudal prong ……………………………… 3
   Epandrial lobe truncated near bifurcated caudal prong and bearing 2–3 acute teeth ………… 4. *Bicaubittacus yangi*
3. Epandrial lobe with a finger-like process directed ventrad near base of ventral margin ……………………………… 1. *Bicaubittacus appendiculatus*
4. Epandrial without such process on ventral margin……………………………3. *Bicaubittacus longiprocessus*

1. 具肢双尾蚊蝎蛉 *Bicaubittacus appendiculatus* (Esben-Petersen, 1927)

【测量】 前翅长 17.0 ~ 19.0 mm。

【形态特征】 头顶黄褐色，触角柄节和梗节黄褐色，鞭节黑褐色，被毛。单眼三角区明显隆起，黑色。喙黑褐色。前胸背板不均匀深褐色。中、后胸背板中部具近似葫芦形的浅色中带。足污黄色，股节末端以及胫节两端黑色。翅透明，微泛污黄色，仅端部有不明显的褐色雾状加深。雄性腹部 1 ~ 8 节黄褐色，第 8 节背板后缘中部具 V 形缺刻。上生殖瓣形状奇特；腹缘近基部有 1 个向下的指形突起，自中部向后有 1 列长刚毛；端部形成 1 根细长的指状尾突。生殖肢基节末端向后向上延伸成 2 根弯曲的鹅颈状突起，其后缘着生 1 列长刚毛；生殖刺突靴状，末端平截。阳茎丝缠绕成小环。尾须短棒状。载肛突上瓣较长，末端分支呈 L 形；下瓣短，向下钩曲。

图 2-1-1 具肢双尾蚊蝎蛉 *Bicaubittacus appendiculatus* 雄成虫背面观（云南香格里拉虎跳峡，NWAFU）

Fig. 2-1-1 Male adult of *Bicaubittacus appendiculatus*, dorsal view (Tiger Leaping Gorge, Shangri-La, Yunnan, NWAFU)

图 2-1-2 具肢双尾蚊蝎蛉 *Bicaubittacus appendiculatus* 雄性外生殖器侧面观

Fig. 2-1-2 Male genitalia of *Bicaubittacus appendiculatus*, lateral view

【发生期】 6 ~ 7 月。

【地理分布】 云南（香格里拉）。

Measurements: Forewing length 17.0–19.0 mm.

Morphology: Vertex yellowish brown. Antennae with yellowish brown scape and pedicel, and blackish brown and ciliated flagellum. Ocellar triangle black and protruded. Rostrum blackish brown. Pronotum unevenly deep brown. Mesonotum and metanotum with a pale, gourd-shaped mesal stripe. Legs sordidly yellow, with apical end of femur and both ends of tibia darkened. Wing membrane hyaline, slightly tinged with sordid yellow. Apex indistinctly tinged with smoky brown. Male A1–A8 yellowish brown. Tergum Ⅷ bearing a V-shaped emargination on posterior margin mesally. Epandrial lobes peculiarly shaped: ventral margin protruding ventrad and forming a finger-like process subbasally, and bearing a row of long setae from middle to end of slender tail distally. Apex of gonocoxites with a pair of long and curved gooseneck-shaped prolongations bearing a row of long setae along their hind margins. Gonostylus boot-shaped with a truncated apex. Penisfilum coiled into a small loop. Cerci short and clavate. Upper branch of proctiger long and slightly branched at apex in a L-shape, while lower branch short and curved ventrad.

Emergence: June to July.

Distribution: Yunnan (Shangri-La).

图 2-1-3 具肢双尾蚊蝎蛉 ***Bicaubittacus appendiculatus*** 地理分布

（中国特有种分布地为红色，全图鉴同）

Fig. 2-1-3 Distribution of ***Bicaubittacus appendiculatus***

(Red indicates species endemic to China, applied in whole atlas)

2. 缅甸双尾蚊蝎蛉 *Bicaubittacus burmanus* (Tjeder, 1973)

【测量】 前翅长 18.0 ~ 19.0 mm。

【形态特征】 头顶黄棕色至红棕色，单眼三角区黑色。触角黑褐色。胸部背板黑褐色，两侧黄褐色，足黄褐色。翅烟黄色透明，无任何斑纹，仅端部有不明显的加深。雄性腹部 1 ~ 8 节黄褐色，第 8 节背板后缘中部具 V 形缺刻。上生殖瓣长矩形，末端分为两根近等长的尾突；背缘内卷，具 3 个向内的突起，其中中间的突起较细长；腹缘近中部向下突出。生殖肢基节末端具 1 对弯曲的长突起，末端分歧；该突起近基部背侧呈叶状突出。生殖刺突狭长，内侧具齿。阳茎叶阔，阳茎丝细长，盘曲成小环。载肛突上瓣鹅颈状，末端尖锐，近末端处膨大并具 1 簇长毛；下瓣较短小，向下弯曲。

图 2-2-2 缅甸双尾蚊蝎蛉 ***Bicaubittacus burmanus*** 雄性外生殖器侧面观

Fig. 2-2-2 Male genitalia of ***Bicaubittacus burmanus***, lateral view

图 2-2-3 缅甸双尾蚊蝎蛉 ***Bicaubittacus burmanus*** 地理分布

（与国外共有种分布地为绿色，全图鉴同）

Fig. 2-2-3 Distribution of ***Bicaubittacus burmanus***

(Green indicates species shared by China and other countries, applied in whole atlas)

图 2-2-1 缅甸双尾蚊蝎蛉 ***Bicaubittacus burmanus*** 成虫侧面观（云南永德棠梨山）

1. 雄虫 2. 雌虫

Fig. 2-2-1 Adults of ***Bicaubittacus burmanus***, lateral view (Mount Tangli, Yongde, Yunnan)

1. Male 2. Female

【发生期】 5 ~ 6 月。

【地理分布】 云南（永德）；缅甸（甘拜地）。

Measurements: Forewing length 18.0–19.0 mm.

Morphology: Vertex yellowish brown to reddish brown. Ocellar triangle black. Antennae blackish brown. Nota blackish brown with yellowish brown sides. Legs yellowish brown.Wing membrane hyaline, tinged with smoky yellow, and devoid of markings, except apex indistinctly darkened. Male A1–A8 yellowish brown. Tergum Ⅷ with a V-shaped emargination on posterior margin mesally. Epandrial lobe long rectangular, and forked apically into a pair of slender prongs; its dorsal margin curled inward and forming three small processes, with middle one more slender than others; its ventral margin protruding ventrad at middle. Gonocoxites with a pair of long and curved distal prolongations bearing a row of long bristles along their hind margin, and divergent with each other. A foliate lobe basal to each prolongation. Gonostylus slender and bearing an inner tooth. Aedeagal lobes broad. Penisfilum coiled into a small loop. Upper branch of proctiger gooseneck-shaped, and bearing a tuft of long setae on enlarged region below its acute apex; lower branch short and curved ventrad.

Emergence: May to June.

Distribution: Yunnan (Yongde); Myanmar (Kambaiti).

3. 长突双尾蚊蝎蛉 *Bicaubittacus longiprocessus*（Huang & Hua, 2005）

【测量】 前翅长 16.5 ~ 17.5 mm。

【形态特征】 头顶黄褐色，单眼三角区色较深，喙黑褐色，触角黄褐色。胸部背板黄褐色。足黄褐色，股节末端、胫节两端颜色稍深。翅透明，微泛黄色，翅面光洁无斑纹，翅短具不明显的棕色加深。翅痣浅黄褐色。雄性腹部 1 ~ 8 节黄褐色。第 8 节背板后缘中部具 V 形缺刻。上生殖瓣基半部近梯形，背、腹缘稍内卷；端半部向后延伸成 1 根向背侧弯曲的长尾突，其背缘具 1 列长刚毛；背缘中部内侧有 1 个大型突起，腹缘端部 2/3 处内侧有 1 个较小突起。生殖肢基节末端向后延伸成 2 根弯曲的颈状突起，其后缘着生稀疏的长毛。阳茎叶窄长，阳茎丝细长并缠绕成小环。尾须棒状。载肛突上瓣长，末端强烈向前弯曲；下瓣短。

图 2-3-1 长突双尾蚊蝎蛉 *Bicaubittacus longiprocessus* 雄成虫侧面观（广东韶关）

Fig.2-3-1 Male adult of *Bicaubittacus longiprocessus* ,lateral view (Shaoguan, Guangdong)

图 2-3-2 长突双尾蚊蝎蛉 *Bicaubittacus longiprocessus* 雄性外生殖器侧面观

Fig. 2-3-2 Male genitalia of *Bicaubittacus longiprocessus*, lateral view

【发生期】 4 ~ 6 月。

【地理分布】 福建（南靖）、江西（井冈山）、广东（韶关、梅州）。

Measurements: Forewing length 16.5–17.5 mm.

Morphology: Vertex yellowish brown, ocellar triangle darkened, rostrum blackish brown, antennae yellowish brown. Nota yellowish brown. Legs yellowish brown, with apical end of femur and both ends of tibia darkened. Pterostigma yellowish brown. Male A1–A8 yellowish brown. Tergum Ⅷ with a V-shaped emargination on posterior margin mesally.

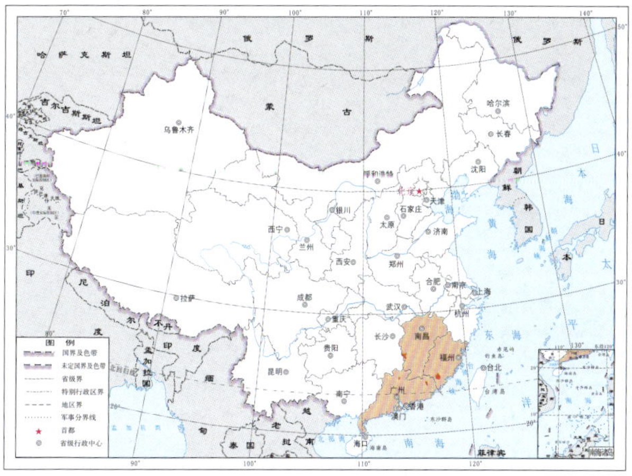

图 2-3-3 长突双尾蚊蝎蛉 *Bicaubittacus longiprocessus* 地理分布

Fig. 2-3-3 Distribution of *Bicaubittacus longiprocessus*

Epandrial lobe with basal half subtrapezoidal, and curled inward at dorsal and ventral margins; apical half extending into a tail-like prong curving dorsad bearing a row of long bristles on dorsal margin; dorsal margin with a large process on inner surface at middle, and ventral margin with a smaller process on inner surface at distal two-thirds. Gonocoxites with a pair of long and curved distal prolongations bearing a row of long sparse bristles along their hind margin. Aedeagal lobes slender. Penisfilum coiled into a small loop. Cerci clavate. Upper branch of proctiger long and greatly curved cephalad at apex, while lower branch short and curved ventrad.

Emergence: April to June.

Distribution: Fujian (Nanjing), Jiangxi (Mount Jinggang) and Guangdong (Shaoguan, Meizhou).

4. 杨氏双尾蚊蝎蛉 *Bicaubittacus yangi* Tan & Hua, 2009

【测量】 前翅长 17.0 mm。

【形态特征】 头顶黄褐色，单眼三角区色较深，触角黄褐色。中、后胸背板前半部分两侧暗褐色，约到翅基处色浅。足黄褐色，股节末端、胫节两端颜色稍深。翅透明，微泛黄色，翅面光洁无斑纹，翅端具不明显的棕色加深。翅痣浅黄褐色。雄性腹部 1 ~ 8 节黄褐色。第 8 节背板后缘中部具 V 形缺刻。上生殖瓣近基部具 1 个向下的指形突起，下缘具 1 列长刚毛；后缘近平截，具 2 ~ 3 个尖锐的短齿，下方具 1 根末端稍分叉的长尾突。生殖肢基节末端向后向上延伸成 2 根弯曲的鹅颈状突起，其后缘着生稀疏的长毛，相互分歧。阳茎叶铲状；阳茎丝细长，缠绕成小环。尾须短棒状。载肛突上瓣较长，末端轻度凹陷成 L 形，被长毛；下瓣短，向腹面弯曲。

【发生期】 5 月。

【地理分布】 广西（田林）。

图 2-4-1 杨氏双尾蚊蝎蛉 *Bicaubittacus yangi* 雄成虫背面观
（广西田林滚平，NWAFU）

Fig. 2-4-1 Male adult of *Bicaubittacus yangi*, dorsal view
(Gunping, Tianlin, Guangxi, NWAFU)

图 2-4-2 杨氏双尾蚊蝎蛉 *Bicaubittacus yangi*
雄性外生殖器侧面观

Fig. 2-4-2 Male genitalia of *Bicaubittacus yangi*, lateral view

Measurements: Forewing length 17.0 mm.

Morphology: Vertex yellowish brown. Ocellar triangle darkened. Antennae blackish brown. Mesonotum and metanotum dark brown anteriorly and paler

图 2-4-3 杨氏双尾蚊蝎蛉 *Bicaubittacus yangi* 地理分布

Fig. 2-4-3 Distribution of *Bicaubittacus yangi*

near wing bases. Legs yellowish brown, with apical end of femur and both ends of tibia darkened. Wing membrane hyaline, and tinged with smoky yellow. Markings absent. Apex tinged with indistinct light brown. Pterostigma light yellowish brown. Male A1–A8 yellowish brown. Tergum Ⅷ with a V-shaped emargination on posterior margin mesally. Epandrial lobe peculiarly shaped: its ventral margin protruding ventrad subbasally and forming a finger-like process; its truncated apex forming 2–3 short and acute teeth, and a tail-like prong slightly bifurcated apically. Apex of gonocoxites with a pair of long and curved gooseneck-shaped prolongations bearing a row of long sparse bristles along their hind margin, and divergent with each other. Aedeagal lobes shovel-like. Penisfilum coiled into a small loop. Cerci short and clavate. Upper branch of proctiger long and slightly branched at apex in a L-shape, while lower branch short and curved ventrad.

Emergence: May.

Distribution: Guangxi (Tianlin).

二、蚊蝎蛉属 *Bittacus* Latreille, 1805

蚊蝎蛉属 *Bittacus* 是蚊蝎蛉科 Bittacidae 中昆虫种类最多的属，全世界已知 100 余种，中国记录 38 种，本图鉴收录 21 种（图 2-d）。

Bittacus is the largest genus in Bittacidae. More than 100 species have been described worldwide. Thirty-eight species have been recorded from China and 21 of them are included in this atlas (Fig. 2-d).

图 2-d　纹翅蚊蝎蛉 ***Bittacus strigatus*** 雄成虫侧面观（湖北房县三座庵）

Fig. 2-d　Male adult of ***Bittacus strigatus***, lateral view (Sanzuo'an, Fangxian, Hubei)

本图鉴蚊蝎蛉属 *Bittacus* 分种检索表

(21 种，雄性)

1. 翅面无斑纹 …… 2
   翅面具各式斑纹 …… 3
2. 上生殖瓣近长矩形；载肛突上瓣约与上生殖瓣等长，末端鹅头状膨大 …… 6. 周氏蚊蝎蛉 *Bittacus choui*
   上生殖瓣短四边形；载肛突上瓣短于上生殖瓣，末端不膨大 …… 17. 净翅蚊蝎蛉 *Bittacus puripennis*
3. 上生殖瓣整体近四边形，无分支，结构简单 …… 4
   上生殖瓣非四边形，末端有时形成各式的分支或突起，结构较复杂 …… 12
4. 触角几乎与身体等长 …… 12. 长角蚊蝎蛉 *Bittacus longantennatus*
   触角远短于体长 …… 5
5. 上生殖瓣粗厚，端半部向腹面弯曲 …… 21. 纹翅蚊蝎蛉 *Bittacus strigatus*
   上生殖瓣扁平，端半部不向腹面弯曲 …… 6
6. 上生殖瓣显著长于生殖肢基节 …… 7
   上生殖瓣短于或几乎等长于生殖肢基节 …… 8
7. 上生殖瓣末端具 1 簇长刚毛；载肛突上瓣仅稍超出上生殖瓣 …… 15. 璧尔蚊蝎蛉 *Bittacus pieli*
   上生殖瓣末端刚毛正常；载肛突上瓣显著超出上生殖瓣 …… 11. 李氏蚊蝎蛉 *Bittacus lii*
8. 上生殖瓣长矩形，长超过宽的 3 倍；载肛突上瓣十分细长 …… 18. 具毛蚊蝎蛉 *Bittacus setigerus*
   上生殖瓣较短，长不及宽的 3 倍；载肛突上瓣相对较短 …… 9
9. 阳茎叶十分宽阔，后面观可见其向下延伸、尖锐的钩状突起 …… 8. 吊罗山蚊蝎蛉 *Bittacus diaoluoshanus*
   阳茎叶较狭窄，后面观无上述突起 …… 10
10. 上生殖瓣末端稍钝圆 …… 16. 扁蚊蝎蛉 *Bittacus planus*
    上生殖瓣末端明显平截 …… 11
11. 上生殖瓣具显著的纵脊 …… 23. 四边蚊蝎蛉 *Bittacus trapezoideus*
    上生殖瓣无明显的纵脊 …… 20. 中国蚊蝎蛉 *Bittacus sinicus*
12. 生殖肢基节末端具 1 对锥形突起 …… 24. 乌苏里蚊蝎蛉 *Bittacus ussuriensis*
    生殖肢基节末端无上述突起 …… 13
13. 上生殖瓣腹缘具 2 个突起，末端延长形成端部平截的尾突 …… 10. 海南蚊蝎蛉 *Bittacus hainanicus*
    上生殖瓣腹缘仅具 1 个突起或无突起，末端不形成端部平截的尾突 …… 14
14. 上生殖瓣近腹缘具 1 条纵沟，并在沟末端处稍内凹；载肛突上瓣近末端背缘具 1 对指形突起 …… 22. 天目山蚊蝎蛉 *Bittacus tienmushana*
    上生殖瓣近腹缘无上述纵沟；载肛突上瓣无上述突起 …… 15
15. 阳茎叶末端分为 2 支 …… 5. 双叉蚊蝎蛉 *Bittacus bifurcatus*
    阳茎叶末端不分支 …… 16
16. 上生殖瓣分为 2 叶 …… 17
    上生殖瓣不分叶 …… 19
17. 上生殖瓣在近基部 1/3 处分为 2 叶 …… 25. 浙江蚊蝎蛉 *Bittacus zhejiangicus*
    上生殖瓣在中部和近端部分为 2 叶 …… 18
18. 上生殖瓣在近中部处分为 2 叶 …… 13. 莫纳斯蚊蝎蛉 *Bittacus monastyrskiyi*
    上生殖瓣在近端部分为 2 叶 …… 19. 中华蚊蝎蛉 *Bittacus sinensis*
19. 上生殖瓣端部延长并尖锐 …… 14. 暗蚊蝎蛉 *Bittacus obscurus*
    上生殖瓣端部延长但较圆钝 …… 20
20. 上生殖瓣近三角形，端部延长成指形尾突 …… 7. 环带蚊蝎蛉 *Bittacus cirratus*
    上生殖瓣长梯形，端部延长成钩状尾突 …… 9. 淡黄蚊蝎蛉 *Bittacus flavidus*

Key to species of *Bittacus* in this atlas

(Twenty-one species, males)

1. Wings devoid of markings …… 2
   Wings with various markings …… 3
2. Epandrial lobe rectangular; upper branch of proctiger nearly as long as epandrial lobe and enlarged apically in a goose head-shape …… 6. *Bittacus choui*

Epandrial lobe short quadrilateral; upper branch of proctiger shorter than epandrial lobe and not enlarged apically ······ 17. *Bittacus puripennis*

3. Epandrial lobe nearly subtrapezoidal, unbranched and simple in structure ······ 4
Epandrial lobe not subtrapezoidal, occasionally branched or with various projections, and complecated in structure ······ 12
4. Antennae almost as long as body ······ 12. *Bittacus longantennatus*
Antennae shorter than body ······ 5
5. Epandrial lobe thick and curved ventrad on apical half ······ 21. *Bittacus strigatus*
Epandrial lobe flat and not curved ventrad on apical half ······ 6
6. Epandrial lobe distinctly longer than gonocoxites ······ 7
Epandrial lobe shorter or nearly as long as gonocoxites ······ 8
7. Epandrial lobe with a tuft of long setae apically; upper branch of proctiger slightly exceeding epandrial lobe ······ 15. *Bittacus pieli*
Epandrial lobe with normal setae apically; upper branch of proctiger distinctly exceeding epandrial lobe for a long distance ······ 11. *Bittacus lii*
8. Epandrial lobe long rectangular, three times as long as wide; upper branch of proctiger extremely elongated ······ 18. *Bittacus setigerus*
Epandrial lobe short, not three times as long as wide; upper branch of proctiger relatively short ······ 9
9. Aedeagal lobe very broad with a hook-like acute projection directing ventrad in caudal view ······ 8. *Bittacus diaoluoshanus*
Aedeagal lobe relatively narrow and devoid of such projection ······ 10
10. Epandrial lobe slightly rounded at apex ······ 16. *Bittacus planus*
Epandrial lobe distinctly truncated at apex ······ 11
11. Epandrial lobe with a distinct longitudinal ridge ······ 23. *Bittacus trapezoideus*
Epandrial lobe without longitudinal ridge ······ 20. *Bittacus sinicus*
12. Gonocoxites with a pair of conical projections apically ······ 24. *Bittacus ussuriensis*
Gonocoxites without such projections apically ······ 13
13. Epandrial lobe with two projections on ventral margin, and elongated apically into a tail with truncated apex ······ 10. *Bittacus hainanicus*
Epandrial lobe with only one, or none projections on ventral margin, and not elongated apically into such a tail ······ 14
14. Epandrial lobe with a longitudinal groove near ventral margin, and emarginated at end of groove; upper branch of proctiger with a pair of finger-like processes on dorsal surface subapically ······ 22. *Bittacus tienmushana*
Epandrial lobe without such groove; upper branch of proctiger without such processes ······ 15
15. Aedeagal lobe bifurcated apically ······ 5. *Bittacus bifurcatus*
Aedeagal lobe not bifurcated apically ······ 16
16. Epandrial lobe bifurcated into two lobes ······ 17
Epandrial lobe not bifurcated ······ 19
17. Epandrial lobe splits at basal third into two lobes ······ 25. *Bittacus zhejiangicus*
Epandrial lobe splits at middle or near its apex ······ 18
18. Epandrial lobe splits at dorsal margin medially into two lobes ······ 13. *Bittacus monastyrskiyi*
Epandrial lobe splits near its apex into two short lobes ······ 19. *Bittacus sinensis*
19. Epandrial lobe elongated and acute apically ······ 14. *Bittacus obscurus*
Epandrial lobe elongated but relatively rounded apically ······ 20
20. Epandrial lobe subtriangular and elongated apically into a finger-like tail ······ 7. *Bittacus cirratus*
Epandrial lobe long trapezoidal and elongated apically into a tail curving dorsad ······ 9. *Bittacus flavidus*

5. 双叉蚊蝎蛉 *Bittacus bifurcatus* Hua & Tan, 2008

【测量】 前翅长 16.0 mm。
【形态特征】 头顶及喙黄褐色。单眼三角区黑褐色，中单眼后方具 2 根鬃毛。触角黑褐色，鞭节线状，被毛。前胸背板黑棕色，前缘两侧各具 1 根粗刚毛。中、后胸背板黑棕色，小盾片黄褐色。侧板及足浅黄褐色，股节及胫节端部加深。翅无色透明，斑纹褐色。前翅 FRs、ORs、CuP 末端具加深的雾状斑；沿横脉加深。翅痣明显，棕色。雄性腹部 1 ~ 8 节黄褐色至黑褐色。上生殖瓣黄褐色，背缘基半部向上隆起，端部形成 1 个向上弯曲的钩状尾突。生殖肢基节末端具 V 形膜质区；生殖刺突细长，似 L 形，基部内侧具 1 个突起。阳茎叶在末端分为 2 支，其中一支圆阔，一支细长。阳茎丝细长，盘绕成环。载肛突上瓣细长，从上生殖瓣之间伸出；下瓣短钩状。

【发生期】 7 ~ 8 月。

【地理分布】 陕西（太白山、鄠邑）。

Measurements: Forewing length 16.0 mm.

Morphology: Vertex and rostrum yellowish brown. Ocellar triangle dark brown and bearing two bristles behind middle ocellus. Antennae blackish brown, with flagellum filiform and ciliated.Pronotum blackish brown and bearing one stout seta on each side at anterior margin. Mesonotum and metanotum dark brown with yellowish brown scutella. Pleura and legs light yellowish brown, with apical end of femur and both ends of tibia darkened. Wing membrane hyaline. Deepened smoky markings along FRs, ORs, and end of CuP in forewings. Pterostigma distinct and brown.

图 2-5-1 双叉蚊蝎蛉 *Bittacus bifurcatus* 雄成虫背面观（腹部已去除，陕西鄠邑朱雀森林公园，NWAFU）

Fig. 2-5-1 Male adult of *Bittacus bifurcatus* with abdomen removed, dorsal view (Zhuque Forest Park, Huyi, Shaanxi, NWAFU)

图 2-5-2 双叉蚊蝎蛉 *Bittacus bifurcatus* 雄性外生殖器侧面观

Fig. 2-5-2 Male genitalia of *Bittacus bifurcatus*, lateral view

Male A1–A8 yellowish brown to blackish brown. Epandrial lobe yellowish brown; half of its dorsal margin protruding dorsad; its ventral margin elongated and forming a hook-like tail at apex. Gonocoxites with a V-shaped membranous area posteriorly. Gonostylus slender and L-shaped, and bearing a small process on inner side. Aedeagal lobe bifurcated into a broad branch and another slender branch. Penisfilum coiled into a small loop. Upper branch of proctiger long and protruding between epandrial lobes, while lower branch short and hook-shaped.

Emergence: July to August.

Distribution: Shaanxi (Taibai Mountain, Huyi).

图 2-5-3 双叉蚊蝎蛉 *Bittacus bifurcatus* 地理分布

Fig. 2-5-3 Distribution of *Bittacus bifurcatus*

6. 周氏蚊蝎蛉 *Bitttacus choui* Hua & Tan, 2007

【测量】 前翅长 15.0 ~ 16.0 mm。

【形态特征】 头顶、喙及触角浅黄褐色，单眼三角区黑褐色，贯穿的黑斑向两侧延伸到复眼边缘。前胸背板黄褐色，中后胸背板不均匀黄褐色。侧板及足浅黄色，股节、胫节末端颜色明显变暗。翅透明，微泛淡黄色，无斑纹。雄性腹部 1 ~ 8 节黑褐色，第 8 节后缘中部有 V 形缺刻。上生殖瓣淡黄色，半透明，与黑褐色的生殖肢形成鲜明对比；近长矩形。生殖肢基节半球形，在末端中央具 V 形膜质区；生殖刺突短，基部阔，顶端钝圆。阳茎叶圆钝；阳茎丝细长，盘绕成环。载肛突上瓣细长，端部鹅头状膨大，具 1 簇细长毛；下瓣长钩状，向腹面弯曲。

【发生期】 7 ~ 8 月。

【地理分布】 陕西（南郑）、四川（峨眉山）、重庆（金佛山）、湖北（神农架）。

Measurements: Forewing length 15.0–16.0 mm.

Morphology: Vertex and rostrum light yellowish brown. Ocellar triangle black. Dark ocellar stripe extending to inner margin of compound eyes laterally. Pronotum yellowish brown. Mesonotum and metanotum unevenly yellowish brown. Pleura and legs light yellowish brown, with apical end of femur and both ends of tibia darkened. Wing membrane hyaline and slightly tinged with light yellow. Markings absent. Male A1–A8 dark brown. Tergum Ⅷ with a deep V-shaped emargination on middle of posterior margin. Epandrial lobe long rectangular, light yellowish brown and translucent, contrast to dark brown gonocoxites. Gonocoxites hemispherical, bearing a membranous V-shaped area at apex. Gonostylus short, broad basally and rounded apically. Aedeagal lobes rounded. Penisfilum coiled into a loop. Upper branch of proctiger slender, enlarged gooshead-shaped and bearing a tuft of dense setae apically, while lower branch long hook-shaped and curved ventrad.

Emergence: July to August.

Distribution: Shaanxi (Nanzheng), Sichuan (Mount Emei), Chongqing (Mount Jinfo) and Hubei (Shennongjia).

图 2-6-2 周氏蚊蝎蛉 *Bitttacus choui* 雄性外生殖器侧面观

Fig. 2-6-2 Male genitalia of *Bitttacus choui*, lateral view

图 2-6-1 周氏蚊蝎蛉 *Bitttacus choui* 雄成虫侧面观（陕西南郑黎坪国家森林公园，NWAFU）

Fig. 2-6-1 Male adult of *Bitttacus choui*, lateral view (Liping National Forest Park, Nanzheng, Shaanxi, NWAFU)

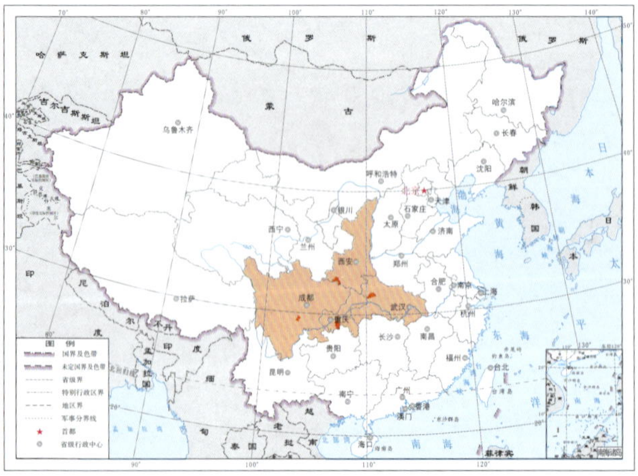

图 2-6-3 周氏蚊蝎蛉 *Bittacus choui* 地理分布

Fig. 2-6-3 Distribution of *Bittacus choui*

7. 环带蚊蝎蛉 *Bittacus cirratus* Tjeder, 1956

【测量】 前翅长 24.0 ~ 26.0 mm。

【形态特征】 头顶、额黄褐色，喙端部稍深。触角黄褐色，鞭节被短毛。单眼三角区黑褐色。前胸背板黄褐色。中胸和后胸背板不均匀黄褐色，小盾片浅。侧板及足黄褐色，股节端部和胫节两端加深。翅透明，微泛黄色。有 4 个褐色小斑分别位于 OM、ORs、FRs 和 CuP 近末端。沿所有横脉褐色雾状加深。雄性腹部 1 ~ 8 节黄褐色。第 8 节背板末端具深 V 形缺刻。上生殖瓣侧面观近似三角形，末端延伸成指形突起；内侧中部有一块区域密生黑色刺毛。生殖肢基节黄褐色，后缘中部浅黄色，向后突出；生殖刺突短，端部内侧有 1 个光滑的突起。阳茎叶较长，端部稍尖；阳茎丝长，盘绕成环。尾须细长，长度约为生殖肢基节之半。载肛突上瓣相当阔，末端有一簇短毛，端半部有 1 对明显的侧叶，侧叶边缘有 1 列刺毛；下瓣短而狭窄。

【发生期】 7 ~ 8 月。

【地理分布】 陕西（佛坪、眉县、周至）、上海（佘山）、浙江（舟山）、吉林（临江）、湖北（武昌）、内蒙古（通辽）。

Measurements: Forewing length 24.0–26.0 mm.

Morphology: Vertex and rostrum yellowish brown, with apex of rostrum slightly darkened. Antennae yellowish brown, with flagellum ciliated. Ocellar triangle dark brown. Pronotum yellowish light brown. Mesonotum and metanotum unevenly yellowish brown with pale scutella. Pleura and legs yellowish brown, with apical end of femur and both ends of tibia darkened. Wing membrane hyaline and slightly tinged with light yellow. Four brown spots on OM, ORs, FRs and near end of CuP. All cross-veins with surrounding blurry stripes. Male A1–A8 yellowish brown. Tergum Ⅷ with a deep V-shaped emargination on posterior margin. Epandrial lobe subtriangular in lateral view, protruding and forming a finger-like prong apically. Dense black spines on inner surface. Gonocoxites yellowish brown, protruding backward apically. Gonostylus short and bearing a smooth process on inner side. Aedeagal lobes long and acute apically. Penisfilum coiled into a loop. Cerci

图2-7-1 环带蚊蝎蛉 *Bittacus cirratus* 雄成虫侧面观（内蒙古通辽，NWAFU）

Fig. 2-7-1 Male adult of *Bittacus cirratus*, lateral view (Tongliao, Inner Mongolia, NWAFU)

图 2-7-2 环带蚊蝎蛉 *Bittacus cirratus* 雄性外生殖器侧面观

Fig. 2-7-2 Male genitalia of *Bittacus cirratus*, lateral view

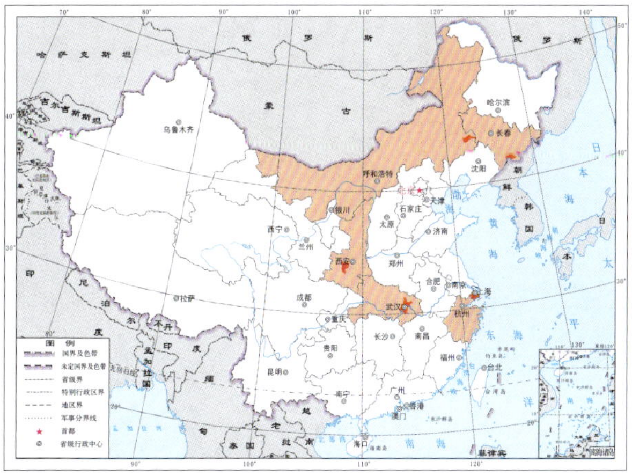

图 2-7-3 环带蚊蝎蛉 *Bittacus cirratus* 地理分布

Fig. 2-7-3 Distribution of *Bittacus cirratus*

slender and approximately half length of gonocoxites. Upper branch of proctiger broad and bearing a tuft of setae at apex, while lower branch shorter and narrower. The upper branch with a lateral lobe on each side, and bearing a row of yellow setae on each lobe marginally.

Emergence: July to August.

Distribution: Shaanxi (Foping, Meixian, Zhouzhi), Shanghai (Sheshan), Zhejiang (Zhoushan), Jilin (Linjiang), Hubei (Wuchang) and Inner Mongolia (Tongliao).

8. 吊罗山蚊蝎蛉 *Bittacus diaoluoshanus* Chen & Hua, 2011

【测量】 前翅长 19.7 ~ 20.2 mm。

【形态特征】 头顶、喙黄褐色；单眼三角区黑色；触角棕色被毛。前胸背板黄褐色。中、后胸背板两侧黑褐色，中间具浅色中带；侧板及足黄褐色，股节末端及胫节两端加深。翅微泛黄色，在 ORs、FRs、OM 等处具褐色斑。翅痣褐色明显。沿翅痣到翅端部褐色加深，沿所有横脉具雾状褐色斑。雄性腹部 1 ~ 8 节黄褐色。第 8 节背板末端具深 V 形缺刻。雄性上生殖瓣近梯形，端部平截。生殖肢基节端部具深 V 形膜质区；生殖刺突短小，内侧具突起。阳茎叶阔，阳茎丝细长盘绕成环。尾须短棒状。载肛突上瓣从上生殖瓣之间伸出，末端尖，具毛簇；下瓣短，向腹面弯曲。

【发生期】 5 月。

【地理分布】 海南（吊罗山）。

Measurements: Forewing length 19.7–20.2 mm.

Morphology: Vertex and rostrum yellowish brown. Ocellar triangle black. Antennae brown and cilia-ted.

图 2-8-1 吊罗山蚊蝎蛉 ***Bittacus diaoluoshanus*** 雄成虫侧面观
（海南陵水吊罗山，NWAFU）

Fig. 2-8-1 Male adult of ***Bittacus diaoluoshanus***, lateral view
(Mount Diaoluo, Lingshui, Hainan, NWAFU)

图 2-8-2 吊罗山蚊蝎蛉 ***Bittacus diaoluoshanus***
雄性外生殖器侧面观

Fig. 2-8-2 Male genitalia of ***Bittacus diaoluoshanus***, lateral view

Pronotum yellowish brown. Mesonotum and meta-notum dark brown with a pale mesal stripe. Pleura and legs yellowish brown, with apical end of femur and both ends of tibia darkened. Wing membrane hyaline and slightly tinged with yellow. Several brown spots on OM, ORs, FRs, respectively. Pterostigma brown and

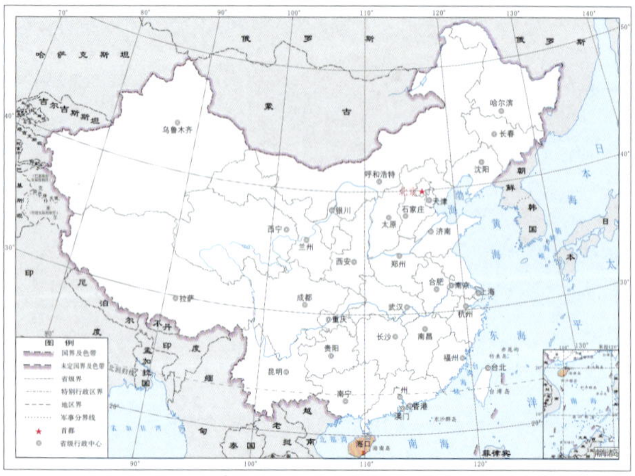

图 2-8-3 吊罗山蚊蝎蛉 ***Bittacus diaoluoshanus*** 地理分布

Fig. 2-8-3 Distribution of ***Bittacus diaoluoshanus***

distinct. A brown stripe extending from pterostigma to wing apex. All crossveins with surrounding blurry stripes. Male A1–A8 yellowish brown. A8 with a deep V-shaped emargination on middle of posterior margin. Male epandrial lobe subtrapezoidal, and truncated at apex. Gonocoxites with a V-shaped membranous area on apex. Gonostylus short and bearing a small process on inner side. Aedeagal lobes broad. Penisfilum coiled into a loop. Cerci short and clavate. Upper branch of proctiger acute and bearing a tuft of setae apically, and stretching out between paired epandrial lobes. Lower branch short and curved ventrad.

Emergence: May.

Distribution: Hainan (Mount Diaoluo).

9. 淡黄蚊蝎蛉 *Bittacus flavidus* Huang & Hua, 2005

【测量】 前翅长 22.0 ~ 25.0 mm。

【形态特征】 头顶不均匀黄褐色，中间浅两侧深。喙黄褐色。单眼三角区黑褐色。触角褐色，被毛。前胸背板两侧前缘具 2 根明显的黑色长刚毛，后缘各 1 根。中、后胸背板黄褐色，中部具约与小盾片等宽的浅色纵带。侧板黄褐色。足黄褐色，股节末端、胫节两端加深。翅透明微泛黄色，翅痣棕色明显，在 ORs、FRs 和 OM 处具褐色斑，沿所有横脉具雾状褐色斑，在 R_5 末端具 1 矩形雾状大斑。雄性腹部 1 ~ 8 节黄褐色，向后逐渐加深。第 8 节背板后缘具深 V 形缺刻。上生殖瓣长梯形，端部具 1 个钩状突起，近腹缘上方具 1 纵脊。生殖肢基节后缘具 V 形缺刻。阳茎叶基部阔，末端尖锐；阳茎丝短，盘绕成环。尾须细长，超过生殖肢基节端部，末端尖。载肛突上瓣细长，末端具长毛，从上生殖瓣之间穿出；下瓣短小，向下钩曲。

【发生期】 7 ~ 8 月。

【地理分布】 陕西（宁陕、南郑、太白、眉县、周至）、湖北（神农架）。

Measurements: Forewing length 22.0–25.0 mm.

Morphology: Vertex unevenly yellowish brown, and darkened laterad. Rostrum yellowish brown. Ocellar triangle dark brown. Antennae brown and ciliated. Pronotum with two long black setae on each side of its anterior margin, and one seta on each side of its posterior margin. Mesonotum and metanotum yellowish brown with a pale mesal stripe almost as wide as scutella. Pleura and legs yellowish brown, with apical end of femur and both ends of tibia darkened. Wing membrane hyaline, slightly tinged with yellow. Pterostigma brown and distinct. Several brown spots on OM, ORs, and FRs, respectively. All crossveins with surrounding blurry stripes. Terminal portion of R_5 with large smoky stripe. Male A1–A8 yellowish

图 2-9-1 淡黄蚊蝎蛉 *Bittacus flavidus* 雄成虫侧面观（陕西南郑黎坪国家森林公园，NWAFU）

Fig. 2-9-1 Male adult of *Bittacus flavidus*, lateral view（Liping National Forest Park, Nanzheng, Shaanxi，NWAFU)

图 2-9-2 淡黄蚊蝎蛉 *Bittacus flavidus* 雄性外生殖器侧面观

Fig. 2-9-2 Male genitalia of *Bittacus flavidus*, lateral view

brown and gradually darkening caudad. Tergum Ⅷ with a deep V-shaped emargination on posterior margin. Epandrial lobe subtrapezoidal, forming a hook-like process on apex, and bearing a longitudinal ridge above ventral margin. Gonocoxites with a V-shaped emargination on posterior margin. Aedeagal lobes broad basally, acute on apex. Penisfilum short and coiled into a loop. Cerci long and extending beyond gonocoxites apically, with acute apexes. Upper branch of proctiger slender and bearing a tuft of setae apically, and stretching out between paired epandrial lobes. Lower branch short and curved ventrad.

Emergence: July to August.

Distribution: Shaanxi (Ningshan, Nanzheng, Taibai, Meixian, Zhouzhi) and Hubei (Shennongjia).

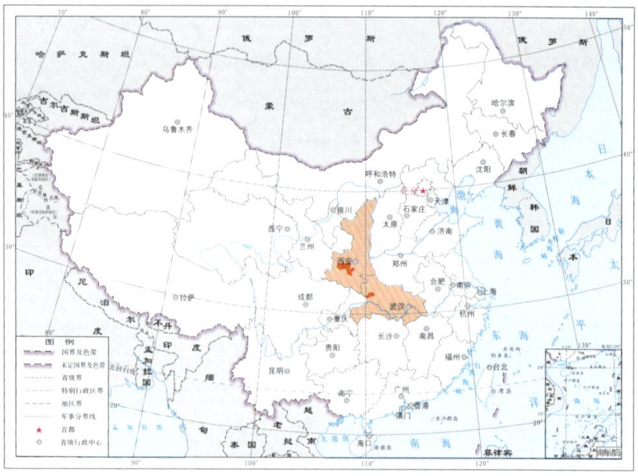

图 2-9-3 淡黄蚊蝎蛉 ***Bittacus flavidus*** 地理分布

Fig. 2-9-3 Distribution of ***Bittacus flavidus***

10. 海南蚊蝎蛉 *Bittacus hainanicus* Tan & Hua, 2008

【测量】 前翅长 18.0 ~ 19.0 mm。

【形态特征】 头顶、额琥珀色，喙黄褐色；单眼三角区黑色；触角棕色被毛。前胸背板黑棕色，前后缘无明显刚毛。中、后胸背板黄棕色，中间具浅色中带；侧面不均匀黄褐色。足黄褐色，股节及胫节端部加深。翅透明微泛黄色，翅痣褐色明显，在 ORs、FRs、OM 等处具褐色斑，沿 R_5 有 1 条长斑，沿所有横脉具雾状褐色斑。雄性腹部黄褐色。第 7、8 节背板末端具深 V 形缺刻。上生殖瓣形状奇特：腹缘内卷，内侧具 2 个突起；背缘中部具 1 个 V 形横裂，其后形成 1 个三角形突起；端部延伸成末端平截的长突起。生殖肢基节顶端具 1 个尖细突起，与阳茎之间由 1 个半透明舌状膜质片连接；生殖刺突近锥状，顶端具 3 根长刚毛。阳茎叶阔，近端部向后方延伸出 1 个臂状突起；阳茎丝盘绕成小环。尾须短棒状。载肛突上瓣细长，端部具毛簇。下瓣短，向下钩曲。

【发生期】 5 月。

【地理分布】 海南（霸王岭、五指山、尖峰岭）。

Measurements: Forewing length 18.0–19.0 mm.

Morphology: Vertex and frons amber. Rostrum yellowish brown. Ocellar triangle black. Antennae brown and ciliated. Pronotum blackish brown, without distinct setae on anterior and posterior margins.

图 2-10-1 海南蚊蝎蛉 ***Bittacus hainanicus*** 雌成虫侧面观（海南乐东尖峰岭，NWAFU）

Fig. 2-10-1 Female adult of ***Bittacus hainanicus***, lateral view (Jianfengling, Ledong, Hainan, NWAFU)

图 2-10-2 海南蚊蝎蛉 ***Bittacus hainanicus*** 雄性外生殖器侧面观

Fig. 2-10-2 Male genitalia of ***Bittacus hainanicus***, lateral view

Mesonotum and metanotum yellowish brown with a pale mesal stripe. Pleura unevenly brown. Legs yellowish brown, with apical end of femur and both ends of tibia darkened.Wing membrane hyaline, slightly tinged with yellow. Pterostigma brown and distinct. Several brown spots on OM, ORs, and FRs, *etc*. A brown stripe extending along R_5. All cross-veins with surrounding blurry stripes. Male abdomen yellowish brown. Terga Ⅶ and Ⅷ each with a deep V-shaped emargination on posterior margin. Epandrial lobe peculiarly shaped: ventral margin curled inward and bearing two small processes mesially; dorsal margin with a V-shaped cleft at middle; a triangular process next to cleft; apex extending caudad and forming a long truncated prolongation. Gonocoxites each with an acute process apically, which connected to aedeagus by a translucent and tongue-shaped plate. Gonostylus subconical, and bearing three long bristles on apex. Aedeagal lobe broad and protruding caudad to form an arm-shaped process. Penisfilum coiled into

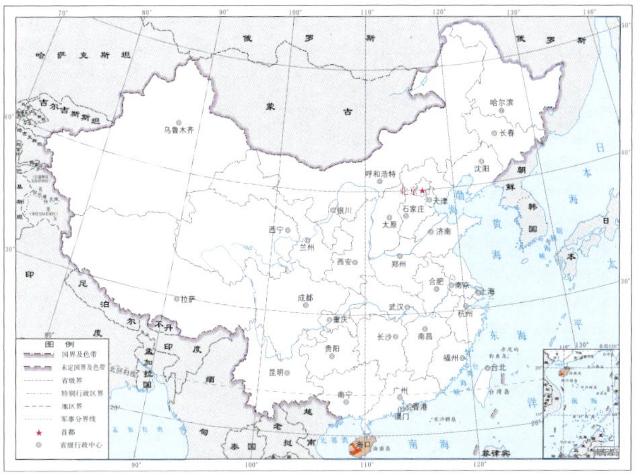

图 2-10-3 海南蚊蝎蛉 ***Bittacus hainanicus*** 地理分布

Fig. 2-10-3 Distribution of ***Bittacus hainanicus***

a small loop. Cerci short and clavate. Upper branch of proctiger slender and bearing a tuft of setae apically. Lower branch of proctiger short and curved ventrad.

Emergence: May.

Distribution: Hainan (Bawangling, Wuzhi Mountain, Jianfengling).

11. 李氏蚊蝎蛉 *Bittacus lii* Zhou, 2003

【测量】 前翅长 19.0 ~ 21.0 mm。

【形态特征】 头顶黄褐色；单眼三角区黑色，1 条黑色带穿过单眼三角区并向两侧延伸至复眼；喙基部深褐色，端部稍浅；触角褐色被毛。前胸背板深褐色，后缘无刚毛。中、后胸背板不均匀深褐色，具浅黄色中带；侧板黄褐色；足黄褐色，股节端部和胫节两端加深。翅透明微泛浅黄褐色，翅痣褐色明显；翅脉分叉处具多个雾状斑；沿几乎所有翅脉具模糊的斑纹。雄性腹部黄褐色。第 8 节背板后缘中部具深 V 形缺刻。上生殖瓣基部愈合区阔，内侧形成 2 个脊状突起，端部圆钝，侧面观近似长梯形，略长于生殖肢基节。生殖肢基节半球形，端部具 V 形膜质区；生殖刺突短。阳茎叶小，末端圆形。载肛突上瓣长，从上生殖瓣之间伸出，端部具毛簇；下瓣短，向下钩曲。

【发生期】 7 ~ 8 月。

【地理分布】 湖南（桑植）、湖北（五峰）、贵州（梵净山）。

Measurements: Forewing length 19.0–21.0 mm.

图 2-11-1 李氏蚊蝎蛉 ***Bittacus lii*** 雄成虫侧面观（湖南桑植八大公山，NWAFU）

Fig. 2-11-1 Male adult of ***Bittacus lii***, lateral view (Badagong Mountains, Sangzhi, Hunan, NWAFU)

Morphology: Vertex yellowish brown. Ocellar triangle black. A black stripe passing through ocellar triangle and extending to compound eyes laterally. Rostrum deep brown basally and paler apically. Antennae brown and ciliated. Pronotum blackish brown, without distinct setae on posterior margin.

图 2-11-2 李氏蚊蝎蛉 *Bittacus lii* 雄性外生殖器侧面观
Fig. 2-11-2 Male genitalia of *Bittacus lii*, lateral view

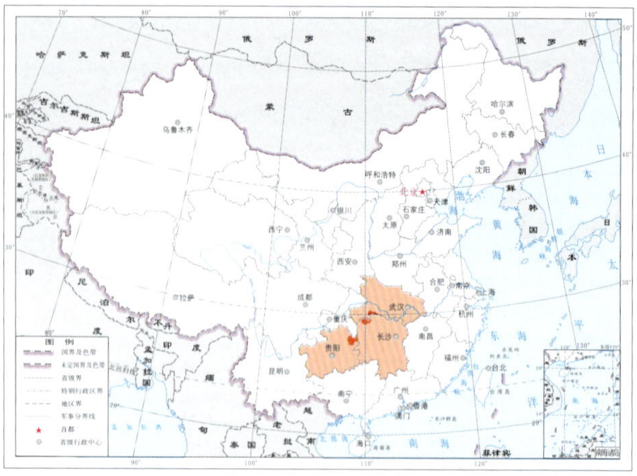

图 2-11-3 李氏蚊蝎蛉 *Bittacus lii* 地理分布
Fig. 2-11-3 Distribution of *Bittacus lii*

Mesonotum and metanotum unevenly yellowish brown with a pale yellow mesal stripe. Pleura yellowish brown. Legs yellowish brown with apical end of femur and both ends of tibia darkened.Wing membrane hyaline, slightly tinged with yellowish brown. Pterostigma brown and distinct. Multiple smoky brown spots on joints of veins. Almost all veins with surrounding blurry stripes. Male abdomen yellowish brown. Tergum Ⅷ with a deep V-shaped emargination on middle of posterior margin. Epandrial lobes broadly fused basally, long subtrapezoidal with rounded apexes, bearing two ridge-like prominences on inner surface, and slightly longer than gonocoxites. Gonocoxites semispherical, with a V-shaped membranous area on apex; gonostylus very short. Aedeagal lobes small with rounded apexes. Upper branch of proctiger long and extending between epandrial lobes, with a tuft of setae on apex; lower branch short and curved ventrad.

Emergence: July to August.

Distribution: Hunan (Sangzhi), Hubei (Wufeng) and Guizhou (Mount Fanjing).

12. 长角蚊蝎蛉 *Bittacus longantennatus* Chen, Tan & Hua, 2013

【测量】 前翅长 19.0 ~ 21.0 mm。

【形态特征】 头顶暗褐色；单眼三角区黑褐色，其黑色斑向两侧延伸到复眼边缘；触角长于足股节。前胸背板黑褐色，前后缘均无明显的长刚毛；侧板黑褐色。中、后胸背板均为不均匀黑褐色，小盾片及后胸背板背中线颜色稍浅，但中胸背板无明显的浅中纵带，侧板黄褐色。足黄褐色，股节端部和胫节两端加深。翅面具多处明显的褐色雾状斑，几乎弥漫整个翅面。翅痣棕色明显。雄性腹部 1 ~ 8 节黑褐色。第 8 节背板末端具浅缺刻。上生殖瓣比生殖肢基节短，侧面观近四边形，端部平截，内缘有密集的黑色刺毛。生殖肢基节末端具 V 形膜质区；生殖刺突短小，末端凹陷，内侧形成 1 个突起。阳茎叶较短，末端钝圆；阳茎丝细长，盘绕成环。载肛突上瓣从上生殖瓣基部中间伸出，端部密生 1 簇长细毛；下瓣短，向下钩曲。

图 2-12-1 长角蚊蝎蛉 *Bittacus longantennatus* 雄成虫侧面观（湖南石门壶瓶山，NWAFU）
Fig. 2-12-1 Male adult of *Bittacus longantennatus*, lateral view (Mount Huping, Shimen, Hunan, NWAFU)

【发生期】 7 ~ 8 月。

【地理分布】 四川（天全、峨眉山）、湖北（巴东）、湖南（石门）。

Measurements: Forewing length 19.0–21.0 mm.

Morphology: Vertex deep brown. Ocellar triangle dark brown. A black stripe passing through ocellar triangle and extending to compound eyes laterally. Antennae longer than femora. Pronotum blackish brown, without distinct setae on anterior and posterior margins. Mesonotum and metanotum unevenly blackish brown, paler at scutella and mesal stripe of metanotum, but devoid of pale mesal stripe on mesonotum. Pleura yellowish brown. Legs yellowish brown with apical end of femur and both ends of tibia darkened. Wing membrane intensively tinged with smoky brown markings. Pterostigma brown and distinct. Male A1–A8 blackish brown. Tergum Ⅷ with a shallow emargination on posterior margin. Epandrial lobes subtrapezoidal in lateral view and slightly shorter than gonocoxites, and bearing dense black spines on inner surface of each truncated apex. Gonocoxites semispherical, with a V-shaped membranous area on apex; gonostylus very short, emarginated apically and bearing a small process on inner surface. Aedeagal lobes short with rounded apexes. Upper branch of proctiger streching out between epandrial lobes, with a tuft of long setae on apex; lower branch short and curved ventrad.

Emergence: July to August.

Distribution: Sichuan (Tianquan, Mount Emei), Hubei (Badong) and Hunan (Shimen).

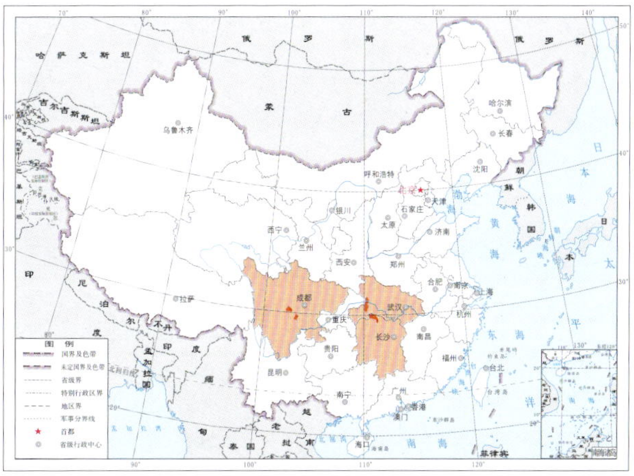

图 2-12-3　长角蚊蝎蛉 *Bittacus longantennatus* 地理分布

Fig. 2-12-3　Distribution of *Bittacus longantennatus*

图 2-12-2　长角蚊蝎蛉 *Bittacus longantennatus* 雄性外生殖器侧面观

Fig. 2-12-2　Male genitalia of *Bittacus longantennatus*, lateral view

13. 莫纳斯蚊蝎蛉 *Bittacus monastyrskiyi* Bicha, 2007

【测量】 前翅长 24.8 mm。

【形态特征】 头顶深褐色。单眼三角区黑褐色，两侧具延伸至复眼的黑色横带。喙黄褐色。前胸背板具不均匀黑褐色。中、后胸背板两侧黑褐色，具不明显的浅色中带。侧板及足黄褐色，股节端部及胫节两端加深。翅黄褐色半透明，斑纹褐色浓厚。翅痣深褐色，窄长明显。前翅外缘强烈向后弯曲近直角形；沿翅外缘从翅痣到 R_5 末端有 1 条带状褐色斑纹，沿 R_5 褐色加深；Av、OM、ORs 及 FRs 处各有 1 明显斑；沿多数横脉均雾状加深。雄性腹部背板中间褐色，两端黄褐色，末端几节背板颜色加深。第 8 节背板后缘具 V 形深缺刻。上生殖瓣长，在近中部处分为 2 叶，形成较长的下臂和较短的上臂；上臂基部下方和腹缘内表面上分别具 1 个尖锐突起。生殖肢基节末端具 V 形膜质区；生殖刺突小，向中间弯曲。阳茎叶半圆形，阳茎丝细长，盘绕成小环。尾须长约为生殖肢基节之半。载肛突上瓣狭长，从上生殖瓣之间伸出，末端具 1 根粗刺；下瓣

图 2-13-1　莫纳斯蚊蝎蛉 ***Bittacus monastyrskiyi*** 雄成虫侧面观
（广西那坡老虎跳，NWAFU）
Fig. 2-13-1　Male adult of ***Bittacus monastyrskiyi***, lateral view
(Laohutiao, Napo, Guangxi, NWAFU)

短，向下钩曲。

【发生期】 7 月。

【地理分布】 广西（那坡）；越南。

Measurements: Forewing length 24.8 mm.

Morphology: Vertex deep brown. Ocellar triangle dark brown. A black stripe passing through ocellar triangle and extending to compound eyes laterally. Rostrum yellowish brown. Pronotum unevenly dark brown. Mesonotum and metanotum unevenly blackish brown without a mesal stripe but only paler at middle. Pleura yellowish brown. Legs yellowish brown with apical end of femur and both ends of tibia darkened. Wing membrane translucent, tinged with yellowish brown, with dense brown markings. Pterostigma deep brown and slender. Apical margin abruptly turning backwards in a right angle. Pterostigma deep brown and slender. A darkened stripe extending from pterostigma to end of R_5. Several distinct spots on Av, OM、ORs and FRs. Almost all crossveins surrounded by smoky and blurred stripes. Male terga brown medially, yellowish brown laterally, and darker on several distal segments. Tergum Ⅷ with a deep V-shaped emargination on posterior margin. Epandrial lobes long and splits at dorsal margin medially into two lobes, divided into a short dorsal and a long ventral branches. Two small acute protuberances under base of upper branch, and on inner surface of ventral margin of epandrial lobe, respectively. Gonocoxites with a V-shaped membranous area on apex. Gonostylus very small and curved mesad. Aedeagal lobes semicircular. Penisfilum very long and coiled into a small loop. Cerci nearly half length of gonocoxites. Upper branch of proctiger slender and protruding between epandrial lobes, and bearing a stout spine apically. Lower branch short and curved ventrad.

Emergence: July.

Distribution: Guangxi (Napo); Vietnam.

图 2-13-2　莫纳斯蚊蝎蛉 ***Bittacus monastyrskiyi*** 地理分布
Fig. 2-13-2　Distribution of ***Bittacus monastyrskiyi***

14. 暗蚊蝎蛉 *Bittacus obscurus* Huang & Hua, 2005

【测量】 前翅长 19.0 ~ 20.0 mm。

【形态特征】 头顶、额黄褐色，单眼三角区黑色，黑色横带向两侧延伸至复眼。触角黄褐色，被毛。前胸背板具不均匀黄褐色，前后缘各具 2 根黑色粗刚毛。中、后胸背板暗褐色，两侧较深。侧板及足基节暗褐色至浅黄色。股节末端及胫节两端黑褐色。翅浅褐色透明，端部边缘雾状加深。沿端部横脉具模糊的条纹。雄性腹部 1 ~ 6 节黄褐色，7 ~ 9 节则加深为黑褐色。第 8 节背板后缘具深 V 形凹陷。上生殖瓣侧面观近斧头状，上后方延伸成尖角，后腹方圆钝。生殖肢基节后缘中部具V形膜质区；生殖刺突极短。阳茎叶相当长，末端处形成很小的突起向内弯曲；阳茎丝极长，盘绕成环。载肛突上瓣延长，鹅颈状，端部略向头部弯曲，末端着生一簇细毛；下瓣相当长，基部稍向腹面弯曲，端半部直且向后渐细。

【发生期】 7～8月。

【地理分布】 陕西（宝鸡天台山、太白、周至、眉县、岚皋、南郑、宁陕）、河南（白云山）、湖北（神农架）。

Measurements: Forewing length 19.0–20.0 mm.

Morphology: Vertex and frons yellowish brown. Ocellar triangle black. A black stripe passing through ocellar triangle and extending to compound eyes laterally. Antennae yellowish brown and ciliated. Pronotum unevenly yellowish brown, bearing two long black setae on anterior and posterior margins. Mesonotum and metanotum dark brown with two sides darkened. Pleura and coxae dark brown to light yellow. Apical end of femur and both ends of tibia dark brown. Wing membrane hyaline and tinged with light brown, and darkened at apex. Some blurred smoky stripes along apical crossveins. Male abdomen A1–A6 yellowish brown. A7–A9 dark brown.Tergum Ⅷ with a deep V-shaped emargination on posterior margin. Epandrial lobe is axe-shapedin lateral view, with an elongated acute apex dorso-posteriorly, and rounded ventro-posteriorly. onocoxites with a V-shaped embranous area

图 2-14-1 暗蚊蝎蛉 *Bittacus obscurus* 雄成虫侧面观

（陕西宝鸡天台山国家森林公园，姜碌摄）

Fig. 2-14-1 Male adult of *Bittacus obscurus*, lateral view

（Tiantaishan National Forest Park, Baoji, Shaanxi，photo by Lu Jiang)

on apex. Gonostylus extremely reduced. Aedeagal lobes very long and narrow, with a small process curving mesad on apex. Penisfilum is very long and coiled into a loop. Upper branch of proctiger is slender and goosenecklike, with its apex curved cephalad and bearing a tuft of thin setae. Lower branch very long, curved ventrad basally and straight distally, and tapering towards apex.

Emergence: July to August.

Distribution: Shaanxi [Baoji (Mount Tiantai), Taibai, Zhouzhi, Meixian, Lan'gao, Nanzheng, Ningshan], Henan (Mount Baiyun) and Hubei (Shennongjia).

图 2-14-2 暗蚊蝎蛉 ***Bittacus obscurus*** 雄性外生殖器侧面观

Fig. 2-14-2 Male genitalia of ***Bittacus obscurus***, lateral view

图 2-14-3 暗蚊蝎蛉 ***Bittacus obscurus*** 地理分布

Fig. 2-14-3 Distribution of ***Bittacus obscurus***

15. 璧尔蚊蝎蛉 *Bittacus pieli* Navás, 1935

【测量】 前翅长 24.0 ~ 26.0 mm。

【形态特征】 头顶黄褐色，单眼三角区黑色，喙褐色。触角具长毛。前胸背板不均匀暗褐色，前后缘无明显刚毛。中、后胸背板黄褐色，两侧加深。足黄褐色，侧板和足基节浅褐色。股节端部及胫节两端黑色。翅透明，浅黄色，翅痣颜色较深；在 Av、OM、ORs、FRs 处具黑褐色浓斑，沿所有横脉具模糊的雾状斑。雄性腹部黄褐色。4 ~ 6 各节背板后部具明显长毛，第 7、8 节背板后缘具 V 形缺刻。上生殖瓣基部愈合面阔；侧面近长方形，背缘中部向内侧弯曲；端部钝圆，具 1 簇长毛。生殖肢基节圆阔，末端具长毛；生殖刺突短小。阳茎叶短阔；阳茎丝细长，盘绕成环。尾须短棒状。载肛突上瓣稍超出上生殖瓣，末端具一束毛；下瓣基部阔，末端变细，向腹面弯曲。

【发生期】 7 ~ 8 月。

【地理分布】 江西（庐山）、浙江（天目山、大盘山）、安徽（黄山）。

Measurements: Forewing length 24.0–26.0 mm.

图 2-15-1 璧尔蚊蝎蛉 ***Bittacus pieli*** 雄成虫侧面观（浙江金华大盘山，NWAFU）

Fig. 2-15-1 Male adult of ***Bittacus pieli***, lateral view (Mount Dapan, Jinhua, Zhejiang, NWAFU)

Morphology: Vertex yellowish brown. Ocellar triangle black. Rostrum brown. Antennae ciliated. Pronotum unevenly dark brown, and without distinct setae on anterior and posterior margins. Mesonota and metanotum yellowish brown with two sides darkened. Pleura and coxae pale brown. Legs yellowish brown, with apical end of femur and both ends of tibia dark black. Wing membrane hyaline and tinged with pale yellow. Pterostigma brown and distinct. Several deep brown spots on Av, OM, ORs and FRs, respectively. Some blurred smoky stripes along apical crossveins. Male abdomen yellowish brown. Terga Ⅳ–Ⅵ each with a tuft of long setae posteriorly. Terga Ⅶ and Ⅷ each with a deep V-shaped emargination on posterior margin. Epandrial lobes broadly fused at base, and oblong in lateral view, with its dorsal margin curled inward. The rounded apex with a tuft of long setae. Gonocoxites rounded and bearing a tuft of long setae apically, gonostylus greatly reduced. Aedeagal lobes short and broad. Penisfilum very long and coiled into a loop. Cerci short and clavate. Upper branch of proctiger slightly extending beyond epandrial lobes, and bearing a tuft of setae apically. Lower branch broad basally and tapering towards acute apex, and curling ventrad.

图 2-15-2 壁尔蚊蝎蛉 *Bittacus pieli* 雄性外生殖器侧面观

Fig. 2-15-2 Male genitalia of *Bittacus pieli*, lateral view

图 2-15-3 壁尔蚊蝎蛉 *Bittacus pieli* 地理分布

Fig. 2-15-3 Distribution of *Bittacus pieli*

Emergence: July to August.

Distribution: Jiangxi (Lushan Mountain), Zhejiang (Tianmu Mountains, Mount Dapan) and Anhui (Mount Huangshan).

16. 扁蚊蝎蛉 *Bittacus planus* Cheng, 1949

【测量】 前翅长 19.0 ~ 21.0 mm。

【形态特征】 头顶黄褐色，喙暗褐色。复眼黑色，单眼三角区棕黑色。触角被毛。前胸背板不均匀黄褐色，前缘有 2 根明显的黑色粗刚毛。中、后胸背板前部两侧黑褐色，中间浅黄色。侧板及足黄褐色。翅透明，黄褐色，翅痣不明显，斑纹较淡。雄性腹部 1 ~ 8 节黄褐色。第 7、8 节末端具深 V 形缺刻。上生殖瓣短于生殖肢基节，侧面观近似四边形，末端内面密生数列黑刺。生殖肢基节圆阔，末端具 V 形膜质区；生殖刺突短小。阳茎叶阔，末端圆。阳茎丝长，缠绕成环。载肛突上瓣指状，从上生殖瓣基部中间伸出，端部被细毛；下瓣较短，基部阔，向末端渐细。

【发生期】 7 ~ 8 月。

【地理分布】 陕西（宁陕、宝鸡天台山、鄠邑、太白山）、河南（白云山）。

Measurements: Forewing length 19.0–21.0 mm.

Morphology: Vertex yellowish brown, rostrum dark brown. Compound eyes black, ocellar triangle dark brown. Antennae ciliated. Pronotum unevenly yellowish brown, bearing two distinct black setae on anterior margin. Mesonotum and metanotum dark brown laterally and pale yellow mesally. Pleura and legs yellowish brown. Wing membrane hyaline, tinged with yellowish-brown. Pterostigma indistinct. No distinct markings. Male abdomen A1–A8 yellowish brown. Tergum Ⅶ and Ⅷ with a deep V-shaped emargination on posterior margin. Epandrial lobes subtrapezoidal in lateral view, shorter than

图 2-16-1　扁蚊蝎蛉 ***Bittacus planus*** 成虫侧面观（陕西宝鸡天台山国家森林公园，姜碌摄）
1. 雄虫　2. 雌虫

Fig. 2-16-1　Adults of ***Bittacus planus***, lateral view (Tiantaishan National Forest Park, Baoji, Shaanxi, photo by Lu Jiang)
1.Male　2. Female

图 2-16-2　扁蚊蝎蛉 ***Bittacus planus*** 雄性外生殖器侧面观
Fig. 2-16-2　Male genitalia of ***Bittacus planus***, lateral view

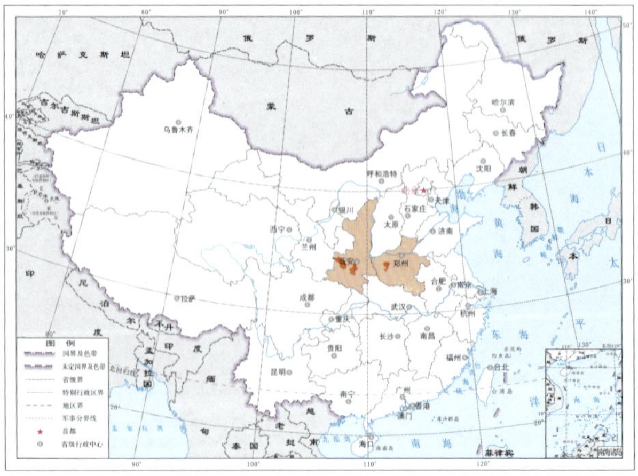

图 2-16-3　扁蚊蝎蛉 ***Bittacus planus*** 地理分布
Fig. 2-16-3　Distribution of ***Bittacus planus***

gonocoxites, and bearing dense black spines on apical inner margins. Gonocoxites rounded and bearing a V-shaped membranous area posteriorly. Gonostylus greatly reduced. Aedeagal lobes broad and rounded apically. Penisfilum long and coiled into a loop. Upper branch of proctiger extending beyond epandrial lobes and bearing a tuft of setae on apex. Lower branch short and tapering towards apex.

Emergence: July to August.

Distribution: Shaanxi [Ningshan, Baoji (Mount Tiantai), Huyi, Taibai Mountain] and Henan (Mount Baiyun).

17. 净翅蚊蝎蛉 *Bittacus puripennis* Cai & Hua, 2006

【测量】 前翅长 15.0 ~ 16.0 mm。

【形态特征】 头顶、喙黄褐色；单眼三角区黑色，其黑斑向两侧延伸至复眼。前胸背板不均匀黑褐色；中、后胸背板黑褐色，沿背中线明显形成 1 条浅黄色带。侧板及足黄色。翅透明，微泛黄色，翅痣阔，浅褐色，翅面光洁无斑纹。雄性腹部 1 ~ 6 节黄褐色。第 7、8 节黑褐色，末端具深 V 形缺刻。上生殖瓣浅黄白色，半透明，侧面形状近长方形，端部渐窄。生殖肢基节深褐色，半球形，末端有 1 块 V 形膜质区。生殖刺突短，末端中部略凹陷，呈 V 形，被细毛。阳茎叶较阔，末端钝圆；阳茎丝较短，盘绕成环。第 10 节背板背上部分马鞍形，基部向两侧延伸并围绕尾须。尾须伸达上瓣的 2/3 处，末端钝圆。载肛突上瓣从两上生殖瓣之间伸出，末端变细，端部有 1 簇长毛；下瓣短，向后渐细。

【发生期】 7 ~ 8 月。

【地理分布】 陕西（南郑）、四川（卧龙）。

Measurements: Forewing length 15.0–16.0 mm.

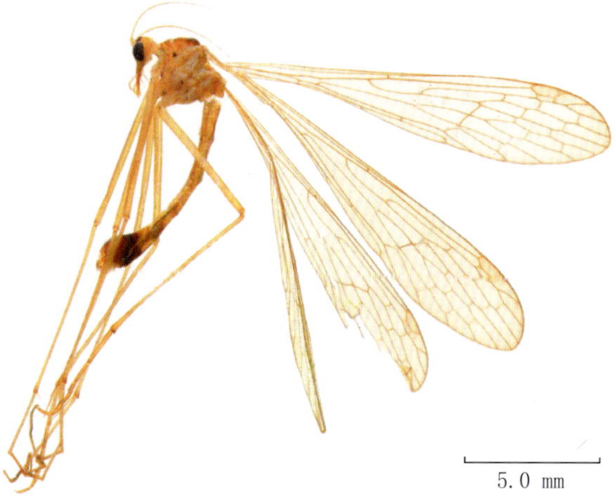

图 2-17-1 净翅蚊蝎蛉 *Bittacus puripennis* 雄成虫侧面观（陕西南郑黎坪国家森林公园，NWAFU）

Fig. 2-17-1 Male adult of *Bittacus puripennis*, lateral view (Liping National Forsest Park, Nanzheng, Shaanxi, NWAFU)

图 2-17-2 净翅蚊蝎蛉 *Bittacus puripennis* 雄性外生殖器侧面观

Fig. 2-17-2 Male genitalia of *Bittacus puripennis*, lateral view

Morphology: Vertex and rostrum yellowish brown. Ocellar triangle black. A black stripe passing through ocellar triangle and extending to compound eyes laterally. Pronotum unevenly dark brown. Mesonotum and metanotum dark brown, with a pale yellow mesal stripe. Pleura and legs yellow.Wing membrane hyaline and slightly tinged with yellow. Pterostigma broad and light brown. No distinct markings. Male A1–A6

图 2-17-3 净翅蚊蝎蛉 *Bittacus puripennis* 地理分布

Fig. 2-17-3 Distribution of *Bittacus puripennis*

yellowish brown. Terga Ⅶ and Ⅷ dark brown and each with a deep V-shaped emargination on posterior margin. Epandrial lobes pale yellow, translucent, subtrapezoidal in lateral view, tapering towards apex, and bearing dense black spines on apical inner margins. Gonocoxites deep brown, semispherical, and bearing a V-shaped membranous area posteriorly. Gonostylus greatly reduced, with apex slightly emarginate on middle of posterior margin and V-shaped, and covered by slender setae. Aedeagal lobes broad and rounded apically. Penisfilum short and coiled into a loop. Tergum Ⅹ saddle-shaped dorsally, and extending laterad at base and enclosing cerci. Cerci extending to two-thirds of epandrial lobes, and rounded at apexs. Upper branch of proctiger extending beyond epandrial lobes and bearing a tuft of long setae on apex. Lower branch short and tapering towards apex.

Emergence: July to August.

Distribution: Shaanxi (Nanzheng) and Sichuan (Wolong).

18. 具毛蚊蝎蛉 *Bittacus setigerus* Chen, Tan & Hua, 2013

【测量】 前翅长 21.5 ~ 22.0 mm。

【形态特征】 头顶黄褐色，单眼三角区黑色。喙基部深褐色，端部稍浅；触角被毛。前胸背板不均匀黄褐色，后缘两侧各具 2 根刚毛。中、后胸背板不均匀黄褐色，中央具与小盾片等宽的浅色中带。侧板及足黄褐色，股节端部及胫节两端加深。翅浅黄色，半透明。翅痣褐色宽阔。前翅 OM、ORs、FRs、CuP 末端以及端半部横脉处均有褐色云斑。雄性腹部黄褐色。第 8 节背板后缘缺刻呈 V 形。上生殖瓣黄褐色，长矩形，略短于生殖基节，末端平截；内侧中部以及腹缘近中部处有许多黑色短刚毛，近端部有 1 排黑褐色长刚毛。生殖肢基节半球状，后方具 1 个 V 形膜质区；生殖刺突短，被长刚毛。阳茎叶末端平截，阳茎丝细长盘绕。尾须棒状，端部略尖。载肛突上瓣细长，仅端部向体前端弯曲，顶端后侧具 1 簇长刚毛；下瓣短钩状。

【发生期】 7 ~ 8 月。

【地理分布】 贵州（梵净山）、湖北（神农架、巴东）。

图 2-18-2 具毛蚊蝎蛉 *Bittacus setigerus* 雄性外生殖器侧面观

Fig. 2-18-2 Male genitalia of *Bittacus setigerus*, lateral view

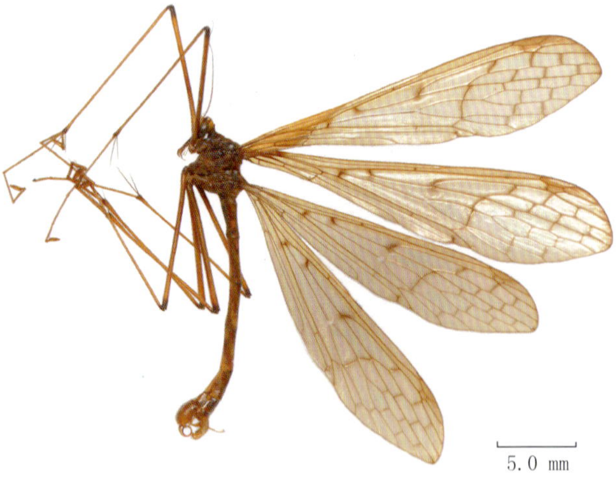

图 2-18-1 具毛蚊蝎蛉 *Bittacus setigerus* 雄成虫侧面观（贵州梵净山，NWAFU）

Fig. 2-18-1 Male adult of *Bittacus setigerus*, lateral view (Mount Fanjing, Guizhou, NWAFU)

Measurements: Forewing length 21.5–22.0 mm.

Morphology: Vertex yellowish brown. Ocellar triangle black. Rostrum dark brown basally and paler apically. Antennae ciliated. Pronotum unevenly dark brown, bearing two stout setae on each side of posterior margin. Mesonotum and metanotum unevenly yellowish brown, with a pale yellow mesal stripe as wide as scutella. Pleura and legs yellowish brown, with apex of femur and both ends of tibia darkened. Wing membrane translucent and tinged with light yellow. Pterostigma broad and brown. Several cloudy markings on OM, ORs, FRs, CuP and along crossveins in apical half of forewings. Male abdomen yellowish brown. Tergum Ⅷ with a V-shaped emargination on posterior margin. Epandrial lobes yellowish brown, oblong with a truncated apex, and slightly longer than gonocoxites, with a series of black short setae on inner surface and ventral margin medially, and a row of dark brown long setae near apex. Gonocoxites semispherical, with a V-shaped membranous area posteriorly; gonostylus greatly reduced and covered with long setae. Aedeagal lobes truncated apically. Penisfilum long and coiled into a loop. Cerci clavate with slightly acute apexes. Upper branch of proctiger slender, with its apical portion curved cephalad and bearing a tuft of setae on posterior margin of apex. Lower branch short and hook-like.

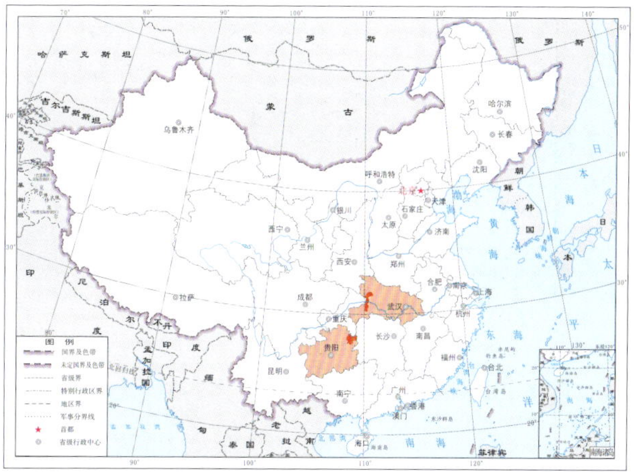

图 2-18-3 具毛蚊蝎蛉 *Bittacus setigerus* 地理分布

Fig. 2-18-3 Distribution of ***Bittacus setigerus***

Emergence: July to August.

Distribution: Guizhou (Mount Fanjing) and Hubei (Shennongjia, Badong).

19. 中华蚊蝎蛉 *Bittacus sinensis* Walker, 1853

【测量】 前翅长 22.0 ~ 23.0 mm。

【形态特征】 头顶、额琥珀色。单眼三角区黑褐色，喙深褐色。触角线状细长，被毛。前胸背板黄褐色，前后缘无明显刚毛；中、后胸背板大部黑褐色，小盾片黄褐色。胸部侧板及足基节黄白色。股节末端及胫节两端黑褐色，其余黄褐色。翅浅黄色。翅痣阔，黄褐色。翅端钝圆。翅面在 OM、ORs、FRs、CuP 和端部横脉处具一系列小斑点。雄性腹部 1 ~ 8 节背、腹板黄褐色。第 8 节背板末端无缺刻。雄性上生殖瓣侧面近似梯形，后缘分裂为上、下 2 叶，下叶长于上叶，末端内表面密生黑色小刺。生殖肢基节后缘中部圆形，密被细毛。阳茎叶阔长，末端稍尖；阳茎丝细长，缠绕成小环。尾须较长，约为上生殖瓣长的 2/3，末端尖。载肛突上瓣较长，从上瓣之间伸出，末端具 1 簇长毛；下瓣弯曲，末端尖。

图 2-19-2 中华蚊蝎蛉 *Bittacus sinensis* 雄性外生殖器侧面观

Fig. 2-19-2 Male genitalia of ***Bittacus sinensis***, lateral view

图 2-19-3 中华蚊蝎蛉 *Bittacus sinensis* 地理分布

Fig. 2-19-3 Distribution of ***Bittacus sinensis***

【发生期】 7～8月。

【地理分布】 陕西（黄陵）、辽宁（清原、沈阳）。

Measurements: Forewing length 22.0–23.0 mm.

Morphology: Vertexand frons amber. Ocellartriangle dark brown. Rostrum deep brown. Antennae thread-like and ciliated.Pronotum yellowish brown, and without distinct setae on anterior and posterior margins. Mesonotum and metanotum mostly dark brown and with yellowish brown scutella. Pleura and coxae yellowish brown. Apex of femur and both ends of tibia darkened, the rest parts of legs yellowish brown. Wing membrane tinged with light yellow. Apex of both forewings and hindwings rounded. Pte-rostigma broad and yellowish brown. Several small spots on OM, ORs, FRs, CuP, and along apical crossveins. Male terga and sterna Ⅰ–Ⅷ yellowish brown. Tergum Ⅷ not emarginate on posterior margin. Epandrial lobes subtra-pezoidal, and divided posteriorly into a short upper branch and a long lower branch. Both branches with dense

图 2-19-1 中华蚊蝎蛉 ***Bittacus sinensis*** 雌成虫侧面观（辽宁沈阳森林公园，姜碌摄）

Fig. 2-19-1 Female adult of ***Bittacus sinensis***, lateral view（Shenyang Forest Park, Liaoning，photo by Lu Jiang)

black setae on their inner margin apically. Gonoco - xites rounded at middle of hind margin and bea- ring dense setae. Aedeagal lobes broad and long, and acute apically. Penisfilum slender and coi-led into a small loop. Cerci two-thirds as long as epandrial lobes, and acute apically. Upper branch of proctiger long, extending beyond epandrial lobes, and bearing a tuft of setae on apex. Lower branch curved ventrad with acute apex.

Emergence: July to August.

Distribution: Shaanxi (Huangling) and Liaoning (Qingyuan, Shenyang).

20. 中国蚊蝎蛉 *Bittacus sinicus* Issiki, 1931

图 2-20-1 中国蚊蝎蛉 ***Bittacus sinicus*** 雄成虫侧面观（四川峨眉山，NWAFU）

Fig. 2-20-1 Male adult of ***Bittacus sinicus***, lateral view (Mount Emei, Sichuan, NWAFU)

【测量】 前翅长 17.5 ~ 19.3 mm。

【形态特征】 头顶褐色。单眼三角区黑色，黑色条带向两侧延伸至复眼边缘。喙黑褐色。触角黑褐色具长毛。前胸背板黄褐色。中、后胸背板黑褐色，具浅黄色中带；中、后胸前侧片黑褐色，后侧片灰白色；足黄褐色不均匀，股节端部和胫节两端黑色。翅透明，浅黄褐色，翅痣褐色明显。沿翅脉具黑褐色雾状斑纹。雄性腹部 2 ~ 6 节背板前部黑褐色，后端黄褐色，明暗相间。第 7、8 节黑褐色，末端具浅缺刻。上生殖瓣浅褐色、半透明，近四边形，末端钝圆。生殖肢基节后缘稍微向上突出，中部色浅；生殖刺突小，近中部有 1 个明显的内突起。阳茎叶短，较窄，末端圆。阳茎丝细长，盘绕成环。尾须短棒状，长约为生殖肢基节之半。载肛突上瓣窄长，端部具毛簇，从上生殖瓣之间伸出；下瓣短小，向腹面钩曲。

【发生期】 7 ~ 8 月。

【地理分布】 四川（峨眉山）、湖北（恩施、兴山）、贵州（梵净山）、陕西（南郑）。

Measurements: Forewing length 17.5–19.3 mm.

图 2-20-2 中国蚊蝎蛉 ***Bittacus sinicus*** 雄性外生殖器侧面观

Fig. 2-20-2 Male genitalia of ***Bittacus sinicus***, lateral view

Morphology: Vertex brown. Ocellar triangle black. A transverse black band passing through ocellar triangle and extending to compound eyes laterally. Rostrum dark brown. Antennae blackish brown and ciliated. Pronotum yellowish brown. Mesonotum and metanotum mostly dark brown and with a pale yellow mesal stripe. Episterna of mesothorax and metathorax dark brown, while epimera pale white. Legs unevenly yellowish brown, with apex of femur and both ends of tibia darkened. Wing membrane hyaline and tinged with light yellowish brown. Pterostigma brown and distinct. Dense smoky brown stripes along veins. Male terga Ⅱ–Ⅵ dark brown medially and yellowish brown posteriorly, forming distinct light and dark patterns. Terga Ⅶ and Ⅷ dark brown and slightly emarginate on posterior margin. Epandrial lobes light brown and translucent, subtrapezoidal, and with a rounded apex. Gonocoxites with hind margin slightly protruded dorsad, and paler at middle. Gonostylus small and bearing a distinct process near middle portion. Aedeagal lobes short and narrow, and rounded apically. Penisfilum slender and coiled into a small loop. Cerci short and approximately half length of gonocoxites. Upper branch of proctiger slender, extending beyond epandrial lobes, and bearing a tuft of setae on apex. Lower branch short and curved ventrad with acute apex.

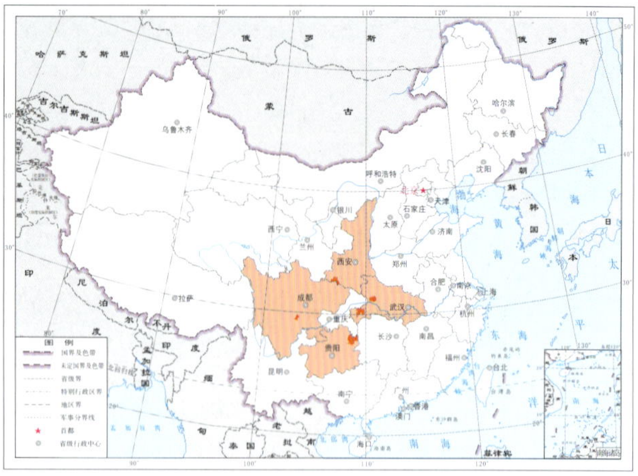

图 2-20-3　中国蚊蝎蛉 *Bittacus sinicus* 地理分布

Fig. 2-20-3　Distribution of ***Bittacus sinicus***

Emergence: July to August.

Distribution: Sichuan (Mount Emei), Hubei (Enshi, Xingshan), Guizhou (Mount Fanjing) and Shaanxi (Nanzheng).

21. 纹翅蚊蝎蛉 *Bittacus strigatus* Hua & Chou, 1998

【测量】 前翅长 20.0 ~ 26.0 mm。

【形态特征】 头顶黄褐色。单眼三角区黑色。喙深褐色。触角褐色被毛。胸部背板黑褐色，无浅色中带；侧面及足黄褐色，股节端部及胫节两端加深。翅透明，浅黄色；翅痣黑褐色明显；斑纹褐色，浓密的黑褐色雾状斑分布成 4 条横带，其中 2 条后部连接成 V 形。雄性腹部 1 ~ 6 节黄褐色，第 7、8 节黑褐色。上生殖瓣黑褐色，宽阔，长于生殖肢

图 2-21-2　纹翅蚊蝎蛉 *Bittacus strigatus* 雄性外生殖器侧面观

Fig. 2-21-2　Male genitalia of ***Bittacus strigatus***, lateral view

图 2-21-3　纹翅蚊蝎蛉 *Bittacus strigatus* 地理分布

Fig. 2-21-3　Distribution of ***Bittacus strigatus***

基节，端半部向腹面弯曲，末端向内突起，突起上密生小刺。生殖肢基节后缘中部微凹。阳茎叶阔；阳茎丝细长，盘绕成环。尾须细长。载肛突上瓣鹅颈状，顶端向头部弯曲，具毛簇；下瓣较短粗，向下钩曲。

【发生期】 7 ~ 8月。

【地理分布】 河南（白云山）、陕西（太白山、南宫山、宁陕）、湖北（神农架、房县）。

Measurements: Forewing length 20.0–26.0 mm.

Morphology: Vertex yellowish brown. Ocellar triangle black. Rostrum deep brown. Antennae brown and ciliated. Nota dark brown without pale mesal stripes. Pleura and legs yellowish brown, with apex of femur and both ends of tibia darkened. Wing membrane hyaline, slightly tinged with light yellow. Pterostigma dark brown and distinct. Dense smoky brown stripes arranged in four transversal lines, with two of them connected pos-teriorly in a V-shape. Male A1–A6 yellowish brown. A7 and A8 dark brown. Epandrial lobes dark brown, very broad and longer than gonocoxites, with their apical halves curved ventrad. The apex of epandrial lobe protruded inward and bearing dense spines. Gonocoxites slightly emarginated apically at

图 2-21-1 纹翅蚊蝎蛉 *Bittacus strigatus* 雄成虫侧面观（湖北房县三座庵）

Fig. 2-21-1 Male adult of ***Bittacus strigatus***, lateral view (Sanzuo'an, Fangxian, Hubei)

middle. Aedeagal lobes broad. Penisfilum slender and coiled into a loop. Cerci slender. Upper branch of proctiger slender and gooseneck-like, with its apical portion curved cephalad and bearing a tuft of setae. Lower branch stout and curved ventrad.

Emergence: July to August.

Distribution: Henan (Mount Baiyun), Shaanxi (Mount Taibai, Mount Nangong, Ningshan) and Hubei (Shennongjia, Fangxian).

22. 天目山蚊蝎蛉 *Bittacus tienmushana* Cheng, 1957

【测量】 前翅长 26.0 ~ 27.0 mm。

【形态特征】 头顶黄褐色，单眼三角区黑色，喙黑褐色。触角黄褐色，丝状被毛。两触角窝中间愈合，连成 1 个近长方形的大窝。前胸背板不均匀暗褐色；中胸背板大部分暗褐色，后部及后胸背板颜色变浅。侧板及足黄褐色，股节端部及胫节两端加深。翅黄色透明。翅痣浅棕色，不明显。在 OM、ORs、FRs、CuP 末端处有明显的黑褐色小斑，沿横脉具雾状加深的条纹。雄性腹部黄褐色。第 8 节背板末端具深 V 形缺刻。雄性上生殖瓣侧面观近三角形；下方近边缘处具 1 条纵沟；外缘在沟末端处稍内凹。生殖肢基节后缘两侧略延伸，形成 2 个小突起；生殖刺突短，靴状，密被细毛。阳茎叶长且阔，末端稍尖；阳茎丝长，盘绕成环。载肛突上瓣从上生殖瓣之间伸出，其背面近端部具 1 对弯曲的指形突起，中央顶部具 1 簇刚毛；下瓣强烈向下钩曲。

【发生期】 5 ~ 7 月。

【地理分布】 浙江（天目山）。

图 2-22-1 天目山蚊蝎蛉 ***Bittacus tienmushana*** 雄成虫侧面观
（浙江临安西天目山，NWAFU）

Fig. 2-22-1 Male adult of ***Bittacus tienmushana***, lateral view
(West Tianmu Mountain, Lin'an, Zhejiang, NWAFU)

图 2-22-2 天目山蚊蝎蛉 ***Bittacus tienmushana*** 雄性外生殖器侧面观

Fig. 2-22-2 Male genitalia of ***Bittacus tienmushana***, lateral view

Measurements: Forewing length 26.0–27.0 mm.

Morphology: Vertex yellowish brown. Ocellar triangle black. Rostrum dark brown. Antennae yellowish brown and ciliated. Two antennal sockets confused mesally and forming an oblong depression. Pronotum unevenly deep brown. Mesonotum mostly dark brown, and paler towards hind portion and metanotum. Pleura and legs yellowish brown, with apex of femur and both ends of tibia darkened. Wing membrane hyaline and slightly tinged with yellow. Pterostigma light brown and indistinct. Several dark brown spots on OM, ORs, FRs, and ending of CuP, respectively. Deepened smoky stripes along cross-veins. Male abdomen yellowish brown. Tergum Ⅷ with a deep V-shaped emargination on posterior margin. Epandrial lobes subtriangular in lateral view; a longitudinal groove slightly above ventral margin; apex slightly emarginate at end of groove. Gonocoxites protruded apically and forming a pair of small processes. Gonostylus short and boot-shaped with dense setae. Aedeagal lobes long and broad with slightly acute apexes. Penisfilum slender and coiled into a loop. Upper branch of proctiger extending beyond epandrial lobes, bearing a pair of curved finger-like processes subapically, and a tuft of setae on central apex. Lower branch intensively curved ventrad.

Emergence: May to July.

Distribution: Zhejiang (Tianmu Mountains).

图 2-22-3　天目山蚊蝎蛉 ***Bittacus tienmushana*** 地理分布

Fig. 2-22-3　Distribution of ***Bittacus tienmushana***

23. 四边蚊蝎蛉 *Bittacus trapezoideus* Huang & Hua, 2005

图 2-23-1　四边蚊蝎蛉 ***Bittacus trapezoideus*** 成虫侧面观（陕西宝鸡天台山国家森林公园，姜碌摄）

1. 雄虫　2. 雌虫

Fig. 2-23-1　Adults of ***Bittacus trapezoideus***, lateral view (Tiantaishan National Forest Park, Baoji, Shaanxi, photo by Lu Jiang)

1. Male　2. Female

【测量】 前翅长 17.0 ~ 19.0 mm。

【形态特征】 头顶褐色，单眼三角区黑褐色，1 条黑色横带向两侧延伸到复眼边缘。触角端部约 2/3 具长毛。前胸背板褐色到黑褐色不均匀。中、后胸背板两侧黑褐色，背中线色浅。侧板、足黄褐色，股节端部及胫节两端加深。翅透明，略带褐色，一系列黑褐色雾状斑纹几乎散布整个翅面。雄性腹部 1 ~ 6 节背板每节前部黑褐色，后部黄褐色。第 7、8 节背板黄褐色。上生殖瓣长度约为生殖肢基节之半，黄白色，侧面观近四边形，端部平截，内侧中部具 1 条纵脊。生殖肢基节末端具 V 形膜质区；生殖刺突短小。阳茎叶圆钝，阳茎丝细长，盘绕成环。尾须稍短于上生殖瓣，细长。载肛突上瓣狭长，从上瓣之间伸出，端部具 1 簇长细毛。

【发生期】 7 ~ 8 月。

【地理分布】 陕西（太白山、天台山、宁陕、南郑）、甘肃（天水）、湖北（神农架）。

Measurements: Forewing length 17.0–19.0 mm.

Morphology: Vertex brown. Ocellar triangle dark brown. A black transversal band passing through ocellar triangle and extending to compound eyes laterally. Antennae ciliated on apical two-thirds. Pronotum unevenly brown to dark brown. Mesonotum and metanotum dark brown with a light mesal stripe. Pleura and legs yellowish brown, with apex of femur and both ends of tibia darkened.Wing membrane hyaline and slightly tinged with brown. A series of smoky dark brown markings covering almost entire wing surface. Male terga Ⅰ–Ⅵ blackish brown anteriorly and yellowish brown posteriorly. Terga Ⅶ and Ⅷ yellowish brown. Epandrial lobes yellowish white, half length of gonocoxites, and subtrapezoidal in lateral view with truncated apex. A longitudinal groove on its middle. Gonocoxites with a V-shaped membranous area posteriorly. Gonostylus short and small. Aedeagal lobes rounded. Penisfilum slender and coiled into a loop. Cerci slender but slightly shorter than epandrial lobes. Upper branch of proctiger slender and protruding between epandrial lobes and bearing a tuft of setae apically.

图 2-23-2 四边蚊蝎蛉 ***Bittacus trapezoideus***
雄性外生殖器侧面观
Fig. 2-23-2 Male genitalia of ***Bittacus trapezoideus***, lateral view

Emergence: July to August.

Distribution: Shaanxi (Mount Taibai, Tiantaishan, Ningshan, Nanzheng), Gansu (Tianshui) and Hubei (Shennongjia).

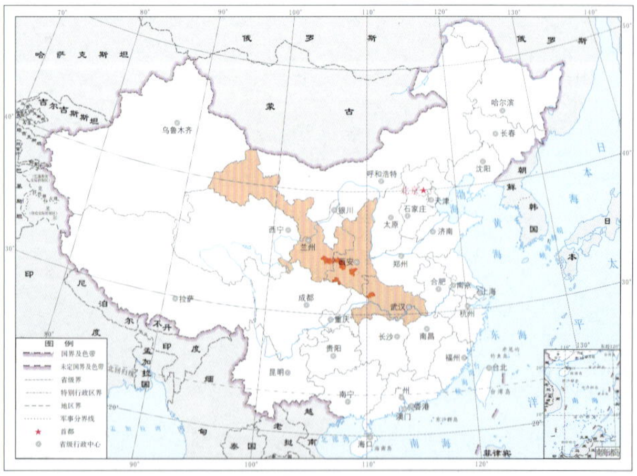

图 2-23-3 四边蚊蝎蛉 ***Bittacus trapezoideus*** 地理分布
Fig. 2-23-3 Distribution of ***Bittacus trapezoideus***

24. 乌苏里蚊蝎蛉 *Bittacus ussuriensis* Plukento, 1985

【测量】 前翅长 16.0 ~ 19.0 mm。

【形态特征】 头顶黑褐色；单眼三角区黑色。喙不均匀褐色。触角线状被毛。前胸背板不均匀褐色；中胸背板基半部深褐色，端半部黄褐色；后胸背板浅黄色。侧板及足浅黄褐色。翅浅黄色透明，翅痣不明显，翅面具零散的黑色斑纹，外缘端部

图 2-24-1 乌苏里蚊蝎蛉 ***Bittacus ussuriensis*** 雌成虫侧面观（辽宁宽甸花脖山，姜碌摄）

Fig. 2-24-1 Female adult of ***Bittacus ussuriensis***, lateral view (Mount Huabo, Kuandian, Liaoning, photo by Lu Jiang)

横脉处雾状加深。雄性腹部 1 ~ 6 节黄褐色，7、8 节暗褐色。上生殖瓣黄白色，形状特别：背缘近基部具 1 个向后的三角形小突起；外缘耳状，内表面中部具 1 条纵脊。生殖肢基节短于上生殖瓣，端部腹面伸出 1 对锥形突起。无明显的阳茎叶，阳茎丝缠绕成小环。载肛突上瓣长，近顶端处略内凹，端部具 1 列长毛。下瓣短粗，向下钩曲。

【发生期】 7 ~ 8 月。

【地理分布】 辽宁（宽甸）；俄罗斯（远东）。

Measurements: Forewing length 16.0–19.0 mm.

Morphology: Vertex dark brown. Ocellar triangle black. Rostrum unevenly brown. Antennae threadlike and ciliated. Pronotum unevenly brown. Mesonotum deep brown basally and yellowish brown distally. Metanotum light yellow. Pleura and legs light yellowish brown.Wing membrane hyaline, slightly tinged with yellow. Pterostigma indistinct. Several black spots present. Apical cross-veins with surrounding smoky stripes. Male A1– A6 yellowish brown, Terga Ⅶ and Ⅷ deep brown. Epandrial lobes yellowish white, and peculiar in shape: dorsal margin with a small triangular process subbasally; outer margin ear-like; inner surface with a longitudinal ridge. Gonocoxites shorter than epandrial lobes and protruded apically to form a pair of cylindrical processes. Aedeagal lobes indistinct. Penisfilum coiled into a small loop. Upper branch of proctiger long and slightly emarginated subapically, and bearing a row of long setae on apex. Lower branch stout and curved ventrad.

Emergence: July to August.

Distribution: Liaoning (Kuandian) and Russia (Far East).

图 2-24-2 乌苏里蚊蝎蛉 ***Bittacus ussuriensis*** 雄性外生殖器侧面观

Fig. 2-24-2 Male genitalia of ***Bittacus ussuriensis***, lateral view

图 2-24-3 乌苏里蚊蝎蛉 ***Bittacus ussuriensis*** 地理分布

Fig. 2-24-3 Distribution of ***Bittacus ussuriensis***

25. 浙江蚊蝎蛉 *Bittacus zhejiangicus* Tan & Hua, 2008

【测量】 前翅长 23.5 mm。

【形态特征】 头顶、额黄褐色，喙褐色。触角被毛。前胸背板不均匀黄褐色，前缘无刚毛。中、后胸背板黑褐色，中央具约与小盾片等宽的灰棕色纵带。侧板及足黄褐色；股节和胫节端部加深。翅略带浅黄色，半透明；翅痣棕色明显。ORs、FRs、OM 和 CuP 末端具加深斑纹；沿横脉均有雾状加深。雄性腹部黄褐色。第 8 节背板后缘具深 V 形缺刻。上生殖瓣长于生殖肢基节；上缘近基部分出 1 个钝分叶；中部延长，端部指状向内弯曲；下缘近基部亦形成 1 个分叶；各分叶及端部内表面具密集的黑刺。生殖肢基节末端钝圆；生殖刺突简单，无分支。尾须狭长，端部尖。载肛突上瓣宽阔，从上生殖瓣之间稍伸出，端部具 1 簇长毛；下瓣短而弯曲。

【发生期】 7 月。

【地理分布】 浙江（凤阳山、乌岩岭、百山祖）。

图 2-25-1 浙江蚊蝎蛉 ***Bittacus zhejiangicus*** 雄成虫侧面观
（浙江龙泉凤阳山，NWAFU）

Fig. 2-25-1 Male adult of ***Bittacus zhejiangicus***, lateral view
(Mount Fengyang, Longquan, Zhejiang, NWAFU)

图 2-25-2 浙江蚊蝎蛉 ***Bittacus zhejiangicus*** 雄性外生殖器侧面观

Fig. 2-25-2 Male genitalia of ***Bittacus zhejiangicus***, lateral view

Measurements: Forewing length 23.5 mm.

Morphology: Vertex and frons yellowish brown. Rostrum brown. Antennae ciliated.Pronotum unevenly yellowish brown and lacking distinct setae on anterior margin. Mesonotum and metanotum dark brown and bearing a greyish brown mesal stripe, almost equal width with scutella. Pleura and legs yellowish brown, with apex of femur and both ends of tibia darkened. Wing membrane transluent and slightly tinged with yellow. Pterostigma brown and distinct. Several black spots on ORs, FRs, OM, and ending of CuP, respectively. Crossveins with surrounding smoky stripes. Male abdomen yellowish brown. Tergum Ⅷ deeply emarginate in a V-shaped on posterior margin. Epandrial lobes longer than gonocoxites. Dorsal margin splits and bearing a rounded short lobe subbasally; middle portion elongated and terminated in a curved finger-like apex; ventral margin with an acute lobe subbasally. All these lobes with dense black spines on inner surfaces. Gonocoxites rounded apically. Gonostylus simple and unbranched. Cerci slender and acute apically. Upper branch of proctiger broad and protruding beyond epandrial lobes, and bearing a tuft of long setae apically. Lower branch short and curved ventrad.

Emergence: July.

Distribution: Zhejiang (Mount Fengyang, Wuyanling, Baishanzu).

图 2-25-3 浙江蚊蝎蛉 ***Bittacus zhejiangicus*** 地理分布

Fig. 2-25-3 Distribution of ***Bittacus zhejiangicus***

三、地蚊蝎蛉属 *Terrobittacus* Tan & Hua, 2009

地蚊蝎蛉属 *Terrobittacus* 与蚊蝎蛉属 *Bittacus* 外形相似，主要区别在于雄性上生殖瓣通常短于生殖肢基节之半。

中国特有属，中国已记录 6 种，本图鉴收录 6 种。

Terrobittacus is similar to *Bittacus* in appearance, but can be differentiated from the latter by the male epandrial lobes shorter than half the length of gonocoxites.

The genus is endemic to China and comprises six species. All of them are included in this atlas.

图 2-e 缠绕地蚊蝎蛉 ***Terrobittacus implicatus*** 雄成虫侧面观

Fig. 2-e Male adult of ***Terrobittacus implicatus***, lateral view

中国地蚊蝎蛉属 *Terrobittacus* 分种检索表

（6 种，雄性）

1. 胸部背板具明显的浅色中带；生殖肢基节端部具 1 ~ 3 对长刚毛；尾须长于上生殖瓣 ………………………………………………… 29. 长毛地蚊蝎蛉 *Terrobittacus longisetus*
   胸部背板无中带；生殖肢基节端部无长刚毛；尾须短于上生殖瓣 ……………… 2
2. 载肛突上瓣较细长，端部简单并具 1 簇刚毛 ……………… 3
   载肛突上瓣膨大，端部呈钩状弯曲 ……………… 4
3. 载肛突上瓣马刀状，稍向腹面弯，末端尖 ……………… 31. 刀地蚊蝎蛉 *Terrobittacus xiphicus*
   载肛突上瓣长棒状，直，末端钝圆 ……………… 27. 具刺地蚊蝎蛉 *Terrobittacus echinatus*
4. 翅面光洁几乎无斑纹，仅端部稍加深 ……………… 30. 钩地蚊蝎蛉 *Terrobittacus rostratus*
   翅沿横脉具雾状加深的斑点 ……………… 5
5. 生殖瓣侧面观近三角形；载肛突上瓣基半部与端半部约等宽 ………28. 缠绕地蚊蝎蛉 *Terrobittacus implicatus*
   上生殖瓣侧面观小棒状；载肛突上瓣基半部窄，端半部十分膨大 …… 26. 狭地蚊蝎蛉 *Terrobittacus angustus*

Key to species of *Terrobittacus* in China

(6 species, males)

1. Nota with a distinct pale mesal stripe; gonocoxites with 1–3 pairs of long bristles apically; cercus longer than epandrial lobe ……………… 29. *Terrobittacus longisetus*
   Nota without mesal stripe; gonocoxites with no such bristles apically; cercus shorter than epandrial lobe ……………… 2
2. Upper branch of proctiger relatively slender and bearing a tuft of setae on apex ……………… 3
   Upper branch of proctiger enlarged and curved apically and hook-like ……………… 4
3. Upper branch of proctiger sabre-shaped and slightly curved ventrad with an acute apex…… 31. *Terrobittacus xiphicus*
   Upper branch of proctiger long stick-shaped and straight with a rounded apex ………… 27. *Terrobittacus echinatus*
4. Wings devoid of markings except faint darkening on apex, and no cloudy spots along crossveins ……………… 30. *Terrobittacus rostratus*
   Wings with cloudy spots along crossveins ……………… 5
5. Epandrial lobe subtriangular in lateral view; upper branch of proctiger same width at basal and distal halves ……………… 28. *Terrobittacus implicatus*
   Epandrial lobe clavate in lateral view; upper branch of proctiger narrow at base and enlarged at distal half ……………… 26. *Terrobittacus angustus*

26. 狭地蚊蝎蛉 *Terrobittacus angustus* Du & Hua, 2017

【测量】 前翅长 18.5 ~ 19.0 mm。

【形态特征】 头顶、额黄褐色。单眼三角区黑色，中单眼上方具 2 根刚毛。喙黄褐色。前胸背板黑褐色，前缘具 2 ~ 4 根粗刚毛。中胸背板两侧深褐色，中间及后部颜色稍浅，小盾片黄褐色；后胸背板黄褐色。侧板及足基节乳黄色或黄褐色。足大部分红褐色，股节端部及胫节两端加深。翅透明，黄褐色。翅痣黄褐色，明显。OM、ORs、FRs 和 CuP 近末端分别具 4 个加深的模糊斑；沿几乎所有横脉呈雾状加深。R_5 末端具向后渐宽的雾状纹。雄性腹部背板黄褐色至黑棕色，腹板黄褐色。第 8 节背板后缘轻微缺刻。上生殖瓣窄，侧面观似小棒状，端部钝圆，内缘着生一系列黑色短刺；中部具 1 条纵脊。生殖肢基节末端中部圆；生殖刺突短小。阳茎叶基部阔，端部尖锐；阳茎丝细长，盘绕成环。第 10 节背板中部退化消失，形成两块环绕尾须的骨片。尾须细长。载肛突上瓣长，强烈骨化，背缘形成两片分瓣，腹缘具 1 条深沟槽，端部略呈鸟嘴状钩曲；下瓣短，向下钩状弯曲。

【发生期】 6 ~ 8 月。

【地理分布】 云南（永德）。

Measurements: Forewing length 18.5–19.0 mm.

Morphology: Vertex and frons yellowish brown.

图 2-26-1 狭地蚊蝎蛉 ***Terrobittacus angustus*** 成虫侧面观（云南永德大雪山）

1. 雄虫 2. 雌虫

Fig. 2-26-1 Adults of ***Terrobittacus angustus***, lateral view (Daxue Mountain, Yongde, Yunnan)

1.Male 2. Female

图 2-26-2 狭地蚊蝎蛉 ***Terrobittacus angustus***

雄性外生殖器侧面观

Fig. 2-26-2 Male genitalia of ***Terrobittacus angustus***, lateral view

图 2-26-3 狭地蚊蝎蛉 ***Terrobittacus angustus*** 地理分布

Fig. 2-26-3 Distribution of ***Terrobittacus angustus***

Ocellar triangle black, with two bristles above median ocellus. Rostrum yellowish brown. Pronotum dark brown with two to four stout setae on anterior margin. Mesonotum deep brown laterally, and paler mesally and posteriorly, with a yellowish brown scutellum. Metanotum yellowish brown. Pleura and coxae cream-colored or yellowish brown. Legs reddish brown with apex of femur and both ends of tibia darkened. Wing membrane hyaline and tinged with yellowish brown. Pterostigma yellowish brown and distinct. Four smoky spots on ORs, FRs, OM and ending of CuP, respectively. Crossveins with surrounding smoky stripes. A smoky stripe along distal portion of R_5 and broadened towards apex. Male terga yellowish brown to dark brown. Sterna yellowish brown. Tergum Ⅷ slightly emarginate posteriorly. Epandrial lobes narrow, clavate in lateral view with rounded apexes; a series of black spines on inner surface of apex; a longitudinal ridge along middle. Gonocoxites rounded apically. Gonostylus short. Aedeagal lobes broad basally with acute apexes. Penisfilum slender and coiled into a loop. Tergum Ⅹ greatly reduced mesally and forming a pair of sclerites surrounding lender cerci. Upper branch of proctiger very long and intensively sclerotized, bilobed dorsally and bearing a deep groove ventrally, and curved ventrad into a hook apically. Lower branch short and curved ventrad.

Emergence: June to August.

Distribution: Yunnan (Yongde).

27. 具刺地蚊蝎蛉 *Terrobittacus echinatus* (Hua & Huang, 2008)

【测量】 前翅长 16.0 ~ 17.0 mm。

【形态特征】 头顶黄褐色。单眼三角区黑色，中单眼上方具 2 根不明显的刚毛。喙深棕色。前胸背板黑褐色。中、后胸背板两侧深褐色，中后部颜色稍浅；侧板及足黄褐色，股节端部及胫节两端加深。翅透明，略带浅黄色。翅痣棕色明显。OM、ORs、FRs 和 CuP 近末端分别具 4 个斑；几乎沿所有横脉具明显的黑褐色雾状加深。雄性腹部背板黄褐色，第 8 节后缘具 V 形缺刻。上生殖瓣黄褐色，端部内缘着生 1 列不规则的黑色短刺，中部具 1 条纵脊。生殖肢基节末端中部圆，生殖刺突短小。阳茎叶基部阔，端部尖锐，阳茎丝细长盘绕成环。第 10 节背板中部退化消失，形成两块环绕尾须的骨片。尾须细长。载肛突上瓣细长，直，端部较尖，具 1 簇长毛；下瓣向下钩状弯曲。

【发生期】 7 ~ 8 月。

图 2-27-1 具刺地蚊蝎蛉 *Terrobittacus echinatus* 雄成虫侧面观（陕西佛坪，NWAFU）

Fig. 2-27-1 Male adult of *Terrobittacus echinatus*, lateral view (Foping, Shaanxi, NWAFU)

图 2-27-2 具刺地蚊蝎蛉 *Terrobittacus echinatus* 雄性外生殖器侧面观

Fig. 2-27-2 Male genitalia of *Terrobittacus echinatus*, lateral view

【地理分布】 河南（白云山）、陕西（佛坪、长安）、福建（邵武）。

Measurements: Forewing length 16.0–17.0 mm .

Morphology: Vertex yellowish brown. Ocellar triangle black, with two indistinct bristles above median ocellus. Rostrum deep brown. Pronotum dark brown. Mesonotum and metanotum deep brown laterally, and paler mesally and posteriorly. Pleura and legs yellowish brown, with apex of femur and both ends of tibia darkened. Wing membrane hyaline and tinged with light yellow. Pterostigma brown and distinct. Four smoky dark spots on ORs, FRs, OM, and ending of CuP, respectively. Crossveins with surrounding smoky stripes. Male terga yellowish brown. Tergum Ⅷ slightly emarginate in a V-shaped posteriorly. Epandrial lobes yellowish brown, bearing a series of black spines on apical inner region, and a longitudinal ridge on middle. Gonocoxites rounded apically. Gonostylus short. Aedeagal lobes broad basally and acute apically. Penisfilum slender and coiled into a loop. Tergum Ⅹ greatly reduced mesally and forming a pair of sclerites surrounding slender cerci. Upper branch of proctiger slender, bearing a tuft of long setae at acute apex. Lower branch hook-like and curved ventrad.

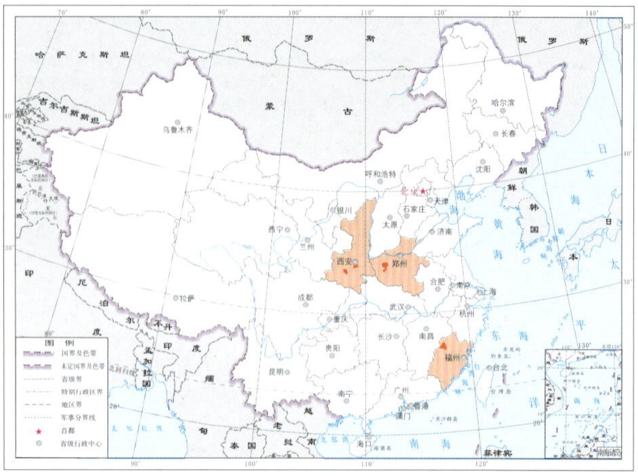

图 2-27-3 具刺地蚊蝎蛉 ***Terrobittacus echinatus*** 地理分布

Fig. 2-27-3 Distribution of ***Terrobittacus echinatus***

Emergence: July to August.

Distribution: Henan (Mount Baiyun), Shaanxi (Foping, Chang'an) and Fujian (Shaowu).

28. 缠绕地蚊蝎蛉 *Terrobittacus implicatus* (Huang & Hua, 2006)

【测量】 前翅长 15.0 ~ 16.0 mm。

【形态特征】 头顶黄褐色。单眼三角区黑色。喙黄褐色，两侧颜色较深。触角被毛。前胸背板黄褐色，前缘有 2 根黑色刚毛，后缘 2 ~ 4 根。中胸背板前半部暗褐色，后半部及后胸背板浅褐色。侧板及足黄褐色，股节末端及胫节两端加深。翅浅黄色透明，翅痣黄褐色明显。在 OM、OR、FRs 和 CuP 末端处分别具小型褐色雾状斑；几乎沿所有横

图 2-28-2 缠绕地蚊蝎蛉 ***Terrobittacus implicatus*** 雄性外生殖器侧面观

Fig. 2-28-2 Male genitalia of ***Terrobittacus implicatus***, lateral view

图 2-28-3 缠绕地蚊蝎蛉 ***Terrobittacus implicatus*** 地理分布

Fig. 2-28-3 Distribution of ***Terrobittacus implicatus***

图 2-28-1 缠绕地蚊蝎蛉 *Terrobittacus implicatus* 成虫侧面观（陕西南郑黎坪国家森林公园）

1. 雄虫 2. 雌虫

Fig. 2-28-1 Adults of *Terrobittacus implicatus*, lateral view (Liping National Forest Park, Nanzheng, Shaanxi)

1. Male 2. Female

脉具雾状加深。雄性腹部背板黄褐色。第 8 节背板后缘中部具 V 形缺刻。上生殖瓣十分短小，侧面观近三角形，外缘内表面密生黑色小刺。生殖肢基节末端具 V 形膜质区；生殖刺突短小。第 10 节背板中部退化，两侧骨片环绕尾须。尾须较短。阳茎叶末端尖锐，阳茎丝细长，盘绕成环。载肛突上瓣高度骨化；腹缘具 1 条深沟槽，内生 1 列短毛；末端向后弯曲成钩状；下瓣较短，向腹面钩曲。

【发生期】 7 ~ 8 月。

【地理分布】 陕西（南郑、岚皋）。

Measurements: Forewing length 15.0–16.0 mm.

Morphology: Vertex yellowish brown. Ocellar triangle black. Rostrum yellowish brown with deepened flanks. Antennae ciliated. Pronotum yellowish brown, bearing two black setae on anterior margin and two to four setae on posterior margin. Mesonotum deep brown anteriorly and light brown posteriorly. Metanotum light brown. Pleura yellowish brown. Legs yellowish brown with apex of femur and both ends of tibia darkened. Wing membrane hyaline and tinged with light yellow. Pterostigma yellowish brown and distinct. Several smoky dark spots on ORs, FRs, OM, and ending of CuP. Almost all crossveins with surrounding smoky stripes. Male terga yellowish brown. Tergum Ⅷ slightly emarginate in a V-shaped posteriorly. Epandrial lobes considerably short, subtriangular in lateral view, and bearing dense black spines on apical inner margin. Gonocoxites with a V-shaped membranous area posteriorly. Gonostylus short. Tergum Ⅹ greatly reduced mesally and forming a pair of sclerites surrounding short cerci. Aedeagal lobes acute apically. Penisfilum slender and coiled into a loop. Upper branch of proctiger intensively sclerotized, with a deep groove ventrally; a row of dense short setae in groove; apex curved caudad and hook-like. Lower branch short and curved ventrad.

Emergence: July to August.

Distribution: Shaanxi (Nanzheng, Lan'gao).

29. 长毛地蚊蝎蛉 *Terrobittacus longisetus* Tan & Hua, 2009

【测量】 前翅长 20.0 ~ 21.0 mm。

【形态特征】 头顶黄褐色。单眼三角区黑色，中单眼上部具 2 根明显的刚毛。喙中部黄棕色，两侧深褐色。触角被毛。前胸背板黄褐色，前缘具 4 根粗刚毛。中、后胸背板黑褐色，具明显的浅色中带。足浅黄褐色。翅透明，微带黄色，无明显斑纹，仅沿端部横脉具雾状加深。翅痣浅黄色，不明显。腹部背板黄褐色，向后逐节加深。第 8 节背板后缘具 V 形缺刻。上生殖瓣短小，铲状，末端圆钝，背侧具 1 簇长毛，端部内缘具 1 列粗大的黑色小刺；中部具 1 条纵脊。生殖肢基节半球形，末端具 V 形膜质区，其上具 1 ~ 3 根显著的长刚毛；生殖刺突短小。阳茎叶末端尖，阳茎丝细长盘绕。第 10 节背板中部退化消失，侧面形成两块环绕尾须的骨片。尾须长，约为生殖肢基节的 2/3，端部略钝圆。载肛突上瓣高度骨化，宽阔，端部镰刀状弯曲；下瓣短，向腹面钩曲。

【发生期】 7 ~ 8 月。

【地理分布】 湖南（张家界国家森林公园、桑植）、贵州（雷公山）。

Measurements: Forewing length 20.0–21.0 mm.

Morphology: Vertex yellowish brown. Ocellar triangle black, bearing two bristles above median ocellus. Rostrum yellowish brown with deepened flanks. Antennae ciliated. Pronotum yellowish brown, bearing four black setae on anterior margin. Mesonotum and metanotum dark brown with a distinct pale mesal stripe. Legs light yellowish brown. Wing membrane hyaline and tinged with light yellow. Pterostigma light yellow and indistinct. Markings absent except apical crossveins with surrounding

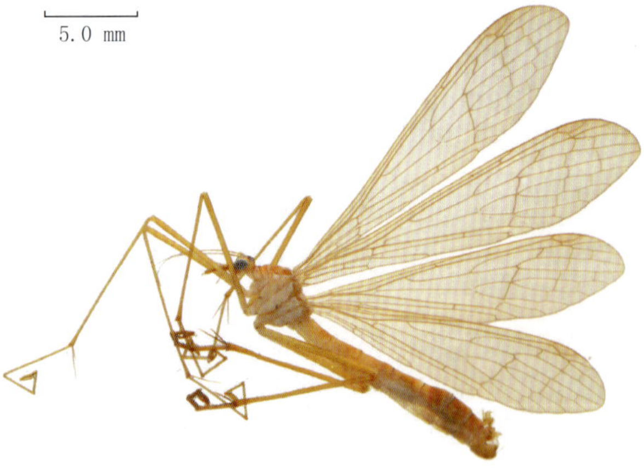

图 2-29-1 长毛地蚊蝎蛉 *Terrobittacus longisetus* 雄成虫侧面观（贵州雷公山，NWAFU）

Fig. 2-29-1 Male adult of *Terrobittacus longisetus*, lateral view (Mount Leigong, Guizhou, NWAFU)

图 2-29-2 长毛地蚊蝎蛉 *Terrobittacus longisetus* 雄性外生殖器侧面观

Fig. 2-29-2 Male genitalia of *Terrobittacus longisetus*, lateral view

图 2-29-3 长毛地蚊蝎蛉 *Terrobittacus longisetus* 地理分布

Fig. 2-29-3 Distribution of *Terrobittacus longisetus*

smoky stripes. Male terga yellowish brown with each segment darkening caudad. Tergum Ⅷ slightly emarginate in a V-shape posteriorly. Epandrial lobes considerably short, shovel-shaped with rounded apexes, and bearing a tuft of long setae on dorsal margins. A series of black and stout spines on inner surface of each lobe apically. A longitudinal ridge on middle of each lobe. Gonocoxites semispherical, bearing a V-shaped membranous area posteriorly, and one to three pairs of rather long setae on membranous area. Gonostylus small. Aedeagal lobes acute apically. Penisfilum slender and coiled into a loop. Tergum Ⅹ greatly reduced mesally and forming a pair of sclerites surrounding cerci. Cerci approximately two-thirds as long as gonocoxites and slightly rounded apically. Upper branch of proctiger intensively sclerotized and broad, strongly curved caudad apically and sickle-like. Lower branch short and curved ventrad.

Emergence: July to August.

Distribution: Hunan (Zhangjiajie National Forest Park, Sangzhi) and Guizhou (Mount Leigong).

30. 钩地蚊蝎蛉 *Terrobittacus rostratus* Du & Hua, 2017

【测量】 前翅长 14.0 ~ 16.0 mm。

【形态特征】 头顶黑棕色，额棕色。单眼三角区黑色，中单眼上方具 2 根不明显的刚毛，喙基部浅褐色，端部黑褐色。前胸背板黑褐色，前缘具 2 根刚毛。中、后胸背板大部分深褐色，中间颜色稍浅；侧板及足红褐色或黄褐色。翅透明，略带浅黄色或灰褐色。翅痣浅褐色，不明显，无明显的斑纹，仅端部稍加深。雄性腹部背板黄褐色至黑褐色。腹板黄褐色。第 8 节背板末端轻微缺刻。上生殖瓣相对较短，黄褐色，侧面观似靴子状，端部钝圆，内缘密生黑色短刺。生殖肢基节末端钝圆，生殖刺突短小。阳茎叶端部尖锐，阳茎丝细长，盘绕成环。第 10 节背板中部退化消失，形成两块环绕尾须的骨片。尾须细长。载肛突上瓣强烈骨化，背缘形成 2 个侧片，腹缘具 1 条深沟槽，内生 1 列短刚毛；端部鸟嘴状钩曲；下瓣短，向腹面钩曲。

【发生期】 7 ~ 8 月。

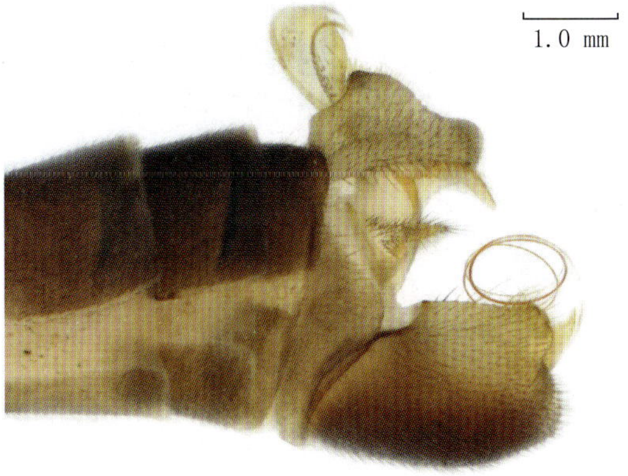

图 2-30-2 钩地蚊蝎蛉 *Terrobittacus rostratus* 雄性外生殖器侧面观

Fig. 2-30-2 Male genitalia of *Terrobittacus rostratus*, lateral view

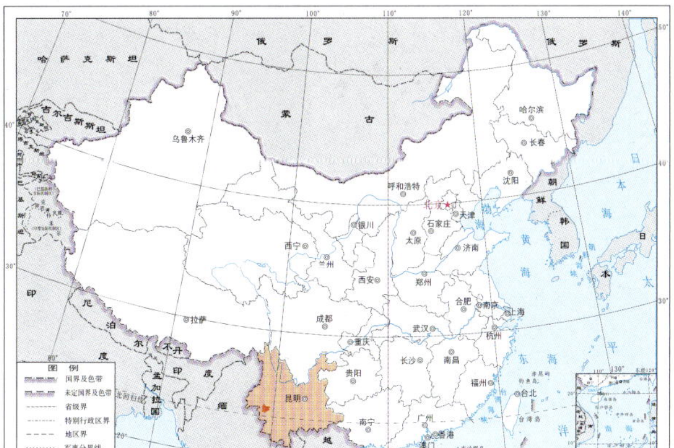

图 2-30-3 钩地蚊蝎蛉 *Terrobittacus rostratus* 地理分布

Fig. 2-30-3 Distribution of *Terrobittacus rostratus*

【地理分布】 云南（永德）。

Measurements: Forewing length 14.0–16.0 mm.

Morphology: Vertex blackish brown. Frons brown. Ocellar triangle black and bearing two bristles above median ocellus. Rostrum light brown basally and blackish brown distally. Pronotum dark brown, bearing two setae on anterior margin. Mesonotum and metanotum mostly dark brown but paler mesally. Pleura and legs reddish or yellowish brown.Wing membrane hyaline and tinged with light yellow or greyish brown. Pterostigma light brown and indistinct. Markings absent but apical portion tinged with smoky brown. Male terga yellowish brown to dark brown. Sterna yellowish brown. Tergum Ⅷ slightly emarginate on posterior margin. Epandrial lobes yellowish brown, relatively short, boot-shaped in lateral view with rounded apexes, and bearing a series of black stout spines on inner surface apically. Gonocoxites rounded apically. Gonostylus short. Aedeagal lobes acute apically. Penisfilum slender and coiled into a loop. Tergum Ⅹ greatly reduced mesally and forming a pair of sclerites surrounding slender cerci. Upper branch of proctiger intensively sclerotized and with

图 2-30-1 钩地蚊蝎蛉 ***Terrobittacus rostratus*** 雄成虫侧面观（云南永德棠梨山）

Fig. 2-30-1 Male adult of ***Terrobittacus rostratus***, lateral view（Mount Tangli, Yongde,Yunnan）

two broad lateral flanks dorsally, a deep groove bearing a row of short setae ventrally and curved caudad into a beak apically. Lower branch short and curved ventrad.

Emergence: July to August.

Distribution: Yunnan (Yongde).

31. 刀地蚊蝎蛉 *Terrobittacus xiphicus* Tan & Hua, 2009

【测量】 翅长 15.0 ~ 16.0 mm。

【形态特征】 头顶红褐色。单眼三角区黑色，黑色横带延伸至复眼边缘。喙黄褐色。触角被毛。前胸背板不均匀黑褐色；中、后胸背板黄棕色，两侧亮黑色。侧板及足黄褐色。翅淡黄色透明，翅痣黑棕色明显。仅在 ORs、FRs、OM 和 CuP 近末端具褐色小斑点，沿横脉具雾状加深。雄性腹部背板黄褐色，向后逐节加深。第 8 节背板后缘具 V 形缺刻。上生殖瓣短于生殖肢基节之半，内缘端部具一些黑色小刺，下缘近基部有 1 个明显的尖齿。生

图 2-31-1 刀地蚊蝎蛉 *Terrobittacus xiphicus* 成虫侧面观（湖南桑植天平山）

1. 雄虫 2. 雌虫

Fig. 2-31-1 Adults of *Terrobittacus xiphicus*, lateral view (Tianping Mountain, Sangzhi, Hunan)

1. Male 2. Female

殖肢基节半球状，末端具疏松的毛簇；生殖刺突短而向内弯曲。阳茎叶末端尖锐；阳茎丝较短，仅盘绕约 3 圈。第 10 节背板中部退化消失，形成两块环绕尾须的骨片。尾须细长，末端尖。载肛突上瓣马刀状，顶端具 1 簇长毛；下瓣短，钩状。

【发生期】 6 ~ 8 月。

【地理分布】 贵州(梵净山)、湖南(八大公山)。

Measurements: Forewing length 15.0–16.0 mm .

Morphology: Vertex reddish brown. Ocellar triangle black and with a transverse black stripe extending laterally to compound eyes. Rostrum yellowish brown. Antennae ciliated.Pronotum unevenly dark brown. Mesonotum and metanotum yellowish brown mesally and shining black laterally. Metanotum yellowish brown. Pleura and legs yellowish brown. Wing membrane hyaline and tinged with light yellow. Pterostigma black brown and distinct. Several brown spots on ORs, FRs, OM, and ending of CuP. Smoky tinges along crossveins. Male terga yellowish brown and gradually darkened caudally. Tergum Ⅷ emarginate in a V-shape posteriorly. Epandrial lobes shorter than half length of gonocoxites, bearing dozens of black spines on inner surfaces apically. An acute tooth protrudes on ventral margin subbasally. Gonocoxites semispherical and bearing a tuft of sparse setae. Gonostylus short and curved inward. Aedeagal lobes acute apically. Penisfilum considerably short and coiled in approximately three circles. Tergum X greatly reduced mesally and forming a pair of sclerites surrounding slender and acute cerci. Upper branch of proctiger sabre-shaped, and bearing a tuft of setae on apex. Lower branch short and hook-like.

Emergence: June to August.

Distribution: Guizhou (Mount Fanjing) and Hunan (Badagong Mountains).

图 2-31-2 刀地蚊蝎蛉 *Terrobittacus xiphicus* 雄性外生殖器侧面观

Fig. 2-31-2 Male genitalia of *Terrobittacus xiphicus*, lateral view

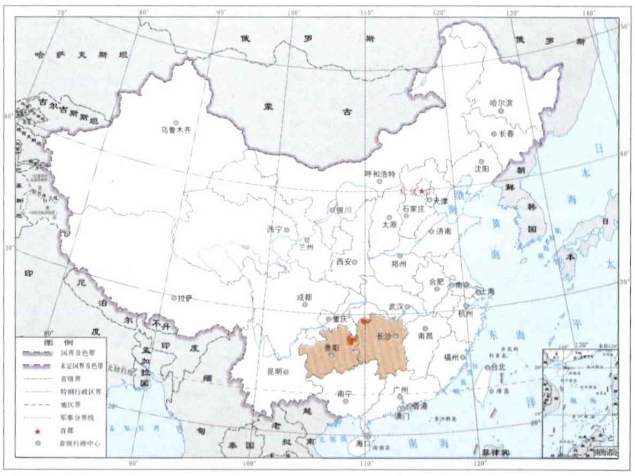

图 2-31-3 刀地蚊蝎蛉 *Terrobittacus xiphicus* 地理分布

Fig. 2-31-3 Distribution of *Terrobittacus xiphicus*

第三部分　蝎蛉科

Part Ⅲ　Panorpidae

图 3-a　蝎蛉 **Panorpidae sp.** 头部正面观（雷波摄）

Fig. 3-a　Head of **Panorpidae sp.**, frontal view (photo by Bo Lei)

蝎蛉科 Panorpidae 昆虫统称蝎蛉（scorpionflies），其头部向下延伸成喙状 (图 3-a)；雄虫外生殖器球形膨大并上举，状如蝎尾。

蝎蛉科 Panorpidae 是长翅目 Mecoptera 中种类最多的科，全世界已知 400 多种，中国已记录 200 余种，本图鉴收录 134 种。

Panorpidae are commonly called scorpionflies. Their heads are prolonged ventrally into a rostrum (Fig. 3-a). Their male genitalia are enlarged and recurved backwards, resembling the stinger of scorpions.

Panorpidae is the largest family in Mecoptera, and more than 400 species have been described worldwide. Over 200 species have been recorded from China, and 134 of them are included in this atlas.

中国蝎蛉科 **Panorpidae** 分属检索表

(6 属)

1. 前翅 1A 脉终于 ORs 以内；1A 和 2A 之间通常仅 1 条横脉 (图 3-b-1)…………………… **新蝎蛉属 *Neopanorpa***
   前翅 1A 脉终止于 ORs 以外，少数刚好到达；1A 和 2A 之间通常具 2 条横脉 (图 3-b-2)……………………… 2

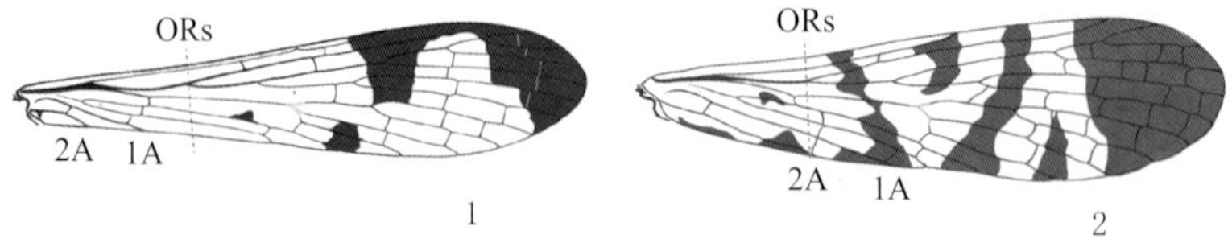

图 3-b　前翅

1. 新蝎蛉属 *Neopanorpa*　2.　蝎蛉属 *Panorpa*

Fig. 3-b　Forewings

1. *Neopanorpa*　2. *Panorpa*)

2. 雄性腹部第 3 节背板无背中突，第 4 节背板无后背突 (图 3-c-1)；雌性生殖板中轴后端延长并分叉 ……………………………………………………………………………… **叉蝎蛉属 *Furcatopanorpa***
   雄性腹部第 3 节背板具背中突，第 4 节背板具后背突 (图 3-c-2)；雌性生殖板中轴末端不延长，且不分叉 … 3

图 3-c　雄性腹部第 3、4 节背板

1. 无背中突、后背突　2. 有背中突、后背突

Fig. 3-c　Male terga Ⅲ and Ⅳ

1. Notal and postnotal organs absent　2. Notal and postnotal organs present

3. 雄性腹部第 6 节具 2 个显著的指状臀角 (图 3-d-1) …………………………… **双角蝎蛉属 *Dicerapanorpa***
   雄性腹部第 6 节无或仅具 1 个臀角 (图 3-d-2，图 3-d-3) ………………………………………… 4

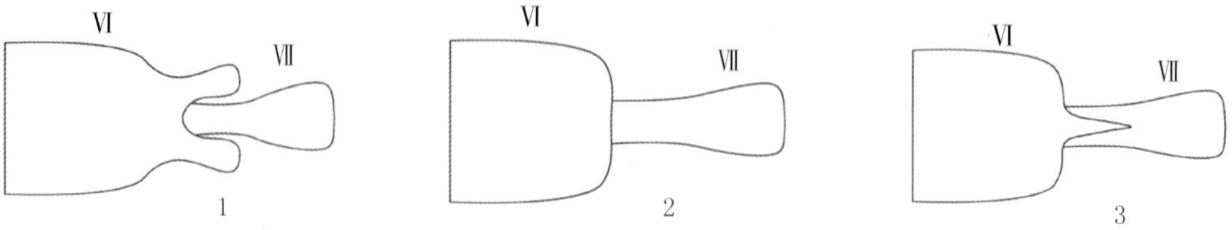

图 3-d　雄性腹部第 6 节

1. 具 2 个臀角　2. 无臀角　3. 具 1 个臀角

Fig. 3-d　Male A6

1. With two anal horns　2. Lacking anal horns　3. With one anal horns

4. 雄性腹部第 6 节具 1 个臀角；雌性生殖板两侧具 2 对侧片 (图 3-e-1) ………………… 单角蝎蛉属 *Cerapanorpa*
雄性腹部第 6 节大多无臀角或仅 1 个臀角；雌性生殖板两侧无侧片或仅 1 对侧片 (图 3-e-2) ………………… 5

1　　2

图 3-e　雌性生殖板

1. 具 2 对侧片 2. 具 1 对侧片

Fig. 3-e　Female genital plate

1. With two pairs of lateral plates 2. With one pairs of lateral plates

5. 翅黄色；前翅 R_2 脉分 3 叉；雄性腹部第 6、7 节背面各具 1 簇长刚毛 (图 3-f-1) ……… 华蝎蛉属 *Sinopanorpa*
翅大多无色，少数黄色或褐色；前翅 R_2 脉一般分 2 叉，少数不分叉或分 3 叉；雄性腹部第 6、7 节背面无特别长的刚毛 (图 3-f -2) …………………………………………………………………………… 蝎蛉属 *Panorpa*

1　　2

图 3-f　雄性腹部

1. 第 6、7 节背面具长刚毛　2. 第 6、7 节背面无刚毛

Fig. 3-f　Male abdomen

1. A6 and A7 with long setae dorsally 2. A6 and A7 without setae dorsally

Key to genera of Chinese Panorpidae

(Six genera)

1. Vein 1A in forewings ended before ORs; usually one crossvein between 1A and 2A (Fig. 3-b-1) …………*Neopanorpa*
Vein 1A in forewings ended beyond ORs; usually two crossveins between 1A and 2A (Fig. 3-b-2) ………………… 2
2. Males lacking notal organ on tergum Ⅲ and postnotal organ on tergum Ⅳ (Fig. 3-c-1); axis in female genital plate elongated and bifurcated posteriorly …………………………………………………………… *Furcatopanorpa*
Males with notal organs on tergum Ⅲ and postnotal organ on tergum Ⅳ (Fig. 3-c-2); axis in female genital plate simple …………………………………………………………………………………………………… 3
3. Males bearing a pair of finger-like anal horns on tergum Ⅵ (Fig. 3-d-1) …………………………… *Dicerapanorpa*
Males bearing a single or no anal horn on tergum Ⅵ (Fig. 3-d-2, Fig. 3-d-3) ……………………………………… 4
4. Males bearing a single anal horn on A6; female genital plate with two pairs of lateral plates (Fig. 3-e-1) …………… ……………………………………………………………………………………………………… *Cerapanorpa*
Males mostly without anal horn or with only one anal horn on A6; female genital plate lacking or with only one pair of lateral plates (Fig. 3-e-2) ……………………………………………………………………………………… 5
5. Wings yellow; R_2 in forewings trifurcated; male A6 and A7 each with a tuft of long setae dorsally (Fig. 3-f-1) ……… ……………………………………………………………………………………………………… *Sinopanorpa*
Wings mostly hyaline but yellow or greyish in some species; R_2 in forewings usually bifurcated, unbranched or trifurcated in a few species; male A6 and A7 without long setae dorsally (Fig. 3-f-2) ……………………… *Panorpa*

四、单角蝎蛉属 *Cerapanorpa* Gao, Ma & Hua, 2016

单角蝎蛉属 *Cerapanorpa* 昆虫主要分布于中国、日本、朝鲜、韩国和俄罗斯远东地区，其主要鉴定特征为雄性腹部第 6 节背板末端具 1 个显著的指状臀角（图 3-g）。

单角蝎蛉属 *Cerapanorpa* 昆虫全世界 22 种，中国已记录 17 种，本图鉴收录 16 种。

Cerapanorpa is mainly distributed in China, Japan, Korean Peninsula and Russian Far East. Its main diagnostic character lies in a distinct finger-like anal horn on the posterior end of tergum Ⅵ in males (Fig. 3-g)。

Twenty-two species have been described, and 16 of them are included in this atlas.

图 3-g 太白单角蝎蛉 ***Cerapanorpa obtusa*** 雄成虫侧面观

Fig. 3-g Male adult of ***Cerapanorpa obtusa***, lateral view

本图鉴单角蝎蛉属 *Cerapanorpa* 分种检索表

（16 种，雄性）

1. 休息时翅呈屋脊状叠于体背；阳茎腹面基部具 1 个棱形突起；阳茎背瓣末端 2 分支 …………………………………………………………………………………………… 32. 尖翅单角蝎蛉 *Cerapanorpa acutipennis*
   休息时翅呈 V 形置于体侧；阳茎近基部无棱形突起；背瓣末端简单 …………………………………… 2
2. 头顶黄色；腹部第 6 节末端臀角锥形；第 7 节基部强烈缢缩成柄状；下瓣短小，不超过生殖肢基节中部 …………………………………………………………………………………………… 40. 埔里单角蝎蛉 *Cerapanorpa horiensis*
   头顶一般黑色；腹部第 6 节末端臀角指形；第 7 节基部缢缩，向端部逐渐变阔，不呈柄状；下瓣长带状，末端超过生殖肢基节中部 …………………………………………………………………………………… 3
3. 腹部第 5 节背板后缘具 1 个尖锐、类似第 6 节臀角的突起 ……… 33. 二角单角蝎蛉 *Cerapanorpa bicornifera*
   腹部第 5 节背板后缘无上述突起 ……………………………………………………………………… 4
4. 臀角长多为第 6 节的 0.2 倍；下瓣仅伸达生殖肢基节中部 ……… 34. 短角单角蝎蛉 *Cerapanorpa brevicornis*
   臀角长少为第 6 节的 0.3 倍；下瓣远超过生殖肢基节中部 …………………………………………… 5
5. 阳基侧突明显远超出生殖肢基节端部 ………………………………………………………………… 6
   阳基侧突不超出，刚好到达或仅稍超出生殖肢基节端部 ……………………………………………… 10
6. 阳基侧突极长，几乎伸达生殖刺突的端部 ………………………… 44. 壮突单角蝎蛉 *Cerapanorpa protrudens*
   阳基侧突不伸达生殖刺突端部 ………………………………………………………………………… 7
7. 阳基侧突端半部较直，向末端渐细 …………………………………………………………………… 8
   阳基侧突端半部呈波状弯曲 …………………………………………………………………………… 9
8. 阳基侧突近平行，末端不超过生殖刺突中部 …………………… 39. 伏牛山单角蝎蛉 *Cerapanorpa funiushana*
   阳基侧突向内靠拢、相互交叉，末端超过生殖刺突中部 ………… 38. 华山单角蝎蛉 *Cerapanorpa emarginata*
9. 翅面几乎无斑纹；阳基侧突仅在端半部内缘具 1 列短刺 ………………37. 拟华山单角蝎蛉 *Cerapanorpa dubia*
   翅面具显著的端带和痣带；阳基侧突内缘从基柄以上具 1 列短刺 …………………………………………………………………………………………… 47. 王屋山单角蝎蛉 *Cerapanorpa wangwushana*
10. 翅面具明显的黑褐色斑纹，端带、痣带、缘斑、基带和基斑完整；阳基侧突在近基部呈 L 形弯曲 ………… …………………………………………………………………………………………… 45. 任氏单角蝎蛉 *Cerapanorpa reni*
    翅面无斑纹或仅具极其退化的端带和痣带；阳基侧突近基部不如上述………………………………… 11
11. 阳基侧突短粗，中部方形膨大，外侧具 1 条尖锐的 L 形分支 …… 35. 拜尔斯单角蝎蛉 *Cerapanorpa byersi*
    阳基侧突较细长，中部不呈方形膨大，外侧无上述分支………………………………………………… 12
12. 阳基侧突直，基半部稍膨大，端半部尖细………………………… 36. 中部单角蝎蛉 *Cerapanorpa centralis*
    阳基侧突不同程度弯曲，基半部与端半部大致等宽……………………………………………………… 13
13. 阳基侧突内缘的刚毛短于其宽度………………………………… 41. 六盘山单角蝎蛉 *Cerapanorpa liupanshana*
    阳基侧突内缘的刺毛长于其宽度………………………………………………………………………… 14
14. 阳基侧突端半部强烈弓形弯曲，内缘着生 1 列极梳齿状长鬃………… 46. 弯曲单角蝎蛉 *Cerapanorpa sinuata*
    阳基侧突端半部轻微膝状弯曲，内缘着生相对较短的梳齿状刺………………………………………… 15
15. 阳基侧突内缘的刺均粗长；阳茎背瓣指状………………………………43. 太白单角蝎蛉 *Cerapanorpa obtusa*
    阳基侧突内缘的刺向后逐渐变短小；阳茎背瓣向末端变阔…… 42. 南五台单角蝎蛉 *Cerapanorpa nanwutaina*

Key to species of *Cerapanorpa* in this atlas

(16 species, males)

1. Wings held roof-like over abdomen at repose; aedeagus with a hamulus on ventral base; dorsal aedeagal valves bifurcated apically…………………………………………………………… 32. *Cerapanorpa acutipennis*
   Wings held V-shape aside abdomen at repose; aedeagus lacking a hamulus; dorsal aedeagal valves not furcated apically …… ………………………………………………………………………………………… 2
2. Vertex yellow; A6 with a conical anal horn apically; A7 intensively stalked at base; hypovalves short and not extending to middle of gonocoxites ……………………………………………… 40. *Cerapanorpa horiensis*
   Vertex generally black; A6 with a finger-like anal horn apically; A7 constricted basally and evenly enlarged towards apex and not stalked; hypovalves long stripe-like and extending to middle of gonocoxites apically ……………………… 3

3. A5 with an acute process posteriorly similar to anal horn on tergum Ⅵ .... 33. ***Cerapanorpa bicornifera***
A5 lacking such a process 4
4. Anal horn at most 0.2 times as long as A6; hypovalves extending to middle of gonocoxites .... 34. ***Cerapanorpa brevicornis***
Anal horn at least 0.3 times as long as A6; hypovalves far exceeding apex of gonocoxites 5
5. Paramere conspicuously extending far beyond apex of gonocoxites.. 6
Paramere not exceeding, or nearly extending to apex of gonocoxites 10
6. Paramere extremely elongated and almost exceeding apex of gonostylus .... 44. ***Cerapanorpa protrudens***
Paramere not exceeding apex of gonostylus 7
7. Paramere straight on apical half and tapering towards apex 8
Paramere curved wave-like on apical half.. 9
8. Parameres nearly parallel, and not exceeding middle of gonocoxites apically .... 39. ***Cerapanorpa funiushana***
Parameres convergent and crossed mesally, and exceeding middle of gonocoxites apically .... 38. ***Cerapanorpa emarginata***
9. Wings nearly devoid of markings; paramere with a row of short spines on inner margin of apical half .... 37. ***Cerapanorpa dubia***
Wings with conspicuous apical and pterostigmal bands; paramere with a row of short spines on inner margin above basal stalk .... 47. ***Cerapanorpa wangwushana***
10. Wings with conspicuous dark brown apical, pterostigmal, basal bands and marginal and basal spots; paramere curved in a L-shape subbasally .... 45. ***Cerapanorpa reni***
Wings devoid of markings or with greatly reduced apical and pterostigmal bands; paramere not curved like above 11
11. Paramere short and stout, swollen rectangular at middle and bearing an acute L-shaped branch on outer margin......... 35. ***Cerapanorpa byersi***
Paramere relatively slender, not swollen rectangular and without such a branch 12
12. Paramere straight and slightly swollen on basal half and slender on distal half................ 36. ***Cerapanorpa centralis***
Paramere variously curved with its basal and distal halves almost equal in width..13
13 .Paramere with spines on inner margin shorter than its width .... 41. ***Cerapanorpa liupanshana***
Paramere with spines on inner margin longer than its width 14
14. Paramere strongly curved and sinuate on distal half, and bearing a row of long comb-like spines on inner margin .... 46. ***Cerapanorpa sinuata***
Paramere slightly geniculate on distal half, and bearing a row of relatively shorter comb-like spines on inner margin 15
15. Paramere with equal-lengthed spines on inner margin; dorsal aedeagal valves finger-like .... 43. ***Cerapanorpa obtusa***
Paramere with spines on inner margin gradually shortened caudad; dorsal aedeagal valves enlarged towards apex ...... 42. ***Cerapanorpa nanwutaina***

32. 尖翅单角蝎蛉 *Cerapanorpa acutipennis* (Hua, 1997)

【测量】 前翅长 14.0 ~ 15.0 mm。

【形态特征】 头顶黑褐色，后头具两个黄褐色斑。单眼三角区黑色。喙红褐色，中央颜色稍浅。胸部背板黑色，侧板黄褐色至黑褐色，足黄褐色。翅灰褐色几乎不透明，翅斑黑褐色，末端略尖；休憩时呈屋脊状叠于体背。前翅端带阔，后缘形成散乱的小斑；痣带基支完整，端支极窄或消失；缘斑小；基带断开成 2 斑；基斑极小或无。雄性腹部 1 ~ 5 节背板黑褐色。第 3 节背中突扁平，盖住第 4 节上尖锐的后背突。第 6 节末端具 1 个褐色臀角。第 7、8 节基部缢缩。雄性外生殖器阔卵形。第 9 节背板末端具深缺刻，两侧形成 1 对细长指形突起。下瓣长带状，仅伸达生殖肢基节约 2/3 处。生殖肢基节端部内缘具数根长刚毛；生殖刺突短于基节，内缘中齿不明显，近基部具大型盘状亚基齿。阳基侧突中部弯曲，端部较直，内缘密生梳齿状短刚毛。阳茎腹瓣基部中央具1个梭形突起；背瓣末端2裂，侧突发达。

【发生期】 5 ~ 8 月。

图 3-32-1 尖翅单角蝎蛉 ***Cerapanorpa acutipennis*** 成虫侧面观（辽宁宽甸花脖山，姜碌摄）

1. 雄虫 2. 雌虫

Fig. 3-32-1 Adults of ***Cerapanorpa acutipennis***, lateral view (Mount Huabo, Kuandian, Liaoning, photo by Lu Jiang)

1. Male 2. Female

【地理分布】 吉林（长白山）、辽宁（宽甸、老秃顶子）。

Measurements: Forewing length 14.0–15.0 mm.

Morphology: Vertex dark brown. Two yellowish brown spots on occiput. Ocellar triangle black. Rostrum reddish brown but paler mesally. Nota black. Pleura yellowish brown to dark brown. Legs mostly yellowish brown. Wing membrane almost opaque and greyish with dark brown markings. Wings slightly acute apically and held roof-like over body at repose. Apical band in forewing broad and forming several separated small spots posteriorly; pterostigmal band with its basal branch present and apical branch extremely reduced or absent; marginal spot small; basal band split into two spots; basal spot very small or absent. Male terga Ⅰ – Ⅴ dark brown. Notal organ on tergum Ⅲ flat and covering acute postnotal organ on tergum Ⅳ. A6 with a single brown anal horn on its dorsal apex. A7 and A8 constricted basally. Male genital bulb broad oval. Epandrium emarginate apically and forming a pair of finger-like processes laterally. Hypovalves stripe-like and nearly two-thirds length of gonocoxites. Gonocoxites with several long bristles on ventral apex. Gonostylus shorter than gonocoxites and bearing an indistinct median tooth, and a large disc-like subbasal process on inner margin. Paramere curved medially and straight apically, and bearing dense comb-like spines along inner margin. Aedeagus with a hamulus basal to ventral valves; dorsal valves bifurcated apically and with stout lateral processes.

Emergence: May to August.

Distribution: Jilin (Changbai Mountains) and Liaoning (Kuandian, Laotudingzi).

图 3-32-2 尖翅单角蝎蛉 ***Cerapanorpa acutipennis*** 雄性外生殖器腹面观

Fig. 3-32-2 Male genitalia of ***Cerapanorpa acutipennis***, ventral view

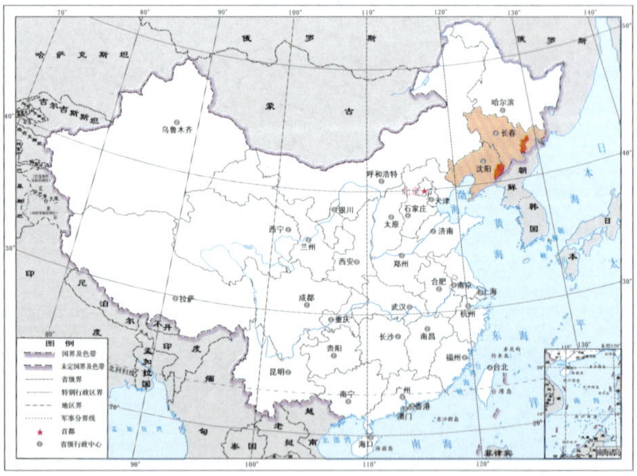

图 3-32-3 尖翅单角蝎蛉 ***Cerapanorpa acutipennis*** 地理分布

Fig. 3-32-3 Distribution of ***Cerapanorpa acutipennis***

33. 二角单角蝎蛉 *Cerapanorpa bicornifera* (Chou & Wang, 1981)

【测量】 前翅长 11.0 mm。

【形态特征】 头顶黑色，喙黄褐色。胸部背板黑色。侧板及足黄褐色。翅无色透明，无明显的斑纹，仅端部稍加深，翅痣浅黄色。雄性腹部 1 ~ 5 节背板黑褐色。第 3 节背中突扁平，盖住第 4 节上尖锐的后背突。第 5 节背板末端具 1 个类似臀角的尖锐突起，但相对较短。第 6 节黑褐色，末端的臀角红褐色，指形。第 7、8 节红褐色，基部缢缩。雄性外生殖器卵形。第 9 节背板末端具深缺刻，两侧形成 1 对宽阔的指形突起。下瓣长带状，仅伸达生殖肢基节约 2/3 处，内缘具长鬃毛。生殖肢基节端部内缘具数根长刚毛；生殖刺突短于基节，内缘

图 3-33-1　二角单角蝎蛉 *Cerapanorpa bicornifera* 雄成虫背面观（外生殖器已去除，陕西秦岭，NWAFU）

Fig. 3-33-1　Male adult of ***Cerapanorpa bicornifera***, dorsal view (With genitalia removed, Qinling Mauntains, Shaanxi, NWAFU)

中齿钝三角形，亚基齿大型盘状。阳基侧突长棒状，端半部具短毛刺，末端尖锐。阳茎腹瓣膜质，短小；背瓣指状，分离。

【发生期】 7 月。

【地理分布】 陕西（秦岭）。

备注：本种自 1981 年以 1 头雄虫为模式标本被发表至今，一直没有被再度发现。我们推测，其模式标本有可能是其他单角蝎蛉的变异个体。

Measurements: Forewing length 11.0 mm.

Morphology: Vertex black. Rostrum yellowish brown.Nota black. Pleura and legs yellowish brown. Wing membrane hyaline, and devoid of markings except darkened apex. Pterostigma light yellow. Male terga Ⅰ–Ⅴ dark brown. Notal organ on tergum Ⅲ flat and covering acute postnotal organ on tergum Ⅳ. Tergum Ⅴ with a small and acute horn-like process on posterior margin. A6 dark brown and bearing a single reddish brown anal horn on its dorsal apex. A7 and A8 reddish brown and constricted basally. Male genital bulb oval. Epandrium emarginate apically and forming a pair of broad finger-like processes laterally. Hypovalves stripe-like, nearly two-thirds length of gonocoxites, and bearing long bristles along inner margin. Gonocoxites with several long bristles on ventral apex. Gonostylus shorter than gonocoxites and bearing an obtuse triangular median tooth and a large disc-like subbasal process on inner

图 3-33-2　二角单角蝎蛉 *Cerapanorpa bicornifera* 雄虫腹部末端侧面观

Fig. 3-33-2　Male teminal abdomen of ***Cerapanorpa bicornifera,*** lateral view

图 3-33-3　二角单角蝎蛉 *Cerapanorpa bicornifera* 雄性外生殖器腹面观

Fig. 3-33-3　Male genitalia of ***Cerapanorpa bicornifera***, ventral view

图 3-33-4　二角单角蝎蛉 *Cerapanorpa bicornifera* 地理分布

Fig. 3-33-4　Distribution of ***Cerapanorpa bicornifera***

margin. Paramere long, stick-like with an acute apex, and bearing numerous spines on apical half. Aedeagus with its ventral valves membranous and short, and dorsal valves finger-like and divergent.

Emergence: July.

Distribution: Shaanxi (Qinling Mountains).

Remarks: This species has not been rediscovered since its publication in 1981 based on a male holotype. We suppose that holotype might be a variant of another *Cerapanorpa* species.

34. 短角单角蝎蛉 *Cerapanorpa brevicornis* (Hua & Li, 2007)

【测量】 前翅长 13.5 ~ 14.0 mm。

【形态特征】 头顶黑褐色。单眼三角区黑色。喙红褐色，中央颜色稍浅。胸部背板黑色，侧板黄褐色，足黄褐色。翅无色透明，无明显斑纹，仅端部稍加深。翅痣黄色。雄性腹部 1 ~ 5 节背板黑褐色。第 3 节背中突扁平，盖住第 4 节上尖锐的后背突。第 6 节末端具 1 个黄褐色臀角。第 7、8 节黄褐色，基部缢缩。雄性外生殖器阔卵形。第 9 节背板末端具深缺刻，两侧形成 1 对指形突起。下瓣长带状，超过生殖肢基节的 2/3 处，内缘各具 1 列整齐的短鬃。生殖肢基节端部内缘具数根长刚毛；生殖刺突短于基节，内缘中齿钝三角形，亚基齿大型盘状。阳基侧突细长，端部向内弯曲，内缘密生短刚毛。阳茎腹瓣短，膜质；背瓣近末端背缘有明显的膜质膨大物，端部指状，侧突起明显。

【发生期】 5 ~ 8 月。

【地理分布】 陕西（南郑、宁陕、宝鸡、佛坪）、湖北（神农架）、四川（平武）。

图 3-34-2 短角单角蝎蛉 *Cerapanorpa brevicornis* 雄性外生殖器腹面观

Fig. 3-34-2 Male genitalia of *Cerapanorpa brevicornis*, ventral view

Measurements: Forewing length 13.5–14.0 mm.

Morphology: Vertex dark brown. Ocellar triangle black. Rostrum reddish brown and pale mesally. Nota black. Pleura and legs yellowish brown. Wing membrane hyaline, and devoid of markings except darkened apex. Pterostigma yellow. Male terga Ⅰ–Ⅴ dark brown. Notal organ on tergum Ⅲ flat and covering acute postnotal organ on tergum Ⅳ. A6 dark brown and bearing a single yellowish brown anal horn on its dorsal apex. A7 and A8 yellowish brown and constricted basally. Male genital bulb broad oval. Epandrium emarginate apically and forming a pair of finger-like processes laterally. Hypovalves stripe-like, exceeding two-thirds length of gonocoxites, and bearing a row of regularly arranged, short bristles along inner margin. Gonocoxites with several long bristles on ventral apex. Gonostylus shorter than gonocoxites and bearing an obtuse triangular median tooth and a large disc-like subbasal process on inner margin. Paramere slender, curved mesad at distal

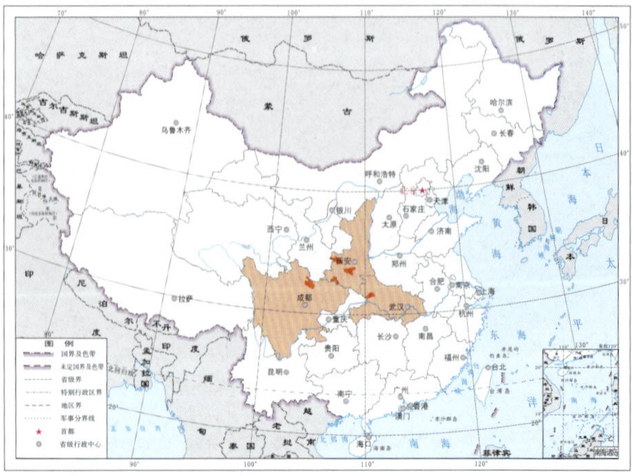

图 3-34-3 短角单角蝎蛉 *Cerapanorpa brevicornis* 地理分布

Fig. 3-34-3 Distribution of *Cerapanorpa brevicornis*

图 3-34-1 短角单角蝎蛉 *Cerapanorpa brevicornis* 成虫侧面观
（陕西宝鸡天台山国家森林公园，姜碌摄）
1. 雄虫 2. 雌虫

Fig. 3-34-1 Adults of *Cerapanorpa brevicornis*, lateral view
(Tiantaishan National Forest Park, Baoji, Shaanxi, photo by Lu Jiang)
1. Male 2. Female

portion, and bearing numerous short spines on apical inner margin. Aedeagus with its ventral valves membranous and short, and dorsal valves finger-like apically. An inflated membranous process on dorsal surface of dorsal valve subapically. Lateral processes conspicuous.

Emergence: May to August.

Distribution: Shaanxi (Nanzheng, Ningshan, Baoji, Foping), Hubei (Shennongjia) and Sichuan (Pingwu).

35. 拜尔斯单角蝎蛉 *Cerapanorpa byersi* (Hua & Huang, 2007)

【测量】 前翅长 12.5 ~ 13.0 mm。

【形态特征】 体型较小的单角蝎蛉。头顶黑色。喙黑棕色但两侧黄褐色。胸部背板黑褐色。侧板及足乳黄色或浅灰褐色。翅无色透明，除极其退化的端带外无斑纹，翅脉黑棕色。雄性腹部 1 ~ 5 节背板黑褐色。第 3 节背中突扁平，盖住第 4 节上尖锐的后背突。第 6 节黑褐色，末端具 1 个黄褐色臀角。第 7、8 节黄褐色，基部缢缩。雄性外生殖器圆球形。第 9 节背板末端具宽 U 形缺刻，两侧形成 1 对指形突起。下瓣长带状，内缘具 1 列长鬃。生殖刺突短于基节，内缘中齿钝三角形，亚基齿大型盘状。阳基侧突短粗，不超过生殖肢基节端部，中部方形膨大，内缘密布梳齿状刚毛；外侧具 1 尖锐的 L 形短支。阳茎背瓣伸达生殖肢基节末端，端部指形。

【发生期】 6 ~ 7 月。

【地理分布】 陕西(太白、眉县)、甘肃(宕昌)。

Measurements: Forewing length 12.5–13.0 mm.

Morphology: Relatively small in size. Vertex black. Rostrum dark brown with yellowish brown flanks. Nota dark brown. Pleura and legs cream-white or pale brown. Wing membrane hyaline, and devoid of markings except greatly reduced apical band. Veins dark brown. Male terga Ⅰ– Ⅴ dark brown. Notal organ on tergum Ⅲ flat and covering acute postnotal organ on tergum Ⅳ. A6 dark brown and bearing a yellowish brown anal horn on its dorsal apex. A7 and A8 yellowish brown and constricted at base. Male genital bulb globular. Epandrium emarginate terminally in a broad U-shape and forming a pair of finger-like processes laterally. Hypovalves stripe-like and bearing a row of long bristles along inner margin. Gonostylus shorter than gonocoxites and bearing an obtuse triangular median tooth and a large disc-like subbasal process on inner margin. Paramere short and stout, not exceeding apex of gonocoxites, swollen trapezoidally

图 3-35-1 拜尔斯单角蝎蛉 *Cerapanorpa byersi* 雄成虫（陕西太白红沟林场）

Fig. 3-35-1 Male adult of *Cerapanorpa byersi* （Honggou Forest Farm, Taibai, Shaanxi）

图 3-35-2 拜尔斯单角蝎蛉 ***Cerapanorpa byersi*** 雌成虫（陕西太白红沟林场）

Fig. 3-35-2 Female adult of ***Cerapanorpa byersi*** (Honggou Forest Farm, Taibai, Shaanxi)

at middle and bearing several comb-like spines along inner margin, and an acute L-shaped short branch laterally. Dorsal valves of aedeagus extending to apex of gonocoxites with finger-like apexes.

图 3-35-3 拜尔斯单角蝎蛉 ***Cerapanorpa byersi*** 雄性外生殖器腹面观（右侧下瓣已去除）

Fig. 3-35-3 Male genitalia of ***Cerapanorpa byersi***, ventral view (with lateral hypovalve removed)

图 3-35-4 拜尔斯单角蝎蛉 ***Cerapanorpa byersi*** 地理分布

Fig. 3-35-4 Distribution of ***Cerapanorpa byersi***

Emergence: June to July.

Distribution: Shaanxi (Taibai, Meixian) and Gansu (Dangchang).

36. 中部单角蝎蛉 *Cerapanorpa centralis* (Tjeder, 1936)

【测量】 前翅长 13.0 mm。

【形态特征】 头顶黑色，喙黄褐色。胸部背板黑色。侧板及足黄褐色。翅无色透明，无斑纹，翅痣浅棕色。雄性腹部 1 ~ 5 节背板黑褐色。第 3 节背中突扁平，盖住第 4 节上尖锐的后背突。第 6 节黑褐色，末端具 1 个黄褐色臀角。第 7、8 节黄褐色，基部缢缩。雄性外生殖器卵形。第 9 节背板末端 U 形缺刻，形成 1 对指形突起。下瓣长带状，内缘具 1 列整齐排列的短鬃。生殖肢基节端部内缘具数根长刚毛；生殖刺突短于基节，内缘中齿钝三角形，亚基齿大型盘状。阳基侧突基半部粗壮，端半部纤细，不超过生殖肢基节端部，内缘着生 1 列狭长的梳齿状刚毛。阳茎腹瓣膜质短小；背瓣伸达生殖肢基节后缘，端部略膨大。

【发生期】 5 ~ 7 月。

【地理分布】 四川（阆中、卧龙）、甘肃（文县）。

Measurements: Forewing length 13.0 mm.

Morphology: Vertex black. Rostrum yellowish brown. Nota black. Pleura and legs yellowish brown. Wing membrane hyaline, and devoid of markings. Pterostigma light brown. Male terga Ⅰ – Ⅴ dark brown. Notal organ on tergum Ⅲ flat and covering acute postnotal organ on tergum Ⅳ. A6 dark brown and bearing a yellowish brown anal horn on its dorsal apex. A7 and A8 yellowish brown and constricted basally. Male genital bulb oval. Epandrium emarginate terminally in a broad U-shape and forming a pair of finger-like processes laterally. Hypovalves stripe-like and bearing a row of long bristles along inner margin.

图 3-36-1 中部单角蝎蛉 *Cerapanorpa centralis* 雄成虫背面观（甘肃文县，IZAS）

Fig. 3-36-1 Male adult of *Cerapanorpa centralis*, dorsal view (Wenxian, Gansu, IZAS)

图 3-36-2 中部单角蝎蛉 *Cerapanorpa centralis* 雄性外生殖器腹面观

Fig. 3-36-2 Male genitalia of *Cerapanorpa centralis*, ventral view

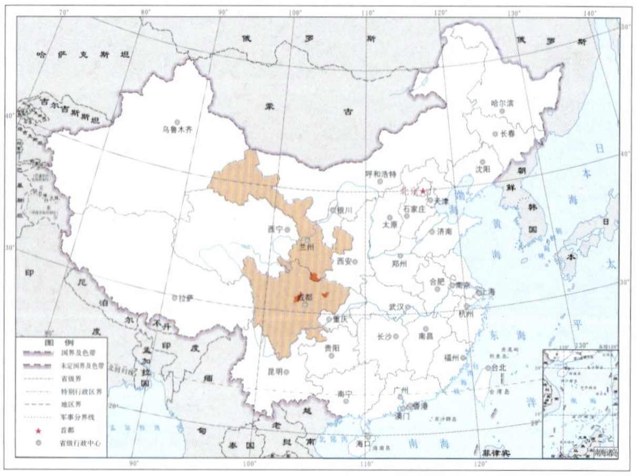

图 3-36-3 中部单角蝎蛉 *Cerapanorpa centralis* 地理分布

Fig 3-36-3 Distribution of *Cerapanorpa centralis*

Gonocoxites with several long setae on ventral apex. Gonostylus shorter than gonocoxites and bearing an obtuse triangular median tooth and a large disc-like subbasal process on inner margin. Paramere stout at basal half and slender at distal half, not exceeding apex of gonocoxites, and bearing several comb-like spines along inner margin. Ventral aedeagal valves membranous and short. Dorsal aedeagal valves exceeding the base of gonocoxites with slightly swollen apexes.

Emergence: May to July.

Distribution: Sichuan (Langzhong, Wolong) and Gansu (Wenxian).

37. 拟华山单角蝎蛉 *Cerapanorpa dubia* (Chou & Wang, 1981)

【测量】 前翅长 13.5 ~ 15.0 mm。

【形态特征】 头顶黑色，喙浅黄色、黄褐色或黑褐色。胸部背板黑色，侧板及足浅黄色。翅透明无斑纹或仅端部略加深，翅痣浅棕色明显。雄性腹部 1 ~ 5 节背板黑褐色。第 3 节背中突扁平，盖住第 4 节上尖锐的后背突。第 6 节末端具 1 个黄褐色臀角。第 7、8 节黄褐色，基部缢缩。雄性外生殖器卵形。第 9 节背板末端 U 形缺刻，两侧形成 1 对指形突起。下瓣长带状，内缘具 1 列鬃毛。生殖肢基节腹面端部具多根长刚毛。生殖刺突短于生殖肢基节，内缘中齿钝三角形，亚基齿大型盘状。阳基侧突细长，直或稍波状弯曲，内侧具 1 列极短刚毛。阳茎腹瓣膜质短小；背瓣细长，超过生殖刺突基部，背面具膜质膨大区。

【发生期】 6 ~ 8 月。

【地理分布】 陕西（眉县、周至、宁陕、华山）、河南（栾川）。

Measurements: Forewing length 13.5–15.0 mm.

Morphology: Vertex black. Rostrum light yellow, yellowish brown or dark brown. Nota black. Pleura and legs light yellow. Wing membrane hyaline, and devoid of markings, or darkened apically. Pterostigma light brown and conspicuous. Male terga Ⅰ–Ⅴ dark brown. Notal organ on tergum Ⅲ flat and covering acute postnotal organ on tergum Ⅳ. A6 dark brown and bearing a yellowish brown anal horn on its dorsal apex. A7 and A8 yellowish brown and constricted at base. Male genital bulb oval. Epandrium emarginate terminally in a broad U-shape and forming a pair of finger-like processes laterally. Hypovalves stripe-like and bearing a row of long bristles along inner margin. Gonocoxites with several long setae on ventral apex. Gonostylus shorter than gonocoxites and bearing an obtuse triangular median tooth and a large disc-

图 3-37-1 拟华山单角蝎蛉 *Cerapanorpa dubia* 成虫背面观（陕西太白山，NWAFU）

1. 雄虫 2. 雌虫

Fig. 3-37-1 Adults of ***Cerapanorpa dubia***, dorsal view (Mount Taibai, Shaanxi, NWAFU)

1. Male 2. Female

图 3-37-2 拟华山单角蝎蛉 *Cerapanorpa dubia* 雄性外生殖器腹面观

Fig. 3-37-2 Male genitalia of *Cerapanorpa dubia*, ventral view

like subbasal process on inner margin. Paramere slender, straight or wave-like, and bearing a row of short spines on inner margin. Ventral aedeagal valves

图 3-37-3 拟华山单角蝎蛉 *Cerapanorpa dubia* 地理分布

Fig. 3-37-3 Distribution of *Cerapanorpa dubia*

membranous and short. Dorsal aedeagal valves slender and exceeding base of gonostylus with membranous swollen processes dorsally.

Emergence: June to August.

Distribution: Shaanxi (Meixian, Zhouzhi, Ningshan, Mount Hua) and Henan (Luanchuan).

38. 华山单角蝎蛉 *Cerapanorpa emarginata* (Cheng, 1949)

【测量】 前翅长 13.5 ~ 14.0 mm。

【形态特征】 头顶黑色，喙浅黄色、黄褐色或黑褐色。胸部背板黑色，侧板及足浅黄色。翅透明，斑纹变异较大，有的个体具完整的痣带和端带，有的个体完全无斑纹。雄性腹部 1 ~ 5 节背板黑褐色。第 3 节背中突扁平，盖住第 4 节上尖锐的后背突。第 6 节黑褐色，末端具 1 个黄褐色臂角。第 7、8 节黄褐色，基部缢缩。雄性外生殖器阔卵形。第 9 节背板末端 U 形缺刻，两侧形成 1 对指形突起。下瓣长带状，内缘具 1 列整齐排列的长鬃。生殖肢

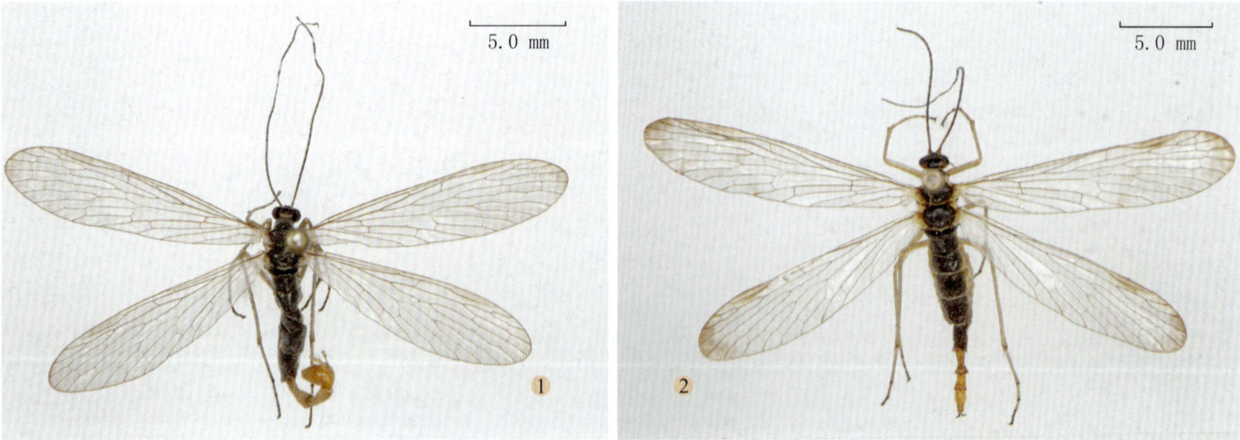

图 3-38-1 华山单角蝎蛉 *Cerapanorpa emarginata* 成虫背面观（陕西太白山，NWAFU）

1. 雄虫 2. 雌虫

Fig. 3-38-1 Adults of *Cerapanorpa emarginata*, dorsal view (Mount Taibai, Songxian, NWAFU)

1. Male 2. Female

基节端部内缘具数根长刚毛；生殖刺突短于基节，内缘中齿钝三角形，亚基齿大型盘状。阳基侧突基半部粗壮，端半部纤细，远超过生殖肢基节端部，内缘着生 1 列梳齿状短刚毛。阳茎腹瓣膜质短小；背瓣伸达生殖肢基节后缘，端部略膨大。

【发生期】 6 ~ 8 月。

【地理分布】 陕西 [华山、宁陕、周至、眉县、太白、宝鸡 (天台山)]、河南 (嵩县)、甘肃 (文县)、湖北 (神农架)。

Measurements: Forewing length 13.5–14.0 mm.

Morphology: Vertex black. Rostrum light yellow, yellowish brown or dark brown. Nota black. Pleura and legs light yellowish . Wing membrane hyaline. Markings highly variable: some individuals with complete apical and pterostigmal bands, while some devoid of markings. Male terga Ⅰ–Ⅴ dark brown. Notal organ on tergum Ⅲ flat and covering acute postnotal organ on tergum Ⅳ. A6 dark brown and bearing a yellowish brown anal horn on its dorsal apex. A7 and A8 yellowish brown and constricted at base. Male genital bulb oval. Epandrium emarginate terminally in a broad U-shape and forming a pair of finger-like processes laterally. Hypovalves stripe-like and bearing a row of long bristles along inner margin. Gonocoxites with several long setae on ventral apex. Gonostylus shorter than gonocoxites and bearing an obtuse triangular median tooth and a large disc-like subbasal process on inner margin. Paramere stout at basical half, slender at apical half, extending far beyond apex of gonocoxites, and bearing a row of short comb-like spines on inner margin. Ventral aedeagal valves membranous and short. Dorsal aedeagal valves slender and extending to apex of gonocoxites with slightly swollen processes dorsally.

Emergence: June to August.

Distribution: Shaanxi [Mount Hua, Ningshan, Zhouzhi, Meixian, Taibai, Baoji(Mount Tiantai)], Henan (Songxian), Gansu (Wenxian) and Hubei (Shennongjia).

图 3-38-2 华山单角蝎蛉 *Cerapanorpa emarginata* 雄性外生殖器腹面观（右侧下瓣已去除）

Fig. 3-38-2 Male genitalia of *Cerapanorpa emarginata*, ventral view (with right hypovalve removed)

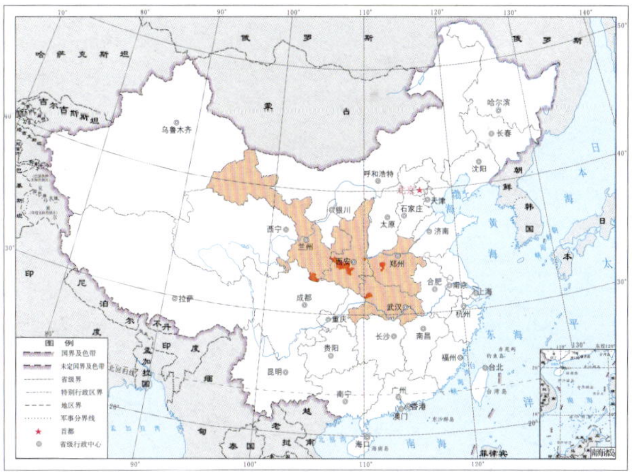

图 3-38-3 华山单角蝎蛉 *Cerapanorpa emarginata* 地理分布

Fig. 3-38-3 Distribution of *Cerapanorpa emarginata*

39. 伏牛山单角蝎蛉 *Cerapanorpa funiushana* (Hua & Chou, 1997)

【测量】 前翅长 13.5 ~ 14.0 mm。

【形态特征】 头顶黑色，喙黄褐色。胸部背板黑褐色，侧板及足黄褐色。翅透明，无斑纹或仅端部稍加深。翅痣浅黄色，不明显。雄性腹部 1 ~ 5 节背板黑褐色。第 3 节背中突扁平，盖住第 4 节上尖锐的后背突。第 6 节黑褐色，末端具 1 个黄褐色臀角。第 7、8 节黄褐色，基部缢缩。雄性外生殖器阔卵形。第 9 节背板末端 U 形缺刻，两侧形成 1 对指形突起。下瓣长带状，内缘鬃毛长而粗壮，较稀疏，排列整齐。生殖肢基节端部腹面具数根长刚

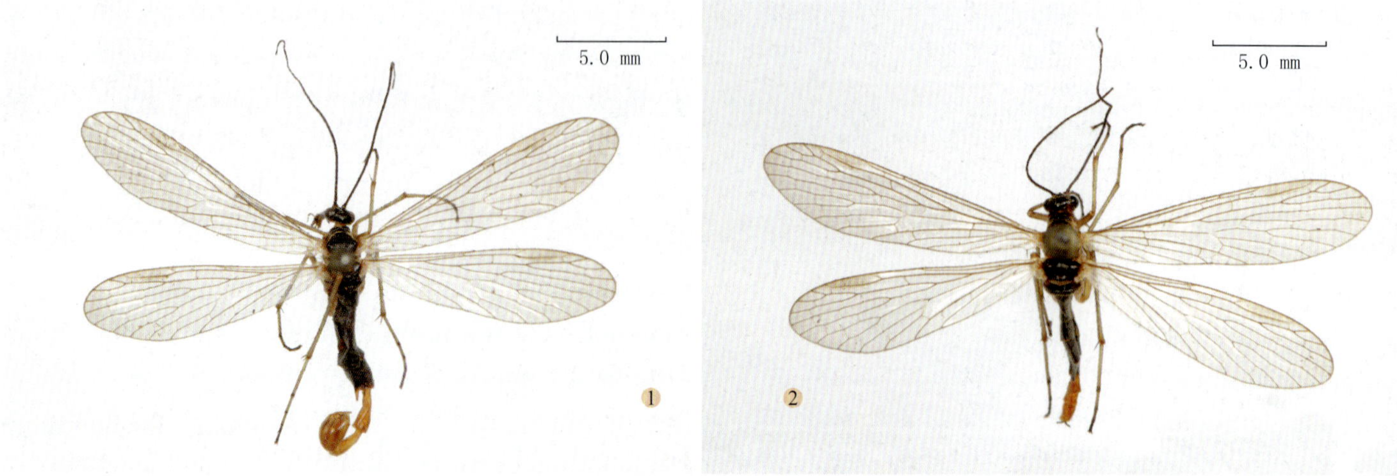

图 3-39-1 伏牛山单角蝎蛉 ***Cerapanorpa funiushana*** 成虫背面观（河南嵩县白云山，NWAFU）

1. 雄虫 2. 雌虫

Fig. 3-39-1 Adults of ***Cerapanorpa funiushana***, dorsal view (Mount Baiyun, Songxian, Henan, NWAFU)

1. Male 2. Female

图 3-39-2 伏牛山单角蝎蛉 ***Cerapanorpa funiushana*** 雄性外生殖器腹面观

Fig. 3-39-2 Male genitalia of ***Cerapanorpa funiushana***, ventral view

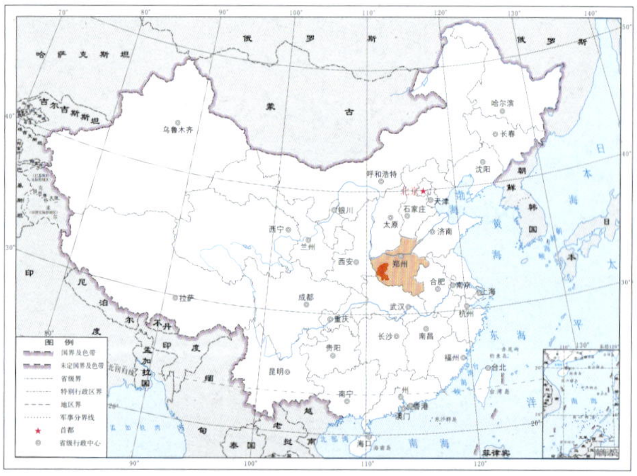

图 3-39-3 伏牛山单角蝎蛉 ***Cerapanorpa funiushana*** 地理分布

Fig. 3-39-3 Distribution of ***Cerapanorpa funiushana***

毛。生殖刺突短于基节，内缘中齿钝三角形，亚基齿大型盘状。阳基侧突近平行，末端不超过生殖刺突中部，内侧具 1 列密集的梳齿状毛。阳茎腹瓣短，膜质；背瓣长，伸达生殖肢基节端部，末端指状，背缘稍膨大。

【发生期】 6～8月。

【地理分布】 河南（伏牛山）。

Measurements: Forewing length 13.5–14.0 mm.

Morphology: Vertex black. Rostrum yellowish brown.Nota dark brown. Pleura and legs yellowish brown. Wing membrane hyaline, devoid of markings or only darkened at apex. Pterostigma light yellow and inconspicuous. Male terga Ⅰ–Ⅴ dark brown. Notal organ on tergum Ⅲ flat and covering acute postnotal organ on tergum Ⅳ. A6 dark brown and bearing a yellowish brown anal horn on its apex. A7 and A8 yellowish brown and constricted at base. Male genital bulb broad oval. Epandrium emarginate terminally in a broad U-shape and forming a pair of finger-like processes laterally. Hypovalves stripe-like and bearing a row of sparsely and regularly arranged long bristles along inner margin. Gonocoxites with several long setae on ventral apex. Gonostylus shorter than gonocoxites and bearing an obtuse triangular median tooth and a large disc-like subbasal process on inner margin. Paramere nearly parallel, and not exceeding middle of gonocoxites apically, and bearing a row of short and dense comb-like spines on inner margin. Ventral aedeagal valves membranous and short. Dorsal aedeagal valves long and extending to apex of gonocoxites, finger-like at apexes, and with slightly swollen processes dorsally.

Emergence: June to August.

Distribution: Henan (Funiu Mountain).

40. 埔里单角蝎蛉 *Cerapanorpa horiensis* (Issiki, 1929)

【测量】 前翅长 11.5 ~ 13.0 mm。

【形态特征】 头顶黄色，单眼三角区具 1 黑色大斑，喙黄褐色。胸部背板黑色，侧板及足浅黄色。翅透明稍带黄色；端部稍加深，痣带退化为一系列不连续的小斑点，无其他斑纹。雄性腹部 1 ~ 5 节背板黑褐色。第 3 节背中突扁平，盖住第 4 节上尖锐的后背突。第 6 节背面黑褐色，腹面黄褐色，末端具 1 个黄褐色锥形臀角。第 7、8 节黄褐色，两侧具黑色纵纹，基部缢缩，稍呈柄状。雄性外生殖器卵形。第 9 节背板末端具深的 U 形缺刻，两侧形成 1 对指形突起。下瓣短小，不伸达生殖肢基节中部，端部钝圆，具稀疏的粗壮鬃毛。生殖刺突短于基节，内缘中齿不明显，基半部具大型碗状凹陷区。阳基侧突刀刃形，超出生殖肢基节末端，基部较直，端部向内侧弯曲，内缘具 1 列梳齿状短刺。阳茎腹瓣短，膜质；背瓣长，末端指状，侧突起十分粗壮。

【发生期】 3 ~ 5 月。

【地理分布】 台湾（南投、台东）。

备注：本种在形态上与同属其他种差异较大，是否真正属于单角蝎蛉，还有待商榷。

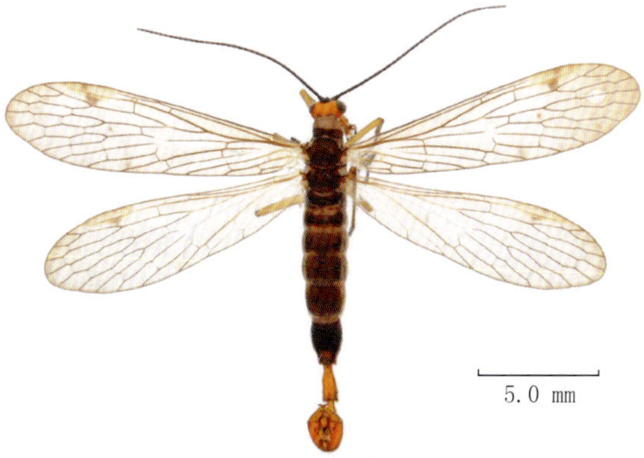

图 3-40-1 埔里单角蝎蛉 *Cerapanorpa horiensis* 雄成虫背面观（台湾台东海端，NWAFU）

Fig. 3-40-1 Male adult of *Cerapanorpa horiensis*, dorsal view (Haituan, Taitung, Taiwan, NWAFU)

图 3-40-2 埔里单角蝎蛉 *Cerapanorpa horiensis* 雄性腹部侧面观

Fig. 3-40-2 Male teminal abdomen of *Cerapanorpa horiensis*, lateral view

Measurements: Forewing length 11.5–13.0 mm.

Morphology: Vertex yellow. Ocellar triangle enclosed by a large black spot. Rostrum yellowish brown. Nota black. Pleura and legs light yellow. Wing membrane hyaline and slightly tinged with yellow. Apex slightly darkened. Pterostigmal band greatly reduced into a series of discontinuous spots. Other markings absent. Male terga Ⅰ–Ⅴ dark brown. Notal organ on tergum Ⅲ flat and covering acute postnotal organ on tergum Ⅳ. A6 dark brown dorsally and yellowish brown ventrally, and bearing a yellowish brown, conical anal horn apically. A7 and A8 yellowish brown and slightly stalked at base, with black lines on lateral surfaces. Male genital bulb oval. Epandrium emarginate terminally in a broad U-shape and forming a pair of finger-like processes laterally. Hypovalves short, not extending to middle of gonocoxites, with sparse, stout setae on rounded apexes. Gonostylus shorter than gonocoxites, bearing an indistinct median tooth, and a large bowl-shaped concavity on basal half of inner margin. Paramere blade-shaped, with its base straight, distal portion curved mesad, exceeding apex of gonocoxites, and bearing a row of comb-like spines along inner margin. Aedeagus with its ventral valves membranous and short; dorsal valves elongated and finger-shaped apically; and lateral processes very stout.

图 3-40-3 埔里单角蝎蛉 ***Cerapanorpa horiensis*** 雄性外生殖器腹面观

Fig. 3-40-3 Male genitalia of ***Cerapanorpa horiensis***, ventral view

Emergence: March to May.

Distribution: Taiwan (Nantou, Taitung).

Remarks: This species is greatly different from its congeners in morphology. Whether it really belongs to *Cerapanorpa* merits further research.

图 3-40-4 埔里单角蝎蛉 ***Cerapanorpa horiensis*** 地理分布

Fig. 3-40-4 Distribution of ***Cerapanorpa horiensis***

41. 六盘山单角蝎蛉 *Cerapanorpa liupanshana* Gao, Ma & Hua, 2016

【测量】 前翅长 12.0 ~ 13.0 mm。

【形态特征】 头顶黑色，喙黄褐色。胸部背板黑色，侧板及足浅黄色。翅透明，无斑纹或仅端部稍加深。翅痣不明显。雄性腹部 1 ~ 5 节背板黑褐色。第 3 节背中突扁平，盖住第 4 节上尖锐的后背突。第 6 节黑褐色，末端具 1 个黄褐色臀角。第 7、8 节黄褐色，基部缢缩。雄性外生殖器阔卵形。第 9 节背板末端具深的 U 形缺刻，并在两侧形成 1 对指形突起。下瓣长带状，超过生殖肢基节内侧突起，内缘鬃毛长而粗壮，较稀疏，排列杂乱。生殖肢基节内缘端部着生 1 簇长刚毛。生殖刺突短于基节，内缘中齿钝三角形，亚基齿大型盘状。阳基侧

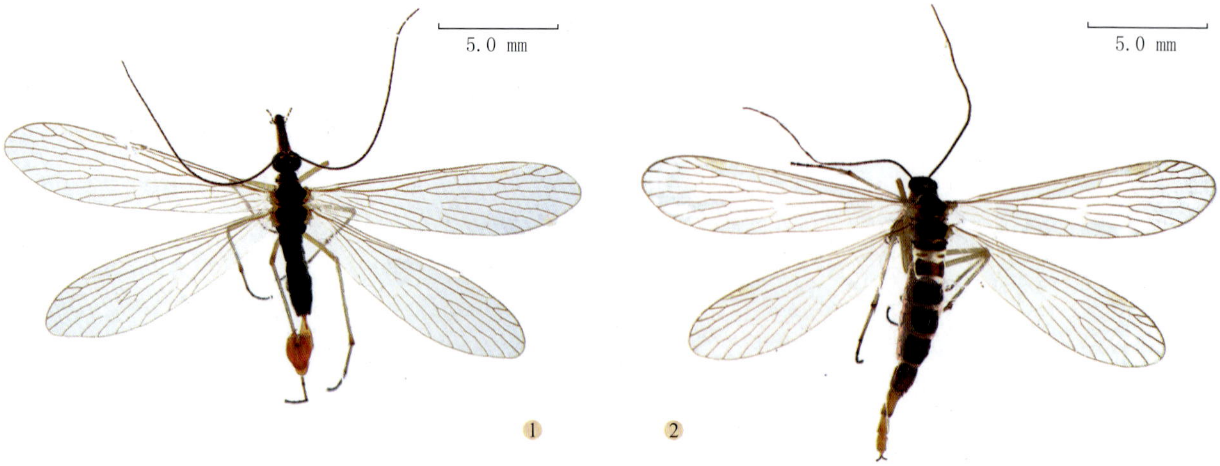

图 3-41-1 六盘山单角蝎蛉 ***Cerapanorpa liupanshana*** 成虫背面观（宁夏六盘山，NWAFU）

1. 雄虫 2. 雌虫

Fig. 3-41-1 Adults of ***Cerapanorpa liupanshana***, dorsal view (Liupan Mountain, Ningxia, NWAFU)

1. Male 2. Female

突不超出生殖肢基节端部，基部较直，端部向内弯曲，背面具 1 列短刚毛；刚毛长度短于阳基侧突的宽度。阳茎腹瓣短，膜质；背瓣长，末端指状。

【发生期】 6 ~ 8 月。

【地理分布】 宁夏（六盘山）。

Measurements: Forewing length 12.0–13.0 mm.

Morphology: Vertex black. Rostrum yellowish brown. Nota black. Pleura and legs light yellow. Wing membrane hyaline, devoid of markings or slightly darkened at apex. Pterostigma inconspicuous. Male terga Ⅰ–Ⅴ dark brown. Notal organ on tergum Ⅲ flat and covering acute postnotal organ on tergum Ⅳ. A6 dark brown and bearing a yellowish brown anal horn on apex. A7 and A8 yellowish brown and constricted at base. Male genital bulb broad oval. Epandrium emarginate terminally in a broad U-shaped and forming a pair of finger-like processes laterally. Hypovalves stripe-like and bearing a row of sparsely and irregularly arranged long bristles along inner margin, with their apex exceeding inner processes of gonocoxites. Gonocoxites with several long setae on ventral apex. Gonostylus shorter than gonocoxites and bearing an obtuse triangular median tooth and a large disc-like subbasal process on inner margin. Paramere short and not exceeding apex of gonocoxites, with its basal portion straight and distal portion curving mesad, and bearing a row of short spines along inner margin, spines on inner margin shorter than width of paramere. Aedeagus with its ventral valves short and membranous; dorsal valves elongated and finger-like apically.

Emergence: June to August.

Distribution: Ningxia (Liupan Mountain).

图 3-41-2 六盘山单角蝎蛉 ***Cerapanorpa liupanshana*** 雄性外生殖器腹面观

Fig. 3-41-2 Male genitalia of ***Cerapanorpa liupanshana***, ventral view

图 3-41-3 六盘山单角蝎蛉 ***Cerapanorpa liupanshana*** 地理分布

Fig. 3-41-3 Distribution of ***Cerapanorpa liupanshana***

42. 南五台单角蝎蛉 *Cerapanorpa nanwutaina* (Chou, 1981)

【测量】 前翅长 13.5 ~ 14.0 mm。

【形态特征】 头顶黑色，喙黑褐色，两侧黄褐色。胸部背板黑色，侧板及足浅黄褐色。翅透明，无斑纹或具端带和痣带，雌性翅斑更为显著。雄性腹部 1 ~ 5 节背板黑褐色。第 3 节背中突扁平，盖住第 4 节上尖锐的后背突。第 6 节黑褐色，末端具 1 个黄褐色臀角。第 7、8 节黄褐色，基部缢缩。雄性外生殖器卵形。第 9 节背板末端具深 U 形缺刻，两侧形成 1 对指状突起。下瓣长带状，超过生殖肢基节内侧突起，内缘鬃毛长而粗壮，较稀疏，排列整齐。生殖肢基节端部内缘具稀疏的长刚毛。生殖刺突短于基节，内缘中齿钝三角形，亚基齿大型盘状。阳基侧突伸达生殖肢基节端部，基部直，端部膝状弯曲，内缘具 1 列密集的短刺毛。阳茎腹

图 3-42-1 南五台单角蝎蛉 *Cerapanorpa nanwutaina* 成虫侧面观（陕西宝鸡天台山国家森林公园，姜碌摄）

1. 雄虫 2. 雌虫

Fig. 3-42-1 Adults of ***Cerapanorpa nanwutaina***, lateral view (Tiantaishan National Forest Park, Baoji, Shaanxi, photo by Lu Jiang)

1. Male 2. Female

瓣短，膜质；背瓣长，侧面观向末端变宽。

【发生期】 6 ~ 8 月。

【地理分布】 陕西（长安、宁陕、宝鸡）、湖北（神农架）、甘肃（宕昌）。

Measurements: Forewing length 13.5–14.0 mm.

Morphology: Vertex black. Rostrum dark brown with yellowish flanks. Nota black. Pleura and legs light yellowish brown. Wing membrane hyaline, without markings or only with apical and pterostigmal bands. Females usually with more markings than males. Male terga Ⅰ–Ⅴ dark brown. Notal organ on tergum Ⅲ flat and covering acute postnotal organ on tergum Ⅳ. A6 dark brown and bearing a yellowish brown anal horn on its dorsal apex. A7 and A8 yellowish brown and constricted at base. Male genital bulb oval. Epandrium emarginate terminally in a broad U-shape and forming a pair of finger-like processes laterally. Hypovalves stripe-like and bearing a row of sparsely and regularly arranged long bristles along inner margin, with their apex exceeding inner processes of gonocoxites. Gonocoxites with sparse long setae on ventral apex. Gonostylus shorter than gonocoxites and bearing an obtuse triangular median tooth and a large disc-like subbasal process on inner margin. Paramere slender and extending to apex of gonocoxites, with its basal portion straight and distal portion geniculate, and bearing a row of short spines along inner margin. Aedeagus with its ventral valves short and membranous; dorsal valves elongated and broaden towards apexes in lateral view.

Emergence: June to August.

Distribution: Shaanxi (Chang'an, Ningshan, Baoji), Hubei (Shennongjia) and Gansu (Dangchang).

图 3-42-2 南五台单角蝎蛉 *Cerapanorpa nanwutaina* 雄性外生殖器腹面观

Fig. 3-42-2 Male genitalia of *Cerapanorpa nanwutaina*, ventral view

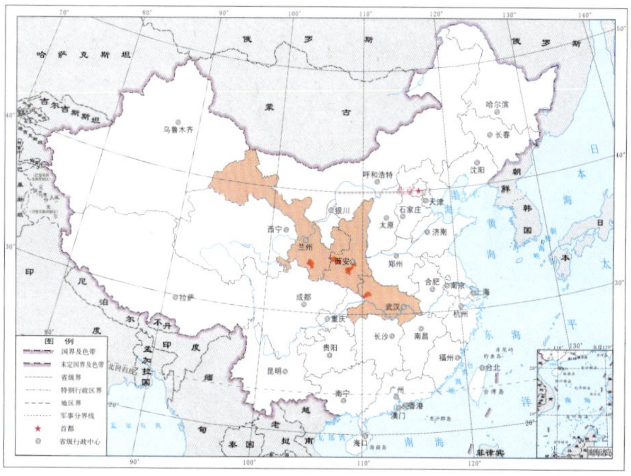

图 3-42-3 南五台单角蝎蛉 *Cerapanorpa nanwutaina* 地理分布

Fig. 3-42-3 Distribution of *Cerapanorpa nanwutaina*

43. 太白单角蝎蛉 *Cerapanorpa obtusa* (Cheng, 1949)

【测量】 前翅长 13.0 ~ 14.5 mm。

【形态特征】 头顶黑褐色，喙黄褐色。胸部背板黑色，侧板及足淡黄褐色。翅透明无色，有或无端带和痣带，变异较大，一般雌性个体斑纹较雄性更为发达。雄性腹部 1 ~ 5 节背板黑褐色。第 3 节背中突扁平，盖住第 4 节上尖锐的后背突。第 6 节黑褐色，背部末端具 1 个黄褐色臀角。第 7、8 节黄褐色，基部缢缩。雄性外生殖器宽卵形。第 9 节背板末端具深 U 形缺刻，两侧形成 1 对指状突起。下瓣长带状，末端超过生殖肢基节内缘突起，内侧边缘具致密而杂乱的长鬃。生殖肢基节端部内缘具稀疏的长刚毛。生殖刺突短于基节，内缘中齿钝三角形，亚基齿大型盘状。阳基侧突伸达生殖肢基节端部，向末端逐渐变细，基半部较直，近端部稍膝

图 3-43-1 太白单角蝎蛉 *Cerapanorpa obtusa* 成虫侧面观（陕西宝鸡天台山国家森林公园，姜禄摄）

1. 雄虫 2. 雌虫

Fig. 3-43-1 Adults of *Cerapanorpa obtusa*, lateral view (Tiantaishan National Forest Park, Baoji, Shaanxi, photo by Lu Jiang)

1. Male 2. Female

状弯曲，内缘具 1 列梳齿状长毛。阳茎腹瓣短，膜质；背瓣长，末端指状。

【发生期】 6 ~ 8 月。

【地理分布】 陕西（太白、眉县、周至、宁陕、宝鸡天台山）、甘肃（宕昌）、宁夏（六盘山）。

Measurements: Forewing length 13.5–14.5 mm.

Morphology: Vertex black. Rostrum yellowish brown. Nota balck. Pleura and legs light yellowish brown. Wing membrane hyaline, with markings highly variable, from totally absent to only apical and pterostigmal bands present. Females usually with denser markings than males. Male terga Ⅰ–Ⅴ dark brown. Notal organ on tergum Ⅲ flat and covering acute postnotal organ on tergum Ⅳ. A6 dark brown and bearing a yellowish brown anal horn on its dorsal apex. A7 and A8 yellowish brown and constricted at base. Male genital bulb broad oval. Epandrium emarginate terminally in a broad U-shape and forming a pair of finger-like processes laterally. Hypovalves stripe-like and bearing a row of sparsely and irregularly arranged long bristles along inner margin, with their apexes exceeding inner processes of gonocoxites. Gonocoxites with sparse long setae on ventral apex. Gonostylus shorter than gonocoxites and bearing an obtuse triangular median tooth and a large disc-like subbasal process on inner margin. Paramere extending to apex of gonocoxites, with its basal portion straight and distal portion geniculate, and bearing a row of long, comb-like spines along inner margin. Aedeagus with its ventral valves short and membranous; dorsal valves long and finger-like apically.

Emergence: June to August.

Distribution: Shaanxi [Taibai, Meixian, Zhouzhi, Ningshan, Baoji (Mount Tiantai)], Gansu (Dangchang) and Ningxia (Liupan Mountain).

图 3-43-2 太白单角蝎蛉 ***Cerapanorpa obtusa***
雄性外生殖器腹面观
Fig. 3-43-2 Male genitalia of ***Cerapanorpa obtusa***, ventral view

图 3-43-3 太白单角蝎蛉 ***Cerapanorpa obtusa*** 地理分布
Fig. 3-43-3 Distribution of ***Cerapanorpa obtusa***

44. 壮突单角蝎蛉 *Cerapanorpa protrudens* Gao, Ma & Hua, 2016

【测量】 前翅长 12.5 ~ 14.2 mm。

【形态特征】 头顶黑褐色，喙黄褐色。胸部背板黑色，侧板及足淡黄褐色。翅透明，翅痣不明显。翅斑多变；无斑纹或仅具端带和痣带，一般雌性斑纹较雄性更为发达。雄性腹部 1 ~ 5 节背板黑褐色。第 3 节背中突扁平，盖住第 4 节上尖锐的后背突。第 6 节黑褐色，末端具 1 个黄褐色臀角。第 7、8 节黄褐色，基部缢缩。雄性外生殖器宽卵形。第 9 节背板末端具深 U 形缺刻，两侧形成 1 对指状突起。下瓣长带状，末端接近生殖基节后缘，内侧边缘具稀疏、整齐排列的长鬃。生殖肢基节端部腹面具稀疏的长刚毛。生殖刺突较短于基节，内缘中齿

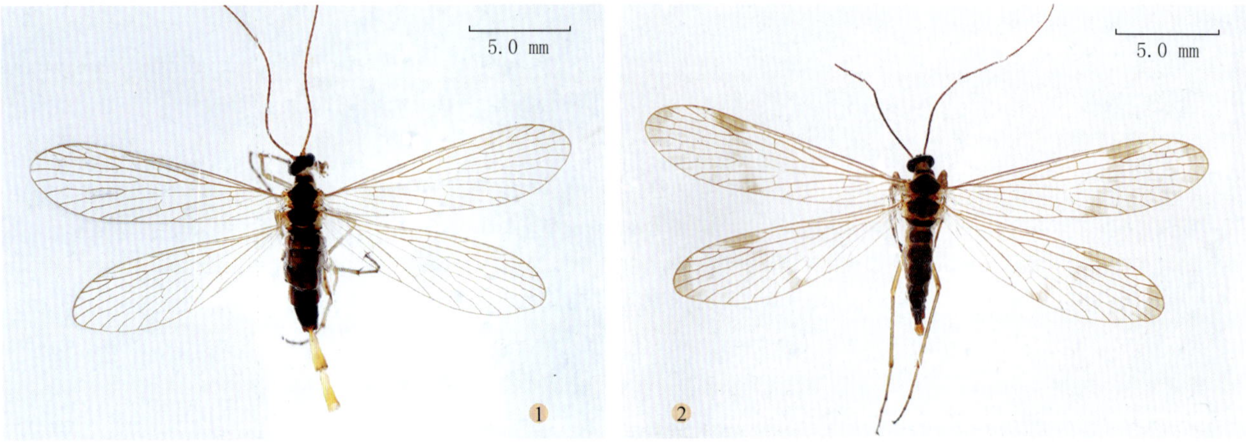

图 3-44-1 壮突单角蝎蛉 ***Cerapanorpa protrudens*** 成虫背面观（陕西安康大巴山，NWAFU）
1. 雄虫（外生殖器已去除） 2. 雌虫

Fig. 3-44-1 Adults of ***Cerapanorpa protrudens***, dorsal view (Daba Mountains, Ankang, Shaanxi, NWAFU)
1. Male (with genitalia removed) 2. Female

钝三角形，亚基齿大型盘状。阳基侧突极长，超过生殖刺突中部，向末端逐渐变细，近基部稍膝状弯曲，内缘具 1 列梳齿状短刺。阳茎腹瓣膜质，短小；背瓣长，末端指状，稍向外弯曲。

【发生期】 7 月。

【地理分布】 陕西（安康）。

Measurements: Forewing length 12.5–14.2 mm.

Morphology: Vertex dark brown. Rostrum yellowish brown.Nota black. Pleura and legs light yellowish brown. Wing membrane hyaline.

图 3-44-2 壮突单角蝎蛉 ***Cerapanorpa protrudens*** 雄性外生殖器腹面观

Fig. 3-44-2 Male genitalia of ***Cerapanorpa protrudens***, ventral view

Pterostigma inconspicuous. Markings highly variable, from totally absent to only apical and pterostigmal bands present. Females usually with denser markings than males. Male terga Ⅰ–Ⅴ dark brown. Notal organ on tergum Ⅲ flat and covering acute postnotal organ on tergum Ⅳ. A6 dark brown and bearing a yellowish brown anal horn on its dorsal apex. A7 and A8 yellowish brown and constricted at base.Male genital bulb broad oval. Epandrium emarginate terminally in a broad U-shape and forming a pair of finger-like processes laterally. Hypovalves stripe-like and bearing a row of sparsely and regularly arranged long bristles along inner margin, with their apexes approximate apex of gonocoxites. Gonocoxites with sparse long setae on ventral apex. Gonostylus shorter

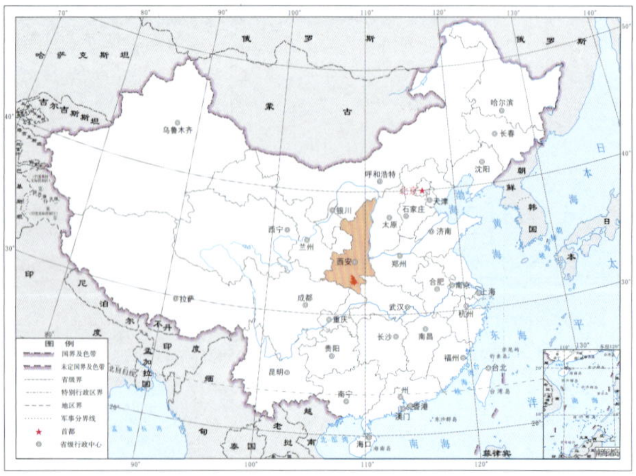

图 3-44-3 壮突单角蝎蛉 ***Cerapanorpa protrudens*** 地理分布

Fig. 3-44-3 Distribution of ***Cerapanorpa protrudens***

than gonocoxites, bearing an obtuse triangular median tooth and a large disc-like subbasal process on inner margin. Paramere extremely elongated, extending beyond middle of gonostylus, with its basal portion slightly geniculate and distal portion curved mesad, and bearing a row of very short, comb-like spines along inner margin. Aedeagus with its ventral valves short and membranous; dorsal valves long and finger-like apically, slightly curved mesad.

Emergence: July.

Distribution: Shaanxi (Ankang).

45. 任氏单角蝎蛉 *Cerapanorpa reni* (Chou, 1981)

【测量】 前翅长 15.0 ~ 16.0 mm。

【形态特征】 体形较大的单角蝎蛉。头顶黑色，喙黄褐色或红褐色。胸部背板黑色，侧板及足基节黑褐色，足大部分红褐色。翅透明无色，斑纹黑褐色，翅痣红棕色。前翅端带阔，后方具 1 个分离的小斑；痣带的端支、基支完整；缘斑大；基带宽阔；基斑极小。雄性腹部 1 ~ 5 节背板黑褐色。第 3 节背中突扁平，盖住第 4 节上尖锐的后背突。第 6 节黑褐色，末端具 1 个红褐色臀角。第 7、8 节红褐色，基部缢缩。雄性外生殖器卵形。第 9 节背板末端具深 U 形缺刻，两侧形成 1 对指状突起。下瓣细长，内侧具数根鬃毛，排列整齐，末端超过生殖肢基节内缘突起。生殖肢基节内缘端部具稀疏的长毛。生殖刺突短于基节，内缘中齿钝三角形，亚基齿大型盘状。阳基侧突 L 形，端部平直，内侧具短刚毛。阳茎腹瓣膜质短小；背瓣长，末端指状。

【发生期】 6 ~ 7 月。

图 3-45-2 任氏单角蝎蛉 ***Cerapanorpa reni*** 雄性外生殖器腹面观（左侧下瓣已去除）

Fig. 3-45-2 Male genitalia of ***Cerapanorpa reni***, ventral view (with left hypovalve removed)

【地理分布】 陕西（鄠邑、宁陕、太白、周至）。

Measurements: Forewing length 15.0–16.0 mm.

Morphology: Large Cerapanorpa species.Vertex black. Rostrum yellowish brown or reddish brown. Nota black. Pleura and coxae dark brown. Legs mostly reddish brown. Wing membrane hyaline. Pterostigma reddish brown. Markings dark brown. Apical band broad, and connected to a small spot near its hind portion; pterostigmal band with complete basal and distal branches; marginal spot large; basal band broad; basal spot very small. Male terga Ⅰ–Ⅴ dark brown. Notal organ on tergum Ⅲ flat and covering acute postnotal organ on tergum Ⅳ. A6 dark brown and bearing a reddish brown anal horn on its dorsal apex. A7 and A8 reddish brown and constricted at base.Male genital bulb broad oval. Epandrium emarginate terminally in a broad U-shape and forming a pair of finger-like processes laterally. Hypovalves slender stripe-like, and bearing a row of sparsely and regularly

图 3-45-3 任氏单角蝎蛉 ***Cerapanorpa reni*** 地理分布

Fig. 3-45-3 Distribution of ***Cerapanorpa reni***

图 3-45-1　任氏单角蝎蛉 ***Cerapanorpa reni*** 成虫侧背面观（陕西秦岭，姜碌摄）

1. 雄虫　2. 雌虫

Fig. 3-45-1　Adults of ***Cerapanorpa reni***, dorso-lateral view (Qinling Mountains, Shaanxi, photo by Lu Jiang)

1. Male　2. Female

arranged long bristles along inner margin, with their apexes exceeding to inner processes of gonocoxites. Gonocoxites with sparse long setae on ventral apex. Gonostylus shorter than gonocoxites, bearing an obtuse triangular median tooth and a large disc-like subbasal process on inner margin. Paramere L-shaped, with its distal portion straight, and bearing a row of short spines along inner margin. Aedeagus with its ventral valves short and membranous; dorsal valves long and finger-like apically.

Emergence: June to July.

Distribution: Shaanxi (Huyi, Ningshan, Taibai, Zhouzhi).

46. 弯曲单角蝎蛉 *Cerapanorpa sinuata* Gao, Ma & Hua, 2016

【测量】 前翅长 12.0 ~ 13.0 mm。

【形态特征】 头顶黑色，喙黄褐色。胸部背板黑色，侧板及足浅黄褐色。翅透明，无明显的斑纹，仅端部稍加深。翅痣不明显。雄性腹部 1 ~ 5 节背板黑褐色。第 3 节背中突扁平，盖住第 4 节上尖锐的后背突。第 6 节黑褐色，末端具 1 个黄褐色臀角。第 7、8 节黄褐色，基部缢缩。雄性外生殖器阔卵形。第 9 节背板末端具深 U 形缺刻，两侧形成 1 对指状突起。下瓣长带状，内缘鬃毛稀疏而排列整齐，末端伸达生殖肢基节端部。生殖肢基节端部内缘具稀疏的长刚毛，生殖刺突短于基节，内缘中齿钝三角形，亚基齿大型盘状。阳基侧突端半部强烈弓状弯曲，内侧具密集的梳齿状长鬃。阳茎腹瓣短，膜质；背瓣长，末端平截，近端部背面具膨大的膜质结构。

【发生期】 6 ~ 7 月。

【地理分布】 陕西（宁陕）。

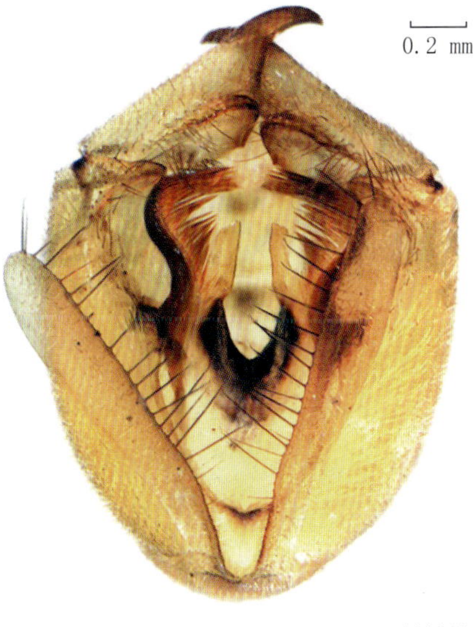

图 3-46-2 弯曲单角蝎蛉 *Cerapanorpa sinuata* 雄性外生殖器腹面观

Fig. 3-46-2 Male genitalia of *Cerapanorpa bicornifera*, ventral view

Measurements: Forewing length 12.0–13.0 mm.

Morphology: Vertex black. Rostrum yellowish brown. Nota black. Pleura and legs light yellowish brown. Wing membrane hyaline and devoid of markings, except apex darkened. Pterostigma inconspicuous. Male tergaⅠ–Ⅴdark brown. Notal organ on tergum Ⅲ flat and covering acute postnotal organ on tergumⅣ. A6 dark brown and bearing a yellowish brown anal horn on its dorsal apex. A7 and A8 yellowish brown and constricted at base.Male genital bulb broad oval. Epandrium emarginate terminally in a broad U-shape and forming a pair of finger-like processes laterally. Hypovalves stripe-like, bearing a row of sparsely regularly arranged long bristles regularly arranged along inner margin, with their apexes extending to apex of gonocoxites. Gonocoxites with sparse long setae on ventral apex. Gonostylus shorter than gonocoxites, bearing an obtuse triangular median tooth and a large disc-like subbasal process on

图 3-46-3 弯曲单角蝎蛉 *Cerapanorpa sinuata* 地理分布

Fig. 3-46-3 Distribution of *Cerapanorpa sinuata*

图 3-46-1 弯曲单角蝎蛉 ***Cerapanorpa sinuata*** 成虫侧背面观（陕西宁陕平河梁，姜碌摄）

1. 雄虫 2. 雌虫

Fig. 3-46-1 Adults of ***Cerapanorpa sinuata***, dorso-lateral view (Pingheliang, Ningshan, Shaanxi, photo by Lu Jiang)

1. Male 2. Female

inner margin. Paramere sinuate and posteriorly curved, and bearing dense comb-like spines along inner margin on distal half. Aedeagus with its ventral valves short and membranous; dorsal valves long and truncated apically and bearing a swollen membranous process on dorsal margin subapically.

Emergence: June to July.

Distribution: Shaanxi (Ningshan).

47. 王屋山单角蝎蛉 *Cerapanorpa wangwushana* (Huang, Hua & Shen, 2004)

【测量】 前翅长 13.0 ~ 14.0 mm。

【形态特征】 头顶黑色，喙黄褐色。胸部背板黑褐色，侧板及足浅黄褐色。翅透明，具明显的端带和痣带，其中痣带无端支。翅痣浅褐色，不明显。雄性腹部 1 ~ 5 节背板黑褐色。第 3 节背中突扁平，盖住第 4 节上尖锐的后背突。第 6 节黑褐色，末端具 1 个黄褐色臀角。第 7、8 节黄褐色，基部缢缩。雄性外生殖器阔卵形。第 9 节背板末端具深 U 形缺刻，两侧形成 1 对指状突起。下瓣长带状，内缘鬃毛密集而排列整齐，末端伸达生殖肢基节端部。生殖肢基节端部内缘具长刚毛，生殖刺突短于基节，内缘中齿钝三角形，亚基齿大型盘状。阳基侧突细长，波状弯曲，内侧棘毛粗壮、梳齿状排列。阳茎腹瓣短，膜质；背瓣长，末端指状。

【发生期】 6 ~ 7 月。

【地理分布】 河南（王屋山）、山西（历山）。

Measurements: Forewing length 13.0–14.0 mm.

Morphology: Vertex black. Rostrum yellowish brown. Nota dark brown. Pleura and legs light yellowish brown. Wing membrane hyaline, with distinct apical and pterostigmal bands. Pterostigmal band only represented by a basal branch. Pterostigma

图 3-47-2 王屋山单角蝎蛉 *Cerapanorpa wangwushana* 雄性外生殖器腹面观（右侧下瓣已去除）

Fig. 3-47-2 Male genitalia of *Cerapanorpa wangwushana*, ventral view (with right hypovalve removed)

图 3-47-1 王屋山单角蝎蛉 *Cerapanorpa wangwushana* 雄成虫背面观（山西历山，NWAFU）

Fig. 3-47-1 Male adult of *Cerapanorpa wangwushana*, dorsal view (Mount Li, Shanxi, NWAFU)

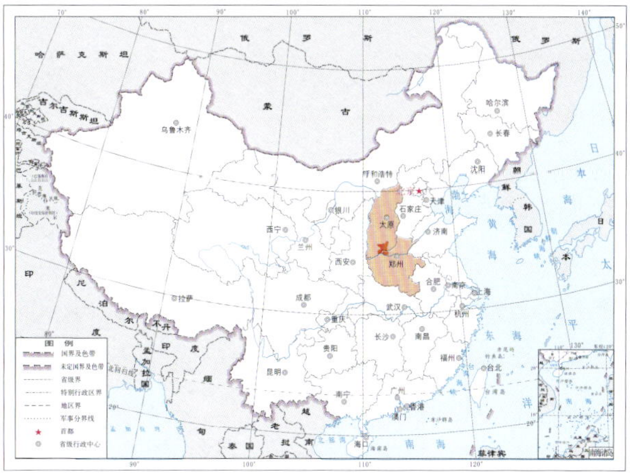

图 3-47-3 王屋山单角蝎蛉 *Cerapanorpa wangwushana* 地理分布

Fig. 3-47-3 Distribution of *Cerapanorpa wangwushana*

light brown and inconspicuous. Male terga Ⅰ–Ⅴ dark brown. Notal organ on tergum Ⅲ flat and covering acute postnotal organ on tergum Ⅳ. A6 dark brown and bearing a yellowish brown anal horn on its dorsal apex. A7 and A8 yellowish brown and constricted at base. Male genital bulb broad oval. Epandrium emarginate terminally in a broad U-shape and forming a pair of finger-like processes laterally. Hypovalves stripe-like, bearing a row of sparsely and regularly arranged long bristles along inner margin, with their apexes extending to apex of gonocoxites. Gonocoxites with sparse long setae on ventral apex. Gonostylus shorter than gonocoxites, bearing an obtuse triangular median tooth and a large disc-like subbasal process on inner margin. Paramere slender and wave-shaped, and bearing a row of comb-like spines along inner margin. Aedeagus with its ventral valves short and membranous; dorsal valves long and finger-like apically.

Emergence: June to July.

Distribution: Henan (Mount Wangwu) and Shanxi (Mount Li).

五、双角蝎蛉属 *Dicerapanorpa* Zhong & Hua, 2013

双角蝎蛉属 *Dicerapanorpa* 的主要识别特征为，雄性腹部第 6 节背板末端具 2 个指状臀角。中国特有属，已知 8 种，本图鉴收录 7 种。

The main diagnostic character of *Dicerapanorpa* is a pair of finger-like anal horns on the apex of male tergum Ⅵ. It is an endemic genus to China. Eight species have been described, and seven species are included in this atlas.

图 3-h 双角蝎蛉 ***Dicerapanorpa* sp.** 雄成虫背面观

Fig. 3-h Male adult of ***Dicerapanorpa* sp.**, dorsal view

本图鉴双角蝎蛉属 *Dicerapanorpa* 分种检索表

（7 种，雄性）

1. 翅面黄色具明显斑纹；喙两侧稍加深，不形成显著的纵带 ······ 2
   翅面无色，斑纹极度退化；喙两侧各具 1 条显著的黑褐色纵带 ······ 4
2. 前翅几乎无缘斑；阳基侧突中支直 ······ 52. 神农双角蝎蛉 ***Dicerapanorpa shennongensis***
   前翅具显著缘斑；阳基侧突中支向中部弯曲 ······ 3
3. 阳基侧突腹支短小，钩状；中支相互交叉 ······ 51. 大双角蝎蛉 ***Dicerapanorpa magna***
   阳基侧突腹支长直；中支不交叉或仅在端部稍接触 ······ 48. 白云山双角蝎蛉 ***Dicerapanorpa baiyunshana***
4. 下瓣末端显著扩大 ······ 5
   下瓣末端不扩大 ······ 6
5. 下瓣末端圆钝 ······ 49. 双角蝎蛉 ***Dicerapanorpa diceras***
   下瓣末端平截 ······ 53. 西康双角蝎蛉 ***Dicerapanorpa stotzneri***
6. 阳基侧突腹支细长而直 ······ 54. 谢德双角蝎蛉 ***Dicerapanorpa tjederi***
   阳基侧突腹支短钩状 ······ 50. 基明斯双角蝎蛉 ***Dicerapanorpa kimminsi***

Key to species of *Dicerapanorpa* in this atlas

(7 species, males)

1. Wings yellowish with conspicuous markings; rostrum darkened laterally but without distinct longitudinal stripes ··· 2
   Wings hyaline with greatly reduced markings; rostrum darkened laterally to form a pair of dark brown longitudinal stripes ······ 4
2. Forewing nearly without markings; paramere with a straight mesal branch ······ 52. ***Dicerapanorpa shennongensis***
   Forewing with a distinct marginal spot; paramere with a convergently curved mesal branch ······ 3
3. Paramere with a short and hook-like ventral branch; median branches of paramere crossed mesally ······ 51. ***Dicerapanorpa magna***
   Paramere with a long and straight ventral branch; mesal branches of paramere not crossed mesally or slightly in contact apically ······ 48. ***Dicerapanorpa baiyunshana***
4. Hypovalve distinctly enlarged on apex ······ 5
   Hypovalve not enlarged on apex ······ 6
5. Hypovalve rounded on apex ······ 49. ***Dicerapanorpa diceras***
   Hypovalve truncated on apex ······ 53. ***Dicerapanorpa stotzneri***
6. Ventral branch of paramere slender and straight ······ 54. ***Dicerapanorpa tjederi***
   Ventral branch of paramere short and hook-like ······ 50. ***Dicerapanorpa kimminsi***

48. 白云山双角蝎蛉 *Dicerapanorpa baiyunshana* Zhong & Hua, 2013

【测量】 前翅长 16.5 ～ 17.5 mm。

【形态特征】 头顶黄色，单眼三角区黑色。喙黄褐色。前胸背板黄褐色。胸部背板两侧黑褐色，中央具 1 条宽阔的浅黄色纵带。侧板及足浅黄褐色。翅浅黄色，斑纹褐色且边缘较模糊。前翅端带中具 1 个大型透明斑；痣带的基支宽阔，端支较狭；基带完整，缘斑三角形，基斑不明显。雄性腹部 1 ～ 5 节背板大部分黄色，两侧边缘黑褐色。第 3 节背中突扁平，盖住第 4 节上尖锐的后背突。第 6 节末端具 2 个指状臀角，第 7 节基部强烈缢缩成柄状。雄性外生殖器阔卵形。第 9 节背板末端具深 U 形缺刻，两侧形成 1 对指形突起。下瓣长带状，内缘具 1 列排列不整齐的长鬃毛，端部不伸达生殖肢基节末端，向内稍弯曲。生殖刺突短于基节，具有发达的中齿和亚基齿，其中亚基齿末端分叉。阳基侧突分 3 叉；腹支长直；中支不交叉或仅在端部稍接触，伸达生殖刺突的中齿；背支仅伸达生殖肢基节端部。阳茎短小，不超过生殖肢基节的端部。

【发生期】 6 ～ 8 月。

【地理分布】 河南（白云山）。

图 3-48-1 白云山双角蝎蛉 ***Dicerapanorpa baiyunshana*** 雄成虫背面观（河南嵩县白云山，NWAFU）

Fig. 3-48-1 Male adult of ***Dicerapanorpa baiyunshana*** , dorsal view (Mount Baiyun, Songxian, Henan, NWAFU)

Measurements: Forewing length 16.5–17.5 mm.

Morphology: Vertex yellow. Ocellar triangle black. Rostrum yellowish brown. Pronotum yellowish brown. Nota dark brown laterally with a broad yellow mesal stripe. Pleura and legs light yellowish brown. Wing membrane hyaline and slightly tinged with yellow, with markings brown and blurred marginally. Apical band broad and with a large hyaline window; pterostigmal band with a broad basal branch and a thin distal branch; basal band complete; marginal spot triangular; and basal spot inconspicuous. Male terga Ⅰ–Ⅴ mostly yellow with dark brown flanks. Notal organ on tergum Ⅲ flat and covering acute postnotal organ on tergum Ⅳ. A6 with a pair of finger-like anal horns on apical margin. A7 intensively stalked at base. Male genital bulb broad oval. Epandrium emarginate terminally in a broad U-shape, with a pair of finger-like processes laterally. Hypovalves stripe -like and bearing a row of sparsely and irregularly arranged long bristles along inner margin, with their apexes not reaching apex of gonocoxites and slightly curved mesad. Gonostylus shorter than gonocoxites and bearing an acute median tooth and a large, bifurcated subbasal process on inner margin. Paramere trifurcated: ventral branch long and straight; median branch of paramere not crossed mesally or slightly in contact apically and extending to median tooth of gonostylus; dorsal branch only extending to apex of gonocoxites. Aedeagus short, and not exceeding apex of gonocoxites.

图 3-48-2 白云山双角蝎蛉 ***Dicerapanorpa baiyunshana*** 雄性外生殖器腹面观

Fig. 3-48-2 Male genitalia of ***Dicerapanorpa baiyunshana***, ventral view

Emergence: June to August.

Distribution: Henan (Mount Baiyun).

图 3-48-3 白云山双角蝎蛉 ***Dicerapanorpa baiyunshana*** 地理分布

Fig. 3-48-3 Distribution of ***Dicerapanorpa baiyunshana***

49. 双角蝎蛉 *Dicerapanorpa diceras* (MacLachlan, 1894)

【测量】 前翅长 14.5 ～ 16.5 mm。

【形态特征】 头顶黄色，单眼三角区具 1 黑色大斑。喙黄色，两侧具明显的深褐色纵带。前胸背板黄色。中、后胸两侧黑褐色，中央具十分宽阔的黄色条带。侧板及足黄色。翅无色透明，无斑纹，仅端部稍加深。翅痣浅黄褐色。雄性腹部 1 ～ 5 节背板中央黄色，两侧边缘黑褐色。第 3 节背中突扁平，盖住第 4 节上尖锐的后背突。第 6 节黄褐色，末端具 2 个指状臀角；第 7 节基部强烈缢缩成柄状。雄性外生殖器宽卵形。第 9 节背板末端具深 U 形凹刻，两侧形成 1 对指形突起。下瓣长带状，内缘具 1 列黑色长鬃毛，端部圆阔，伸达生殖肢基节端部。生殖刺突短于基节，内缘具发达的中齿和亚基齿，其中亚基齿末端分叉。阳基侧突分 3 叉；腹支平直；中支、背支末端稍向内弯曲。阳茎短小，不超过生殖肢基节的端部；腹瓣膜质；背瓣具长毛。

【发生期】 5 ～ 6 月。

【地理分布】 四川（康定）。

Measurements: Forewing length 14.5 –16.5 mm.

Morphology: Vertex yellow. Ocellar triangle enclosed by a large black spot. Rostrum yellow, with a pair of conspicuous dark brown stripes laterally. Pronotum yellow. Mesonotum and metanotum dark brown laterally. Pleura and legs yellow. Wing membrane hyaline, and devoid of markings, except for apex slightly darkened. Pterostigma light yellowish brown. Male terga Ⅰ–Ⅴ mostly yellow with dark brown flanks.

图 3-49-1 双角蝎蛉 *Dicerapanorpa diceras* 雄成虫背面观（四川康定，NWAFU）

Fig. 3-49-1 Male adult of ***Dicerapanorpa diceras***, dorsal view (Kangding, Sichuan, NWAFU)

图 3-49-2 双角蝎蛉 ***Dicerapanorpa diceras*** 雄性外生殖器腹面观

Fig. 3-49-2 Male genitalia of ***Dicerapanorpa diceras***, ventral view

Notal organ on tergum Ⅲ flat and covering acute postnotal organ on tergum Ⅳ. A6 yellowish brown, bearing a pair of finger-like anal horns on dorsal apex. A7 intensively stalked at base. Male genital bulb broad oval. Epandrium emarginate terminally in a broad U-shape to form a pair of finger-like processes laterally. Hypovalves stripe-like and bearing a row of sparsely and irregularly arranged long bristles along inner margin, with their expanded rounded apexes extending to apex of gonocoxites. Gonostylus shorter than gonocoxites and bearing an acute median tooth and a large, bifurca-

图 3-49-3 双角蝎蛉 ***Dicerapanorpa diceras*** 地理分布

Fig. 3-49-3 Distribution of ***Dicerapanorpa diceras***

ted subbasal process on inner margin. Paramere rifurcated: ventral branch straight; median and dorsal branches curved mesad apically. Aedeagus short, not exceeding apex of gonocoxites, with membranous ventral valves, and bearing several long setae on dorsal valves.

Emergence: May to June.

Distribution: Sichuan (Kangding).

50. 基明斯双角蝎蛉 *Dicerapanorpa kimminsi* (Carpenter, 1948)

【测量】 前翅长 15.0 ~ 18.0 mm。

【形态特征】 头顶大部分黄色，单眼三角区被 1 个黑色大斑包围。喙黄色，两侧具深褐色纵带。前胸背板黄褐色。中、后胸背板两侧黑褐色，中央具 1 条宽阔的亮黄色纵带。侧板及足黄褐色。翅无色透明，几乎无斑纹，仅端部加深。翅痣浅黄褐色。雄性腹部 1 ~ 5 节背板大部分黄色，两侧边缘黑褐色。第 3 节背中突扁平，盖住第 4 节上尖锐的后背突。第 6 节黄褐色，末端具 2 个指状臀角。第 7 节基部强烈缢缩成柄状。雄性外生殖器宽卵形。第 9 节背板末端具深 U 形缺刻，两侧形成 1 对指形突起。下瓣长带状，内缘具 1 列黑色长鬃毛，端部钝圆，延伸至生殖肢基节端部。生殖刺突短于基节，内缘具发达尖锐的中齿和亚基齿。阳基侧突分 3 叉；腹支短小、钩状；中支近平直，伸达生殖刺突中齿；背支稍短，向内弯曲。阳茎短小，不超过生殖肢基节的端部。

【发生期】 5 ~ 7 月。

【地理分布】 四川（峨眉山）。

Measurements: Forewing length 15.0–18.0 mm.

Morphology: Vertex mostly yellow. Ocellar triangle enclosed by a large black spot. Rostrum yellow and with a pair of conspicuous dark brown stripes laterally. Pronotum yellowish brown. Mesonotum and metanotum dark brown laterally and bearing a broad

图 3-50-1 基明斯双角蝎蛉 *Dicerapanorpa kimminsi* 雄成虫背面观（四川峨眉山，NWAFU）

Fig. 3-50-1 Male adult of *Dicerapanorpa kimminsi*, dorsal view (Mount Emei, Sichuan, NWAFU)

图 3-50-2 基明斯双角蝎蛉 *Dicerapanorpa kimminsi* 雄性外生殖器腹面观

Fig. 3-50-2 Male genitalia of *Dicerapanorpa kimminsi*, ventral view

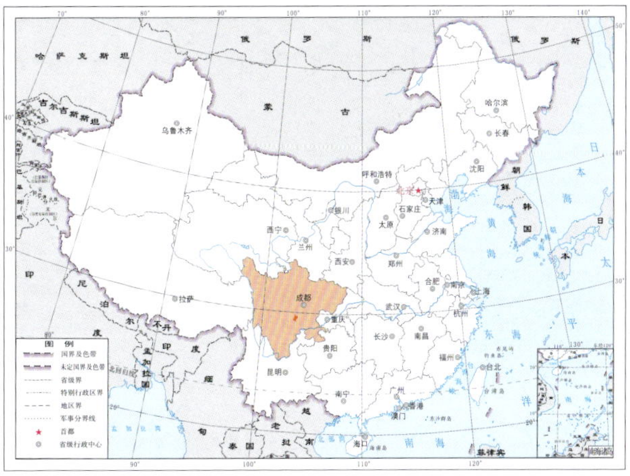

图 3-50-3 基明斯双角蝎蛉 *Dicerapanorpa kimminsi* 地理分布

Fig. 3-50-3 Distribution of *Dicerapanorpa kimminsi*

yellow mesal stripe. Pleura and legs yellowish brown. Wing membrane hyaline, and devoid of markings, except for apex slightly darkened. Pterostigma light yellowish brown. Male terga Ⅰ–Ⅴ mostly yellow with dark brown flanks. Notal organ on tergum Ⅲ flat and covering acute postnotal organ on tergum Ⅳ. A6 yellowish brown and bearing a pair of finger-like anal horns on dorsal apex. A7 intensively stalked at base. Male genital bulb broad oval. Epandrium emarginate terminally in a broad U-shape to form a pair of finger-like processes laterally. Hypovalves stripe-like and with a row of sparsely and irregularly arranged long bristles along inner margin, with their rounded apexes extending to apex of gonocoxites. Gonostylus shorter than gonocoxites and bearing an acute median tooth and a large subbasal process on inner margin. Paramere trifurcated: ventral branch short and hook-like; median branch slender and straight, extending to median tooth of gonostylus; dorsal branch short and curving mesad. Aedeagus short and not exceeding apex of gonocoxites.

Emergence: May to July.

Distribution: Sichuan (Mount Emei).

51. 大双角蝎蛉 *Dicerapanorpa magna* (Chou, 1981)

【测量】 前翅长 16.8 ~ 17.5 mm。

【形态特征】 头顶大部分黄色，单眼三角区被1个黑色大斑包围。喙黄褐色。前胸背板黄色。中、后胸背板两侧黑褐色，中央具1条宽阔的浅黄色纵带。侧板及足黄褐色。翅浅黄色，斑纹褐色且边缘较模糊。前翅端带中具透明斑，后侧有1个分离的狭斑；痣带的基支阔，端支较狭；基带完整，缘斑小，基斑不明显。雄性腹部1 ~ 5节背板大部分黄白色，两侧边缘黑褐色。第3节背中突扁平，盖住第4节上尖锐的后背突。第6节黄褐色，筒形，末端具2个指状臀角；第7节基部强烈缢缩成柄状。雄性外生殖器卵形。第9节背板末端具深U形缺刻，两侧形成1对指形突起。下瓣长带状，内缘具1列黑色长鬃毛，端部钝圆，延伸至生殖肢基节端部。生殖刺突短于基节，内缘具发达的尖锐中齿和亚基齿，其中亚基齿末端分叉。阳基侧突分3叉：

图 3-51-3 大双角蝎蛉 *Dicerapanorpa magna* 雄性腹部末端侧背面观

Fig. 3-51-3 Male teminal abdomen of *Dicerapanorpa magna*, dorsal view

图 3-51-2 大双角蝎蛉 *Dicerapanorpa magna* 雄性外生殖器腹面观

Fig. 3-51-2 Male genitalia of *Dicerapanorpa magna*, ventral view

图 3-51-4 大双角蝎蛉 *Dicerapanorpa magna* 地理分布

Fig. 3-51-4 Distribution of *Dicerapanorpa magna*

图 3-51-1 大双角蝎蛉 ***Dicerapanorpa magna*** 成虫侧背面观（陕西鄠邑朱雀森林公园）

1. 雄虫 2. 雌虫

Fig. 3-51-1 Adults of ***Dicerapanorpa magna***, dorso-lateral view (Zhuque Forest Park, Huyi, Shaanxi)

1. Male 2. Female

腹支短小、钩状，中支长且弯曲并左右交叉，背支略短于中支。阳茎短小，不超过生殖肢基节的端部，背瓣具长毛。

【发生期】 5 ~ 8 月。

【地理分布】 陕西 [宁陕、鄠邑、周至、眉县、太白、宝鸡（天台山）]、甘肃（文县）。

Measurements: Forewing length 16.8–17.5 mm.

Morphology: Vertex mostly yellow. Ocellar triangle enclosed by a large black spot. Rostrum yellowish brown. Pronotum yellowish brown. Mesonotum and metanotum dark brown laterally and bearing a broad yellow mesal stripe. Pleura and legs yellowish brown. Wing membrane tinged with yellow, with markings brown and blurred marginally. In forewings, apical band broad, with a large hyaline window innerly, and a separated slender spot posteriorly; pterostigmal band with a broad basal branch and a thin distal branch; basal band complete; marginal and basal spots small. Male terga Ⅰ–Ⅴ mostly yellowish white with dark brown flanks. Notal organ on tergum Ⅲ flat and covering acute postnotal organ on tergum Ⅳ. A6 yellowish brown, cylindrical with a pair of finger-like anal horns on dorsal apex. A7 intensively stalked at base.Male genital bulb broad oval. Epandrium emarginate terminally in a broad U-shape and forming a pair of finger-like processes laterally. Hypovalves stripe-like and bearing a row of sparsely and irregularly arranged long bristles along inner margin, with their rounded apexes extending to apex of gonocoxites. Gonostylus shorter than gonocoxites and bearing an acute median tooth and a large, bifurcated subbasal process on inner margin. Paramere trifurcated: ventral branch short and hook-like; median branch long slender, curving mesad and crossed messaly; dorsal branch shightly shorter than median branch. Aedeagus short and not exceeding apex of gonocoxites, with long setae on dorsal valves.

Emergence: May to August.

Distribution: Shaanxi [Ningshan, Huyi, Zhouzhi, Meixian, Taibai, Baoji (Mount Tiantai)] and Gansu (Wenxian).

52. 神农双角蝎蛉 *Dicerapanorpa shennongensis* Zhong & Hua, 2013

【测量】 前翅长 17.0 ~ 17.5 mm。

【形态特征】 头大部分黄色，单眼三角区被 1 个黑色大斑包围。喙黄褐色。前胸背板黄褐色。中、后胸背板两侧黑褐色，中央具 1 条宽阔的浅黄色纵带。侧板及足黄褐色。翅浅黄色、透明，斑纹褐色且边缘较模糊。前翅端带阔，具透明斑；痣带的基支阔，端支较狭；基带不完整或退化，前后不与翅缘相接；缘斑、基斑退化，不明显。雄性腹部 1 ~ 5 节背板大部分浅黄色，两侧边缘黑褐色。第 3 节背中突扁平，盖住第 4 节上尖锐的后背突。第 6 节黄

图 3-52-2 神农双角蝎蛉 *Dicerapanorpa shennongensis* 雄性外生殖器腹面观

Fig. 3-52-2 Male genitalia of *Dicerapanorpa shennongensis*, ventral view

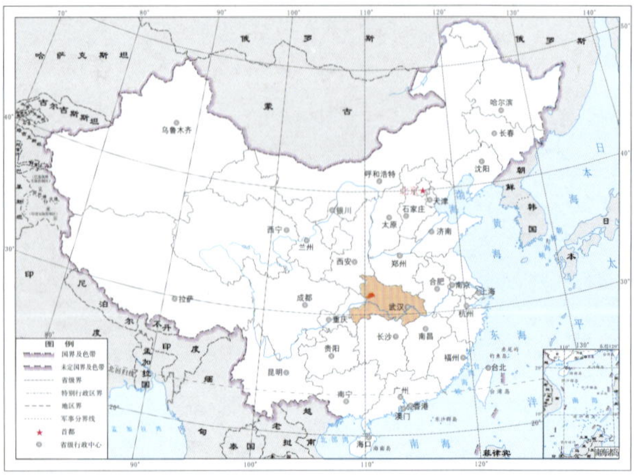

图 3-52-3 神农双角蝎蛉 *Dicerapanorpa shennongensis* 地理分布

Fig. 3-52-3 Distribution of *Dicerapanorpa shennongensis*

图 3-52-1 神农双角蝎蛉 ***Dicerapanorpa shennongensis*** 成虫背面观（湖北神农架）

1. 雄虫 2. 雌虫

Fig. 3-52-1 Adults of ***Dicerapanorpa shennongensis***, dorsal view (Shennongjia, Hubei)

1. Male 2. Female

褐色，末端具 2 个指状臀角；第 7 节基部强烈缢缩成柄状。雄性外生殖器卵形。第 9 节背板末端具深 U 形缺刻，两侧形成 1 对指形突起。下瓣长带状，内缘具 1 列黑色长鬃毛，端部钝圆，延伸至生殖肢基节端部。生殖刺突短于基节，内缘具发达的中齿和亚基齿，其中亚基齿末端分叉。阳基侧突分 3 叉，其腹支较短，基部弯曲；中支长，相对较直；背支略短且弯曲。阳茎短，不超出生殖肢基节端部。

【发生期】 6 ~ 8 月。

【地理分布】 湖北（神农架）。

Measurements: Forewing length 17.0–17.5 mm.

Morphology: Vertex mostly yellow. Ocellar triangle black. Rostrum yellowish brown. Pronotum yellowish brown. Mesonotum and metanotum dark brown laterally and bearing a broad light yellow stripe mesally. Pleura and legs yellowish brown. Wing membrane hyaline and slightly tinged with pale yellow, with markings brown and blurred marginally. Apical band broad, with a large hyaline window spot; pterostigmal band with a broad basal branch and a thin distal branch; basal band incomplete or reduced, and disconnected with wing margins anteriorly and posteriorly; marginal and basal spots reduced and inconspicuous. Male terga Ⅰ–Ⅴ mostly pale yellow with dark brown flanks. Notal organ on tergum Ⅲ flat and covering acute postnotal organ on tergum Ⅳ. A6 yellowish brown with a pair of finger-like anal horns on dorsal apex. A7 intensively stalked at base. Male genital bulb broad oval. Epandrium emarginate terminally in a broad U-shape and forming a pair of finger-like processes laterally. Hypovalves stripe-like and bearing a row of sparsely and irregularly arranged long bristles along inner margin, with their rounded apexes extending to apex of gonocoxites. Gonostylus shorter than gonocoxites and bearing an acute median tooth and a large, bifurcated subbasal process on inner margin. Paramere trifurcated: ventral branch short and curved basally; median branch long and relatively straight; and dorsal branch slightly shorter and curved. Aedeagus short and not exceeding apex of gonocoxites.

Emergence: June to August.

Distribution: Hubei (Shennongjia).

53. 西康双角蝎蛉 *Dicerapanorpa stotzneri* (Esben-Petersen, 1934)

【测量】 前翅长 17.0 ~ 17.5 mm。

【形态特征】 头大部分黄色，单眼三角区被 1 个黑色大斑包围，喙黄褐色，具两条黑色纵带。前胸背板黄褐色。中、后胸背板黑褐色，中央具 1 条宽阔的浅黄色纵带。侧板及足黄褐色。翅近无色，

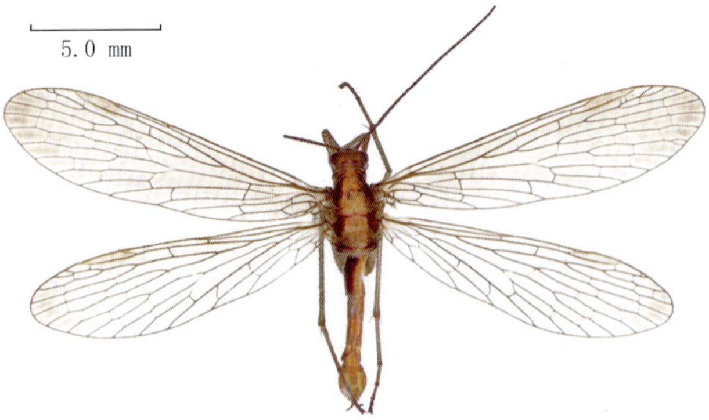

图 3-53-1 西康双角蝎蛉 *Dicerapanorpa stotzneri* 雌成虫背面观（四川康定，NWAFU）

Fig. 3-53-1 Female adult of *Dicerapanorpa stotzneri*, dorsal view (Kangding, Sichuan, NWAFU)

图 3-53-2 西康双角蝎蛉 *Dicerapanorpa stotzneri* 雄性外生殖器腹面观

Fig. 3-53-2 Male genitalia of *Dicerapanorpa stotzneri*, ventral view

无明显的斑纹，仅端部颜色稍加深。雄性腹部 1 ~ 5 节背板大部分黄色，两侧边缘黑褐色。第 3 节背中突扁平，盖住第 4 节上尖锐的后背突。第 6 节黄褐色，末端具 2 个指状臀角；第 7 节基部强烈缢缩成柄状。雄性外生殖器卵形。第 9 节背板末端具深 U 形缺刻，两侧形成 1 对指形突起。下瓣长带状，内缘具 1 列稀疏且不规则排列的黑色长鬃毛，末端宽阔且平截，延伸至生殖肢基节端部。生殖刺突短于基节，内缘具发达的中齿和亚基齿。阳基侧突分 3 叉，其腹支几乎伸达阳茎末端，中支、背支超过生殖肢基节端部，向内弯曲。阳茎短小，不超过生殖肢基节端部。

【发生期】 6 ~ 8 月。

【地理分布】 四川（都江堰、康定）。

图 3-52-3　西康双角蝎蛉 ***Dicerapanorpa stotzneri*** 地理分布

Fig. 3-52-3　Distribution of ***Dicerapanorpa stotzneri***

Measurements: Forewing length 17.0–17.5 mm.

Morphology: Vertex mostly yellow. Ocellar triangle enclosed by a large black spot. Rostrum yellowish brown, with two black stripes laterally. Pronotum yellowish brown. Mesonotum and metanotum dark brown laterally and bearing a broad light yellow stripe mesally. Pleura and legs yellowish brown. Wing membrane hyaline nearly colorless, and devoid of markings except for slight tinge on apex. Male terga Ⅰ–Ⅴ mostly yellow with dark brown flanks. Notal organ on tergum Ⅲ flat and covering acute postnotal organ on tergum Ⅳ. A6 yellowish brown, bearing a pair of finger-like anal horns on dorsal apex. A7 intensively stalked at base. Male genital bulb broad oval. Epandrium emarginate terminally in a broad U-shape and forming a pair of finger-like processes laterally. Hypovalves stripe-like, bearing a row of sparsely and irregularly arranged long bristles along inner margin, with their expanded and truncated apexes extending to apex of gonocoxites. Gonostylus shorter than gonocoxites, bearing an acute median tooth and a large subbasal process on inner margin. Paramere trifurcated: ventral branch almost extending to apex of aedeagus; median and dorsal branches exceeding apex of gonocoxites and curved mesad apically. Aedeagus short and not exceeding apex of gonocoxites.

Emergence: June to August.

Distribution: Sichuan (Dujiangyan, Kangding).

54. 谢德双角蝎蛉 *Dicerapanorpa tjederi* (Carpenter, 1938)

【测量】 前翅长 15.0 ~ 16.0 mm。

【形态特征】 头大部分黄色，单眼三角区被 1 个黑色大斑包围，喙黄褐色，前方具 2 条黑色纵带。前胸背板黄褐色。中、后胸背板两侧黑褐色，中央具 1 条宽阔的浅黄色纵带。侧板及足黄褐色。翅无色透明，无明显斑纹，仅端部稍加深。雄性腹部 1 ~ 5 节背板大部分浅黄色，两侧边缘黑褐色。第 3 节背中突扁平，盖住第 4 节上尖锐的后背突。第 6 节黄褐色，末端具 2 个指状臀角；第 7 节基部强烈缢缩成柄状。雄性外生殖器卵形。第 9 节背板末端仅具浅 U 形缺刻，两侧形成 1 对指形突起。下瓣长带状，内缘具 1 列黑色长鬃毛，端部钝圆，延

图 3-54-1　谢德双角蝎蛉 ***Dicerapanorpa tjederi*** 雄成虫背面观（云南丽江玉龙雪山，NWAFU）

Fig. 3-54-1　Male adult of ***Dicerapanorpa tjederi***, dorsal view (Yulong Snow Mountain, Lijiang, Yunnan, NWAFU)

图 3-54-2 谢德双角蝎蛉 ***Dicerapanorpa tjederi*** 雄性外生殖器腹面观

Fig. 3-54-2 Male genitalia of ***Dicerapanorpa tjederi***, ventral view

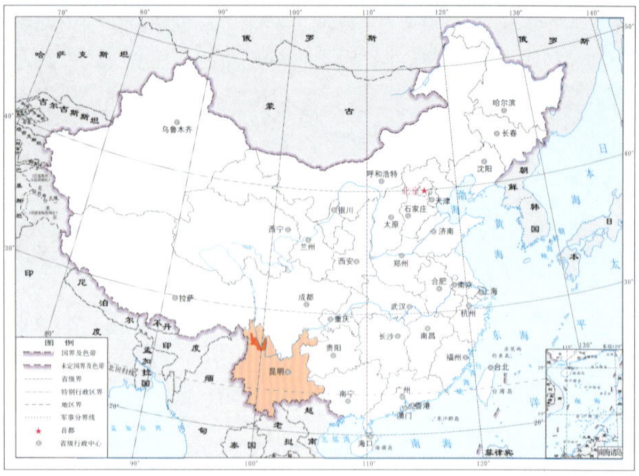

图 3-54-3 谢德双角蝎蛉 ***Dicerapanorpa tjederi*** 地理分布

Fig. 3-54-3 Distribution of ***Dicerapanorpa tjederi***

伸至生殖肢基节端部。生殖刺突短于基节，内缘具发达的中齿和大型亚基齿。阳基侧突分 3 叉，其腹支几乎伸达阳茎末端，中支、背支较平直，伸达生殖刺突的中齿。阳茎短小，不超出生殖肢基节端部。

【发生期】 6 ~ 8 月。

【地理分布】 云南（丽江、维西）。

Measurements: Forewing length 15.0–16.0 mm.

Morphology: Vertex mostly yellow. Ocellar triangle enclosed by a large black spot. Rostrum yellowish brown, with two dark brown stripes frontally. Pronotum yellowish brown. Mesonotum and metanotum dark brown laterally and bearing a broad light yellow stripe mesally. Pleura and legs yellowish brown. Wing membrane hyaline and colorless, devoid of instinct markings. Male terga I–V mostly pale yellow with dark brown flanks. Notal organ on tergum III flat and covering acute postnotal organ on tergum IV. A6 yellowish brown and bearing a pair of finger-like anal horns on dorsal apex. A7 intensively stalked at base. Male genital bulb broad oval. Epandrium emarginate terminally in a shallow U-shape and forming a pair of finger-like processes laterally. Hypovalves stripe-like and bearing a row of sparsely and irregularly arranged long bristles along inner margin, with their rounded apexes extending to apex of gonocoxites. Gonostylus shorter than gonocoxites and bearing an acute median tooth and a large subbasal process on inner margin. Paramere trifurcated: ventral branch straight and exceeding apex of aedeagus; median and dorsal branches long and straight, extending to median teeth of gonostylus. Aedeagus short and not exceeding apex of gonocoxites.

Emergence: June to August.

Distribution: Yunnan (Lijiang, Weixi).

六、叉蝎蛉属 *Furcatopanorpa* Ma & Hua, 2011

叉蝎蛉属 *Furcatopanorpa* 昆虫雄性腹部第 3 节背中突和第 4 节后背突完全消失，第 6 ~ 8 节不像大多数蝎蛉那样延长。雌性生殖板中轴后方相当延长，并在末端分为 2 叉。叉蝎蛉在停歇时，翅呈屋脊状叠放于体背。它们较短的腹部时常被翅盖住，因此从背面看去较难分辨雌雄（图 3-i）。

叉蝎蛉属 *Furcatopanorpa* 全世界仅 1 种，中国记录 1 种，本图鉴收录 1 种。

The notal organ on tergum Ⅲ and the postnotal organ on tergum Ⅳ in males are completely lost. The A6–A8 not elongated as the common scorpionflies. The females have the axis of genital plate extremely elongated posteriorly, and bifurcated terminally. The wings are held roof-like at repose. Their relatively short abdomens are often covered by the wings. Therefore, their genders are hard to determine from dorsal aspect (Fig. 3-i).

A monotypic genus.

图 3-i 长瓣叉蝎蛉 *Furcatopanorpa longihypovalva* 雌成虫背面观（徐晗摄）

Fig. 3-i Female adult of *Furcatopanorpa longihypovalva*, dorsal view (photo by Han Xu)

55. 长瓣叉蝎蛉 *Furcatopanorpa longihypovalva* (Hua & Cai, 2009)

【测量】 前翅长 15.5 ~ 18.5 mm。

【形态特征】 头顶黑色，单眼三角区黑色。喙黄褐色。胸部背板黑色，侧板及足浅灰褐色。翅无色透明，休息时呈屋脊状叠于体背，斑纹褐色，翅

图 3-55-1 长瓣叉蝎蛉 *Furcatopanorpa longihypovalva* 雄成虫侧面观（陕西宝鸡天台山国家森林公园）

Fig. 3-55-1 Male adult of *Furcatopanorpa longihypovalva*, dorsal view (Tiantaishan National Forest Park, Baoji, Shaanxi)

图 3-55-2 长瓣叉蝎蛉 *Furcatopanorpa longihypovalva* 雄性外生殖器腹面观

Fig. 3-55-2 Male genitalia of *Furcatopanorpa longihypovalva*, ventral view

痣褐色。前翅端带较窄；痣带宽阔，无端支；无缘斑、基斑；从 CuP 末端起向翅基部，沿后缘具 1 条不规则长斑。雄性腹部 1 ~ 5 节背、腹板黑色；第 3 节无背中突；第 4 节无后背突。第 6 节短筒形，短于第 5 节，无臀角，第 7、8 节短于第 6 节，基部不缢缩。雄性外生殖器长椭圆形。第 9 节背板末端具浅缺刻。下瓣窄长，明显超过生殖肢基节的端部。生殖刺突短于生殖基节之半，内缘的亚基齿大，中齿具锯齿状边缘。阳基侧突结构复杂：近基部分成两支，其中腹支细长弯曲；背支末端扩大盖住阳茎。阳茎较短，腹瓣膜质；背瓣指状，分离。

【发生期】 5 ~ 8 月。

【地理分布】 陕西（宝鸡天台山、宁陕、太白、周至、眉县、留坝）、四川（平武）、甘肃（文县）、湖北（神农架）。

Measurements: Forewing length 15.5–18.5 mm.

Morphology: Vertex black. Ocellar triangle black. Rostrum yellowish brown. Nota black. Pleura and legs light greyish brown. Wing membrane hyaline, with a brown pterostigma. Wings held roof-like over body at repose. Markings brown. In forewings, apical band narrow; pterostigmal band broad but without a distal branch; marginal and basal spots absent; an irregular stripe extending form end of CuP to wing base along hind margin. Male terga and sterna Ⅰ–Ⅴ black. Tergu Ⅲ without a notal organ. Tergum Ⅳ without a postnotal organ. A6 shorter than A5, cylindrical and devoid of anal horns. A7 and A8 shorter than A6 and not constricted basally. Male genital bulb long elliptical. Epandrium slightly emarginate on apical margin. Hypovalves narrow and very long, extending beyond apex of gonocoxites. Gonostylus shorter than half length of gonocoxites, with a large subbasal process and a mesal tooth surrounded by a serrate margin. Paramere complicated in structure: bifurcated subbasally; ventral branch slender and curved; dorsal branch enlarged distally and enclosing aedeagus. Aedeagus short, with its ventral valves membranous, and dorsal valves finger-like and divergent.

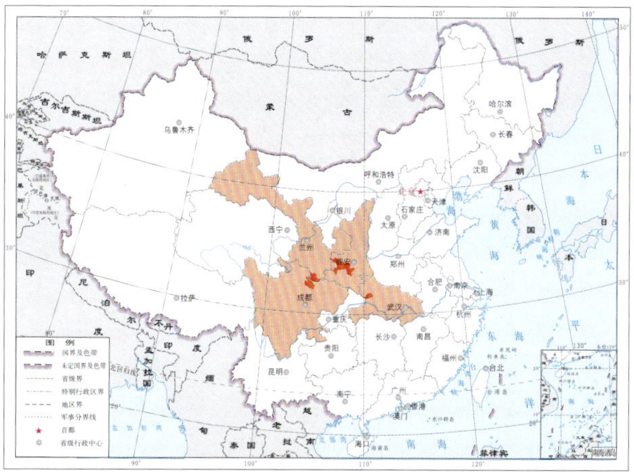

图 3-55-3 长瓣叉蝎蛉 ***Furcatopanorpa longihypovalva*** 地理分布

Fig. 3-55-3 Distribution of ***Furcatopanorpa longihypovalva***

Emergence: May to August.

Distribution: Shaanxi [Baoji (Mount Tiantai), Ningshan, Taibai, Zhouzhi, Meixian, Liuba], Sichuan (Pingwu), Gansu (Wenxian) and Hubei (Shennongjia).

七、新蝎蛉属 *Neopanorpa* van der Weele, 1909

相对于中国其他的属，新蝎蛉属 *Neopanorpa* 喙和翅都较为狭长。
全世界共 170 余种，中国已记录 100 余种，本图鉴收录 49 种。

The genus *Neopanorpa* has relatively longer rostrums and wings compared with other Chinese genera.

More than 170 species have been described worldwide. Over 100 species are recorded from China. Forty-nine species are included in this atlas.

图 3-j　新蝎蛉 ***Neopanorpa* sp.** 雄成虫背面观（雷波摄）

Fig. 3-j　Male adult of ***Neopanorpa* sp.**, dorsal view (photo by Bo Lei)

本图鉴新蝎蛉属 *Neopanorpa* 分种检索表

（49 种，雄性）

1. 腹部第 3 节背板后缘的背中突极其发达，超过第 4 节背板的末端 …………………… 2
   腹部第 3 节背板后缘的背中突不超过第 4 节背板末端 …………………… 11
2. 腹部远长于翅；第 6 节长约为第 5 节的 2 倍 …………………… 63. 契氏新蝎蛉 ***Neopanorpa chillcotti***
   腹部仅稍长于翅；第 6 节长不超过第 5 节的 2 倍 …………………… 3
3. 背中突极长，伸达第 7 节末端；第 4 节背板长约为第 3 节的 2 倍 ………… 64. 周氏新蝎蛉 ***Neopanorpa choui***
   背中突不伸达第 7 节末端；第 4 节背板约与第 3 节等长 …………………… 4
4. 背中突超过第 6 节末端 …………………… 5
   背中突不超过第 5 节末端 …………………… 6
5. 无翅斑；腹部 1 ~ 5 节背板及 6 ~ 8 节全为黑色 …………… 86. 莫干山新蝎蛉 ***Neopanorpa moganshanensis***
   翅斑发达；腹部 1 ~ 5 节背板向后从黑色渐变为黄褐色；6 ~ 8 节黄褐色 …………………… 97. 绍德新蝎蛉 ***Neopanorpa sauteri***
6. 第 9 节腹板具极短的基柄 …………………… 74. 胡氏新蝎蛉 ***Neopanorpa hushengchangi***
   第 9 节腹板具阔长的基柄 …………………… 7
7. 下瓣内缘近基部具 1 个尖锐的指形突起 …………………… 8
   下瓣内缘无上述突起，或仅具不明显突起 …………………… 9
8. 下瓣细长，两瓣不接触 …………………… 75. 鸡公山新蝎蛉 ***Neopanorpa jigongshanensis***
   下瓣宽阔，两瓣内缘相互重叠 …………………… 80. 河南新蝎蛉 ***Neopanorpa longiprocessa***
9. 翅无缘斑、基带 …………………… 10
   翅具发达的缘斑和基带 …………………… 11
10. 端带和痣带发达；痣带具完整的基支和端支；腹部 7 ~ 9 节黄褐色 …………………… 100. 西藏新蝎蛉 ***Neopanorpa tibetensis***
    端带和痣带退化；痣带仅为前缘的 1 个大斑；腹部全黑色 …………………… 98. 具毛新蝎蛉 ***Neopanorpa setigera***
11. 翅端带与痣带之间具 1 条细长的横带；下瓣窄长 …………………… 96. 网纹新蝎蛉 ***Neopanorpa retina***
    翅端带与痣带之间仅在后缘具 1 条分离的短横带；下瓣宽、近矩形 …………………… 84. 莽山新蝎蛉 ***Neopanorpa mangshanensis***
12. 无翅斑 …………………… 13
    有发达的翅斑 …………………… 21
13. 第 3 节背板后缘在背中突两侧有 1 对小突起；背中突下方还有 1 个尖锐的小突起 …………………… 14
    第 3 节背板无上述突起 …………………… 15
14. 腹部第 6 节长筒形；第 7 节约与第 6 节等长；第 9 节背板末端形成 1 对向中部弯曲的尖锐突起 …………………… 58. 布里斯新蝎蛉 ***Neopanorpa brisi***
    腹部第 6 节背板末端呈片状向上隆起；第 7 节长约为第 6 节之半；第 9 节背板末端向两侧呈角状突出 …………………… 92. 垂齿新蝎蛉 ***Neopanorpa pendula***
15. 外生殖器黑褐色至黑色，下瓣基半部黑褐色，端半部灰白色 …………………… 16
    外生殖器黄褐色，下瓣颜色均一 …………………… 18
16. 腹部第 7、8 节基半部黄白色，端半部黑褐色；第 9 节腹板的基柄伸达生殖肢基节的中部 …………………… 78. 李氏新蝎蛉 ***Neopanorpa lifashengi***
    腹部第 7、8 节全为黑色；第 9 节腹板的基柄较短，不伸达生殖肢基节中部 …………………… 17
17. 下瓣末端向中部稍弯曲；生殖刺突内缘具大型柱状亚基齿 …………………… 89. 黑色新蝎蛉 ***Neopanorpa nigritis***
    下瓣末端向腹面卷曲；生殖刺突内缘的亚基齿圆钝 …………………… 91. 淡瓣新蝎蛉 ***Neopanorpa pallivalva***
18. 下瓣内缘互相重叠 …………………… 18
    下瓣内缘仅相接触而不重叠 …………………… 19
19. 下瓣末端较圆钝；生殖刺突内缘的亚基齿钝圆 …………………… 81. 路氏新蝎蛉 ***Neopanorpa lui***
    下瓣末端较平截；生殖刺突内缘的亚基齿尖 …………………… 66. 明翅新蝎蛉 ***Neopanorpa claripennis***
20. 下瓣子叶状，端半部比基半部宽；两瓣在端部稍分歧 …………………… 95. 净翅新蝎蛉 ***Neopanorpa puripennis***
    下瓣端半部与基半部几乎等宽；两瓣平行 …………………… 85. 小新蝎蛉 ***Neopanorpa minuta***
21. 下瓣短于第 9 节腹板基柄宽度，向内呈钩状弯曲 …………………… 57. 短瓣新蝎蛉 ***Neopanorpa brevivalvae***

下瓣长于第 9 节腹板基柄宽度 ……………………………………………………………………………………22
22. 腹部第 6 节长远超过第 5 节的 2 倍 …………………………………………………………………………23
腹部第 6 节长不超过第 5 节的 2 倍 ……………………………………………………………………………24
23. 翅缘斑与痣带相连，形成 K 形斑纹；下瓣内缘近基部具 1 个尖锐的突起 ……………………………………………… 76. K 纹新蝎蛉 *Neopanorpa k-maculata*
翅缘斑小，不与痣带相连；下瓣内缘无上述突起 ………………………… 83. 大新蝎蛉 *Neopanorpa magna*
24. 翅面超过一半的面积被斑纹覆盖 ……………………………… 71. 哈曼德新蝎蛉 *Neopanorpa harmandi*
翅面小于一半的面积被斑纹覆盖 ………………………………………………………………………………25
25. 生殖刺突短于基节之半 …………………………………………………………………………………………26
生殖刺突长于基节之半 …………………………………………………………………………………………27
26. 外生殖器长椭圆形；第 9 节背板末端缺刻 …………………………… 65. 显斑新蝎蛉 *Neopanorpa clara*
外生殖器阔卵形；第 9 节背板末端两侧强烈突出 ……………………68. 方祥新蝎蛉 *Neopanorpa fangxianga*
27. 第 9 节腹板基柄极短，下瓣长带状 ………………………………… 82. 马氏新蝎蛉 *Neopanorpa maai*
第 9 节腹板具阔长的基柄，下瓣形状多样 ……………………………………………………………………28
28. 生殖刺突内缘有 1 个大型碗状亚基齿，上生长毛 ……………… 102. 天平山新蝎蛉 *Neopanorpa tienpingshana*
生殖刺突内缘的亚基齿不如上述 ………………………………………………………………………………29
29. 下瓣纤细，长超过宽的 5 倍 ……………………………………………………………………………………30
下瓣较阔，长不达宽的 5 倍 ……………………………………………………………………………………31
30. 前翅端带内侧有 1 条分离的短带 ……………………………… 72. 华氏新蝎蛉 *Neopanorpa hualizhongi*
前翅端带内侧有 1 条不分离的短带 …………………………… 70. 海南新蝎蛉 *Neopanorpa hainanica*
31. 下瓣内侧不重叠，或仅相互接触 ………………………………………………………………………………32
下瓣内侧多少相互重叠 …………………………………………………………………………………………36
32. 下瓣向末端渐尖 …………………………………………………………………………………………………33
下瓣端部圆钝或平截 ……………………………………………………………………………………………35
33. 下瓣基部最宽，呈锐角三角形 …………………………………… 67. 尖齿新蝎蛉 *Neopanorpa cuspidata*
下瓣基部稍缢缩，端部渐窄，中部最宽 ………………………………………………………………………34
34. 第 9 节腹板基柄约与下瓣等长 ……………………………………… 93. 璧尔新蝎蛉 *Neopanorpa pielina*
第 9 节腹板基柄明显短于下瓣 ……………………………………… 87. 异新蝎蛉 *Neopanorpa mutabilis*
35. 翅斑黄褐色较淡；雄性下瓣末端近平截，两瓣在基部距离较远 …… 103. 透翅新蝎蛉 *Neopanorpa translucida*
翅斑黑褐色明显；雄性下瓣末端圆钝并向腹面翻卷，两瓣在基部距离较近 ………………………………………… 77. 雷公山新蝎蛉 *Neopanorpa leigongshana*
36. 下瓣外缘向背面卷曲，呈长筒状 ………………………………………………………………………………37
下瓣其他形状 ……………………………………………………………………………………………………42
37. 前、后翅端带内各具 1 个大型透明斑 …………………………… 69. 台湾新蝎蛉 *Neopanorpa formosana*
前、后翅端带内无透明斑 ………………………………………………………………………………………38
38. 前翅端带内缘具 1 个向内突出的钝角 …………………………………………………………………………39
前翅端带内缘平直 ………………………………………………………………………………………………41
39. 前翅端带内缘向内突出的尖角，长度约等于端带的宽度…………………60. 卡氏新蝎蛉 *Neopanorpa cavaleriei*
前翅端带内缘向内突出的角短而钝，长度短于端带的宽度……………………………………………… 40
40. 生殖刺突的亚基齿有 1 个尖锐的齿突 ………………………… 88. 尼尔森新蝎蛉 *Neopanorpa nielseni*
生殖刺突内缘的亚基齿无上述齿突 …………………………… 56. 锚形新蝎蛉 *Neopanorpa anchoroides*
41. 胸部背板黑褐色，仅两侧角稍浅 ……………………………… 73. 湖南新蝎蛉 *Neopanorpa hunanensis*
胸部背板两侧黄褐色，中央具明显的黑色中带 ………………………… 94. 丽新蝎蛉 *Neopanorpa pulchra*
42. 腹部第 6 节长超过第 5 节的 1.7 倍；第 7、8 节约与第 6 节等长；生殖刺突约与基节等长 ……………………………………………………………………… 62. 爪新蝎蛉 *Neopanorpa chelata*
腹部第 6 节不及第 5 节 1.7 倍；第 7、8 节短于第 6 节；生殖刺突短于基节……………………………… 43
43. 缘斑细长，向前延伸与痣带的基支相连………………………………………………………………………… 44
缘斑不与痣带基支相连…………………………………………………………………………………………… 45
44. 翅斑网状：端带伸出 1 条细纹与痣带相连；缘斑向前伸出 1 条细纹与痣带相连，向后与基带相连 ……………………………………………………………………… 61. 网翅新蝎蛉 *Neopanorpa caveata*
端带后缘伸出的细纹不与痣带相连；缘斑不与基带相连………………59. 卡本特新蝎蛉 *Neopanorpa carpenteri*

45. 下瓣末端尖，内侧近基部具 1 个耳状突起 ………………………… 104. 杨氏新蝎蛉 *Neopanorpa youngi*
下瓣末端较钝，内侧近基部无上述突起 …………………………………………………………… 46
46. 翅缘斑和基带极度退化，仅为数个散点 ………………………………………………………… 47
翅缘斑和基带发达 …………………………………………………………………………………… 48
47. 生殖突内缘具发达的方形亚基齿 ……………………………… 79. 黎坪新蝎蛉 *Neopanorpa lipingensis*
生殖突内缘亚基齿大，片状 ……………………………………………… 99. 申氏新蝎蛉 *Neopanorpa sheni*
48. 前翅端带内缘形成 2 ～ 3 个透明斑；雄性下瓣向末端稍渐变窄 …… 90. 细纹新蝎蛉 *Neopanorpa ophthalmica*
前翅端带内缘具 1 ～ 2 个透明斑，或分离出 1 条短带；雄性下瓣末端圆钝，与基部约等宽 ……………………
……………………………………………………………………… 101. 天目山新蝎蛉 *Neopanorpa tienmushana*

Key to species of *Neopanorpa* in this atlas
(49 species, males)

1. Notal organ on hind border of tergum Ⅲ extremely developed, and exceeding hind border of tergum Ⅳ …………… 2
Notal organ on hind border of tergum Ⅲ not exceeding hind border of tergum Ⅳ ………………………………… 11
2. Abdomen much longer than wings; A6 approximately twice as long as A5 ………………… 63. *Neopanorpa chillcotti*
Abdomen slightly longer than wings; A6 shorter than twice length of A5 ………………………………………… 3
3. Notal organ extremely elongated, exceeding hind border of A7; tergum Ⅳ twice length of tergum Ⅲ ………………
……………………………………………………………………………………………… 64. *Neopanorpa choui*
Notal organ not exceeding hind border of A7; tergum Ⅳ as long as tergum Ⅲ ………………………………… 4
4. Notal organ exceeding hind border of A6 ……………………………………………………………………… 5
Notal organ not exceeding hind border of A5 ………………………………………………………………… 6
5. Wing markings absent; terga Ⅰ–Ⅴ and A6–A8 entirely black ……………………… 86. *Neopanorpa moganshanensis*
Wing markings well-developed; terga Ⅰ–Ⅴ gradually turned from black to yellowish brown caudad; A6–A8 yellowish brown ……………………………………………………………………………… 97. *Neopanorpa sauteri*
6. Hypandrium with a greatly shortened basal stalk ………………………………… 74. *Neopanorpa hushengchangi*
Hypandrium with a long broad basal stalk …………………………………………………………………… 7
7. Hypovalve with an acute finger-like process on inner margin subbasally ………………………………………… 8
Hypovalve lacking such process …………………………………………………………………………… 9
8. Hypovalves slender and divergent ……………………………………………… 75. *Neopanorpa jigongshanensis*
Hypovalves broad and overlapped mesally …………………………………………… 80. *Neopanorpa longiprocessa*
9. Wings lacking marginal spot and basal band ………………………………………………………………… 10
Wings with well-developed marginal spot and basal band ………………………………………………………… 11
10. Apical and pterostigmal bands well-developed; pterostigmal band with complete basal and distal branches; A7–A9 yellowish brown(Wing markings only represented by apical and pterostigmal bands) … 100. *Neopanorpa tibetensis*
Apical and pterostigmal bands reduced; pterostigmal band represented by a large spot on anterior margin; abdomen entirely black ……………………………………………………………………98. *Neopanorpa setigera*
11. Apical and pterostigmal bands connected by a slender band; hypovalves slender (Wings with a slender transverse band between apical and pterostigmal bands; hypovalves slender) ……………………………… 96. *Neopanorpa retina*
Apical and pterostigmal bands with a separated short transverse band on posterior margin; hypovalves broad and subtrapezoidal (Wings with a separated short transverse band on hind margin between apical and pterostigmal bands; hypovalves broad and subtrapezoidal) ……………………………………………………… 81. *Neopanorpa mangshanensis*
12. Wings devoid of markings ………………………………………………………………………………… 13
Wings with well-developed markings ……………………………………………………………………… 21
13. Tergum Ⅲ with a pair of small teeth aside notal organ on hind margin; a small acute tooth under notal organ ……………… 14
Tergum Ⅲ lacking such teeth …………………………………………………………………………… 15
14. A6 long cylindrical; A7 as long as A6; A9 forming a pair of acute processes curved mesad on apex …………………
……………………………………………………………………………………………… 58. *Neopanorpa brisi*
A6 projected dorsad and plate-like terminally; A7 half as long as A6; A9 acutely projected laterad on apex …………
……………………………………………………………………………………… 92. *Neopanorpa pendula*
15. Genitalia dark brown to black; hypovalves dark brown at basal half and greyish white on distal half ……………… 16
Genitalia yellowish brown; hypovalves uniformly colored …………………………………………………… 18

16. A7 and A8 yellowish white at basal half, and dark brown at distal half; basal stalk of hypandrium extending to middle of gonocoxites··················78. ***Neopanorpa lifashengi***
A7 and A8 black; basal stalk of hypandrium short, not extending to middle of gonocoxites··················17
17. Hypovalves slightly curved mesad on apexes; gonostylus with a large columnar subbasal process on inner margin. ··················89. ***Neopanorpa nigritis***
Hypovalves curved ventrad on apexes; gonostylus with a rounded subbasal process on inner margin ··················91. ***Neopanorpa pallivalva***
18. Hypovalves overlapped mesally ··················18
Hypovalves merely in contact but not overlapped mesally ··················19
19. Hypovalves rounded apically; gonostylus with a rounded subbasal process on inner margin ··················81. ***Neopanorpa lui***
Hypovalves truncated apically; gonostylus with an acute subbasal process on inner margin ··················66. ***Neopanorpa claripennis***
20. Hypovalves cotyledon-shaped and broader at distal portion than basal; two valves slightly divergent apically ··················95. ***Neopanorpa puripennis***
Hypovalves equal in width at basal and distal portions; two valves almost parallel ··················85. ***Neopanorpa minuta***
21. Hypovalves shorter than width of basal stalk of hypandrium, hook-like and curved mesad ··················57. ***Neopanorpa brevivalvae***
Hypovalves longer than width of basal stalk of hypandrium ··················22
22. A6 much longer than twice length of A5 ··················23
A6 not exceeding twice length of A5 ··················24
23. Marginal spot connected with pterostigmal band and forming a K-shaped pattern; hypovalve with an acute process on inner margin subbasally ··················76. ***Neopanorpa k-maculata***
Marginal spot small and not connected with pterostigmal band; hypovalve without such process ··················83. ***Neopanorpa magna***
24. Wings covered with dense markings over half its surface ··················71. ***Neopanorpa harmandi***
Wings covered with markings less than half its surface ··················25
25. Gonostylus shorter than half length of gonocoxites··················26
Gonostylus longer than half length of gonocoxites ··················27
26 .Genitalia long elliptical; epandrium slightly emarginate terminally ··················65. ***Neopanorpa clara***
Genitalia broad oval; epandrium strongly projected on lateral corners of apex ··················68. ***Neopanorpa fangxianga***
27. Hypandrium with a greatly shortened basal stalk; hypovalves long stripe-like ··················82. ***Neopanorpa maai***
Hypandrium with a long and broad basal stalk; hypovalves variously shaped ··················28
28. Gonostylus with a large bowl-shaped subbasal process bearing long setae on inner margin ··················102. ***Neopanorpa tienpingshana***
Gonostylus with a subbasal process not as above ··················29
29. Hypovalves slender, more that five times as long as wide ··················30
Hypovalves broad, not as five times as long as wide ··················31
30. Forewing with a separate short band on inner side of apical band ··················72. ***Neopanorpa hualizhongi***
Forewing with a short band connected to apical band··················70. ***Neopanorpa hainanica***
31. Hypovalves not overlapped, or in contact mesally ··················32
Hypovalves more or less overlapped mesally ··················36
32. Hypovalves tapering towards apexes ··················33
Hypovalves rounded or truncated apically ··················35
33. Hypovalves broadest at base, sharp triangular··················67. ***Neopanorpa cuspidata***
Hypovalves constricted basally and tapering towards apexes, broadest at middle ··················34
34. Basal stalk of tergum Ⅸ as long as hypovalves ··················93. ***Neopanorpa pielina***
Basal stalk of tergum Ⅸ distinctly shorter than hypovalves ··················87. ***Neopanorpa mutabilis***
35. Wing markings faint yellowish brown; hypovalves truncated on apexes, and divergent at base ··················103. ***Neopanorpa translucida***
Wing markings dark brown and distinct; hypovalves rounded apically and curled ventrad apically, and close to each other basally ··················77. ***Neopanorpa leigongshana***

36. Hypovalves curled dorsad on outer margin, and long cylindrical ………………………………………… 37
Hypovalves in other shape …………………………………………………………………………………… 42
37. Apical band of both forewings and hindwings with a large hyaline spot ……………… 69. *Neopanorpa formosana*
Apical band of both forewings and hindwings without such a hyaline spot ……………………………… 38
38 .Apical band of forewing with a projected and rounded angle on inner margin ………………………… 39
Apical band of forewing straight on inner margin ………………………………………………………… 41
39. Projected angle on inner margin of apical band acute, with its length approximate to width of apical band ……………
……………………………………………………………………………………… 60. *Neopanorpa cavaleriei*
Projected angle on inner margin of apical band obtuse and short, with its length shorter than width of apical band … 40
40. Gonostylus with an acute tooth on subbasal process ……………………………………… 88. *Neopanorpa nielseni*
Gonostylus without such a tooth on subbasal process ………………………………… 56. *Neopanorpa anchoroides*
41. Nota dark brown and slightly paler at lateral corners ……………………………… 73. *Neopanorpa hunanensis*
Nota yellowish brown laterally and bearing a distinct black mesal stripe ………………… 94. *Neopanorpa pulchra*
42. A6 longer than 1.7 times of length of A5; A7 and A8 as long as A6; gonostylus nearly as long as gonocoxites ………
……………………………………………………………………………………… 62. *Neopanorpa chelata*
A6 shorter than 1.7 times of length of A5; A7 and A8 shorter than A6; gonostylus shorter than gonocoxites ……… 43
43. Marginal spot slender, extending anteriorly and connected with basal branch of pterostigmal band ………………… 44
Marginal spot not connected with pterostigmal band ……………………………………………………………… 45
44. Wing markings reticulate: apical band as a slender stripe connected to pterostigmal band; marginal spot as a slender stripe connected to pterostigmal band anteriorly and to basal band posteriorly …………………… 61. *Neopanorpa caveata*
Apical band with an elongated slender band not connected with pterostigmal band; marginal spot not connected with basal band……………………………………………………………………………… 59. *Neopanorpa carpenteri*
Wing markings not reticulate as above ……………………………………………………………………… 43
45. Hypovalves acute apically, with an auriculate process on inner margin subbasally ……… 104. *Neopanorpa youngi*
Hypovalves rounded apically and lacking such a process ……………………………………………………… 46
46. Wings with greatly reduced marginal spot and basal band, represented by several scattered spots ……………… 47
Wings with well-developed marginal spot and basal band ………………………………………………………… 48
47. Gonostylus with a stout trapezoidal subbasal process on inner margin ……………… 79. *Neopanorpa lipingensis*
Gonostylus with a large plate-like subbasal process on inner margin ……………………… 99. *Neopanorpa sheni*
48. Apical band of forewing with 2–3 hyaline spots on inner margin; hypovalves tapering towards apex …………………
……………………………………………………………………………… 90. *Neopanorpa ophthalmica*
Apical band of forewing with 1–2 hyaline spots, or a separated short stripe on inner margin; hypovalves rounded apically, with distal portion almost as wide as basal part ……………………………… 101. *Neopanorpa tienmushana*

56. 锚形新蝎蛉 *Neopanorpa anchoroides* Zhou, 2003

【测量】 前翅长 18.0 mm。

【形态特征】 头顶黑色，单眼三角区黑色。喙黄褐色，前方具 2 条棕色纵带。前胸背板黑色。中、后胸背板大部分黑色，两侧后角黄褐色。侧板及足黄褐色。翅无色透明，斑纹褐色。前翅端带完整，内缘具 1 伸出的直角；痣带完整，在前缘与端带相连，基支阔，端支窄；缘斑极小；基带仅为后缘处的 1 个不规则斑；无基斑。雄性腹部 1 ~ 5 节背板黑色。第 3 节背中突长棒状，不伸达第 4 节背板后缘。第 6 节黑色，长筒形，端部稍狭；第 7、8 节黄褐色，基部缢缩。雄性外生殖器长卵形。第 9 节背板阔长，端部平截。第 9 节腹板具阔长的基柄，下瓣阔长，超出生殖肢基节端部，外缘向背面卷折，内侧相互重叠，内缘近基部形成 1 个近圆形窗口。生殖刺突细长，内缘具三角形中齿和粗壮的亚基齿。阳基侧突短细，大部分与阳茎愈合。阳茎稍超过生殖肢基节端部，侧突起发达。

【发生期】 7 月。

【地理分布】 贵州（梵净山、雷公山）。

Measurements: Forewing length 18.0 mm.

Morphology: Vertex black. Ocellar triangle black. Rostrum yellowish brown with a pair of brown longitudinal stripes frontally. Pronotum black. Mesonotum

图 3-56-1　锚形新蝎蛉 *Neopanorpa anchoroides* 雄成虫背面观（贵州雷公山，NWAFU）

Fig. 3-56-1　Male adult of *Neopanorpa anchoroides*, dorsal view (Mount Leigong, Guizhou, NWAFU)

and metanotum mostly black but yellowish brown at each latero-posterior corner. Pleura and legs yellowish brown. Wing membrane hyaline with brown markings. Apical band complete with a protruding right angle on inner margin; pterostigmal band complete and connected to apical band anteriorly, with a broad basal branch and a thin apical branch; marginal spot very small; basal band represented by an irregular spot on hind margin; basal spot absent. Male terga Ⅰ–Ⅴ black. Notal organ on tergum Ⅲ stick-like and not exceeding hind border of tergum Ⅳ. A6 black and long cylindrical, and slightly tapering towards apex; A7 and A8 yellowish brown and constricted basally. Male genital bulb long oval. Epandrium long, broad, and truncated terminally. Hypandrium with a long and broad basal stalk. Hypovalves long, extending beyond apex of gonocoxites, with their outer and inner margins curling dorsad, and inner margins overlapped and forming a nearly circular window subbasally. Gonostylus slender and bearing a triangular median tooth and a stout subbasal process on inner margin. Paramere short and thin, and mostly fused with aedeagus. Aedeagus slightly exceeding apex of gonocoxites and with stout lateral processes.

图 3-56-2　锚形新蝎蛉 *Neopanorpa anchoroides* 雄性外生殖器腹面观

Fig. 3-56-2　Male genitalia of *Neopanorpa anchoroides*, ventral view

图 3-56-3　锚形新蝎蛉 *Neopanoroa anchoroides* 地理分布

Fig. 3-56-3　Distribution of *Neopanorpa anchoroides*

Emergence: July.

Distribution: Guizhou (Mount Fanjing, Mount Leigong).

57. 短瓣新蝎蛉 *Neopanorpa brevivalvae* Chou & Wang, 1988

【测量】 前翅长 12.0 ~ 13.5 mm。

【形态特征】 头顶黑色，单眼三角区黑色。喙黄褐色。前胸背板黑褐色。中、后胸背板黄褐色，中央具黑色纵带。侧板及足黄褐色。翅无色透明，斑纹褐色。前翅端带完整，后缘具 1 个透明窗斑，后方内侧具 1 个分离的条斑；痣带完整，有时端支

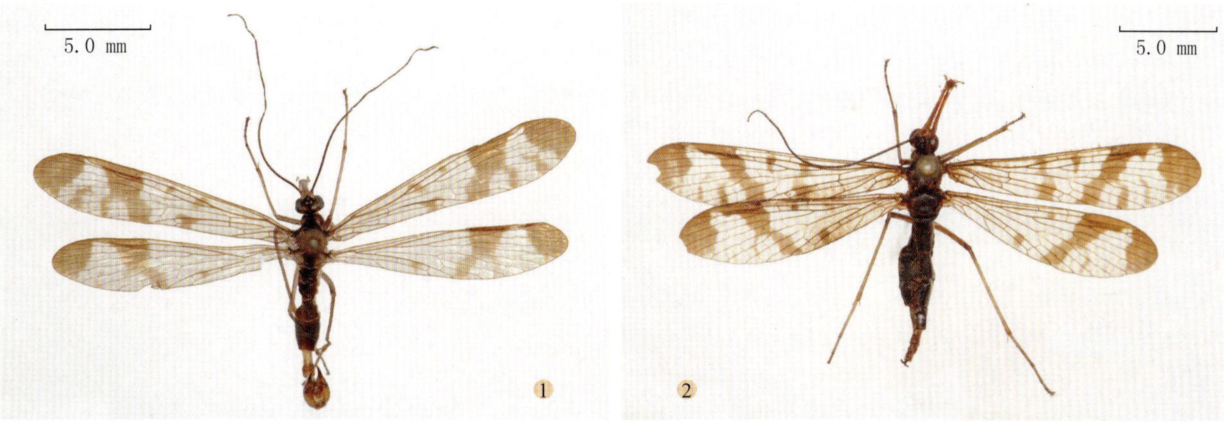

图 3-57-1 短瓣新蝎蛉 *Neopanorpa brevivalvae* 成虫背面观（广东连州大东山，NWAFU）

1. 雄虫 2. 雌虫

Fig. 3-56-1 Adults of *Neopanorpa brevivalvae*, dorsal view (Mount Dadong, Lianzhou, Guangdong, NWAFU)

1. Male 2. Female

图 3-57-2 短瓣新蝎蛉 *Neopanorpa brevivalvae* 雄性外生殖器腹面观

Fig. 3-57-2 Male genitalia of *Neopanorpa brevivalvae*, ventral view

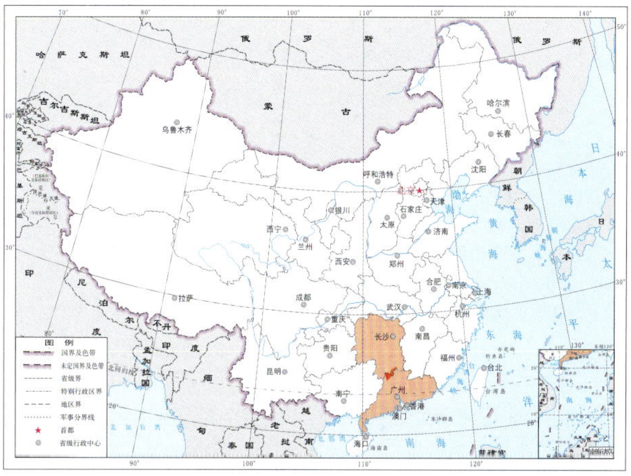

图 3-57-3 短瓣新蝎蛉 *Neopanorpa brevivalvae* 地理分布

Fig. 3-57-3 Distribution of *Neopanorpa brevivalvae*

分离；缘斑小；基带分离成 2 个斑点；无基斑。雄性腹部 1 ~ 5 节背板黑褐色。第 3 节背中突弯曲，长棒状，伸达第 4 节背板中部；第 4 节后背突粗壮，强烈隆起。雄性外生殖器卵形。第 9 节背板阔长，末端平截。第 9 节腹板具阔长的基柄；下瓣黑褐色，极短，末端向内弯曲，内缘基部形成 1 个圆形窗口。生殖刺突内缘具发达的亚基齿和钝三角形中齿。阳基侧突大部分与阳茎愈合，相互靠拢，端部强烈分歧。阳茎腹瓣灰白色，膜质，左右各形成 1 个叉状突起；背瓣较短；侧突起钩状发达。

【发生期】 5 ~ 6 月。

【地理分布】 湖南（莽山）、广东（连州）。

Measurements: Forewing length 12.0–13.5 mm.

Morphology: Vertex black. Ocellar triangle black. Rostrum yellowish brown. Pronotum dark brown. Mesonotum and metanotum mostly yellowish brown with a black mesal stripe. Pleura and legs yellowish brown. Wing membrane hyaline with brown markings. In forewings, apical band complete with a large hyaline window innerly and a separated stripe aside its hind portion; pterostigmal band complete or occasionally with a separate apical branch; marginal spots small; basal band split into two spots; basal spot absent. Male terga Ⅰ–Ⅴ dark brown. Notal organ on tergum Ⅲ slender and clavate, extending to middle of tergum Ⅳ; postnotal organ on tergum Ⅳ stout and strongly raised. Male genital bulb oval. Epandrium long and broad, and truncated terminally. Hypandrium with a long broad basal stalk. Hypovalves dark brown and extremely short, curved mesad and forming a circular window innerly. Gonostylus with an obtuse triangular median tooth and a stout subbasal process on inner margin. Parameres largely fused with aedeagus and close to each other but abruptly divergent apically. Aedeagus with its ventral valves greyish white, membranous, and forming a pair of hook-like processes laterally; dorsal valves relatively short and with stout and hook-like lateral processes.

Emergence: May to June.

Distribution: Hunan (Mount Mangshan) and Guangdong (Lianzhou).

58. 布里斯新蝎蛉 *Neopanorpa brisi* (Navás, 1930)

【测量】 前翅长 16.5 ~ 17.5 mm。

【形态特征】 头顶黑褐色，单眼三角区黑色。喙黄褐色，中间颜色较深。胸部背板黑色，侧板及足基节乳黄色，足大部棕绿色，跗节黑褐色。翅透明，稍带烟灰色，翅脉黑褐色，无任何斑纹，或仅端部略暗。雄性腹部 1 ~ 5 节背板黑褐色。第 3 节背中突发达，超过第 4 节背板的后背突，其腹面还有 1 个尖锐突起；其两侧的背板后缘各具 1 个小突

图 3-58-1 布里斯新蝎蛉 *Neopanorpa brisi* 雄成虫背面观（云南昆明西山）

Fig. 3-58-1 Male adult of *Neopanorpa brisi*, dorsal view (West Hill, Kunming, Yunnan)

图 3-58-2　布里斯新蝎蛉 ***Neopanorpa brisi*** 雄虫腹部侧面观

Fig 3-58-2　Male abdomen of ***Neopanorpa brisi***, lateral view

起。第 6 节长圆筒形，大部分黑褐色，端部红褐色；第 7、8 节红褐色；第 7 节基部稍缢缩，端部侧面具 2 个圆钝的齿突。雄性外生殖器长椭圆形。第 9 节背板十分阔长，端部两侧形成 1 对向中部弯曲的尖锐突起。第 9 节腹板基柄阔长；下瓣阔长，两瓣相互重叠，端部超过生殖刺突的中部，内缘近基部形成 1 圆形窗口。生殖刺突短于基节之半，外缘近基部稍凹陷。阳基侧突刀片状；阳茎锥形，侧突起发达。

【发生期】 5 ~ 8 月。

【地理分布】 云南（永胜、丽江玉龙雪山、昆明）、四川（螺髻山）。

Measurements: Forewing length 16.5–17.5 mm .

Morphology: Vertex dark brown. Ocellar triangle black. Rostrum yellowish brown and darkened mesally. Nota black. Pleura and coxae creamy yellow. Legs mostly greenish brown with dark brown tarsi. Wing membrane hyaline, tinged with smoky grey, and devoid of markings or only darkened at apex. Male tergaⅠ–Ⅴ blackish brown. Notal organ on tergumⅢstout and exceeding postnotal organ on tergum Ⅳ. A smaller acute tooth on ventral side of notal organ. Beside notal organ two small teeth on posterior margin of tergum Ⅲ laterally. A6 mostly dark brown but reddish brown apically, long and cylindrical; A7 and A8 reddish brown. A7 slightly constricted at base and bearing a pair of rounded teeth on each side of its apex. Male genital bulb long elliptical. Epandrium long and broad, and forming a pair of acute processes curved mesad. Hypandrium with a long broad basal stalk. Hypovalves

图 3-58-3　布里斯新蝎蛉 ***Neopanorpa brisi*** 雄性外生殖器腹面观

Fig. 3-58-3　Male genitalia of ***Neopanorpa brisi***, ventral view

图 3-58-4　布里斯新蝎蛉 ***Neopanorpa brisi*** 地理分布

Fig. 3-58-4　Distribution of ***Neopanorpa brisi***

long and broad, overlapped mesally, exceeding middle of gonostylus and forming a circular window subbasally on inner margin. Gonostylus shorter than half length of gonocoxites, and concaved on outer margin subbasally. Paramere blade-like. Aedeagus conical with a pair of stout lateral processes.

Emergence: May to August.

Distribution: Yunnan [Yongsheng , Lijiang (Yulong Snow Mountain), Kunming] and Sichuan (Mount Luoji).

59. 卡本特新蝎蛉 *Neopanorpa carpenteri* Cheng, 1957

1　2

图 3-59-1　卡本特新蝎蛉 *Neopanorpa carpenteri* 成虫背面观（广东梅州）

1. 雄虫　2. 雌虫

Fig. 3-59-1　Adults of *Neopanorpa carpenteri*, dorsal view (Meizhou, Guangdong)

1. Male　2. Female

图 3-59-2　卡本特新蝎蛉 *Neopanorpa carpenteri* 雄性外生殖器腹面观

Fig. 3-59-2　Male genitalia of *Neopanorpa carpenteri*, ventral view

图 3-59-3　卡本特新蝎蛉 *Neopanorpa carpenteri* 地理分布

Fig. 3-59-3　Distribution of *Neopanorpa carpenteri*

【测量】 前翅长 15.0 ~ 16.0 mm。

【形态特征】 头顶黄褐色，1 条宽阔的黑色带横贯单眼三角区；喙黄褐色。前胸背板黑褐色；中、后胸背板前缘黑色，中部具 1 条黑色宽中带，两侧后角黄褐色。侧板及足黄褐色。翅略带黄色，斑纹褐色。前翅端带内缘后侧具 1 条分离的细带，与端带之间形成 1 个透明窗；痣带具完整的端支和基支；基带前端窄，后端阔；缘斑窄长，向后与痣带的基支相连；无基斑；翅痣红褐色明显。雄性腹部 1 ~ 5 节背板黑色，腹板黄白色，侧面黄色。第 3 节背中突棒状，不超过第 4 节后缘。第 6 节黑色。第 7、8 节黄褐色，基部稍缢缩。雄性外生殖器卵形。第 9 节背板阔长，末端平截；腹板具阔长的基柄；下瓣黄白色，内侧向腹面翻折。生殖刺突稍短于生殖肢基节，内缘中齿钝三角形，近基部具圆阔的亚基齿。阳基侧突短，端部膨大；阳茎锥形，侧突起发达。

【发生期】 4 ~ 6 月。

【地理分布】 广东（梅州）。

Measurements: Forewing length 15.0–16.0 mm.

Morphology: Vertex yellowish brown, with a broad black stripe passing through ocellar triangle; rostrum yellowish brown. Pronotum dark brown; mesonotum and metanotum black at anterior margin, with a broad black mesal stripe and yellowish brown at latero-posterior corners. Pleura and legs yellowish brown. Wings slightly tinged with yellow, markings brown. Forewing apical band with a separated thin stripe posteirorly, which formed a hyaline window with opical band; pterostigmal band with complete basal and apical branches; basal branch thin anteriorly and thick posteriorly; marginal spot elongated and connected to basal branch of pterostigmal band;basal spot absent; pterostigmal reddish brown and distinct. Male terga Ⅰ–Ⅴ black; sterna yellowish white; pleura yellow. Notal organ on tergum Ⅲ short clavate, not exceeding hind border of tergumⅣ; A6 black. A7 and A8 yellowish brown and slightly constricted at base. Male genital bulb oval. Epandrium long broad, with a truncated apex. Hypandrium with a long broad basal stalk. Hypovalves yellowish white and curled ventrad at inner margins. Gonostylus slightly shorter than gonocoxites, with an obtuse triangular median tooth and a rounded subbasal process on inner margin. Paramere short, enlarged apically; aedeagus conical with stout lateral processes.

Emergence: April to June.

Distribution: Guangdong (Meizhou).

60. 卡氏新蝎蛉 *Neopanorpa cavaleriei* (Navás, 1908)

【测量】 前翅长 16.0 ~ 17.0 mm。

【形态特征】 头顶黑褐色，单眼三角区黑色；喙黄褐色。前胸背板黑褐色；中、后胸背板前缘黑色，中部具 1 条黑色宽中带，两侧后角黄褐色。侧板及足浅黄褐色。翅无色透明，斑纹褐色。前翅端带内缘中部具 1 个斜尖角；痣带具完整的端支和基支；基带中部弱化，形成 2 个大斑；缘斑窄长；无基斑。雄性腹部 1 ~ 5 节背板黑色，腹板黄白色，侧面黄色。第 3 节背中突短棒状，不超过第 4 节后缘。第 6 节黑色。第 7、8 节黄褐色，基部稍缢缩。

图 3-60-1 卡氏新蝎蛉 *Neopanorpa cavaleriei* 雄成虫背面观（贵州荔波，NWAFU）

Fig. 3-60-1 Male adult of *Neopanorpa cavaleriei*, dorsal view (Libo, Guizhou, Fujian, NWAFU)

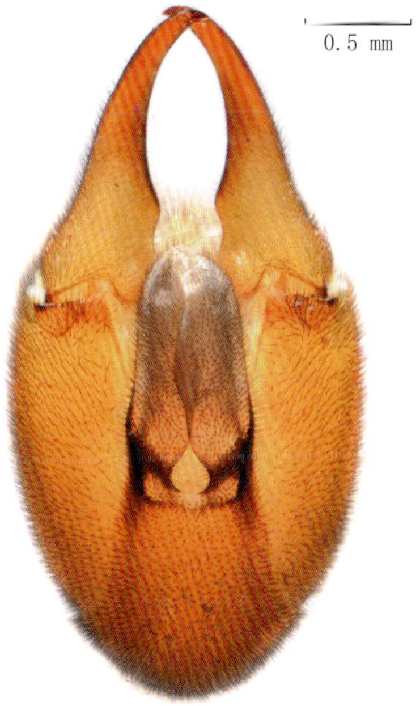

图 3-60-2 卡氏新蝎蛉 *Neopanorpa cavaleriei* 雄性外生殖器腹面观

Fig. 3-60-2 Male genitalia of *Neopanorpa cavaleriei*, ventral view

雄性外生殖器长卵形。第9节背板阔长，末端平截；腹板具阔长的基柄；下瓣侧缘卷曲，相互套叠，内缘近基部具1个近圆形窗口。生殖刺突稍短于生殖肢基节，内缘中齿钝三角形，近基部具圆阔的亚基齿，亚基齿下方延伸出1个弯曲的尖齿。阳基侧突大部分与阳茎愈合；阳茎锥形，侧突起发达。

【发生期】 4 ~ 6月。

【地理分布】 贵州（荔波）。

Measurements: Forewing length 16.0–17.0 mm.

Morphology: Vertex dark brown, ocellar triangle black; rostrum yellowish brown. Pronotum dark brown; mesonotum and metanotum dark brown at anterior margin, with a broad black mesal stripe, and yellowish brown at latero-posterior corners. Pleura and legs light yellowish brown. Wing membrane hyaline, with brown markings. Apical band with an oblique prolongation at middle of inner margin; pterostigmal band with complete apical and basal branches; basal band reduced at middle and forming two large spots; marginal spot slender; basal spot absent. Male terga Ⅰ–Ⅴ black; sterna yellowish white ; pleura yellow. Notal organ on tergum Ⅲ short clavate, not exceeding hind border of tergum Ⅳ; A6 black. A7 and A8 yellowish brown and slightly constricted at base.

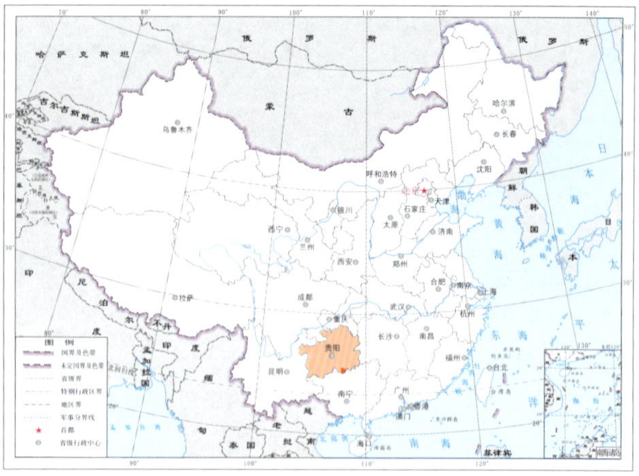

图 3-60-3 卡氏新蝎蛉 *Neopanorpa cavaleriei* 地理分布

Fig. 3-60-3 Distribution of *Neopanorpa cavaleriei*

Male genital bulb long oval. Epandrium long broad, with a truncated apex. Hypandrium with a long broad basal stalk. Hypovalves curled along lateral margins and overlapped mesally, forming an subcircular window subbasally. Gonostylus slightly shorter than gonocoxites, with an obtuse triangular median tooth and a rounded subbasal process bearing an curved acute tooth. Paramere almost fused with aedeagus; aedeagus conical with stout lateral processes.

Emergence: April to June.

Distribution: Guizhou (Libo).

61. 网翅新蝎蛉 *Neopanorpa caveata* Cheng, 1957

【测量】 前翅长 15.0 ~ 16.0 mm。

【形态特征】 头顶黄褐色，单眼三角区黑色。1条黑色纵带向后穿过单眼三角区。喙红褐色。前胸背板黑褐色。中、后胸背板黄褐色，中央具显著的黑色中纵带。侧板及足黄褐色。翅浅黄色，透明，斑纹黄褐色、网状。前翅端带阔，与痣带之间具1条细长的横带，向内形成网状斑纹与痣带的端支相连；痣带完整；缘斑向外与痣带基支相连，向内与基带相连；基带完整。无基斑。雄性腹部1 ~ 5节背板黑褐色。第3节背中突长棒状，不超过第4节背板后缘。第6节大部分黑褐色，端部黄褐色。雄性外生殖器阔卵形。第9节背板阔长，末端平截。第9节腹板具阔的基柄；下瓣十分宽阔，左右互相重叠，超过生殖肢基节端部，内缘近基部形成1个近心形窗口。生殖刺突短于生殖基节之半，外缘稍凹陷，内缘具发达的亚基齿和钝三角形中齿。阳基侧突细长，端部尖锐；阳茎锥状，侧突起发达。

【发生期】 5 ~ 9月。

【地理分布】 福建（邵武、建宁、武夷山）、浙江（龙王山）。

Measurements: Forewing length 15.0–16.0 mm.

Morphology: Vertex yellowish brown. Ocellar triangle black. A longitudinal black stripe passing through ocellar triangle. Rostrum reddish brown. Pronotum dark brown. Mesonotum and metanotum yellowish brown, with a black longitudinal mesal stripe. Pleura and legs yellowish brown. Wing membrane hyaline and tinged with light yellow. Markings yellowish brown and reticulate: apical band broad, with a slender stripe connected to pterostigmal band, extending inward and connected to apical

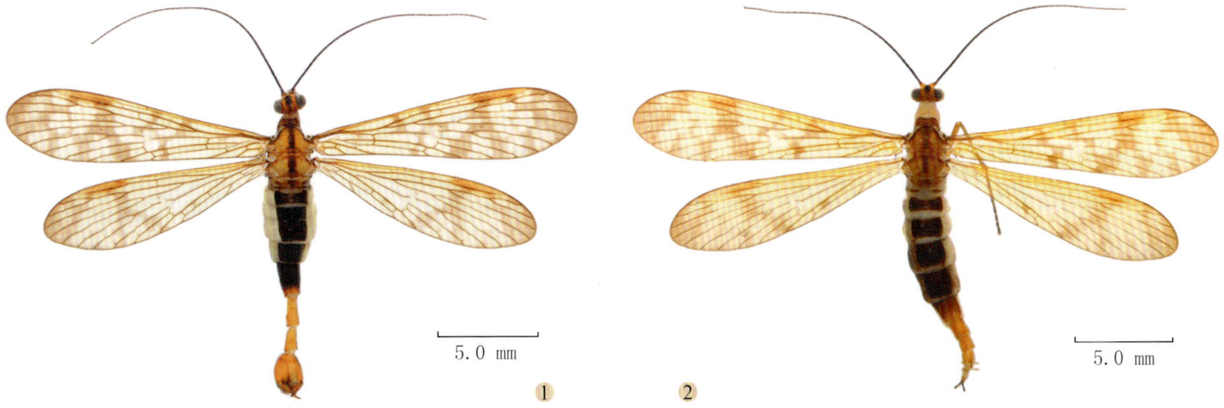

图 3-61-1 网翅新蝎蛉 ***Neopanorpa caveata*** 成虫背面观（福建武夷山，NWAFU）

1. 雄虫 2. 雌虫

Fig. 3-61-1 Adults of ***Neopanorpa caveata***, dorsal view (Wuyi Mountain, Fujian, NWAFU)

1. Male 2. Female

图 3-61-2 网翅新蝎蛉 ***Neopanorpa caveata*** 雄性外生殖器腹面观

Fig. 3-61-2 Male genitalia of ***Neopanorpa caveata***, ventral view

图 3-61-3 网翅新蝎蛉 ***Neopanorpa caveata*** 地理分布

Fig. 3-61-3 Distribution of ***Neopanorpa caveata***

branch of pterostigmal band; pterostigmal band complete; marginal spot extending outward to connect with basal branch of pterostigmal band, and inward to connect with complete basal band; basal spot absent. Male terga Ⅰ – Ⅴ black. Notal organ on tergum Ⅲ long stick-like and not exceeding posterior margin of tergum Ⅳ. A6 mostly dark brown with a yellowish brown apex. Male genital bulb broad oval. Epandrium long, broad and truncated apically. Hypandrium with a long and broad basal stalk. Hypovalves specially broad, overlapped mesally, exceeding apex of gonocoxites and forming a heart-shaped window subbasally. Gonostylus shorter than half length of gonocoxites, slightly concaved on outer margin, and bearing an obtuse triangular median tooth and a stout subbasal process on inner margin. Paramere slender and acute apically. Aedeagus conical with a pair of stout lateral processes.

Emergence: May to September.

Distribution: Fujian (Shaowu, Jianning, Wuyi Mountains) and Zhejiang (Mount Longwang).

62. 爪新蝎蛉 *Neopanorpa chelata* Carpenter, 1938

图 3-62-1　爪新蝎蛉 *Neopanorpa chelata* 雄成虫正在取食其他昆虫尸体（四川崇州鞍子河，柴义君摄）
1. 侧面观　2. 背面观

Fig. 3-62-1　Male adult of *Neopanorpa chelata* feeding on a dead insect (Anzihe, Chongzhou, Sichuan, photo by Yijun Chai)
1. Lateral view　2. Dorsal view

【测量】 前翅长 15.0 ~ 17.0 mm。

【形态特征】 头顶黑色，单眼三角区黑色。喙黄褐色至暗褐色。前胸背板黑褐色。中、后胸背板大部分黑褐色，两侧黄褐色。侧板及足黄褐色。翅略带淡褐色，透明，斑纹黑褐色。前翅端带半圆形，较阔，后缘延伸出 1 条细纹；痣带阔，基支阔，端支细，有时断开；缘斑细长；基带只在 R_1 近基部和翅后缘呈 2 处不规则斑；无基斑。雄性腹部

图 3-62-2 爪新蝎蛉 *Neopanorpa chelata* 雄性外生殖器腹面观

Fig. 3-62-2 Male genitalia of ***Neopanorpa chelata***, ventral view

1 ~ 5 节背板黑褐色。第 3 节背中突发达，短棒状，盖住第 4 节上圆钝的后背突。第 6 节黑褐色，长筒形。第 7、8 节红褐色，长筒形，十分粗壮，基部缢缩不明显。雄性外生殖器长卵形。第 9 节背板阔长，端部平截。第 9 节腹板基柄阔长；下瓣阔长，稍超过生殖肢基节端部，外缘末端向背侧卷曲，内缘相互重叠。生殖刺突约与基节等长，内缘亚基齿梯形，中齿钝三角形。阳基侧突基半部与阳茎愈合，端半部向外弯曲，叶状；阳茎锥形，侧突起发达。

【发生期】 5 ~ 8 月。

【地理分布】 四川（峨眉山、雅安、都江堰、鞍子河）。

Measurements: Forewing length 15.0 –17.0 mm.

Morphology: Vertex black. Ocellar triangle black.

图 3-62-3 爪新蝎蛉 *Neopanorpa chelata* 地理分布

Fig. 3-62-3 Distribution of ***Neopanorpa chelata***

Rostrum yellowish brown to dark brown. Pronotum dark brown. Mesonotum and metanotum mostly dark brown but yellowish brown laterally. Pleura and legs yellowish brown. Wing membrane hyaline and slightly tinged with light brown with markings dark brown. In forewing, apical band semicircular and broad, and with a separate small spot posteriorly; pterostigmal band broad, with broad basal branch, and thin or detached apical branch; marginal spot slender; basal band represented by two irregular spots near base of R_1 and on hind margin, respectively; basal spot absent. Male terga Ⅰ–Ⅴ dark brown. Notal organ on tergum Ⅲ short and clavate, covering rounded postnotal organ on tergum Ⅳ. A6 dark brown and long cylindrical. A7 and A8 reddish brown and long cylindrical, considerably stout and inconspicuously constricted basally. Male genital bulb long oval. Epandrium slightly truncated on apical margin. Hypandrium with a broad basal stalk. Hypovalves broad and long, slightly exceeding apex of gonocoxites, with their outer margins curled dorsad, and inner margins overlapped mesally. Gonostylus nearly as long as gonocoxites, with a trapezoidal subbasal process and an obtuse triangular mesal tooth on inner margin. Paramere fused with aedeagus basally, and curved laterad and foliate apically. Aedeagus conical, with a pair of stout lateral processes.

Emergence: May to August.

Distribution: Sichuan (Mount Emei, Ya′an, Dujiangyan, Anzihe).

63. 契氏新蝎蛉 *Neopanorpa chillcotti* Byers, 1971

【测量】 前翅长 15.0 ~ 16.0 mm。

【形态特征】 头顶黑色，单眼三角区黑色。喙黄褐色至暗褐色，前侧具 2 条弥散的深色纵带。前胸背板黑褐色。中、后胸背板、侧板及足黄褐色。翅透明略带黄色，斑纹黄褐色。前翅端带半圆形，后缘不完整，有时向内伸出 1 个尖角；痣带细长，前半部常不超过 R_1；基支常断开；缘斑不明显；基带仅为小斑点。雄性腹部 1 ~ 5 节背板污黄棕色。第 3 节背中突发达，长棒状，末端稍膨大，延伸达第 5 节中部，盖住第 4 节上圆钝的后背突。第 6 节筒形，长约第 5 节的 2 倍，末端具黑色边缘。第 7、8 节约与第 6 节等长，基部稍缢缩。雄性外生殖器

图 3-63-1 契氏新蝎蛉 *Neopanorpa chillcotti* 成虫背面观（西藏吉隆，NWAFU）

1. 雄虫 2. 雌虫

Fig. 3-63-1 Adults of *Neopanorpa chillcotti*, dorsal view (Gyirong, Tibet, NWAFU)

1. Male 2. Female

图 3-63-2 契氏新蝎蛉 *Neopanorpa chillcotti* 雄性外生殖器腹面观

Fig. 3-63-2 Male genitalia of *Neopanorpa chillcotti*, ventral view

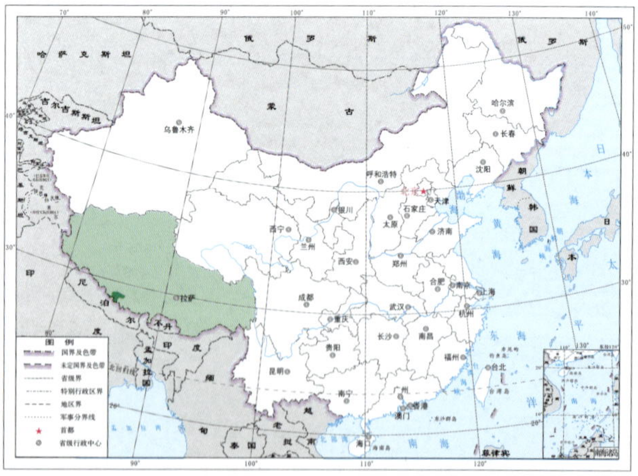

图 3-63-3 契氏新蝎蛉 *Neopanorpa chillcotti* 地理分布

Fig. 3-63-3 Distribution of *Neopanorpa chillcotti*

卵形。第 9 节背板阔长，端部钝圆。第 9 节腹板基柄短；下瓣窄长，基部分离；端半部阔，向内弯曲，着生许多长刚毛。生殖刺突细长，内缘近基部具大的方形亚基齿，中齿钝三角形。阳基侧突线状，基

半部与阳茎愈合，端半部游离。阳茎锥形，背瓣细长指状，向后延伸；侧突起发达。

【发生期】 5 ~ 8 月。

【地理分布】 西藏（吉隆）；尼泊尔（加德满都）。

Measurements: Forewing length 15.0–16.0 mm.

Morphology: Vertex black. Ocellar triangle black. Rostrum yellowish brown to deep brown with two diffuse brown longitudinal stripes frontally. Pronotum dark brown. Mesonotum and metanotum, pleura and legs yellowish brown. Wing membrane hyaline and slightly tinged with light yellow. Markings yellowish brown. Forewing apical band semicircular with incomplete hind border, usually with an acute elongation innerly; pterostigmal band slender and usually not exceeding R_1 anteriorly; basal branch of pterostigmal band usually separated; marginal spot inconspicuous; basal spot greatly reduced into a small spot. Male terga Ⅰ–Ⅴ sordid yellowish brown. Notal organ on tergum Ⅲ greatly developed, long clavate, slightly enlarged terminally and extending to middle of tergum Ⅴ, and covering rounded postnotal organ on tergum Ⅳ. A6 cylindrical, approximately twice as long as A5, and with black margin terminally. A7 and A8 approximately as long as A6 and slightly constricted basally. Male genital bulb long oval. Epandrium long and broad, and rounded at apex. Hypandrium with a short basal stalk. Hypovalves slender, divergent basally and convergent apically, with numerous long setae on expanded apical portion. Gonostylus slender, bearing an obtuse triangular median tooth and a large oblong subbasal process on inner margin. Paramere thread-like, fused to aedeagus basally and free distally. Aedeagus conical, bearing a pair of stout lateral processes, with its finger-like dorsal valves greatly elongated caudad.

Emergence: May to August.

Distribution: Tibet (Gyirong) and Nepal (Kathmandu).

64. 周氏新蝎蛉 *Neopanorpa choui* Cheng, 1949

【测量】 前翅长 13.5 ~ 14.0 mm。

【形态特征】 头顶黑色，单眼三角区黑色。喙黄褐色。前胸背板黑褐色。中、后胸背板黄褐色，具宽阔的黑色中带；前胸背板前缘黑褐色。侧板及足黄褐色。翅略带黄色，透明，斑纹褐色。前翅端带呈散乱的一系列斑点；痣带的基支、端支均断开；基带断为 2 点，或仅为 CuP 末端的 1 点；缘斑不明显；无基斑。雄性腹部 1 ~ 5 节背板黄褐色。第 3 节背中突极其发达，向后延伸达第 7 节末端，下缘着生密集、向前弯曲的黑色短毛，基半部向背面拱起，端部略上翘。第 4 节背板长约为第 3 节背板的 2 倍，中央背中线上着生 1 列向前弯曲的黑色短

图 3-64-1 周氏新蝎蛉 *Neopanorpa choui* 成虫背面观（四川青城山，NWAFU）

1. 雄虫 2. 雌虫

Fig. 3-64-1 Adults of *Neopanorpa choui*, dorsal view (Mount Qingcheng, Sichuan, NWAFU)

1. Male 2. Female

毛。第6节筒形；第7、8节基部稍缢缩。雄性外生殖器阔卵形。第9节背板向后渐窄，末端稍缺刻。第9节腹板具短宽阔的基柄；下瓣基部较窄，端部2/3处最宽，末端尖，内、外缘均着生稀疏粗长的鬃毛。生殖刺突略短于基节，内缘具钝三角形中齿和近方形亚基齿。阳基侧突基部细长，端部叶片状，末端具2短尖；阳茎短粗，不超过生殖肢基节端部，侧突起发达。

【发生期】 6～7月。

【地理分布】 四川（周公山、青城山）。

Measurements: Forewing length 13.5–14.0 mm.

Morphology: Vertex black. Ocellar triangle black. Rostrum yellowish brown. Pronotum dark brown. Mesonotum and metanotum yellowish brown, with broad black mesal stripe. Pleura and legs yellowish brown. Wing membrane hyaline, slightly tinged with light yellow, markings brown. In forewing, apical band represented by series of scattered spots; pterostigmal with detached basal and apical branches; basal band split into two spots, or only represented by spot on terminal portion of CuP; marginal spot inconspicuous; basal spot absent. Male terga Ⅰ–Ⅴ yellowish brown. Notal organ on tergum Ⅲ extremely elongated and extending beyond apex of A7 and bearing numerous cephalad-curling setae on ventral surface, with its basal half arched dorsad and apex curved dorsad. Tergum Ⅳ with a row of short and cephalad-curling setae on its dorso-mesal line, and nearly twice length of tergum Ⅲ. A6 cylindrical. A7 and A8 slightly constricted basally. Male genital bulb broad oval. Epandrium tapering caudad, and slightly emarginate terminally. Hypandrium with a short broad basal stalk. Hypovalves narrow at base and acute at apex, broadest at distal two-thirds, with sparse and long bristles on inner and outer margins. Gonostylus slightly shorter than gonocoxites, with an obtuse triangular median tooth and an oblong subbasal process on inner margin. Paramere thin basally and foliate distally, with its apex split into two acute tips. Aedeagus short and stout, not exceeding apex of gonocoxites, with stout lateral processes.

Emergence: June to July.

Distribution: Sichuan (Mount Zhougong, Mount Qingcheng).

图 3-64-2 周氏新蝎蛉 ***Neopanorpa choui*** 雄虫腹部侧面观

Fig. 3-64-2 Male teminal abdomen of ***Neopanorpa choui*** , lateral view

图 3-64-3 周氏新蝎蛉 ***Neopanorpa choui*** 雄性外生殖器腹面观

Fig. 3-64-3 Male genitalia of ***Neopanorpa choui***, ventral view

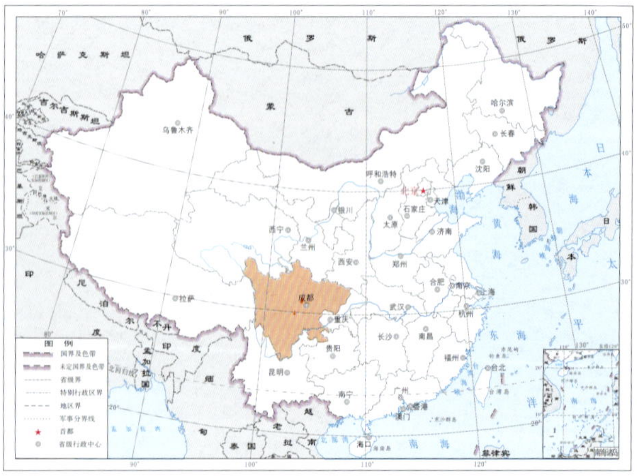

图 3-64-4 周氏新蝎蛉 ***Neopanorpa choui*** 地理分布

Fig. 3-64-4 Distribution of ***Neopanorpa choui***

65. 显斑新蝎蛉 *Neopanorpa clara* Chou & Wang, 1988

【测量】 前翅长 14.0 mm。

【形态特征】 头顶黑色，单眼三角区黑色。喙黄褐色。前胸背板黑色。中、后胸背板黑褐色。侧板及足黄褐色。翅无色透明。端带完整，后缘具 1 分离的小斑；痣带完整，端支极窄，基支宽阔；缘斑小；基带中部模糊，两端形状不规则，后端宽；基斑小。雄性腹部 1 ~ 5 节背板黑褐色。第 3 节背板后缘的背中突长棒状，稍超过第 4 节背板后背突。第 6 节大部分黑色，末端黄褐色，长筒形。第 7、8 节黄褐色，基部稍缢缩。雄性外生殖器长椭圆形。第 9 节背板阔长，末端稍缺刻。第 9 节腹板基柄阔长，下瓣长，内缘相互重叠。生殖肢基节长约为生殖刺突 3 倍；生殖刺突粗壮，内缘中齿钝三角形，亚基齿短柱形。阳基侧突叶片状。阳茎锥形，超出生殖肢基节端部，侧突起发达。

【发生期】 4 ~ 7 月。

【地理分布】 湖南（汝城、绥宁、八面山、双牌）。

Measurements: Forewing length 14.0 mm.

Morphology: Vertex black. Ocellar triangle black. Rostrum yellowish brown. Pronotum black. Mesonotum and metanotum dark brown. Pleura and legs yellowish brown. Wing membrane hyaline. In forewings, apical band complete; a separated spot on hind border near apical band; pterostigmal band complete, with a thin apical branch and a broad basal branch; marginal spot small; basal band blurred medially and irregular at anterior and posterior

图 3-65-1 显斑新蝎蛉 *Neopanorpa clara* 雄成虫背面观（外生殖器已除去，湖南汝城，NWAFU）

Fig. 3-65-1 Male adult of *Neopanorpa clara*, dorsal view (With genitala removed, Rucheng, Hunan, NWAFU)

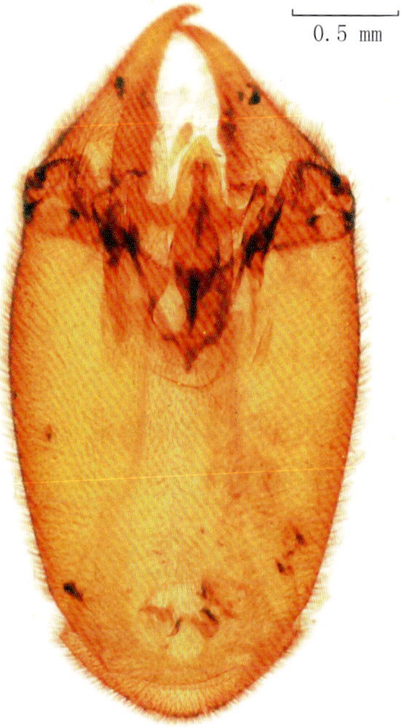

图 3-65-2 显斑新蝎蛉 *Neopanorpa clara* 雄性外生殖器腹面观

Fig. 3-65-2 Male genitalia of *Neopanorpa clara*, ventral view

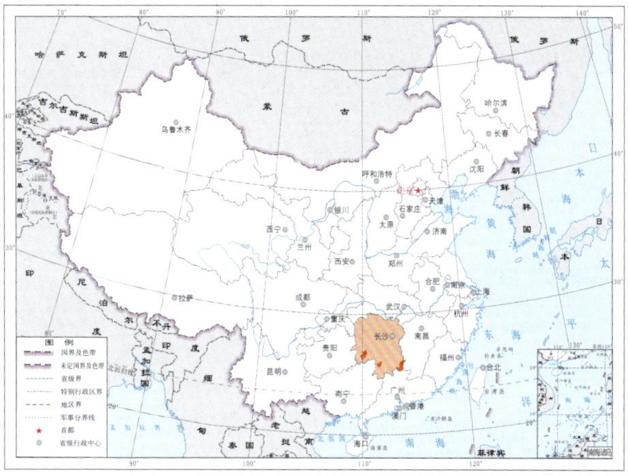

图 3-65-3 显斑新蝎蛉 *Neopanorpa clara* 地理分布

Fig. 3-65-3 Distribution of *Neopanorpa clara*

ends; basal spot small. Male terga Ⅰ–Ⅴ dark brown. Notal organ on tergum Ⅲ long and stick-like, slightly exceeding postnotal organ on tergum Ⅳ. A6 mostly black with a yellowish brown apex, cylindrical; A7 and A8 yellowish brown and slightly constricted basally. Male genital bulb long elliptical. Epandrium long and broad, and slightly emarginate terminally. Hypandrium with a long and broad basal stalk. Hypovalves long and overlapped mesally. Gonostylus nearly third as long as gonocoxites, with an obtuse triangular median tooth and a cylindrical subbasal process on inner margin. Paramere foliate. Aedeagus conical, slightly exceeding apex of gonocoxites, with stout lateral processes.

Emergence: April to July.

Distribution: Hunan (Rucheng, Suining, Mount Bamian, Shuangpai).

66. 明翅新蝎蛉 *Neopanorpa claripennis* Carpenter, 1938

【测量】 前翅长 13.0 mm。

【形态特征】 头顶黑色，单眼三角区黑色。喙不均匀黄褐色。前胸背板黑棕色。中、后胸背板黄褐色，中央具十分宽大的深褐色中带。侧板及足黄褐色。翅无色透明，无明显的斑纹，仅端部具不明显的加深。翅痣不明显。雄性腹部 1 ~ 5 节背板黑色。第 3 节背板后缘的背中突极短，仅盖住第 4 节上的后背突。第 6 节黑色，筒形。第 7、8 节黄褐色，基部稍缢缩。雄性外生殖器卵形。第 9 节背板阔长，末端平截。第 9 节腹板具宽阔的基柄，下瓣不超过生殖肢基节端部，内缘端半部相互重叠，近基部形成 1 个卵形窗口。生殖刺突纤细，内缘中齿钝三角形，亚基齿发达。阳基侧突纤细；阳茎锥形，略超出生殖肢基节端部，侧突起发达。

【发生期】 7 月。

【地理分布】 四川（白鹿顶、平武）。

Measurements: Forewing length 13.0 mm.

Morphology: Vertex black. Ocellar triangle black. Rostrum unevenly yellowish brown.Pronotum dark brown. Mesonotum and metanotum yellowish brown, bearing a broad deep brown mesal stripe. Pleura and legs yellowish brown. Wing membrane hyaline and devoid of markings, except for indistinct tinge on apex. Pterostigma inconspicuous. Male terga Ⅰ–Ⅴ black. Notal organ on tergum Ⅲ short and covering postnotal organ on tergum Ⅳ. A6 black, cylindrical.

图 3-66-1 明翅新蝎蛉 *Neopanorpa claripennis* 雄成虫背面观（四川平武老河沟，NWAFU）

Fig. 3-66-1 Male adult of *Neopanorpa claripennis*, dorsal view (Laohegou, Pingwu, Sichuan, NWAFU)

图 3-66-2 明翅新蝎蛉 *Neopanorpa claripennis* 雄性外生殖器腹面观

Fig. 3-66-2 Male genitalia of *Neopanorpa claripennis*, ventral view

A7 and A8 yellowish brown and slightly constricted basally. Male genital bulb oval. Epandrium long and broad, and truncated terminally. Hypandrium with a broad basal stalk, with hypovalves not exceeding apex of gonocoxites. Hypovalves overlapped mesally and forming an oval window subbasally. Gonostylus slender and bearing an obtuse triangular median tooth and a stout subbasal process on inner margin. Paramere slender. Aedeagus conical, and slightly exceeding apex of gonocoxites, with stout lateral processes.

Emergence: July.

Distribution: Sichuan (Bailuding, Pingwu).

图 3-66-3 明翅新蝎蛉 *Neopanorpa claripennis* 地理分布

Fig. 3-66-3 Distribution of *Neopanorpa claripennis*

67. 尖齿新蝎蛉 *Neopanorpa cuspidata* Byers, 1965

【测量】 前翅长 11.0 ~ 13.0 mm。

【形态特征】 头顶黑色，单眼三角区黑色。喙黄褐色。胸部背板黑褐色，中、后胸背板两侧黄褐色。侧板及足黄褐色。翅透明无色，斑纹褐色。前翅端带宽，后缘具透明小圆斑；痣带完整；缘斑较阔；基带近中部断开，形成 2 个小斑；基斑不明显。雄性腹部 1 ~ 5 节背板黑褐色。第 3 节背中突发达，锐三角形，盖住第 4 节上圆钝的后背突。第 6 节黑褐色，长筒形。第 7、8 节黄褐色，基部缢缩。雄性外生殖器长卵形。第 9 节背板阔长，端部平截；腹板基柄阔长；下瓣超过生殖肢基节端部，末端尖，内缘近基部形成 1 个圆形窗口。生殖刺突内缘亚基齿杯状，中齿钝三角形。阳基侧突基半部与阳茎愈合，端半部游离；阳茎锥形，侧突起发达。

【发生期】 5 ~ 8 月。

【地理分布】 云南（纳板河、勐龙、勐往）；泰国。

Measurements: Forewing length 11.0–13.0 mm.

图 3-67-1 尖齿新蝎蛉 *Neopanorpa cuspidata* 雄成虫背面观
（云南西双版纳纳板河，IZAS）

Fig. 3-67-1 Male adult of *Neopanorpa cuspidata*, dorsal view
(Nabanhe, Xishuangbanna, Yunnan, IZAS)

图 3-67-2 尖齿新蝎蛉 *Neopanorpa cuspidata* 雄性外生殖器腹面观

Fig. 3-67-2 Male genitalia of *Neopanorpa cuspidata*, ventral view

Morphology: Vertex black. Ocellar triangle black. Rostrum yellowish brown. Pronotum dark brown. Mesonotum and metanotum dark brown but yellowish brown laterally. Pleura and legs yellowish brown. Wing membrane hyaline, with markings dark brown. In forewings, apical band broad with a small hyaline window posteriorly; pterostigmal band complete; marginal spot broad; basal band split medially and forming two spots; basal spots inconspicuous. Male terga Ⅰ–Ⅴ dark brown. Notal organ on tergum Ⅲ well-developed acute triangular, covering rounded postnotal organ on tergum Ⅳ. A6 dark brown and long cylindrical. A7 and A8 yellowish brown and slightly constricted basally. Male genital bulb long oval. Epandrium long and broad, and truncated terminally. Hypandrium with a broad basal stalk. Hypovalves exceeding apex of gonocoxites, acute apically, and forming an oval window at inner margin subbasally. Gonostylus slender, bearing an obtuse triangular median tooth and a cup-like subbasal process on inner margin. Paramere fused with aedeagus at basal half, and free at distal half. Aedeagus conical, with stout lateral processes.

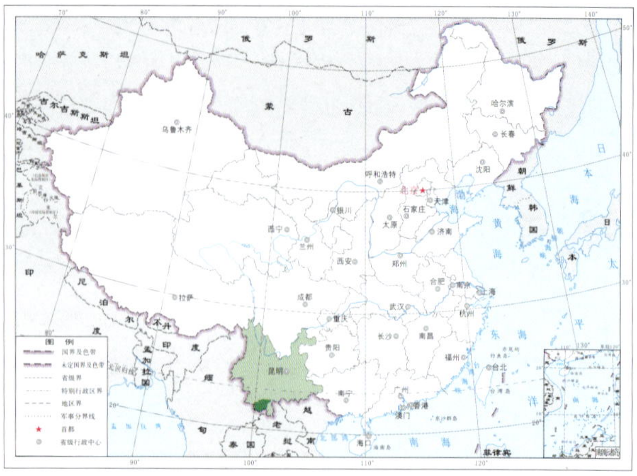

图 3-67-3 尖齿新蝎蛉 *Neopanorpa cuspidata* 地理分布

Fig. 3-67-3 Distribution of *Neopanorpa cuspidata*

Emergence: May to August.

Distribution: Yunnan (Naban River, Menglong, Mengwang) ; Thailand.

68. 方祥新蝎蛉 *Neopanorpa fangxianga* Zhou & Zhou, 2007

【测量】 前翅长 15.0 mm。

【形态特征】 头顶黑色；单眼三角区黑色。喙黄褐色，两侧具深褐色纵带。胸部背板黑色，侧板及足黄褐色。翅略带黄色、透明；翅痣黄褐色；斑纹黑褐色。前翅端带宽，后方较模糊；痣带宽阔，基支和端支退化、不明显；无其他斑纹。雄性腹部1 ~ 5 节背板黑褐色。第 3 节背板后缘的背中突短

图 3-68-1 方祥新蝎蛉 *Neopanorpa fangxianga* 雄成虫背面观

（贵州雷公山，NWAFU）

Fig. 3-68-1 Male adult of *Neopanorpa fangxianga*, dorsal view

(Mount Leigong, Guizhou, NWAFU)

图 3-68-2 方祥新蝎蛉 *Neopanorpa fangxianga*

雄性外生殖器腹面观

Fig. 3-68-2 Male genitalia of *Neopanorpa fangxianga*,

ventral view

棒状，不超过第 4 节背板中部，其下方还有 1 个尖锐的小突起。第 6 节黑色，长筒形，长于第 5 节，末端背缘具 1 半圆形缺刻。第 7、8 节黄褐色，基部稍缢缩。雄性外生殖器阔卵形。第 9 节背板末端强烈向两侧突出；腹板具宽阔的基柄；下瓣黑色，阔长，外缘向背面翻卷，内缘相互重叠，端部伸达生殖刺突中部。生殖肢基节端部内缘具 1 簇长刚毛；生殖刺突粗壮，短于基节之半，基半部具浓密的刚毛；中齿不明显，亚基齿筒形。阳茎锥形，侧突起发达。

【发生期】 6 月。

【地理分布】 贵州（雷公山）。

图 3-68-3 方祥新蝎蛉 *Neopanorpa fangxianga* 地理分布

Fig. 3-68-3 Distribution of *Neopanorpa fangxianga*

Measurements: Forewing length 15.0 mm.

Morphology: Vertex black. Ocellar triangle black. Rostrum yellowish brown, with two brown longitudinal stripes laterally. Nota black. Pleura and legs yellowish brown. Wing membrane hyaline and slightly tinged with yellow. Pterostigma yellowish brown. Markings dark brown. In forewings, apical band broad and blurred posteriorly; pterostigmal band broad, with its basal and apical branches reduced and inconspicuous; other markings absent. Male terga Ⅰ–Ⅴ dark brown. Notal organ on tergum Ⅲ short clavate, not exceeding middle of tergum Ⅳ. A smaller acute tooth beneath notal organ. A6 black, cylindrical, longer than A5, and bearing a semicircular emargination on apex dorsally. A7 and A8 yellowish brown and slightly constricted basally. Male genital bulb broad oval. Epandrium greatly protruded laterad on latero-apical corners. Hypandrium with a broad basal stalk. Hypovalves black, long broad, curling dorsad along outer margins, and overlapping along inner margins, with their apexes extending to middle of gonostylus. Gonocoxites with a tuft of long bristles on ventral apex. Gonostylus stout and shorter than half length of gonocoxites, and bearing dense setae on basal half; median tooth inconspicuous; subbasal process cylindrical. Aedeagus conical with stout lateral processes.

Emergence: June.

Distribution: Guizhou (Mount Leigong).

69. 台湾新蝎蛉 *Neopanorpa formosana* (Navás, 1911)

【测量】 前翅长 14.0 ~ 15.0 mm。

【形态特征】 头顶黑褐色，单眼三角区黑色。喙黄褐色。胸部背板几乎全部黑褐色，仅中、后胸背板后侧角处稍浅。侧板及足黄褐色。翅无色透明，斑纹褐色。前翅端带阔，后角处有 1 个分离的小斑，内部还有 1 个大型透明斑；痣带完整，具有较宽的基支和窄的端支，在翅前缘处与端带相接；基带断开呈 2 个小斑；缘斑明显；基斑极小，模糊。雄性腹部 1 ~ 5 节背板黑褐色。第 3 节背中突长棒状，伸达第 4 节背板中部，盖住第 4 节上圆钝的后背突。第 6 节黑褐色，筒形。第 7、8 节黄褐色，基部缢缩。雄性外生殖器长椭圆形。第 9 节背板阔长，端部平截。第 9 节腹板具阔长的基柄。下瓣黑且长，中部相互重叠，内缘稍重叠，近基部形成 1 个近圆形窗口。生殖刺突短于基节之半，内缘亚基齿大，中齿钝三角形。阳茎锥形，侧突起十分发达。

【发生期】 3 ~ 5 月。

【地理分布】 台湾（新北、南投、屏东）。

Measurements: Forewing length 14.0–15.0 mm.

Morphology: Vertex dark brown. Ocellar triangle black. Rostrum yellowish brown. Nota mostly dark brown except latero-posterior corners of mesonotum and metanotum slightly paler. Pleura

图 3-69-1 台湾新蝎蛉 ***Neopanorpa formosana*** 雄成虫背面观（台湾新北三峡，NWAFU）

Fig. 3-69-1 Male adult of ***Neopanorpa formosana***, dorsal view (Sanxia, Xinbei, Taiwan, NWAFU)

图 3-69-2 台湾新蝎蛉 ***Neopanorpa formosana*** 雄性外生殖器腹面观

Fig. 3-69-2 Male genitalia of ***Neopanorpa formosana***, ventral view

and legs yellowish brown. Wing membrane hyaline, with markings brown. In forewings, apical band broad with a small separate spot near hind margin,

图 3-69-3 台湾新蝎蛉 ***Neopanorpa formosana*** 地理分布

Fig. 3-69-3 Distribution of ***Neopanorpa formosana***

and with a large hyaline spot innerly; pterostigmal band complete, with a thin apical branch and a broad basal branch, and connected to apical band on anterior margin; basal band split into two small spots; marginal spot conspicuous; basal spot very small and blurred. Male terga Ⅰ–Ⅴ dark brown. Notal organ on tergum Ⅲ long stick-like, extending to middle of tergum Ⅳ and covering rounded postnotal organ on tergum Ⅳ. A6 dark brown and cylindrical. A7 and A8 yellowish brown and constricted basally. Male genital bulb long elliptical. Epandrium long and broad with a truncated apex. Hypandrium with a long broad basal stalk. Hypovalves black and long, with inner margin overlapped mesally and forming a subcircular window subbasally. Gonostylus shorter than half length of gonocoxites, with an obtuse triangular median tooth and a stout subbasal process on inner margin. Aedeagus conical with stout lateral processes.

Emergence: March to May.

Distribution: Taiwan (New Taipei, Nantou, Pingtung).

70. 海南新蝎蛉 *Neopanorpa hainanica* Hua & Chou, 1998

【测量】 前翅长 12.0 mm。

【形态特征】 头顶黑色，单眼三角区黑色。喙黄褐色。胸部背板黄褐色，中间有显著的黑色纵带。侧板及足黄褐色。翅透明，斑纹褐色。前翅端带阔，后角处有 1 个长透明斑；痣带阔，完整，具有较宽的基支和窄的端支，在翅前缘处与端带几乎相接；基带断成 2 斑；缘斑明显；无基斑。雄性腹部 1 ~ 5 节背板黑褐色。第 3 节背中突三角形，盖住第 4 节上圆钝的后背突。第 6 节黑褐色，长筒形。第 7、8 节黄褐色，基部缢缩。雄性外生殖器长卵形。第 9 节背板阔长，端部平截；腹板具较短的基柄；下瓣窄长，中部稍靠拢，末端分离，超过

图 3-70-1 海南新蝎蛉 *Neopanorpa hainanica* 成虫背面观（海南尖峰岭，NWAFU）
1. 雄虫 2. 雌虫
Fig. 3-70-1 Adults of ***Neopanorpa hainanica***, dorsal view (Jianfengling, Hainan, NWAFU)
1. Male 2. Female

生殖肢基节端部。生殖刺突内缘亚基齿，中齿钝三角形。阳茎锥状，侧突起发达。

【发生期】 2 ~ 10 月。

【地理分布】 海南（尖峰岭、五指山）。

Measurements: Forewing length 12.0 mm.

Morphology: Vertex black. Ocellar triangle black. Rostrum yellowish brown.Nota mostly yellowish brown, with a black mesal stripe. Pleura and legs yellowish brown. Wing membrane hyaline, with markings brown. In forewings, apical band broad with a long hyaline spot at posterior portion; pterostigmal band broad with a thin apical branch and a broad basal branch and complete, and nearly connected to apical band on anterior margin; basal band split into two spots; marginal spot conspicuous; basal spot absent. Male terga Ⅰ–Ⅴ black. Notal organ on tergum Ⅲ triangular, covering rounded postnotal organ on tergum Ⅳ. A6 dark brown and long cylindrical. A7 and A8 yellowish brown and constricted basally. Male genital bulb long oval. Epandrium long broad with a truncated apex. Hypandrium with a shorter basal stalk. Hypovalves slender and slightly convergent medially, divergent distally, and exceeding apex of gonocoxites.

图 3-70-2 海南新蝎蛉 *Neopanorpa hainanica* 雄性外生殖器腹面观
Fig. 3-70-2 Male genitalia of ***Neopanorpa hainanica***, ventral view

图 3-70-3 海南新蝎蛉 *Neopanorpa hainanica* 地理分布
Fig. 3-70-3 Distribution of ***Neopanorpa hainanica***

Gonostylus with an obtuse triangular median tooth and a stout subbasal process on inner margin. Aedeagus conical, with a pair of stout lateral processes.

Emergence: February to October.

Distribution: Hainan (Jianfengling, Mount Wuzhi).

71. 哈曼德新蝎蛉 *Neopanorpa harmandi* (Navás, 1908)

【测量】 前翅长 14.0 ~ 15.0 mm。

【形态特征】 头顶黑色，单眼三角区黑色。喙黄褐色。前胸背板黑褐色。中、后胸背板黄褐色，中部具明显的黑色纵带；侧板及足黄褐色。翅面褐色斑纹显著。前翅端带宽阔；痣带宽阔，在前缘处与端带相连；基支宽，端支与痣带断开，呈一三角形小斑，有时与端带相连；基带阔，后方与痣带基支相连；缘斑相当阔，近三角形，末端与痣带相连；基斑延长成带状。雄性腹部 1 ~ 5 节背板黑褐色。第 3 节背板后缘背中突锥形。第 6 节黑褐色，向后渐变为黄褐色。第 7、8 节黄褐色，基部稍缢缩。

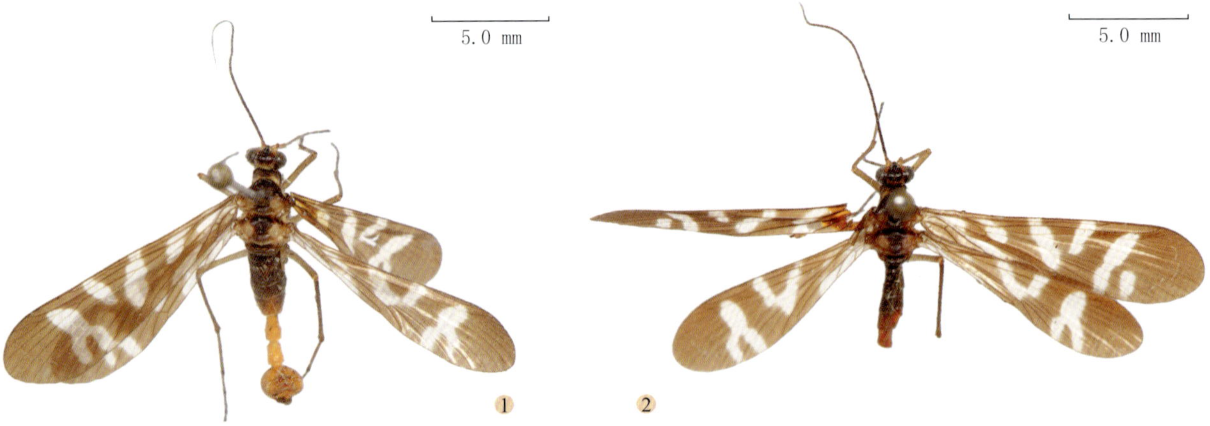

图 3-71-1 哈曼德新蝎蛉 *Neopanorpa harmandi* 成虫背面观（云南西双版纳，NWAFU）

1. 雄虫 2. 雌虫

Fig. 3-71-1 Adults of *Neopanorpa harmandi*, dorsal view (Xishuangbanna, Yunnan, NWAFU)

1. Male 2. Female

图 3-71-2 哈曼德新蝎蛉 *Neopanorpa harmandi* 雄性外生殖器腹面观

Fig. 3-71-2 Male genitalia of *Neopanorpa harmandi*, ventral view

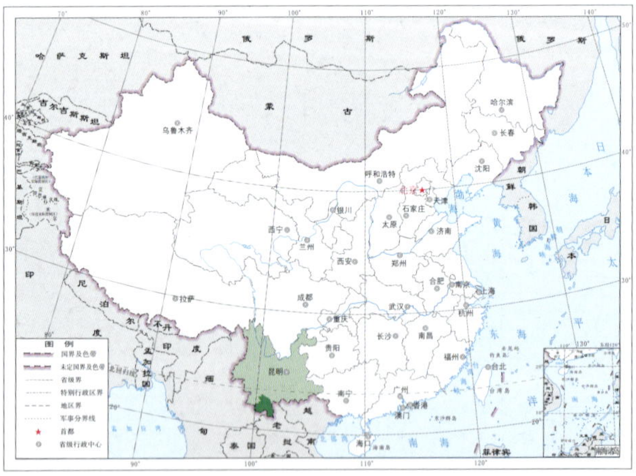

图 3-71-3 哈曼德新蝎蛉 *Neopanorpa harmandi* 地理分布

Fig. 3-71-3 Distribution of *Neopanorpa harmandi*

雄性外生殖器卵形。第9节背板阔长，末端平截。腹板基柄相当阔长，下瓣近三角形，内侧稍重叠，内缘近基部形成1圆形窗口。生殖刺突内缘亚基齿2分叉；中齿钝三角形。阳茎锥形，侧突起发达。

【发生期】 5 ~ 6月。

【地理分布】 云南（普洱、西双版纳）；泰国。

Measurements: Forewing length 14.0–15.0 mm.

Morphology: Vertex black. Ocellar triangle black. Rostrum yellowish brown. Pronotum dark brown. Mesonotum and metanotum yellowish brown with a distinct black mesal stripe. Pleura and legs yellowish brown. Wing membrane hyaline, with dense brown markings. In forewings, apical band broad; pterostigmal band very broad and connected to apical band anteriorly, with its basal branch broad, and apical branch separated as a small triangular spot connected to apical band occasionally; basal band very broad and connected to basal branch of pterostigmal band posteriorly; marginal band broad triangular and connected to pterostigmal band posteriorly; basal spot greatly elongated into a stripe. Male terga Ⅰ–Ⅴ dark brown. Notal organ on tergumⅢ conical. A6 dark brown and gradually turned to yellowish brown caudad. A7 and A8 yellowish brown and constricted basally. Male genital bulb oval. Epandrium long broad, with a truncated apex. Hypandrium with a long broad basal stalk. Hypovalves subtriangular and slightly overlapped mesally, forming a circular window subbasally. Gonostylus with an obtuse triangular median tooth and a bifurcated subbasal process on inner margin. Aedeagus conical with a pair of stout lateral processes.

Emergence: May to June.

Distribution: Yunnan (Pu'er, Xishuangbanna) and Thailand.

72. 华氏新蝎蛉 *Neopanorpa hualizhongi* Hua & Chou, 1998

【测量】 前翅长 12.0 mm。

【形态特征】 头顶黑色，单眼三角区黑色；喙黄褐色，两侧加深。前胸背板黑褐色。中、后胸背板浅褐色，中央具黑色纵带。侧板及足黄褐色。翅透明，斑纹褐色。前翅端带阔，后部具1个透明窗；痣带阔，基支细，端支宽；在端带和痣带之间，还有额外1条横带；缘斑明显；基带断成2斑；无基斑。雄性腹部1 ~ 5节背板黑褐色。第3节背中突短，三角形。第6节黑褐色。第7、8节黄褐色，基部稍缢缩。雄性外生殖器卵形。第9节背板阔长，端

图 3-72-1 华氏新蝎蛉 *Neopanorpa hualizhongi* 雄成虫背面观
（广西金秀，NWAFU）

Fig. 3-72-1 Male adult of *Neopanorpa hualizhongi*, dorsal view
(Jinxiu, Guangxi, NWAFU)

图 3-72-2 华氏新蝎蛉 *Neopanorpa hualizhongi* 雄性外生殖器腹面观

Fig. 3-72-2 Male genitalia of *Neopanorpa hualizhongi*, ventral view

部平截；腹板的基柄阔长；下瓣细长，内缘略重叠。生殖刺突内缘具有1个大的亚基齿和1个钝三角形中齿。阳茎锥形，侧突起发达。

【发生期】 3～7月。

【地理分布】 海南（尖峰岭）、广西（阳朔、金秀）、广东（连州）。

Measurements: Forewing length 12.0 mm.

Morphology: Vertex black. Ocellar triangle black. Rostrum yellowish brown with darkened flanks. Pronotum dark brown. Mesonotum and metanotum light brown, with a black mesal stripe. Pleura and legs yellowish brown. Wing membrane hyaline, with brown markings. In forewings, apical band broad and with a hyaline window posteriorly; pterostigmal band broad and with a broad apical branch and a thin basal branch; an additional stripe between apical and pterostigmal bands; marginal spot conspicuous; basal band split into two spots; basal spot absent. Male terga Ⅰ–Ⅴ dark brown. Notal organ on tergum Ⅲ short triangular. A6 dark brown. A7 and A8 yellowish brown and slightly constricted basally. Male genital bulb oval. Epandrium long and broad with a truncated apex. Hypandrium with a long and broad basal stalk. Hypovalves slender and slightly overlapped mesally. Gonostylus with an obtuse triangular median tooth and a large subbasal process on inner margin. Aedeagus conical, with a pair of stout lateral processes.

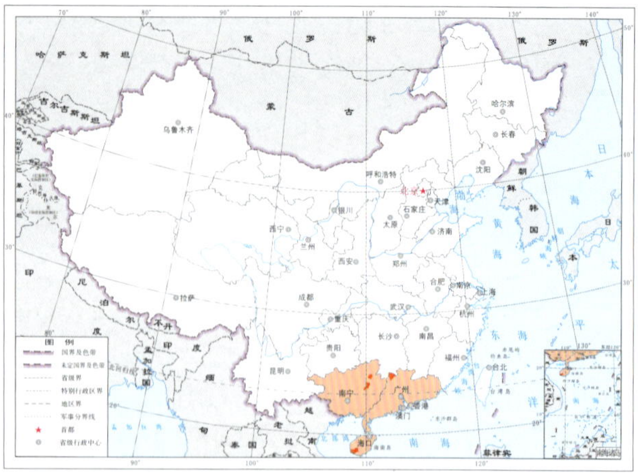

图 3-72-3 华氏新蝎蛉 *Neopanorpa hualizhongi* 地理分布

Fig. 3-72-3 Distribution of *Neopanorpa hualizhongi*

Emergence: March to July.

Distribution: Hainan (Jianfengling), Guangxi (Yangshuo, Jinxiu) and Guangdong (Lianzhou).

73. 湖南新蝎蛉 *Neopanorpa hunanensis* Hua, 2002

【测量】 前翅长 17.0～19.0 mm。

【形态特征】 头顶中部黑色，两侧黄褐色；单眼三角区黑色。喙黄褐色，两侧加深。前胸背板黑褐色。中、后胸背板黑褐色，两侧稍浅；侧板及足黄褐色。翅无色透明，斑纹黑褐色。前翅端带完整，内缘凹；痣带阔，前缘与端带连接；缘斑小；基带完整，后缘阔；基斑不明显。雄性腹部1～5节背板黑褐色。第3节背中突较短，三角形。第6节黑褐色。第7、8节黄褐色，基部缢缩。雄性外生殖器长卵形。第9节背板阔长，末端平截；腹板具阔长的基柄；下瓣阔长，内外缘背向卷曲，内缘重叠，近基部形成1个圆形窗口。阳基侧突短细，端部互相分歧。阳茎锥形，稍超过生殖肢基节端部，侧突起发达。

【发生期】 5～8月。

【地理分布】 湖南（莽山、东安）、广东（连州）、广西（金秀、龙胜、田林、兴安）、贵州（梵净山）。

Measurements: Forewing length 17.0–19.0 mm.

Morphology: Vertex black mesally but yellowish brown laterally. Ocellar triangle black. Rostrum yellowish brown with darkened flanks. Pronotum dark brown. Mesonotum and metanotum dark brown with paler flanks. Pleura and legs yellowish brown. Wing membrane hyaline, with dark brown markings. In forewings, apical band complete with a notched inner margin; pterostigmal band broad and connected to apical band on anterior margin; marginal spot small; basal band complete and broad posteriorly; basal spot inconspicuous. Male terga Ⅰ–Ⅴ dark brown. Notal organ on tergum Ⅲ short and triangular. A6 dark brown. A7 and A8 yellowish brown and slightly constricted basally. Male genital bulb long oval. Epandrium long and broad with a truncated apex. Hypandrium with a long broad basal stalk. Hypovalves long and

图 3-73-1 湖南新蝎蛉 ***Neopanorpa hunanensis*** 成虫背面观（广西兴安猫儿山）

1. 雄虫 2. 雌虫

Fig. 3-73-1 Adults of ***Neopanorpa hunanensis***, dorsal view (Mount Mao'er, Xing'an, Guangxi)

1. Male 2. Female

图 3-73-2 湖南新蝎蛉 *Neopanorpa hunanensis* 雄性外生殖器腹面观

Fig. 3-73-2 Male genitalia of *Neopanorpa hunanensis*, ventral view

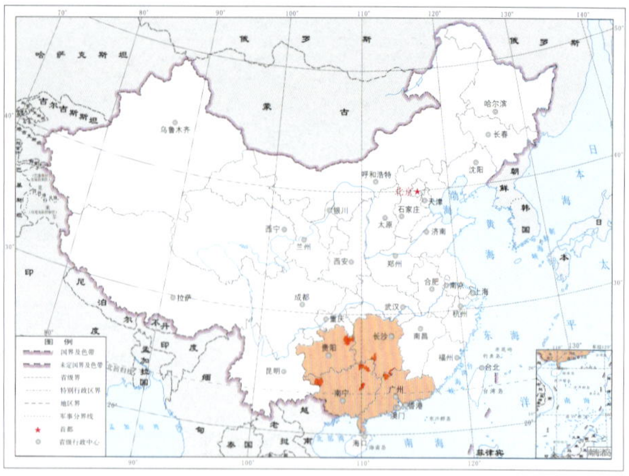

图 3-73-3 湖南新蝎蛉 *Neopanorpa hunanensis* 地理分布

Fig. 3-73-3 Distribution of *Neopanorpa hunanensis*

broad, curled dorsad along outer and inner margins, overlapped mesally, and forming a subcircular window subbasally. Paramere short, thin and divergent apically. Aedeagus conical, slightly exceeding apex of gonocoxites and bearing a pair of stout lateral processes.

Emergence: May to August.

Distribution: Hunan (Mount Mangshan, Dong′an), Guangdong (Lianzhou), Guangxi (Jinxiu, Longsheng, Tianlin, Xing′an) and Guizhou (Mount Fanjing).

74. 胡氏新蝎蛉 *Neopanorpa hushengchangi* Hua & Chou, 1999

【测量】 前翅长 13.5 mm。

【形态特征】 头顶中部黑色，两侧黄褐色。单眼三角区黑色；喙黄褐色，两侧具深色条带。前胸背板黑色。中、后胸背板黄褐色，中央具窄的黑色中带；侧板及足黄褐色。翅透明，略带浅黄色，斑纹浅褐色。前翅端带不完整，只在 R_3 与 M_1 间存在一小斑；痣带窄，基支弱化，端支发达；缘斑明显；基带不完整，只在 CuP 近末端处有 1 个小斑；无基斑。雄性腹部 1 ~ 5 节背板不均匀黄褐色。第 3 节背中突长，伸达第 5 节背板 2/3 处。第 6 节黄褐色。第 7、8 节黄褐色，基部缢缩。雄性外生殖器卵形。第 9 节背板阔长，端部平截。腹板的基柄极短，下瓣细长，两瓣分叉大，呈 V 形；生殖刺突短，长度约为基节之半，内缘亚基齿大，中齿钝三角形。阳茎锥形，侧突起发达。

【发生期】 7 ~ 8 月。

【地理分布】 西藏（樟木）。

图 3-74-1 胡氏新蝎蛉 *Neopanorpa hushengchangi* 雄成虫背面观（西藏樟木，NWAFU）

Fig. 3-74-1 Male adult of *Neopanorpa hushengchangi*, dorsal view (Zham, Tibet, NWAFU)

Measurements: Forewing length 13.5 mm.

Morphology: Vertex black mesally but yellowish brown laterally. Ocellar triangle black. Rostrum yellowish brown with two darkened stripes. Pronotum dark brown. Mesonotum and metanotum yellowish

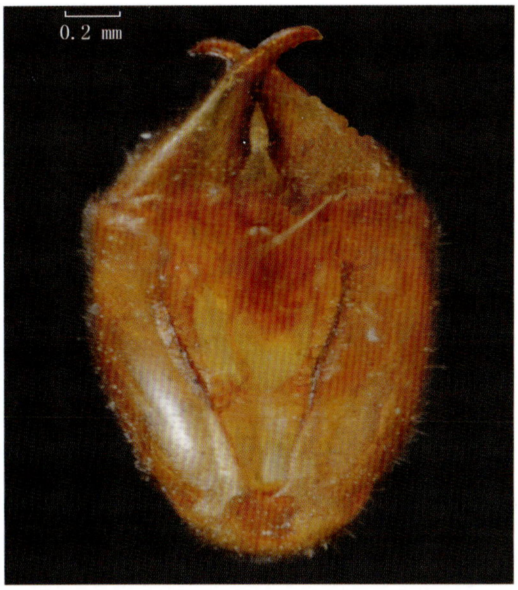

图 3-74-2 胡氏新蝎蛉 *Neopanorpa hushengchangi* 雄性外生殖器腹面观

Fig. 3-74-2 Male genitalia of *Neopanorpa hushengchangi*, ventral view

图 3-74-3 胡氏新蝎蛉 *Neopanorpa hushengchangi* 地理分布

Fig. 3-74-3 Distribution of *Neopanorpa hushengchangi*

brown with a black mesal stripe. Pleura and legs light yellowish brown. Wing membrane hyaline and tinged with light yellow, with light brown markings. In forewings, apical band incomplete and represented by a small spot between R_3 and M_1; pterostigmal band narrow and with a reduced basal branch and a stout apical branch; marginal spot conspicuous; basal band only represented by a small spot near end of CuP; basal spot absent. Male terga Ⅰ–Ⅴ unevenly yellowish brown. Notal organ on tergum Ⅲ long, extending to two-thirds of tergum Ⅴ. A6 yellowish brown. A7 and A8 yellowish brown and constricted basally. Male genital bulb oval. Epandrium long and broad with a truncated apex. Hypandrium with a greatly shortened basal stalk. Hypovalves slender and divergent in a V-shape. Gonostylus approximately half as long as gonocoxites, and bearing an obtuse triangular median tooth and a large subbasal process on inner margin. Aedeagus conical with stout lateral processes.

Emergence: July to August.

Distribution: Tibet (Zham).

75. 鸡公山新蝎蛉 *Neopanorpa jigongshanensis* Hua, 1998

【测量】 前翅长 12.0 mm。

【形态特征】 头顶黄褐色，从额后缘至后头具 1 条黑褐色的纵带，贯穿黑色单眼三角区。喙黄褐色，两侧具深色纵带。前胸背板黑褐色。中、后胸背板黄褐色，中部具明显的黑色纵带。侧板及足黄褐色。翅略带黄色，透明，斑纹褐色。前翅端带宽，后缘具大型透明窗斑；痣带完整，具发达的端支和基支；缘斑大；基带断成两斑；无基斑。雄性腹部 1 ~ 5 节背板黑褐色。第 3 节背中突长棒状，向后延伸超过第 4 节背板后缘；第 4 节背板前方具 1 个方形膜质区。第 6 节基半部黑褐色，向端半部渐变为黄褐色。第 7、8 节黄褐色，基部缢缩。雄性外生殖器卵形。第 9 节背板阔长，端部平截。腹板具宽阔的基柄；下瓣细长，两瓣似 V 形，内缘亚基部具 1 个尖锐的三角形突起。生殖刺突内缘具钝三角形中齿和发达的碗形亚基齿。阳茎侧扁，端部尖锐；腹瓣向基部呈钩状突出；侧突起发达。

【发生期】 7 月。

【地理分布】 河南（鸡公山、桐柏山）。

Measurements: Forewing length 12.0 mm.

Morphology: Vertex yellowish brown. A dark brown stripe extending from hind border of frons to occiput, and passing through black ocellar triangle. Rostrum yellowish brown with two darkened longitudinal stripes laterally. Pronotum dark brown. Mesonotum and metanotum yellowish brown, with a distinct black mesal stripe. Pleura and legs yellowish

图 3-75-1 鸡公山新蝎蛉 ***Neopanorpa jigongshanensis*** 成虫背面观（河南桐柏山，NWAFU）

1. 雄虫 2. 雌虫

Fig. 3-75-1 Adults of ***Neopanorpa jigongshanensis***, dorsal view (Tongbai Mountain, Henan, NWAFU)

1. Male 2. Female

brown. Wing membrane hyaline and tinged with light yellow, with brown markings. In forewings, apical band broad with a large hyaline window posteriorly; pterostigmal band complete with well-developed apical and basal branches; marginal spot large; basal band split into two spots; basal spot absent. Male terga Ⅰ–Ⅴ dark brown. Notal organ on tergum Ⅲ long stick-like, extending beyond hind border of tergum Ⅳ. A trapezoidal membranous area on anterior portion of tergum Ⅳ. A6 dark brown at basal half and gradually becoming yellowish brown towards apex. A7 and A8 yellowish brown and constricted basally. Male genital bulb oval. Epandrium long and broad, with a truncated apex. Hypandrium with a broad basal stalk. Hypovalves slender and divergent in a V-shape, and with an acute triangular tooth on inner margin subbasally. Gonostylus with an obtuse triangular median tooth, and a large bowl-shaped subbasal process on inner margin. Aedeagus laterally flattened with an acute apex; its ventral valves protruding basad and hook-like; lateral processes stout.

Emergence: July.

Distribution: Henan (Mount Jigong, Tongbai Mountain).

图 3-75-2 鸡公山新蝎蛉 ***Neopanorpa jigongshanensis*** 雄性外生殖器腹面观

Fig. 3-75-2 Male genitalia of ***Neopanorpa jigongshanensis***, ventral view

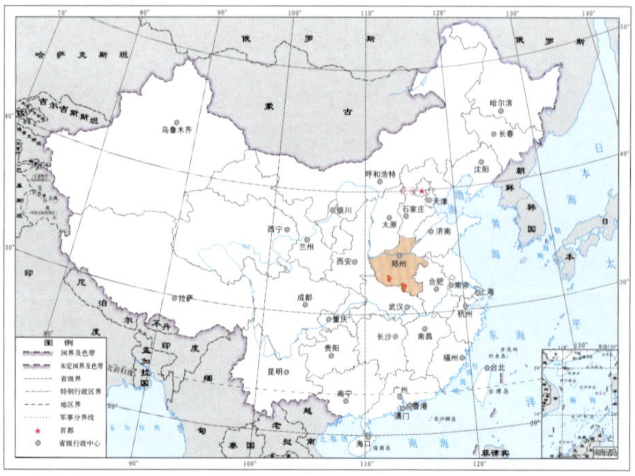

图 3-75-3 鸡公山新蝎蛉 ***Neopanorpa jigongshanensis*** 地理分布

Fig. 3-75-3 Distribution of ***Neopanorpa jigongshanensis***

76. K 纹新蝎蛉 *Neopanorpa k-maculata* Cheng, 1952

【测量】 前翅长 18.5 ~ 20.0 mm。

【形态特征】 头顶黄褐色，中央 1 条黑色纵带延伸至后头区。单眼三角区黑色。喙黄褐色。前胸背板黑褐色；中胸背板大部分黄褐色，前缘黑褐色，具 1 条黑色中带；后胸背板前缘具 2 个黑褐色斑点，并具 1 条黑色中带。翅无色透明，翅斑褐色。前翅端带宽，内缘散乱，近内缘具 2 ~ 3 个透明小斑；痣带阔，具发达的基支和端支；缘斑阔，向外后方延伸并与痣带的基支相接，与痣带形成 1 个明显的 K 形斑纹；在 ORs 附近及翅后缘还有一系列散乱的小斑。雄性腹部 1 ~ 5 节背板不均匀污黄色。第 3 节背板后缘的背中突发达，指状，向后延伸稍超过第 4 节背板中部，盖住第 4 节上圆钝的后背突。6 ~ 8 节显著延长。雄性外生殖器长卵形。第 9 节背板末端具 V 形缺刻；腹板具宽阔的基柄；下瓣阔长，超过生殖肢基节端部，内缘近基部具 1 尖锐的齿突，端半部向内弯曲，内缘着生许多长刚毛。生殖刺突长于基节，内缘具三角形中齿和梯形亚基齿。阳茎锥状，阳基侧突尖锐。

【发生期】 4 ~ 5 月。

【地理分布】 台湾（高雄、屏东）。

Measurements: Forewing length 18.5–20.0 mm.

Morphology: Vertex yellowish brown, with a black longitudinal stripe extending to occiput. Ocellar triangle black. Rostrum yellowish brown. Pronotum dark brown. Mesonotum mostly yellowish brown, with a dark brown anterior margin and a black mesal stripe. Metanotum with two dark brown spots on anterior margin and a black mesal stripe. Metanotum with two dark brown spots on anterior margin and a black mesal stripe. Wing membrane hyaline, with brown markings. In forewings, apical band broad

图 3-76-2 K 纹新蝎蛉 *Neopanorpa k-maculata* 雄性外生殖器腹面观

Fig. 3-76-2 Male genitalia of ***Neopanorpa k-maculata***, ventral view

图 3-76-1 K 纹新蝎蛉 *Neopanorpa k-maculata* 雄成虫背面观

（台湾屏东大汉山，NWAFU）

Fig. 3-76-1 Male adult of ***Neopanorpa k-maculata***, dorsal view

(Mount Tahan, Pingtung, Taiwan, NWAFU)

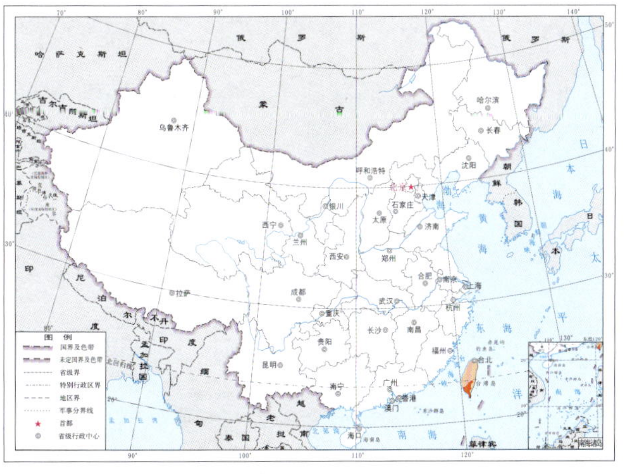

图 3-76-3 K 纹新蝎蛉 *Neopanorpa k-maculata* 地理分布

Fig. 3-76-3 Distribution of ***Neopanorpa k-maculata***

with interrupted inner margin, and 2–3 small hyaline spots innerly; pterostigmal band broad with stout basal and apical branches; marginal spot broad, extending postero-apicad to connect basal branch of pterostigmal band, thereby forming a distinct K-shaped marking with pterostigmal band; a series of interrupted markings near ORs and along hind margin. Male terga Ⅰ–Ⅴ unevenly and sordidly yellowish. Notal organ on tergum Ⅲ developed and finger-like, extending slightly beyond middle of tergum Ⅳ. A6–A8 distinctly elongated. Male genital bulb long oval. Epandrium emarginate terminally in a V-shape. Hypandrium with a broad basal stalk. Hypovalves broad, exceeding apex of gonocoxites, and bearing an acute tooth on inner margin subbasally, with their distal portions curved mesally and numerous long bristles along inner margin. Gonostylus longer than gonocoxites, bearing a triangular median tooth and a trapezoidal subbasal process on inner margin. Aedeagus conical, with acute parameres.

Emergence: April to May.

Distribution: Taiwan (Kaohsiung, Pingtung).

77. 雷公山新蝎蛉 *Nepanorpa leigongshana* Zhou & Zhou, 2007

【测量】 前翅长 13.0 mm。

【形态特征】 头顶黑褐色，单眼三角区黑色。喙黄褐色。前胸背板黑色。中、后胸背板黑褐色，两侧稍黄褐色。侧板及足黄褐色。翅无色透明，斑纹褐色。前翅端带下方不完整，终止于 R_5 末端；痣带不完整，前方与端支相连，基支宽，端支断开为 1 小斑或消失；缘斑小；基带分离为两个小斑；无基斑。雄性腹部 1 ~ 5 节背板深褐色。第 3 节背中突短，近三角形。第 6 节黑褐色。第 7、8 节黄褐色，基部缢缩。雄性外生殖器卵形。第 9 节背板阔长，末端平截；腹板具宽阔的基柄；下瓣端部卷曲，部分重叠，内缘近基部形成 1 个近圆形窗口。阳基侧突端部平行，靠拢。阳茎锥形，侧突起发达。

【发生期】 6 ~ 8 月。

【地理分布】 贵州（雷公山）。

Measurements: Forewing length 13.0 mm.

Morphology: Vertex dark brown. Ocellar triangle black. Rostrum yellowish brown. Pronotum black. Mesonotum and metanotum dark brown, with slightly yellowish brown flanks. Pleura and legs yellowish brown. Wing membrane hyaline, with brown markings. In forewings, apical band incomplete posteriorly and ending at termination of R_5; pterostigmal band incomplete, connected to apical band anteriorly and with a broad basal branch and a separate reduced apical branch; marginal spot small; basal band split into two small spots; basal spot absent. Male terga Ⅰ–Ⅴ dark

图 3-77-1 雷公山新蝎蛉 *Nepanorpa leigongshana* 成虫背面观（贵州雷公山，NWAFU）

1. 雄虫 2. 雌虫

Fig. 3-77-1 Adults of *Nepanorpa leigongshana*, dorsal view (Mount Leigong, Guizhou, NWAFU)

1. Male 2. Female

图 3-77-2 雷公山新蝎蛉 *Nepanorpa leigongshana* 雄性外生殖器腹面观

Fig. 3-77-2 Male genitalia of ***Nepanorpa leigongshana***, ventral view

brown. Notal organ on tergum Ⅲ short subtriangular. A6 dark brown. A7 and A8 yellowish brown and

图 3-77-3 雷公山新蝎蛉 *Neopanorpa leigongshana* 地理分布

Fig. 3-77-3 Distribution of ***Neopanorpa leigongshana***

constricted basally. Male genital bulb oval. Epandrium long broad, with a truncated apex. Hypandrium with a broad basal stalk. Hypovalves curled apically and overlapped mesally, forming a subcircular window subbasally. Parameres parallel and convergent apically. Aedeagus conical, with stout lateral processes.

Emergence: June to August.

Distribution: Guizhou (Mount Leigong).

78. 李氏新蝎蛉 *Nepanorpa lifashengi* Hua & Chou, 1999

【测量】 前翅长 15.0 mm。

【形态特征】 头顶黑褐色，单眼三角区黑色。喙黄褐色。胸部背板大部分黑褐色，后胸后侧缘黄褐色；侧板及足黄褐色。翅透明，无斑纹，或仅存呈三角形的退化痣带。翅痣棕色明显。雄性腹部 1 ~ 5 节背板黑褐色。第 3 节背中突短。第 6 节黑

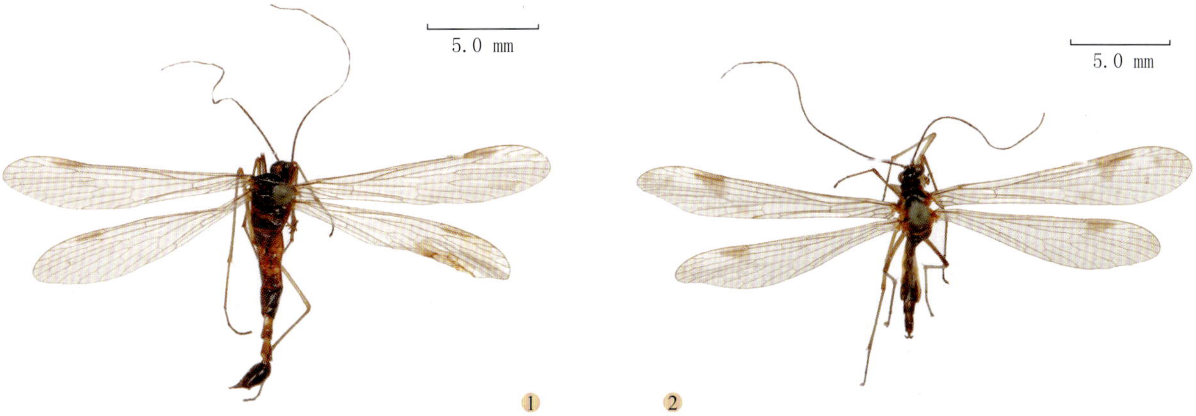

图 3-78-1 李氏新蝎蛉 *Neopanorpa lifashengi* 成虫背面观（西藏波密易贡，NWAFU）

1. 雄虫 2. 雌虫

Fig. 3-78-1 Adults of ***Neopanorpa lifashengi***, dorsal view (Yigong, Bomê, Tibet, NWAFU)

1. Male 2. Female

图 3-78-2 李氏新蝎蛉 *Neopanorpa lifashengi* 雄性外生殖器腹面观
Fig. 3-78-2 Male genitalia of *Neopanorpa lifashengi*, ventral view

褐色。7 ~ 8 节基半部黄色，端半部黑褐色。雄性外生殖器长卵形。第 9 节背板阔长，端部平截。腹板的基柄阔长，伸达生殖肢基节的中部，下瓣端部黄白色，向腹面卷曲。生殖刺突细长，内缘基齿大，中齿钝三角形；阳茎锥形，侧突起发达。

【发生期】 4 ~ 5 月。

【地理分布】 西藏（波密）。

Measurements: Forewing length 15.0 mm.

Morphology: Vertex dark brown. Ocellar triangle black. Rostrum yellowish brown. Nota mostly dark

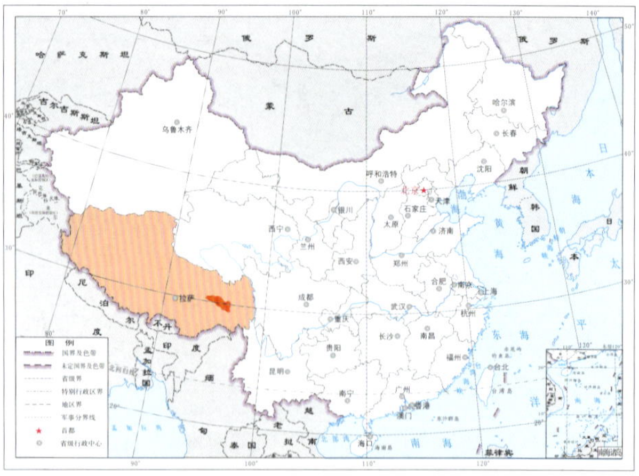

图 3-78-3 李氏新蝎蛉 *Neopanorpa lifashengi* 地理分布
Fig. 3-78-3 Distribution of *Neopanorpa lifashengi*

brown, but latero-posterior corners of metanotum yellowish brown. Pleura and legs yellowish brown. Wing membrane hyaline, devoid of markings or only with reduced and triangular pterostigmal band. Pterostigma brown and distinct. Male terga Ⅰ–Ⅴ dark brown. Notal organ on tergum Ⅲ short. A6 dark brown. A7 and A8 yellow in basal half and yellowish brown apical half. Male genital bulb oval. Epandrium long broad with a truncated apex. Hypandrium with a broad basal stalk, extending to middle of gonocoxites. Hypovalves yellowish white brown and curled ventrad distally. Gonostylus slender with an obtuse triangular median tooth and a large subbasal process. Aedeagus conical, with stout lateral processes.

Emergence: April to May.

Distribution: Tibet (Bomê).

79. 黎坪新蝎蛉 *Neopanorpa lipingensis* Cai & Hua, 2009

【测量】 前翅长 12.0 ~ 12.5 mm。

【形态特征】 头顶、单眼三角区黑褐色。喙黄褐色。前胸背板黑褐色。中、后胸背板两侧黄棕色，中部具黑色纵带。侧板及足黄白色至浅黄褐色。翅无色透明，斑纹黑褐色。前翅端带后缘不完整；痣带基支发达，端支细小，有时断开；缘斑和基带十分退化；无基斑。雄性腹部 1 ~ 5 节背板黑色。第 3 节背中突短，不超过第 4 节背板的中部。第 6 节黑色。第 7、8 节黄褐色，基部稍缢缩。雄性外生殖器卵形。第 9 节背板阔长，末端平截；腹板具阔长的基柄；下瓣宽阔，中部互相重叠，基部内缘稍缢缩形成 1 圆形窗口。生殖刺突近基部具 1 个方形大的亚基齿，中齿钝三角形。阳基侧突基部与阳茎愈合，端部伸达阳茎中部。阳茎粗壮，腹瓣尖长，侧突起细长，向后延伸。

【发生期】 6 月。

【地理分布】 陕西（南郑）。

Measurements: Forewing length 12.0 – 12.5 mm.

Morphology: Vertex and ocellar triangle dark brown. Rostrum yellowish brown. Pronotum dark brown. Mesonotum and metanotum

图 3-79-1 黎坪新蝎蛉 *Neopanorpa lipingensis* 成虫侧面观（陕西南郑黎坪国家森林公园，姜碌摄）

1. 雄虫 2. 雌虫

Fig. 3-79-1 Adults of *Neopanorpa lipingensis*,lateral view (Liping National Forest Park, Nanzheng, Shaanxi, photo by Lu Jiang)

1. Male 2. Female

图 3-79-2 黎坪新蝎蛉 *Neopanorpa lipingensis* 雄性外生殖器腹面观

Fig. 3-79-2 Male genitalia of *Neopanorpa lipingensis*, ventral view

yellowish brown laterally and with a black mesal stripe. Pleura and legs yellowish white to light yellowish brown. Wing membrane hyaline with dark brown markings. In forewings, apical band incomplete posteriorly; pterostigmal band with a stout basal branch and a slender or separated apical branch; marginal spot and basal band greatly reduced; basal spot absent. Male terga Ⅰ–Ⅴ black. Notal organ on tergum Ⅲ short, not exceeding middle of tergum

图 3-79-3 黎坪新蝎蛉 *Neopanorpa lipingensis* 地理分布

Fig. 3-79-3 Distribution of *Neopanorpa lipingensis*

Ⅳ. A6 black. A7 and A8 yellowish brown and slightly constricted basally. Male genital bulb oval. Epandrium long and broad with a truncated apex. Hypandrium with a broad basal stalk. Hypovalves flat and slightly overlapped mesally, forming a circular window subbasally. Gonostylus with an obtuse triangular median tooth and a large trapezoidal subbasal process on inner margin. Paramere fused basally with aedeagus and extending to middle of aedeagus apically. Aedeagus stout with a pair of slender lateral processes directing caudad.

Emergence: June.

Distribution: Shaanxi (Nanzheng).

80. 河南新蝎蛉 *Neopanorpa longiprocessa* Hua & Chou, 1997

【测量】 前翅长 13.5 ~ 15.0 mm。

【形态特征】 头顶深褐色，单眼三角区黑色。喙黄褐色。胸部背板黑褐色，侧板乳黄色，足大部浅黄色。翅无色透明，斑纹黑褐色。前翅端带阔，前缘有 2 个透明窗斑；痣带基支阔，端支窄；痣带与端带之间具 1 条额外短横带；缘斑大；基带断裂呈 2 斑；无基斑。雄性腹部 1 ~ 5 节背板黑褐色；第 3 节背中突发达，棒状，向后延伸几乎达第 5 节后缘。第 6 节黑褐色，圆筒形，向后渐细。7 ~ 8 节黄褐色，基部稍缢缩。雄性外生殖器长卵形。第 9 节背板阔长，端部平截，腹面近端部具 1 对膨大的突起；腹板具阔长的基柄；下瓣宽阔，扁平，内缘中部略有重叠，近基部形成 1 个圆形窗口，下瓣内缘近基部具 1 个指状突起。生殖刺突内缘的亚基齿大，方形；中齿钝三角形。阳基侧突基半部与阳茎愈合，端半部相互分离。阳茎锥形，具发达的侧突起。

【发生期】 5 ~ 8 月。

【地理分布】 河南（嵩县、栾川）、陕西（宁陕、太白、周至、鄠邑）。

Measurements: Forewing length 13.5–15.0 mm.

Morphology: Vertex deep brown. Ocellar triangle black. Rostrum yellowish brown. Nota dark brown. Pleura creamy yellow. Legs mostly light yellow. Wing membrane hyaline, with dark brown markings. In forewings, apical band broad, with two hyaline windows frontally; pterostigmal band with a broad

图 3-80-1 河南新蝎蛉 ***Neopanorpa longiprocessa*** 成虫侧面观（陕西秦岭，姜碌摄）

1. 雄虫 2. 雌虫

Fig. 3-80-1 Adults of ***Neopanorpa longiprocessa***, lateral view (Qinling Mountains, Shaanxi, photo by Lu Jiang)

1. Male 2. Female

图 3-80-2 河南新蝎蛉 *Neopanorpa longiprocessa* 雄成虫侧面观

Fig. 3-80-2 Male adult of *Neopanorpa longiprocessa*, lateral view

图 3-80-3 河南新蝎蛉 *Neopanorpa longiprocessa* 雄性外生殖器腹面观

Fig. 3-80-3 Male genitalia of *Neopanorpa longiprocessa*, ventral view

图 3-80-4 河南新蝎蛉 *Neopanorpa longiprocessa* 地理分布

Fig. 3-80-4 Distribution of *Neopanorpa longiprocessa*

basal branch and a thin apical branch; an additional stripe between apical band and pterostigmal band; marginal spot large; basal band split into two spots; basal spot absent. Male terga Ⅰ–Ⅴ dark brown. Notal organ on tergum Ⅲ stout, clavate, and almost extending to hind border of tergum Ⅴ. A6 dark brown, cylindrical and tapering posteriorly. A7 and A8 yellowish brown, slightly constricted basally. Male genital bulb long oval. Epandrium long and broad, with a truncated apex and a pair of enlarged processes on ventral surface subapically. Hypandrium with a broad basal stalk. Hypovalves broad, flat and slightly overlapped mesally, forming a circular window subbasally. Each hypovalve with a finger-like subbasal process on inner margin. Gonostylus with an obtuse triangular median tooth and a large trapezoidal subbasal process on inner margin. Parameres fused at basal half with aedeagus and divergent at apical half. Aedeagus conical, with a pair of stout lateral processes.

Emergence: May to August.

Distribution: Henan (Songxian, Luanchuan) and Shaanxi (Ningshan, Taibai, Zhouzhi, Huyi).

81. 路氏新蝎蛉 *Neopanorpa lui* Chou & Ran, 1981

【测量】 前翅长 13.5 ~ 14.0 mm。

【形态特征】 头顶黑色，单眼三角区黑色。喙黄褐色。胸部背板黑褐色，中后胸两侧黄褐色，中间有 1 条黑色条斑。侧板及足乳黄色，足大部浅黄褐色。翅透明，无任何斑纹，或仅具极度退化的痣带。翅痣明显，浅褐色。雄性腹部 1 ~ 5 节背板黑褐色。第 3 节背中突较短。第 6 节黑褐色。第 7、8 节黄褐色，基部稍缢缩。雄性外生殖器卵形。第 9 节背板阔长，末端平截；腹板具阔长的基柄；下瓣扁平，基部和端部窄，中部最宽，内缘略有重叠，近基部形成 1 个卵形窗口。生殖刺突内缘中部具钝三角形中齿和较大的亚基齿。阳基侧突大部分与阳茎愈合，末端尖。阳茎锥状，侧突起尖长。

【发生期】 5 ~ 8 月。

【地理分布】 河南（嵩县、栾川）、陕西（宁陕、太白、周至、鄠邑）。

图 3-81-1 路氏新蝎蛉 ***Neopanorpa lui*** 成虫侧面观（陕西宝鸡天台山国家森林公园，姜碌摄）

1. 雄虫 2. 雌虫

Fig. 3-81-1 Adults of ***Neopanorpa lui***, lateral view (Tiantaishan National Forest Park, Baoji,Shaanxi, photo by Lu Jiang)

1. Male 2. Female

图 3-81-2 路氏新蝎蛉 *Neopanorpa lui* 雄性外生殖器腹面观
Fig. 3-81-2 Male genitalia of *Neopanorpa lui,* ventral view

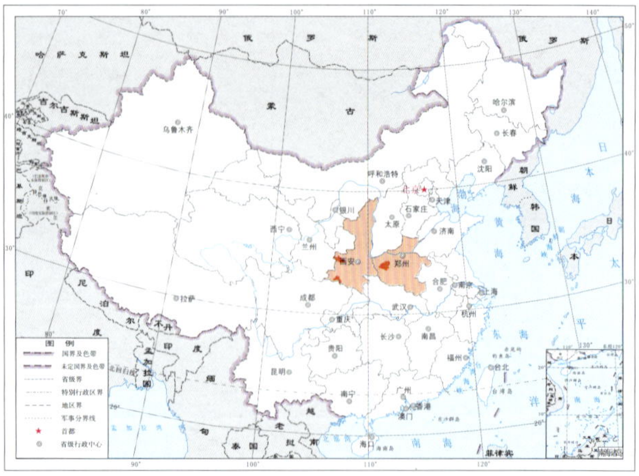

图 3-81-3 路氏新蝎蛉 *Neopanorpa lui* 地理分布
Fig. 3-81-3 Distribution of *Neopanorpa lui*

Measurements: Forewing length 13.5–14.0 mm.

Morphology: Vertex black. Ocellar triangle black. Rostrum yellowish brown. Nota dark brown. Mesonotum and metanotum yellowish brown laterally and with a black mesal stripe. Pleura and legs creamy yellow. Legs mostly light yellow. Wing membrane hyaline, devoid of markings or only with a greatly reduced pterostigmal band. Pterostigma distinct, light brown. Male terga Ⅰ–Ⅴ dark brown. Notal organ on tergumⅢ short. A6 dark brown. A7 and A8 yellowish brown and slightly constricted basally.Male genital bulb oval. Epandrium long and broad, with a truncated apex. Hypandrium with a broad basal stalk. Hypovalves flat, narrow basally and apically, broad medially and slightly overlapped mesally, forming an oval window subbasally. Gonostylus with an obtuse triangular median tooth and a large subbasal process on inner margin. Parameres fused basally with aedeagus and acute apically. Aedeagus conical with a pair of slender and acute lateral processes.

Emergence: May to August.

Distribution: Henan (Songxian, Luanchan), Shaanxi (Ningshan, Taibai, Zhouzhi, Huyi) .

82. 马氏新蝎蛉 *Neopanorpa maai* Cheng, 1957

【测量】 前翅长 14.0 mm。

【形态特征】 头顶红褐色，单眼三角区周围有1个大型黑色斑。喙红棕色，两侧具深色纵带。前胸背板黑褐色。中、后胸背板大部分黑褐色，两侧后角红褐色。侧板及足黄褐色。翅无色透明，斑纹灰褐色。前翅端带阔，内部具1个模糊的窗斑，后方模糊；痣带阔，端支断开；端带与痣带之间的后缘具1个额外的斑；基带散乱，与较小缘斑相连；无基斑。雄性腹部1～5节背板黑色。第3节背板后缘背中突棒状，不超过第4节背板后缘；第4节背板中央具1个方形膜质区。第6节黑色，向端部渐窄。第7节及以后红褐色。雄性外生殖器长卵形。第9节背板阔长，末端平截；腹板具短阔的基柄；下瓣长带状，超出生殖肢基节端部。生殖刺突内缘具钝三角形中齿和碟形亚基齿。阳基侧突线状，不超出阳茎端部；阳茎锥状，不超出生殖肢基节端部，侧突起发达。

【发生期】 4～7月。

【地理分布】 福建（邵武、武夷山、建阳）。

Measurements: Forewing length 14.0 mm.

Morphology: Vertex reddish brown. Ocellar triangle enclosed by a large black spot. Rostrum reddish brown, with two deep brown longitudinal stripes. Pronotum dark brown. Mesonotum and metanotum mostly dark brown, with latero-posterior

图 3-82-1 马氏新蝎蛉 ***Neopanorpa maai*** 雄成虫背面观（福建邵武大竹岚，NWAFU）

Fig. 3-82-1 Male adult of ***Neopanorpa maai***, dorsal view (Dazhulan, Shaowu, Fujian, NWAFU)

图 3-82-2 马氏新蝎蛉 ***Neopanorpa maai*** 雄性外生殖器腹面观

Fig. 3-82-2 Male genitalia of ***Neopanorpa maai***, ventral view

图 3-82-3 马氏新蝎蛉 ***Neopanorpa maai*** 地理分布

Fig. 3-82-3 Distribution of ***Neopanorpa maai***

corners reddish brown. Pleura and legs mostly yellowish brown. Wing membrane hyaline, slightly tinged with greyish brown. In forewings, apical band broad, with a large hyaline window innerly, and blurred posteriorly; pterostigmal band broad, with a separated apical branch; an additional spot between apical and pterostigmal bands; basal band sacttered and connected to small marginal spot; basal spot absent. Male terga Ⅰ–Ⅴ black. Notal organ on hind border of tergum Ⅲ long clavate, not exceeding hind border of tergum Ⅳ; tergum Ⅳ with trapezoidal membranous region on anterior portion mesally. A6 black and tapering towards apex. A7 and succeeding segments reddish brown. Male genital bulb oval. Epandrium long and broad, with a truncated apex. Hypandrium with a broad but short basal stalk. Hypovalves slender and stripe-like, exceeding apex of gonocoxites apically. Gonostylus with an obtuse triangular median tooth and a disc-shaped subbasal process on inner margin. Paramere thread-like, not exceeding apex of aedeagus; aedeagus conical, not exceeding apex of gonocoxites; lateral processes of aedeagus stout.

Emergence: April to July.

Distribution: Fujian (Shaowu, Wuyi Mountains, Jianyang).

83. 大新蝎蛉 *Neopanorpa magna* Issiki, 1927

【测量】 前翅长 20.0 ~ 21.0 mm。

【形态特征】 头顶黑色，单眼三角区黑色。喙红棕色。前胸背板黑色。中、后胸背板大部分黑色，后侧角处红褐色。侧板及足红褐色。翅狭长，无色透明，斑纹黑色。前翅端带完整；痣带完整具基支和端支；缘斑和基带退化；无基斑。雄性腹

图 3-83-1 大新蝎蛉 *Neopanorpa magna* 雄成虫侧背面观（台湾，林义祥摄）
Fig. 3-83-1 Male adult of *Neopanorpa magna*, lateral view (Taiwan, photo by Yechan Lin)

部 1 ~ 5 节背板黑棕色。第 3 节背中突棒状，伸达第 4 节背板中部，盖于第 4 节的后背突之上。第 6 节黑褐色，长筒形；第 7、8 节基部强烈缢缩成柄状。雄性外生殖器阔卵形；第 9 节背板阔长，末端稍缺刻；腹板具较短的基柄；下瓣内缘中部具 1 折角；生殖刺突长于基节，向内弧形弯曲，内缘具发达的亚基齿和三角形中齿。

【发生期】 7 ~ 9 月。

【地理分布】 台湾。

Measurements: Forewing length 20.0–21.0 mm.

Morphology: Vertex black. Ocellar triangle black. Rostrum reddish brown. Pronotum black. Mesonotum and metanotum mostly black, with latero-posterior corners reddish brown. Pleura and legs mostly reddish brown. Wing membrane hyaline, slender, with black markings. In forewings, apical band complete; pterostigmal band with basal and apical branches; marginal spot and basal band reduced; basal

图 3-83-2 大新蝎蛉 *Neopanorpa magna* 地理分布
Fig. 3-83-2 Distribution of *Neopanorpa magna*

spot absent. Male tergaⅠ–Ⅴ dark brown. Notal organ on tergumⅢ clavate, extending to middle of tergum Ⅳ and covering postnotal organ. A6 dark brown, long cylindrical. A7 and A8 intensively constricted and stalk-like basally. Male genital bulb oval. Epandrium long and broad, slightly emarginate terminally. Hypandrium with a short basal stalk. Hypovalve with a projection on inner margin medially. Gonostylus longer than gonocoxites, greatly curved mesad, with a triangular median tooth and a stout subbasal process on inner margin.

Emergence: July to September.

Distribution: Taiwan.

84. 莽山新蝎蛉 *Neopanorpa mangshanensis* Chou & Wang, 1988

图 3-84-1 莽山新蝎蛉 *Neopanorpa mangshanensis* 成虫背面观（湖南宜章莽山，NWAFU）

1. 雄虫 2. 雌虫

Fig. 3-84-1 Adults of *Neopanorpa mangshanensis*, dorsal view (Mount Mangshan, Yizhang, Hunan, NWAFU)

1. Male 2. Female

图 3-84-2 莽山新蝎蛉 *Neopanorpa mangshanensis* 雄性外生殖器腹面观

Fig. 3-84-2 Male genitalia of *Neopanorpa mangshanensis*, ventral view

【测量】 前翅长 13.0 mm。

【形态特征】 头顶黑褐色，单眼三角区黑色。喙黄褐色。前胸背板黑褐色。中、后胸背板黑褐色，两侧后角黄褐色；侧板及足黄褐色。翅无色透明，斑纹棕色。前翅端带完整，内缘中部向内突出，后缘分离出 1 块小斑；痣带基支宽，端支极窄或断开；基带断开为 2 个小斑；缘斑小；基斑不明显。雄性腹部 1 ~ 5 节背板黑褐色。第 3 节背中突长棒状，超出第 4 节背板末端，盖住第 4 节上的后背突。第 6 节黑褐色，长筒形。第 7、8 节黄褐色，较短小，基部缢缩。雄性外生殖器卵形。第 9 节背板阔长，近端部两侧各具 1 个小突起，末端平截；腹板基柄阔；下瓣基半部边缘黑色，其余灰白色；近矩形，近基部内缘形成 1 个菱形窗口，两瓣端半部内缘相互重叠。生殖刺突内缘中齿钝三角形，亚基齿粗大。阳基侧突末端阔叶状，左右分歧。阳茎锥状，侧突起发达。

【发生期】 5 ~ 6 月。

【地理分布】 湖南（莽山、八面山）。

Measurements: Forewing length 13.0 mm.

Morphology: Vertex dark brown. Ocellar triangle black. Rostrum yellowish brown. Pronotum dark brown. Mesonotum and metanotum mostly dark brown, with latero-posterior corners yellowish brown. Pleura and legs yellowish brown.Wing membrane hyaline, with brown markings. In forewings, apical band complete and protruding on inner margin; a thin stripe near apical band posteriorly; pterostigmal band with a broad basal branch and a thin or seperated apical branch; basal band split into two small spots; marginal spot small; basal spot inconspicuous. Male tergaⅠ–Ⅴ dark brown. Notal organ on tergum Ⅲ long stick-like, exceeding hind border of tergum Ⅳ and covering postnotal organ. A6 dark brown, long cylindrical. A7 and A8 yellowish brown, relatively short and constricted basally.Male genital bulb oval. Epandrium long and broad, protruding laterad and forming two small processes subapically, and truncated terminally. Hypandrium with a broad basal stalk. Hypovalve mostly greyish white, blackish marginally on basal half, and almost oblong in shape. Two hypovalves overlapped mesally, forming a diamond-shaped window subbasally. Gonostylus with an obtuse triangular median tooth and a stout subbasal process on inner margin. Parameres broad foliate and divergent apically. Aedeagus conical and with stout lateral processes.

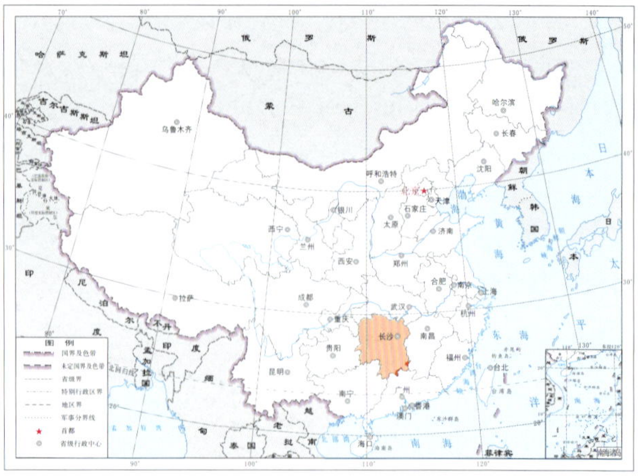

图 3-84-3 莽山新蝎蛉 *Neopanorpa mangshanensis* 地理分布

Fig. 3-84-3 Distribution of *Neopanorpa mangshanensis*

Emergence: May to June.

Distribution: Hunan (Mount Mangshan, Mount Bamian).

85. 小新蝎蛉 *Neopanorpa minuta* Chou & Wang, 1988

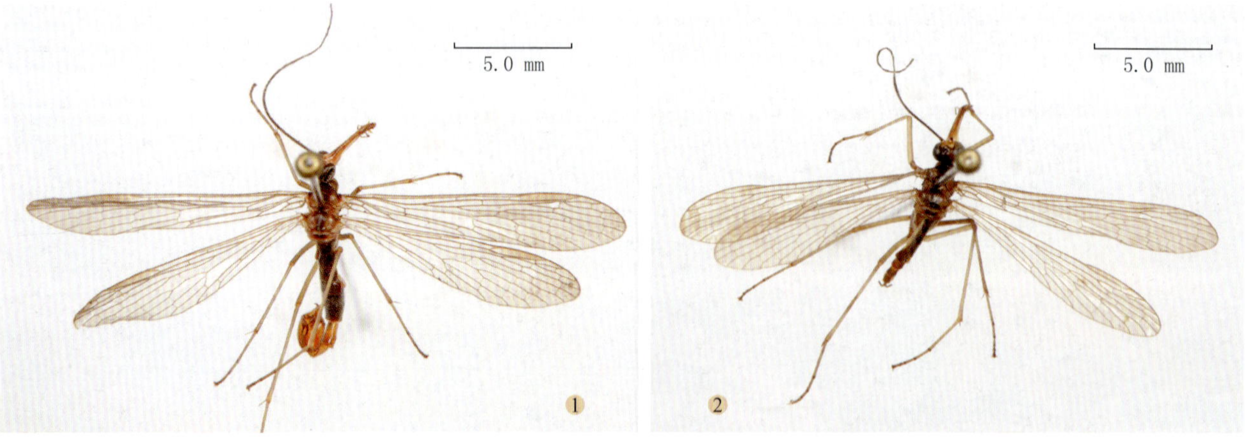

图 3-85-1 小新蝎蛉 *Neopanorpa minuta* 成虫背面观（湖南宜章莽山，NWAFU）

1. 雄虫 2. 雌虫

Fig. 3-85-1 Adults of *Neopanorpa minuta*, dorsal view (Mount Mangshan, Yizhang, Hunan, NWAFU)

1. Male 2. Female

【测量】 前翅长 11.0 ~ 12.0 mm。

【形态特征】 头顶黑褐色，单眼三角区黑色。喙黄褐色。前胸背板黑褐色。中、后胸背板黄褐色，中央有一条黑色纵带。侧板和足黄色。翅略带黄色透明，翅痣黄色，无任何明显斑纹。雄性腹部 1 ~ 5 节背板黑褐色。腹部第 3 节背中突较短，不超过第 4 节背板中部，盖住第 4 节上的后背突。第 6 节黑褐色，筒形。第 7、8 节黄褐色，基部缢缩。雄性

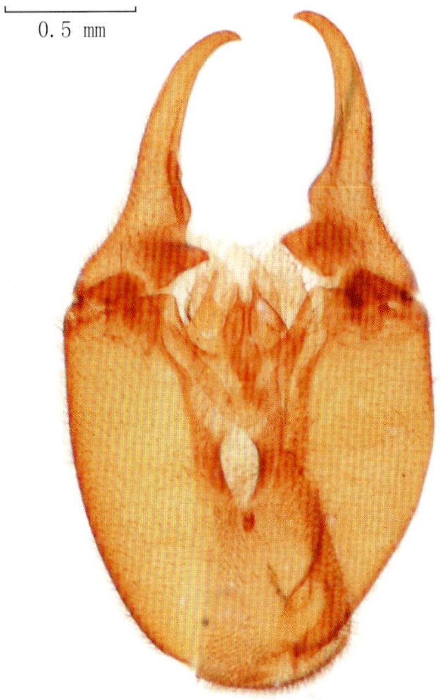

图 3-85-2 小新蝎蛉 *Neopanorpa minuta* 雄性外生殖器腹面观

Fig. 3-85-2 Male genitalia of ***Neopanorpa minuta***, ventral view

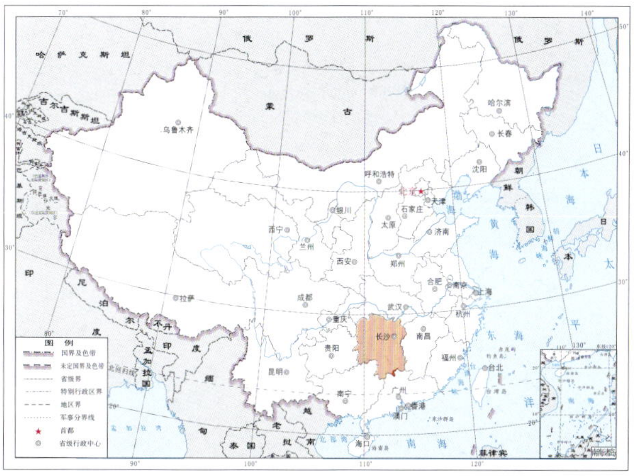

图 3-85-3 小新蝎蛉 *Neopanorpa minuta* 地理分布

Fig. 3-85-3 Distribution of ***Neopanorpa minuta***

外生殖器长卵形。第 9 节背板阔长，末端平截；腹板具阔长的基柄；下瓣阔长，末端圆钝。生殖刺突尖细，外缘内凹，内缘中齿钝三角形，亚基齿尖桃形。阳基侧突粗短，末端尖；阳茎锥形，侧突起发达。

【发生期】 5 ~ 6 月。

【地理分布】 湖南（莽山）。

Measurements: Forewing length 11.0 –12.0 mm.

Morphology: Vertex dark brown. Ocellar triangle black. Rostrum yellowish brown. Pronotum dark brown. Mesonotum and metanotum mostly yellowish brown, with a black mesal stripe. Pleura and legs yellowish brown. Wing membrane hyaline, tinged with yellow, devoid of markings. Pterostigma yellow. Male terga I–V dark brown. Notal organ on tergum III relatively short, not exceeding middle of tergum IV and covering postnotal organ on tergum IV. A6 dark brown, cylindrical. A7 and A8 yellowish brown, relatively short and constricted basally. Male genital bulb long oval. Epandrium long and broad, truncated terminally. Hypandrium with a broad basal stalk. Hypovalve long and broad, with rounded apex. Gonostylus slender, concaved on outer margin, and bearing an obtuse triangular median tooth and an acute peach-shaped subbasal process on inner margin. Paramere short and stout with an acute apex. Aedeagus conical and with stout lateral processes.

Emergence: May to June.

Distribution: Hunan (Mount Mangshan).

86. 莫干山新蝎蛉 *Neopanorpa moganshanensis* Zhou & Wu, 1993

【测量】 前翅长 12.5 ~ 13.0 mm。

【形态特征】 头顶黑色，单眼三角区黑色；喙深褐色，两侧具黑褐色纵带。胸部背板深褐色。侧板和腹部黄褐色。翅狭长，透明，略带烟褐色，无斑纹，仅翅端色稍暗；翅痣褐色。腹部 1 ~ 5 节背板黑色。第 3 节背中突极长，超过第 6 节末端。第 6 节黑色筒形，向后渐细。第 7、8 节黑褐色，基部仅稍缢缩。雄性外生殖器卵形。第 9 节背板阔长，末端具深 U 形缺刻，两侧形成 1 对指形突起；腹板基柄阔长；下瓣短，末端较狭，基部分开。生殖刺突与生殖肢基节约等长，尖细，内缘具 1 三角形亚基齿。阳基侧突基部游离，端部叶片状。阳茎锥状，侧突起粗壮。

【发生期】 6 ~ 7 月。

【地理分布】 浙江（莫干山、凤阳山）、福建（武夷山）。

Measurements: Forewing length 12.5–13.0 mm.

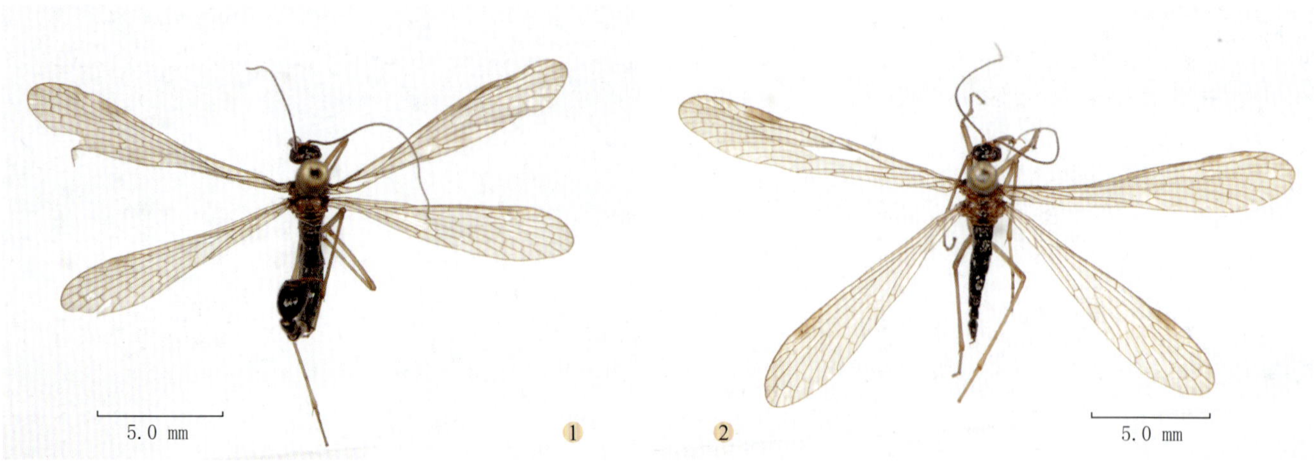

图 3-86-1　莫干山新蝎蛉 ***Neopanorpa moganshanensis*** 成虫背面观（浙江龙泉凤阳山，NWAFU）
1. 雄虫　2. 雌虫

Fig. 3-86-1　Adults of ***Neopanorpa moganshanensis***, dorsal view (Mount Fengyang, Longquan, Zhejiang, NWAFU)
1. Male　2. Female

图 3-86-2　莫干山新蝎蛉 ***Neopanorpa moganshanensis*** 雄性腹部侧面观

Fig. 3-86-2　Male abdomen of ***Neopanorpa moganshanensis*** , lateral view

图 3-86-3　莫干山新蝎蛉 ***Neopanorpa moganshanensis*** 雄性外生殖器腹面观

Fig. 3-86-3　Male genitalia of ***Neopanorpa moganshanensis***, ventral view

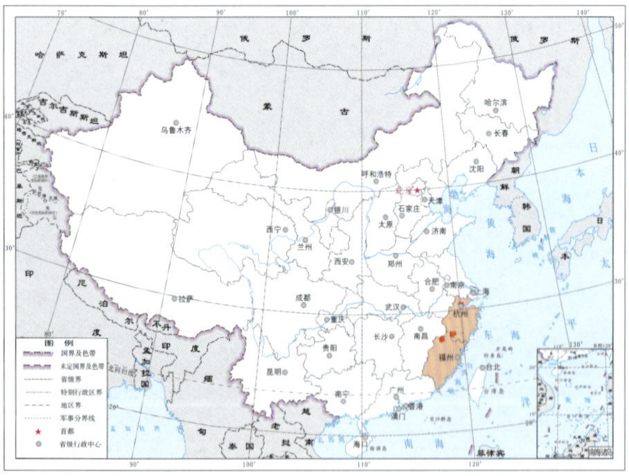

图 3-86-4　莫干山新蝎蛉 ***Neopanorpa moganshanensis*** 地理分布

Fig. 3-86-4　Distribution of ***Neopanorpa moganshanensis***

Morphology: Vertex black. Ocellar triangle black. Rostrum deep brown with two dark brown longitudinal stripes laterally. Nota dark brown. Pleura and legs mostly yellowish brown. Wing membrane hyaline, slender, slightly tinged with greyish, and devoid of markings, except for darkened tinge on apex. Pterostigma brown. Male terga Ⅰ–Ⅴ black. Notal organ on tergum Ⅲ greatly elongated, exceeding hind border of A6. A6 black, cylindrical and tapering toward apex. A7 and A8 dark brown, slightly constricted basally. Male genital bulb oval. Epandrium long and broad, deeply emarginate in a U-shape terminally and forming a pair of finger-like processes laterally. Hypandrium with a broad basal stalk. Hypovalve short and thin apically, and separated basally. Gonostylus almost as long as gonocoxites, and bearing a large triangular subbasal process on inner margin. Paramere free and foliate apically. Aedeagus conical and with stout lateral processes.

Emergence: June to July.

Distribution: Zhejiang (Mount Mogan, Mount Fengyang) and Fujian (Wuyi Mountains).

87. 异新蝎蛉 *Neopanorpa mutabilis* Cheng, 1957

【测量】 前翅长 12.5 ~ 13.0 mm。

【形态特征】 头顶黑色，单眼三角区黑色。喙黄褐色。前胸背板黑褐色。中、后胸背板两侧黄褐色，中央具宽阔的黑色纵纹。侧板及足黄褐色。翅无色，透明，斑纹褐色。前翅端带阔，后角具1个开放的透明窗口；痣带完整，基支较宽，端支较窄；在痣带与端带之间具1条额外斜带；缘斑狭长；基带呈2个不规则斑；无基斑。雄性腹部1 ~ 5节背板黑褐色。第3节背中突长棒状，不超过第4节背板的后缘，并盖住第4节的后背突。第6节黑褐色。第7、8节黄褐色，基部缢缩。雄性外生殖器长椭圆形。第9节背板阔长，末端平截；腹板具宽阔的基柄；下瓣细长，中间稍阔，基部和端部较窄，伸达生殖肢基节末端。生殖刺突内缘具大型的碗形亚基齿和钝三角形中齿。阳基侧突分为2叉，腹支纤细而游离，背支与阳茎愈合。阳茎锥状，侧突起发达。

【发生期】 4 ~ 7月。

【地理分布】 福建（邵武、长汀、武夷山）、浙江（天目山）、安徽（黄山）。

Measurements: Forewing length 12.5–13.0 mm.

Morphology: Vertex black. Ocellar triangle black. Rostrum yellowish brown.Pronotum dark brown. Mesonotum and metanotum yellowish brown laterally, with a broad black mesal stripe. Pleura and legs mostly yellowish brown. Wing membrane hyaline, with brown markings. In forewings, apical band broad, with a large hyaline open window posteriorly; pterostigmal band complete with a broad basal branch and a thin apical branch; an additional oblique stripe between apical and pterostigmal bands; marginal spot slender; basal band split into two irregular spots; basal spot absent. Male terga Ⅰ–Ⅴ dark brown. Notal organ on tergum Ⅲ long stick-like, not exceeding hind border of tergum Ⅳ and covering postnotal organ. A6 dark brown. A7 and A8 yellowish brown, and constricted basally. Male genital bulb long oval. Epandrium long and broad, and truncated terminally. Hypandrium with

图 3-87-1 异新蝎蛉 *Neopanorpa mutabilis* 成虫背面观（安徽黄山，NWAFU）

1. 雄虫 2. 雌虫

Fig. 3-87-1 Adults of *Neopanorpa mutabilis*, dorsal view (Mount Huangshan, Anhui, NWAFU)

1. Male 2. Female

图 3-87-2 异新蝎蛉 *Neopanorpa mutabilis* 雄性外生殖器腹面观
Fig. 3-87-2 Male genitalia of ***Neopanorpa mutabilis***, ventral view

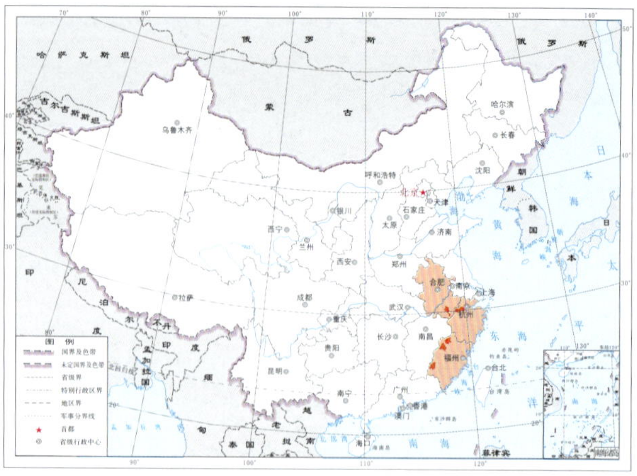

图 3-87-3 异新蝎蛉 *Neopanorpa mutabilis* 地理分布
Fig. 3-87-3 Distribution of ***Neopanorpa mutabilis***

a broad basal stalk. Hypovalve slender, broad medially and thin basally, with its tapering apex extending to apex of gonocoxites. Gonostylus with an obtuse triangular median tooth and a bowl-shaped subbasal process on inner margin. Paramere bifurcated, with its ventral branch thin and free, and dorsal branch fused with aedeagus. Aedeagus conical, with stout lateral processes.

Emergence: April to July.

Distribution: Fujian (Shaowu, Changting, Wuyi Mountains), Zhejiang (Tianmu Mountains) and Anhui (Mount Huangshan).

88. 尼尔森新蝎蛉 *Neopanorpa nielseni* Byers, 1965

【测量】 前翅长 16.5 ~ 17.0 mm。

【形态特征】 头顶黑褐色，单眼三角区黑褐色；喙深褐色，两侧具深色纵条带。前胸背板黑褐色。中、后胸背板黑褐色，两侧后角稍黄褐色。侧板亮黄色，足黄褐色。翅无色透明，斑纹黑褐色。前翅端带内缘中部有个 1 直角；痣带具完整的端支和基支，在雄性个体上端支和基支常不同程度弱化；基带仅在 CuP 末端附近呈 1 个不规则斑；无缘斑和基斑。雄性腹部 1 ~ 5 节背板黑褐色，腹板黄白色，侧板黄色。第 3 节背中突长棒状，不超过第 4 节。第 6 节黑褐色。第 7、8 节橙红色，基部缢缩。雄性外生殖器长卵形。第 9 节背板阔长，末端平截；腹板具阔长的基柄；下瓣侧缘卷曲，相互套叠，内缘近基部形成 1 个近圆形窗口。生殖刺突细长，内缘中齿钝三角形，近基部具显著的亚基齿，亚基齿下方延伸出 1 个尖锐的突起。阳基侧突几乎完全与阳茎愈合；阳茎锥形，侧突起发达。

【发生期】 4 ~ 8 月。

【地理分布】 云南（永德、西双版纳）；越南。

Measurements: Forewing length 16.5–17.0 mm.

Morphology: Vertex dark brown. Ocellar triangle dark brown. Rostrum dark brown, with two darkened longitudinal stripes laterally. Pronotum dark brown. Mesonotum and metanotum mostly dark brown, with latero-posterior corners yellowish brown. Pleura shining yellow. Legs mostly yellowish brown.Wing membrane hyaline, with dark brown markings. In forewings, apical band broad and with a right angle on inner margin, pterostigmal band with a basal branch and an apical branch, which is variously reduced in males; basal band reduced into an irregular spot near end of CuP; marginal and basal spots absent.Male terga Ⅰ–Ⅴ dark brown; sterna yellowish white; pleura yellow. Notal organ on tergum Ⅲ long clavate, not exceeding hind border of tergum Ⅳ. A6 dark brown. A7

图 3-88-1 尼尔森新蝎蛉 ***Neopanorpa nielseni*** 成虫背侧面观（云南永德棠梨山）

1. 雄虫 2. 雌虫

Fig. 3-88-1 Adults of ***Neopanorpa nielseni***, dorso-lateral view (Mount Tangli, Yongde, Yunnan)

1. Male 2. Female

图 3-88-2 尼尔森新蝎蛉 *Neopanorpa nielseni* 雄性外生殖器腹面观

Fig. 3-88-2 Male genitalia of *Neopanorpa nielseni*, ventral view

图 3-88-3 尼尔森新蝎蛉 *Neopanorpa nielseni* 地理分布

Fig. 3-88-3 Distribution of *Neopanorpa nielseni*

and A8 reddish brown and constricted basally. Male genital bulb long oval. Epandrium long broad, with a truncated apex. Hypandrium with a long and broad basal stalk. Hypovalves curled along lateral margins and overlapped mesally, forming a subcircular window subbasally. Gonostylus slender, bearing an obtuse triangular median tooth, and a large subbasal process extending basad and forming an acute tooth. Paramere almost fused with aedeagus. Aedeagus conical with stout lateral processes.

Emergence: April to August.

Distribution: Yunnan (Yongde, Xishuangbanna) and Vietnam.

89. 黑色新蝎蛉 *Neopanorpa nigritis* Carpenter, 1938

【测量】 前翅长 13.0 mm。

【形态特征】 头顶黑色，单眼三角区黑色；喙黑褐色。胸部背板黑色。侧板及足灰褐色。翅灰褐色，透明；无斑纹。翅痣明显，浅红棕色。雄性腹部 1 ~ 5 节背板黑色。第 3 节背中突棒状，几乎伸达第 4 节末端。第 6 节黑色，筒形。第 7、8 节黑色，基部缢缩。雄性外生殖器卵形。第 9 节背板阔长，末端稍缺刻，腹面亚端部具 1 对突起；腹板的基柄较短；下瓣细长，端半部灰白色，末端向内侧弯曲。生殖刺突内缘亚基齿发达，筒形，中齿钝三角形。阳基侧突叶片状；阳茎锥形，侧突起发达。

【发生期】 6 ~ 7 月。

【地理分布】 四川（峨眉山）、重庆（金佛山）。

Measurements: Forewing length 13.0 mm.

Morphology: Vertex black. Ocellar triangle black. Rostrum dark brown. Nota black. Pleura and legs greyish brown. Wing membrane hyaline, tinged with greyish brown, and devoid of markings. Pterostigma distinct, light reddish brown. Male terga Ⅰ – Ⅴ black. Notal organ on tergum Ⅲ long clavate, almost exceeding hind border of tergum Ⅳ. A6 black, cylindrical. A7 and A8 black and constricted basally. Male genital bulb oval. Epandrium long and broad, slightly emarginate terminally, and bearing a pair of subapical processes on ventral surface. Hypandrium with a relatively short basal stalk. Hypovalves slender, whitish at apical half, and curving mesad apically. Gonostylus with an obtuse triangular median tooth and a very large cylindrical subbasal process on inner margin. Paramere foliate. Aedeagus conical, with stout lateral processes.

图 3-89-1 黑色新蝎蛉 ***Neopanorpa nigritis*** 雄成虫背面观（重庆南川金佛山）
Fig. 3-89-1 Male adult of ***Neopanorpa nigritis***, dorsal view (Mount Jinfo, Nanchuan, Chongqing)

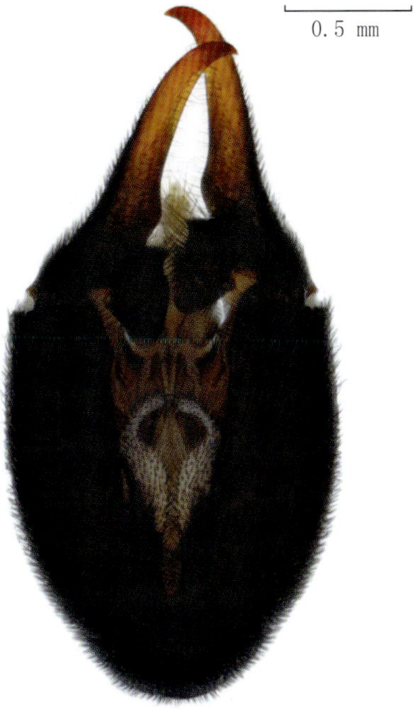

图 3-89-2 黑色新蝎蛉 ***Neopanorpa nigritis*** 雄性外生殖器腹面观
Fig. 3-89-2 Male genitalia of ***Neopanorpa nigritis***, ventral view

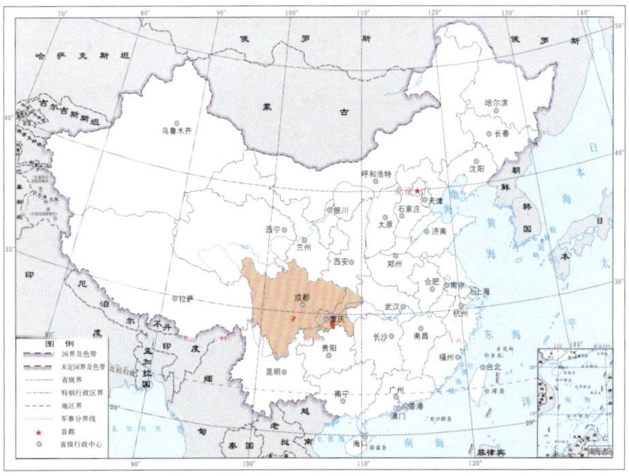

图 3-89-3 黑色新蝎蛉 ***Neopanorpa nigritis*** 地理分布
Fig. 3-89-3 Distribution of ***Neopanorpa nigritis***

Emergence: June to July.

Distribution: Sichuan (Mount Emei) and Chongqing (Mount Jinfo).

90. 细纹新蝎蛉 *Neopanorpa ophthalmica* (Navás, 1911)

【测量】 前翅长 13.5 ~ 14.5 mm。

【形态特征】 头顶黑褐色，单眼三角区黑色。喙黄褐色。前胸背板黑褐色。中、后胸背板黄褐色，中央具显著的黑色纵带。侧板及足黄褐色。翅略带黄色，透明，斑纹褐色。前翅端带阔，具有几个透明斑，在翅前缘处与痣带相连；痣带基支宽，端支窄；缘斑窄长；基带完整或断开成 2 斑；基斑极小。雄性腹部 1 ~ 5 节背板黑褐色；第 3 节背中突长棒状，不超过第 4 节背板后缘；第 6 节黑褐色。第 7、8 节黄褐色，基部缢缩。雄性外生殖器长椭圆形。第 9 节背板阔长，末端平截；腹板具阔长的基柄；下瓣阔长，基部分离，内缘相互重叠。生殖刺突内缘中齿钝三角形，亚基齿大。阳茎锥状，侧突起发达。

【发生期】 6 ~ 7 月。

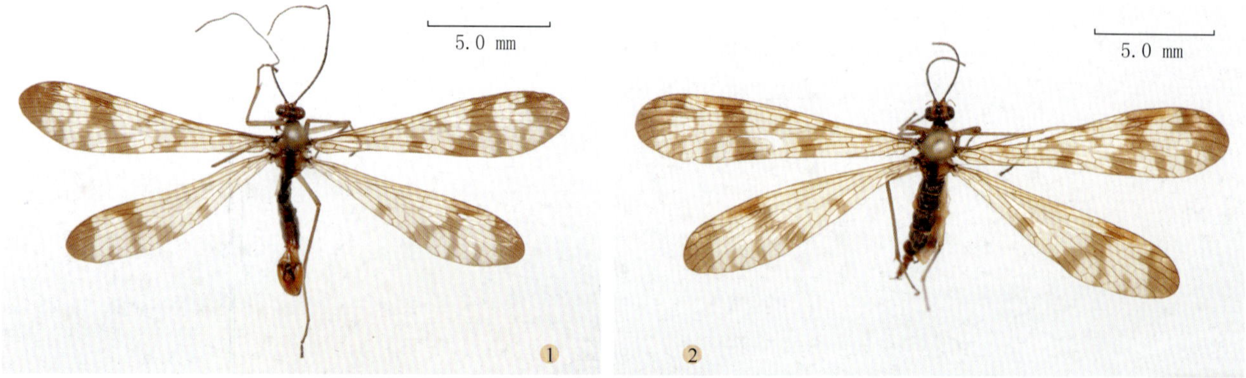

图 3-90-1 细纹新蝎蛉 *Neopanorpa ophthalmica* 成虫背面观（浙江龙泉凤阳山，NWAFU）

1. 雄虫 2. 雌虫

Fig. 3-90-1 Adults of *Neopanorpa ophthalmica*, dorsal view

(Mount Fengyang, Longquan, Zhejiang, NWAFU)

1. Male 2. Female

图 3-90-2 细纹新蝎蛉 *Neopanorpa ophthalmica* 雄性外生殖器腹面观

Fig. 3-90-2 Male genitalia of *Neopanorpa ophthalmica*, ventral view

图 3-90-3 细纹新蝎蛉 *Neopanorpa ophthalmica* 地理分布

Fig. 3-90-3 Distribution of *Neopanorpa ophthalmica*

【地理分布】 浙江（凤阳山）、台湾。

Measurements: Forewing length 13.5–14.5 mm.

Morphology: Vertex dark brown. Ocellar triangle black. Rostrum yellowish brown. Pronotum dark brown. Mesonotum and metanotum yellowish brown, with a distinct black mesal stripe. Pleura and legs yellowish brown. Wing membrane hyaline with brown markings, and tinged with yellow. In forewings, apical band broad with several hyaline spots and connected to pterostigmal band anteriorly; pterostigmal band with a broad basal band and a thin apical band; marginal spot slender; basal band complete or split into two spots; basal spot small. Male terga Ⅰ–Ⅴ dark brown. Notal organ on tergum Ⅲ long stick-like, not exceeding hind border of tergum Ⅳ. A6 dark brown. A7 and A8 yellowish brown and constricted basally. Male genital bulb long elliptical. Epandrium long and broad, truncated terminally. Hypandrium with a long broad basal stalk. Hypovalves long and broad, separated basally and overlapped mesally. Gonostylus with an obtuse triangular median tooth and a large subbasal process on inner margin. Aedeagus conical, with stout lateral processes.

Emergence: June to July.

Distribution: Zhejiang (Mount Fengyang) and Taiwan.

91. 淡瓣新蝎蛉 *Neopanorpa pallivalva* Zhou, 2003

【测量】 前翅长 12.5 ~ 13.0 mm。

【形态特征】 头顶黑色，单眼三角区黑色；喙黑棕色，侧缘黄褐色。胸部背板黑色，侧板及足深褐色。翅透明，略带褐色，无斑纹。翅痣浅黄褐色。雄性腹部 1 ~ 5 节背板黑色。第 3 节背中突短棒状，不超过第 4 节背板中部，盖住第 4 节的后背突。第 6 节黑褐色，筒形。第 7、8 节黑褐色，基部缢缩。雄性外生殖器卵形，黑色。第 9 节背板阔长，末端稍缺刻；腹板具宽阔的基柄；下瓣细长，端半部灰白色，向腹面扭曲。生殖刺突细长，内缘具钝三角形中齿和大型亚基齿。阳基侧突叶片状。阳茎腹瓣厚，稍分离；侧突起发达。

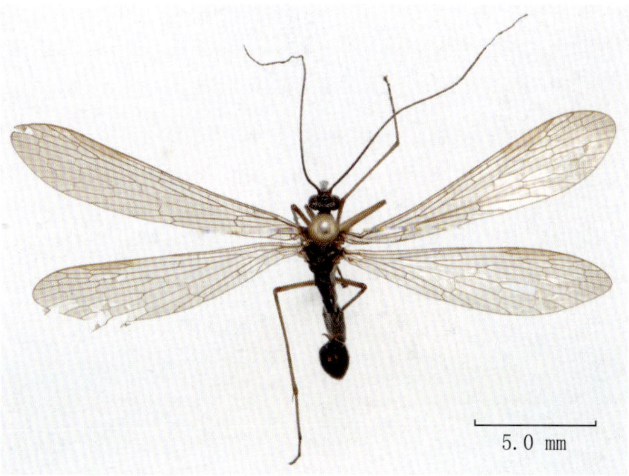

图 3-91-1 淡瓣新蝎蛉 *Neopanorpa pallivalva* 雄成虫背面观（贵州梵净山，NWAFU）

Fig. 3-91-1 Male adult of *Neopanorpa pallivalva*, dorsal view (Mount Fanjing, Guizhou, NWAFU)

【发生期】 7 月。

【地理分布】 贵州（梵净山）。

Measurements: Forewing length 12.5–13.0 mm.

Morphology: Vertex black. Ocellar triangle black. Rostrum dark brown, with yellowish brown flanks.Nota black. Pleura and legs dark brown. Wing membrane hyaline, tinged with light brown and devoid of markings. Pterostigma light yellowish brown. Male terga Ⅰ – Ⅴ black. Notal organ on tergum Ⅲ short

图 3-91-2 淡瓣新蝎蛉 *Neopanorpa pallivalva* 雄性外生殖器腹面观

Fig. 3-91-2 Male genitalia of *Neopanorpa pallivalva*, ventral view

clavate, not exceeding middle of tergum Ⅳ and covering postnotal organ on tergum Ⅳ. A6 dark brown, cylindrical. A7 and A8 dark brown, constricted basally.Male genital bulb black oval. Epandrium long broad, slightly emarginate terminally. Hypandrium with a broad basal stalk. Hypovalves slender with whitish at apical half, and curved ventrad apically. Gonostylus slender, with an obtuse triangle median tooth and a large subbasal process on inner margin. Paramere foliate. Aedeagus with its ventral valves thick and slightly divergent; lateral processes stout.

Emergence: July.

Distribution: Guizhou (Mount Fanjing).

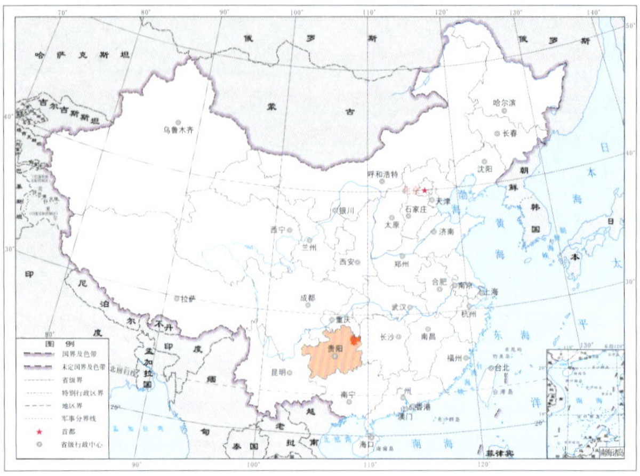

图 3-91-3　淡瓣新蝎蛉 ***Neopanorpa pallivalva*** 地理分布

Fig. 3-91-3　Distribution of ***Neopanorpa pallivalva***

92. 垂齿新蝎蛉 *Neopanorpa pendula* Qian & Zhou, 2001

【测量】 前翅长 17.0 ~ 17.5 mm。

【形态特征】 头顶黑色，单眼三角区黑色；喙黄褐色，两侧各具 1 黑褐色纵条纹。前胸背板黑色。中、后胸背板黑色，两侧后角黄褐色。侧板及足浅黄褐色。翅无色透明，无明显斑纹。翅痣浅褐色。雄性腹部 1 ~ 5 节背板黑色。第 3 节背板后缘的背中突长指状，超过第 4 节背板中部，其两侧各具 1 个小齿突；其下还有 1 个尖锐的突起。第 6 节大部分黑色，端部渐变为黄褐色；末端成片状向上隆起并具密集的刚毛。第 7 节短于第 6 节之半，

图 3-92-1　垂齿新蝎蛉 ***Neopanorpa pendula*** 雄成虫背面观（云南大理苍山，NWAFU）

Fig. 3-92-1　Male adult of ***Neopanorpa pendula***, dorsal view (Mount Cangshan, Dali, Yunnan, NWAFU)

图 3-92-3　垂齿新蝎蛉 ***Neopanorpa pendula*** 雄性外生殖器腹面观（左侧生殖刺突末端残缺）

Fig. 3-92-3　Male genitalia of ***Neopanorpa pendula***, ventral view (with the left gonostylus broke apically)

图 3-92-2　垂齿新蝎蛉 ***Neopanorpa pendula*** 雄性腹部侧面观

Fig. 3-92-2　Male abdomen of ***Neopanorpa pendula*** , lateral view

基部缢缩。第 8 节长于第 7 节。雄性外生殖器长椭圆形。第 9 节背板阔长，末端两侧呈角状向两侧突出；腹板具阔长的基柄；下瓣宽阔，内缘互相重叠，近基部形成 1 个倒心形窗口。生殖刺突短于生殖肢基节之半，内缘着生 1 簇弯曲的粗毛，中齿不明显，亚基齿呈半圆形突出。阳茎锥形，侧突起发达。

【发生期】 6 ~ 7 月。

【地理分布】 云南（苍山）。

Measurements: Forewing length 17.0–17.5 mm.

Morphology: Vertex black. Ocellar triangle black. Rostrum yellowish brown, with a pair of dark brown longitudinal stripes laterally. Pronotum black. Mesonotum and metanotum black, with latero-posterior corners yellowish brown. Pleura and legs light yellowish brown. Wing membrane hyaline, devoid of markings. Pterostigma light brown. Male terga Ⅰ–Ⅴ black. Notal organ on tergum Ⅲ long finger-like, exceeding middle of tergum Ⅳ, accompanied by a pair of small teeth laterally and an acute tooth on ventral surface. A6 mostly black and gradually turning into yellowish brown towards apex, and raised dorsally into a broad plate bearing dense setae along hind margin. A7 shorter than half length of A6 and constricted basally. A8 slightly longer than A7. Male

图 3-92-4 垂齿新蝎蛉 *Neopanorpa pendula* 地理分布

Fig. 3-92-4 Distribution of *Neopanorpa pendula*

genital bulb long elliptical. Epandrium long broad, with latero-apical corners protruding laterad and acute. Hypandrium with a long and broad basal stalk. Hypovalves broad, overlapped mesally and forming an inverted heart-shaped window subbasally. Gonostylus shorter than half length of gonocoxites, bearing a tuft of long curled bristles, an inconspicuous median tooth and a stout semicircular subbasal process on inner margin. Aedeagus conical with stout lateral processes.

Emergence: June to July.

Distribution: Yunnan (Mount Cangshan).

93. 璧尔新蝎蛉 *Neopanorpa pielina* Navás, 1936

【测量】 前翅长 11.5 mm。

【形态特征】 头顶黄褐色，单眼三角区黑色。喙黄褐色。前胸背板黑褐色。中、后胸背板黄褐色，中央具宽阔的黑色纵带。侧板及足浅黄褐色。翅无色透明，翅痣及斑纹褐色。前翅端带阔，内部具一系列透明斑，后缘分离出 1 个条形斑；痣带完整，基支阔，端支窄；缘斑小；基带成 2 斑；无基斑。雄性腹部1 ~ 5节背板黑褐色。第3节背中突长，超过第 4 节背板末端。第 6 节黑褐色。第 7、8 节黄褐色，基部稍缢缩。雄性外生殖器长卵形。第 9 节背板阔长，末端平截；腹板具宽阔的基柄；下瓣窄长，外缘向背面卷折，端部稍尖，内缘近基部形成一个近圆形窗口。生殖刺突内缘具钝三角形中齿，近基部具碗形亚基齿。阳基侧突 Y 形，末端分歧。阳茎锥状，侧突起发达。

【发生期】 4 ~ 8 月。

【地理分布】 江西（庐山）。

Measurements: Forewing length 11.5 mm.

Morphology: Vertex yellowish brown. Ocellar triangle black. Rostrum yellowish brown. Pronotum dark brown. Mesonotum and metanotum yellowish brown with a broad black mesal stripe. Pleura and legs light yellowish brown. Wing membrane hyaline, with pterostigma and markings brown. Wing membrane hyaline, with brown markings. In forewings, apical band broad with a series of hyaline windows and a separated slender stripe posteriorly; pterostigmal band complete, with a broad basal branch and a thin apical branch; marginal spot small; basal band split into a pair of small spots; basal spot absent. Male terga Ⅰ–Ⅴ dark brown. Notal organ on tergum Ⅲ long, exceeding

图 3-93-1 璧尔新蝎蛉 *Neopanorpa pielina* 成虫背面观（江西庐山，NWAFU）
1. 雄虫 2. 雌虫
Fig. 3-93-1 Adults of *Neopanorpa pielina*, dorsal view (Lushan Mountain, Jiangxi, NWAFU)
1.Male 2. Female

hind border of tergum Ⅳ. A6 dark brown. A7 and A8 yellowish brown, slightly constricted basally. Male genital bulb long oval. Epandrium long and broad, with a truncated apex. Hypandrium with a broad basal stalk. Hypovalves thin and long, with lateral margins curled dorsad and apexes slightly acute, forming a subcircular window subbasally. Gonostylus with an obtuse triangular median tooth and a bowl-shaped subbasal process on inner margin. Parameres Y-shaped and divergent. Aedeagus conical with stout lateral processes.

Emergence: April to August.

Distribution: Jiangxi (Lushan Mountain).

图 3-93-2 璧尔新蝎蛉 *Neopanorpa pielina* 雄性外生殖器腹面观
Fig. 3-93-2 Male genitalia of *Neopanorpa pielina*, ventral view

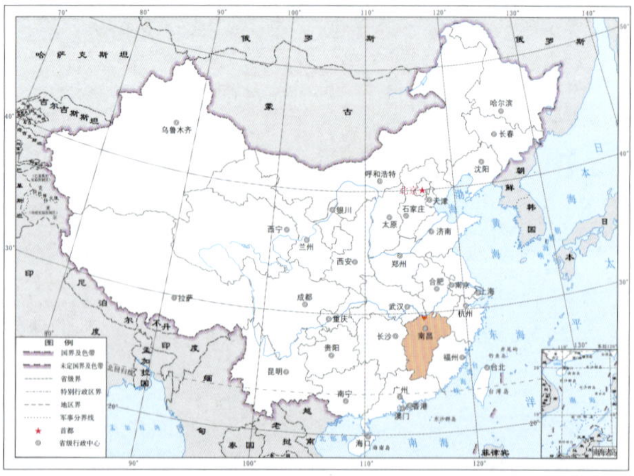

图 3-93-3 璧尔新蝎蛉 *Neopanorpa pielina* 地理分布
Fig. 3-93-3 Distribution of *Neopanorpa pielina*

94. 丽新蝎蛉 *Neopanorpa pulchra* Carpenter, 1945

【测量】 前翅长 14.0 ~ 16.0 mm。

【形态特征】 头顶黑褐色，单眼三角区黑色。喙黄褐色。前胸背板黑褐色。中、后胸背板黄褐色，中央具显著的黑色纵带。侧板及足黄褐色。翅无色

图 3-94-1 丽新蝎蛉 ***Neopanorpa pulchra*** 成虫背面观（海南尖峰岭，NWAFU）

1. 雄虫 2. 雌虫

Fig. 3-94-1 Adults of ***Neopanorpa pulchra***, dorsal view (Jianfengling, Hainan, NWAFU)

1. Male 2. Female

图 3-94-2 丽新蝎蛉 ***Neopanorpa pulchra*** 雄性外生殖器腹面观

Fig. 3-94-2 Male genitalia of ***Neopanorpa pulchra***, ventral view

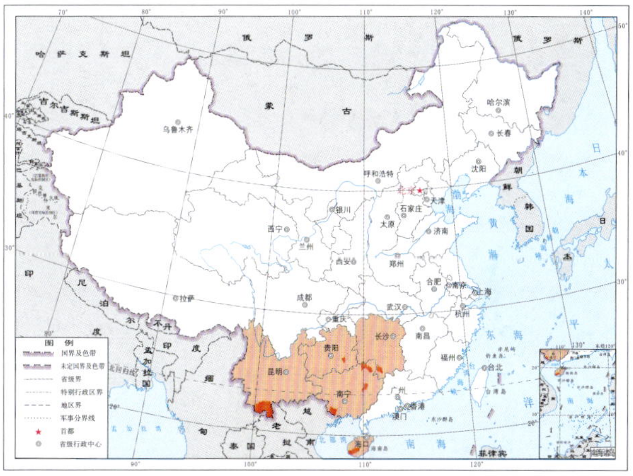

图 3-94-3 丽新蝎蛉 ***Neopanorpa pulchra*** 地理分布

Fig. 3-94-3 Distribution of ***Neopanorpa pulchra***

透明，斑纹褐色。前翅端带宽阔，内缘近平直，前缘与痣带相接；痣带基支和端支完整，端支窄，基支阔；缘斑细长；基带断裂成 2 斑；无基斑。雄性腹部 1 ~ 5 节背板黑褐色。第 3 节背板后缘背中突发达，近三角形。第 6 节黑褐色，长圆筒形。第 7、8 节黄褐色，基部缢缩。雄性外生殖器长椭圆形。第 9 节背板阔长，末端平截；腹板基柄阔长；下瓣长，明显超出生殖基节末端，侧缘卷曲，内缘相互套叠，近基部形成 1 近圆形小窗口。生殖刺突内缘中齿钝三角形，亚基齿大。阳茎短，侧突起发达。

【发生期】 4 ~ 9 月。

【地理分布】 广西（宁明、龙州、金秀、猫儿山）、湖南（道县）、贵州（雷公山）、海南（尖峰岭、五指山）、云南（勐海、勐养、勐龙、勐仑）。

Measurements: Forewing length 14.0–16.0 mm.

Morphology: Vertex dark brown. Ocellar triangle black. Rostrum yellowish brown. Pronotum dark brown. Mesonotum and metanotum yellowish brown, with a broad black mesal stripe. Pleura and legs light yellowish brown. Wing membrane hyaline, with brown markings. In forewings, apical band broad with a straight inner margin, and connected to pterostigmal band anteriorly; pterostigmal band complete, with a broad basal branch and a thin apical branch; marginal spot slender; basal band split into two spots; basal spot absent. Male terga Ⅰ–Ⅴ dark brown. Notal organ on tergum Ⅲ stout subtriangular. A6 dark brown, long cylindrical. A7 and A8 yellowish brown and constricted basally. Male genital bulb long elliptical. Epandrium long and broad with a truncated apex. Hypandrium with a long broad basal stalk. Hypovalves long, distinctly extending beyong apex of gonocoxites, curled on lateral margins, overlapped mesally, and forming a subcircular window subbasally. Gonostylus with an obtuse triangular median tooth and a large subbasal process on inner margin. Aedeagus short with stout lateral processes.

Emergence: April to September.

Distribution: Guangxi (Ningming, Longzhou, Jinxiu, Mount Mao'er), Hunan (Daoxian), Guizhou (Mount Leigong), Hainan (Jianfengling, Mount Wuzhi) and Yunnan (Menghai, Mengyang, Menglong, Menglun).

95. 净翅新蝎蛉 *Neopanorpa puripennis* Chou & Wang, 1988

【测量】 前翅长 12.0 mm。

【形态特征】 头顶黄褐色，单眼三角区黑色。喙黄褐色。前胸背板黄褐色。中、后胸背板黄褐色，中央具宽阔的黑褐色纵带。侧板及足浅黄褐色。翅无色透明，翅痣浅棕色，无明显的斑纹。雄性腹部 1 ~ 5 节背板黑褐色。第 3 节背板后缘的背中突短，略呈三角形。第 6 节大部分棕色，末端黄色。第 7、8 节黄褐色，基部缢缩。雄性外生殖器卵形。第 9 节背板阔长，端部平截；腹板具阔长的基柄；下瓣端部圆阔，相互分歧，超过生殖肢基节端部。生殖刺突内缘具钝三角形中齿和圆钝的亚基齿。阳基侧突不明显；阳茎侧突起十分发达；背瓣侧方具弯曲的角状突起。

【发生期】 7 ~ 8 月。

【地理分布】 湖南（道县）、贵州（雷公山）。

Measurements: Forewing length 12.0 mm.

Morphology: Vertex yellowish brown. Ocellar triangle black. Rostrum yellowish brown. Pronotum yellowish brown. Mesonotum and metanotum yellowish brown, with a broad dark brown mesal stripe. Pleura and legs light yellowish brown. Wing membrane hyaline, devoid of markings. Pterostigma light brown. Male terga Ⅰ–Ⅴ dark brown. Notal organ

图 3-95-1 净翅新蝎蛉 *Neopanorpa puripennis* 雄成虫背面观（贵州雷公山，NWAFU）

Fig. 3-95-1 Male adult of *Neopanorpa puripennis*, dorsal view (Mount Leigong, Guizhou, NWAFU)

图 3-95-2 净翅新蝎蛉 *Neopanorpa puripennis* 雄性外生殖器腹面观

Fig. 3-95-2 Male genitalia of *Neopanorpa puripennis*, ventral view

on tergum Ⅲ short and slightly triangular. A6 mostly brown with an yellow apex. A7 and A8 yellowish brown, and constricted basally. Male genital bulb oval. Epandrium long and broad, with a truncated apex. Hypandrium with a long broad basal stalk. Hypovalves rounded and divergent apically, exceeding apex of gonocoxites. Gonostylus with an obtuse triangular median tooth and a rounded subbasal process. Paramere indistinct. Aedeagus with very strong lateral processes; dorsal valves with curled horn-like processes laterally.

Emergence: July to August.

Distribution: Hunan (Daoxian) and Guizhou (Mount Leigong).

图 3-95-3　净翅新蝎蛉 *Neopanorpa puripennis* 地理分布

Fig. 3-95-3　Distribution of *Neopanorpa puripennis*

96. 网纹新蝎蛉 *Neopanorpa retina* Chou & Li, 1988

【测量】 前翅长 13.0 ~ 14.0 mm。

【形态特征】 头顶黄褐色，单眼三角区被 1 大黑斑包围。喙黄褐色。前胸背板黑褐色。中、后胸背板黑褐色，两侧后角黄褐色。侧板及足黄褐色。翅无色透明，斑纹褐色。前翅端带较阔，中间常有 2 个透明斑；痣带基支和端支完整，端支基部较狭窄；在翅痣的外侧另有 1 条和端带内缘平行的斑纹；基带中间断开；缘斑较大；无基斑。雄性腹部 1 ~ 5 节背板黑褐色。腹部第 3 节背中突超过第 4 节背板后缘。第 6 节黑褐色，筒形。第 7、8 节黄褐色，基部缢缩。雄性外生殖器卵形。第 9 节背板阔长，末端圆钝；腹板基柄阔长；下瓣末端圆钝，内缘相互重叠。生殖刺突内缘亚基齿碗形，中齿钝三角形。阳茎短小，不达生殖肢基节端部，侧突起发达。

【发生期】 4 ~ 10 月。

【地理分布】 湖南（汝城、平江、安仁、江永、临湘）。

Measurements: Forewing length 13.0–14.0 mm.

Morphology: Vertex yellowish brown. Ocellar triangle enclosed by a large black spot. Rostrum yellowish brown. Pronotum dark brown. Mesonotum

图 3-96-1　网纹新蝎蛉 *Neopanorpa retina* 雄成虫背面观（湖南汝城，NWAFU）

Fig. 3-96-1　Male adult of *Neopanorpa retina*, dorsal view (Rucheng, Hunan, NWAFU)

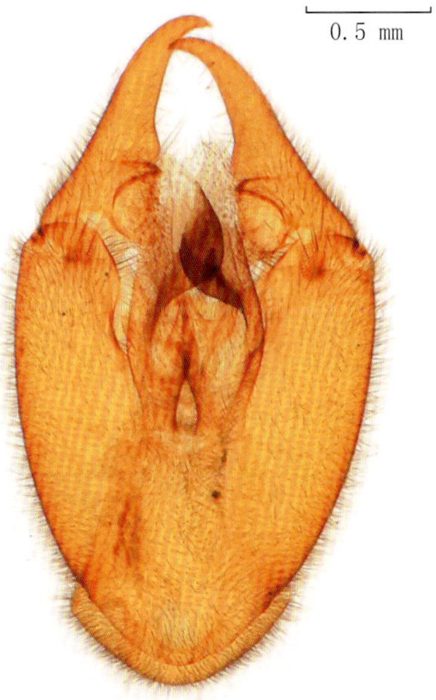

图 3-96-2　网纹新蝎蛉 *Neopanorpa retina* 雄性外生殖器腹面观

Fig. 3-96-2　Male genitalia of *Neopanorpa retina*, ventral view

and metanotum mostly dark brown, with latero-posterior corners yellowish brown. Pleura and legs yellowish brown. Wing membrane hyaline, with brown markings. In forewings, apical band broad with two hyaline windows; pterostigmal band with a basal branch and an apical branch; base of apical branch thin; an additional stripe beyond pterostigmal band and parallel to inner margin of apical band; basal band split medially; marginal spot large; basal spot absent. Male terga I–V dark brown. Notal organ on tergum III exceeding hind border of tergum IV. A6 dark brown and cylindrical. A7 and A8 yellowish brown and constricted basally. Male genital bulb oval. Epandrium long and broad with a rounded apex. Hypandrium with a long broad basal stalk. Hypovalves rounded, overlapping mesally. Gonostylus with an obtuse triangular median tooth and a bowl-shaped subbasal process. Aedeagus short, not exceeding apex of

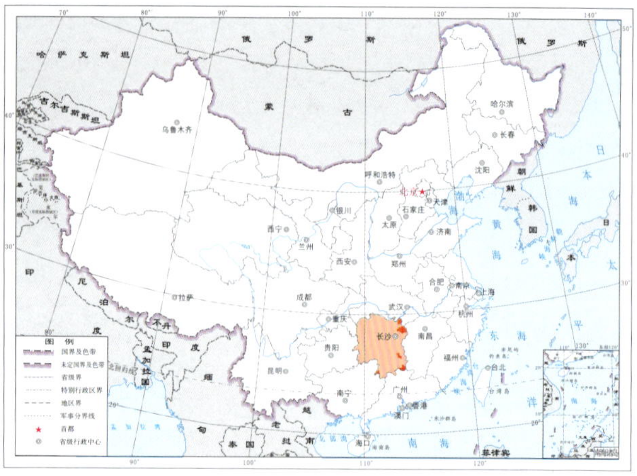

图 3-96-3 网纹新蝎蛉 *Neopanorpa retina* 地理分布

Fig. 3-96-3 Distribution of ***Neopanorpa retina***

gonocoxites, with stout lateral processes.

Emergence: April to October.

Distribution: Hunan (Rucheng, Pingjiang, Anren, Jiangyong, Linxiang).

97. 绍德新蝎蛉 *Neopanorpa sauteri* (Esben-Petersen, 1912)

【测量】 前翅长 13.5 mm。

【形态特征】 头顶黄褐色，单眼三角区周围具大型黑色斑。喙黄褐色。前胸背板黄褐色。中、后胸背板黄褐色，具 1 条狭窄的黑色中纵带。侧板及足黄褐色。翅黄色透明；斑纹黑褐色。前翅端带内缘不规则，具 1 ~ 2 个大型窗斑；痣带完整；缘斑伸出 1 条带与痣带基支相连；基带前半部窄，后半部宽；基斑断为 2 个小斑。雄性腹部 1 ~ 5 节背板从黑褐色向后渐变为黄褐色。第 3 节背中突极长，棒状，几乎伸达第 6 节末端。第 6 节黄褐色，筒形。第 7、8 节黄褐色，基部稍缢缩。雄性外生殖器卵形。第 9 节背板阔长，末端具深 U 形缺刻；腹板具阔长的基柄；下瓣近三角形，相互重叠，近基部形成 1 个大型近圆窗口。生殖刺突内缘具大型基齿，中齿钝三角形。阳茎锥状，侧突起发达。

【发生期】 6 ~ 7 月。

【地理分布】 台湾（玉山、屏东、嘉义、宜兰）。

图 3-97-1 绍德新蝎蛉 *Neopanorpa sauteri* 雄成虫背面观

（台湾屏东大汉山，CAU）

Fig, 3-97-1 Male adult of ***Neopanorpa sauteri***, dorsal view

(Mount Tahan, Pingtung, Taiwan, CAU)

Measurements: Forewing length 13.5 mm.

Morphology: Vertex yellowish brown. Ocellar triangle enclosed by a large black spot. Rostrum yellowish brown. Pronotum yellowish brown. Mesonotum and metanotum mostly yellowish brown, with a narrow black mesal stripe. Pleura and legs yellowish brown.Wing membrane hyaline, tinged with yellow, with dark brown markings. In forewings, apical band irregular along inner margin and with 1–2 large hyaline windows; pterostigmal band complete; marginal spot extending outwards and connected to basal branch of pterostigmal band; basal band narrow at anterior half and broad at posterior half; basal spot split into two small spots. Male terga I–V dark brown

图 3-97-2 绍德新蝎蛉 *Neopanorpa sauteri* 雄性外生殖器腹面观
Fig. 3-97-2 Male genitalia of ***Neopanorpa sauteri***, ventral view

图 3-97-3 绍德新蝎蛉 *Neopanorpa sauteri* 地理分布
Fig. 3-97-3 Distribution of ***Neopanorpa sauteri***

and gradually turning into yellowish brown caudad. Notal organ on tergum Ⅲ extremely elongated, clavate, almost extending to hind border of A6. A6 yellowish brown and cylindrical. A7 and A8 yellowish brown and slightly constricted basally. Male genital bulb oval. Epandrium long and broad, with a deep U-shaped terminal emargination. Hypandrium with a long broad basal stalk. Hypovalves almost subtriangular, overlapped mesally and forming a large subcircular window subbasally. Gonostylus with an obtuse triangular median tooth and a large subbasal process on inner margin. Aedeagus conical, with stout lateral processes.

Emergence: June to July.

Distribution: Taiwan (Mount Yushan, Pingtung, Chiayi, Yilan).

98. 具毛新蝎蛉 *Neopanorpa setigera* Wang & Hua, 2018

【测量】 前翅长 13.0 ~ 14.0 mm。

【形态特征】 头顶黑色，单眼三角区黑色；喙黄褐色，具 2 条黑褐色纵带。胸部背板黑褐色，中、后胸背板两侧近翅基部渐浅。侧板黄白色，足黄褐色。翅略带浅棕色，透明，翅痣褐色明显。翅斑不明显，仅有模糊的端带和极度退化的痣带。雄性腹部 1 ~ 5 节背板黑色。第 3 节背中突长棒状，向后延伸超过第 5 节但不超过第 6 节。4 ~ 6 节背板中央各具 1 列密集的短毛。第 6 节黑色。第 7、8 节黑色，基部强烈缢缩成柄状。雄性外生殖器黑色，长椭圆形。第 9 节背板末端平截；腹板具阔长的基柄；下瓣狭长，侧缘向背面卷曲，内侧近基部形成 1 个水滴形窗口，背侧近基部具 1 个指形突起。阳基侧突细长；阳茎锥状，侧突起发达、宽大。

【发生期】 6 ~ 8 月。

【地理分布】 四川（石棉）。

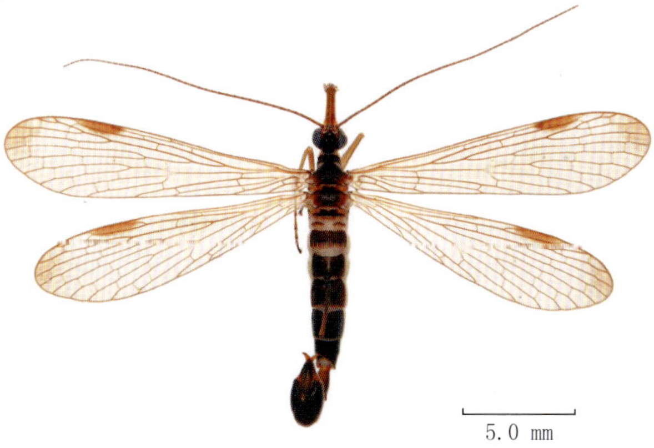

图 3-98-1 具毛新蝎蛉 *Neopanorpa setigera* 成虫背面观（四川石棉，NWAFU）
Fig. 3-98-1 Adults of ***Neopanorpa setigera***, dorsal view (Shimian, Sichuan, NWAFU)

图 3-98-2 具毛新蝎蛉 *Neopanorpa setigera* 雄性外生殖器腹面观
Fig. 3-98-2 Male genitalia of ***Neopanorpa setigera***, ventral view

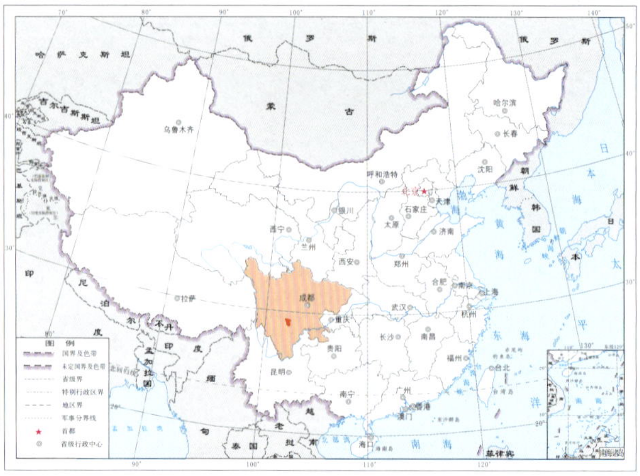

图 3-98-3 具毛新蝎蛉 ***Neopanorpa setigera*** 地理分布
Fig. 3-98-3 Distribution of ***Neopanorpa setigera***

Measurements: Forewing length 13.0–14.0 mm.

Morphology: Vertex black, ocellar triangle black; rostrum yellowish brown, with two dark brown longitudinal stripes. Nota dark brown, Mesonotum and metanotum grading to yellowish brown towards wing base. Pleura yellowish white, legs yellowish brown. Wing slightly tinged with light brown, hyaline, with brown distinct pterostigma. Markings indistinct, represented only by vague apical band and greatly reduced pterostigmal band. Male terga I–V black. Notal organ on tergum III long clavate, extending beyond A5 but not exceeding A6. Terga IV – VI with a row of dense short setae mesally. A6 black. A7 and A8 black, greatly constricted stalk-like at base. Male genital bulb black, long elliptical. Epandrium truncated at apex. Hypandrium with long broad basal stalk; hypovalves slender, curled dorsad at lateral margin, forming a water drop-shaped window subbasally, and with a finger-like process on dorsal surface subbasally. Paramere slender; aedeagus conical, with stout and broad lateral processes.

Emergence: June to August.

Distribution: Sichuan (Shimian).

99. 申氏新蝎蛉 *Neopanorpa sheni* Hua & Chou, 1997

【测量】 前翅长 11.0 ~ 12.0 mm。

【形态特征】 头顶黑色，单眼三角区黑色。喙黄褐色。前胸背板黑色。中、后胸背板黑褐色，两侧后角黄褐色。侧板及足黄褐色。翅无色透明，斑纹褐色。前翅端带宽；痣带宽，基支宽，发达，端支细且与痣带分离；缘斑小，为 1 小圆点；基带断开成 2 小点；无基斑。雄性腹部 1 ~ 5 节背板黑褐色。第 3 节背板后缘背中突较短，近三角形。第 6 节黑褐色，长圆筒形。第 7、8 节黄褐色，基部缢缩。雄性外生殖器卵形。第 9 节背板阔长，末端平截；腹板具阔长的基柄；下瓣细长，内缘相互重叠，近基部形成 1 长椭圆形窗口；生殖刺突内缘具钝三角形中齿和大型亚基齿。阳茎锥状，侧突起发达。

【发生期】 7 ~ 8 月。

【地理分布】 河南（嵩县、栾川）、陕西（宁陕）。

Measurements: Forewing length 11.0–12.0 mm.

Morphology: Vertex black. Ocellar triangle black. Rostrum yellowish brown. Pronotum black. Mesonotum and metanotum blackish brown, and yellowish brown at latero-posterior corners. Pleura and legs yellowish brown. Wing membrane colorless and hyaline, with brown markings. In forewings,

图 3-99-1 申氏新蝎蛉 *Neopanorpa sheni* 成虫背面观（河南嵩县白云山，NWAFU）

1. 雄虫 2. 雌虫

Fig. 3-99-1 Adults of ***Neopanorpa sheni***, dorsal view (Mount Baiyun, Songxian, Henan, NWAFU)

1. Male 2. Female

图 3-99-2 申氏新蝎蛉 *Neopanorpa sheni* 雄性外生殖器腹面观

Fig. 3-99-2 Male genitalia of ***Neopanorpa sheni***, ventral view

图 3-99-3 申氏新蝎蛉 *Neopanorpa sheni* 地理分布

Fig. 3-99-3 Distribution of ***Neopanorpa sheni***

apical band broad; pterostigmal band broad with well-developed basal branch and separated thin apical branch; marginal spot very small, represented by a small circular spot; basal band split into two small spots; basal spot absent. Male terga Ⅰ—Ⅴ dark brown. Notal organ on tergum Ⅲ short subtriangular. A6 dark brown, cylindrical; A7 and A8 yellowish brown, constricted basally. Male genital bulb oval. Epandrium long and broad, with a truncated apex. Hypandrium with a long broad basal stalk. Hypovalves slender, overlapped mesally and forming a long elliptical window subbasally. Gonostylus with an obtuse triangular tooth and a large subbasal process on inner margin. Aedeagus conical, with stout lateral processes.

Emergence: July to August.

Distribution: Henan (Songxian, Luanchuan) and Shaanxi (Ningshan).

100. 西藏新蝎蛉 *Neopanorpa tibetensis* Hua & Chou, 1999

【测量】 前翅长 14.5 ~ 15.5 mm。

【形态特征】 头顶黑色，单眼三角区黑色。喙黄褐色，两侧具黑褐色纵带。前胸背板黑褐色。中、后胸背板大部黑褐色，后侧角黄褐色。侧板及足黄褐色。翅透明，略带黄色，斑纹黄褐色。前翅端带完整，与痣带在翅前缘相连；痣带具完整的端支和基支；基带不完整，仅在翅后缘处呈 1 小斑；无缘斑和基斑。雄性腹部 1 ~ 5 节背板黑褐色，向

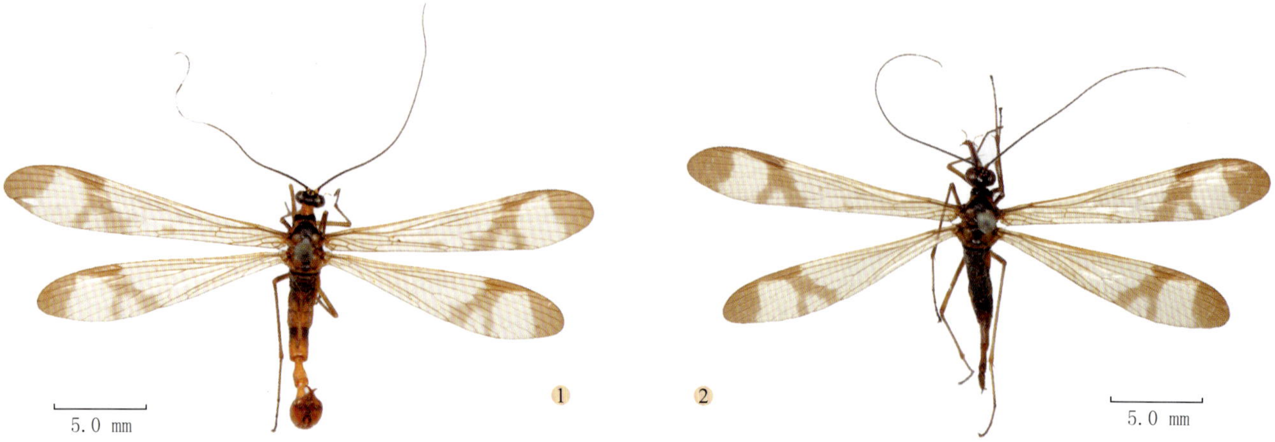

图 3-100-1 西藏新蝎蛉 *Neopanorpa tibetensis* 成虫背面观（西藏通麦，NWAFU）
1. 雄虫 2. 雌虫
Fig. 3-100-1 Adults of ***Neopanorpa tibetensis***, dorsal view (Tangmai, Tibet, NWAFU)
1. Male 2. Female

图 3-100-2 西藏新蝎蛉 *Neopanorpa tibetensis* 雄性腹部侧面观
Fig. 3-100-2 Male teminal abdomen of ***Neopanorpa tibetensis,*** lateral view

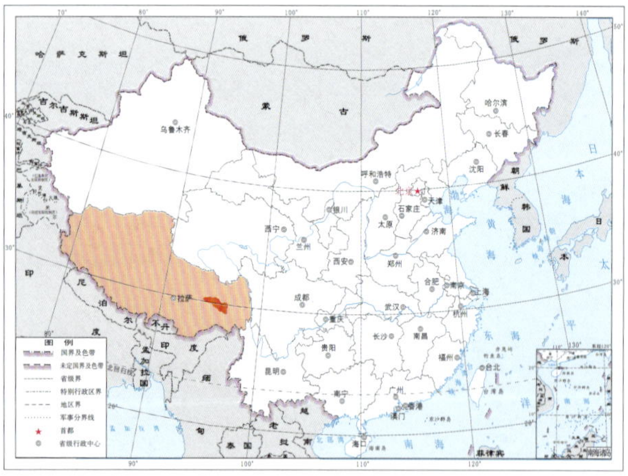

图 3-100-4 西藏新蝎蛉 *Neopanorpa tibetensis* 地理分布
Fig. 3-100-4 Distribution of ***Neopanorpa tibetensis***

图 3-100-3 西藏新蝎蛉 *Neopanorpa tibetensis* 雄性外生殖器腹面观
Fig. 3-100-3 Male genitalia of ***Neopanorpa tibetensis,*** ventral view

后逐渐变为黄褐色。第 3 节背中突发达，长棒状，伸达第 6 节背板中央。第 6 节圆筒形，基半部黑褐色，端半部黄褐色。第 7、8 节黄褐色，基部缢缩。雄性外生殖器卵形。第 9 节背板阔长，端部平截。第 9 节腹板基柄阔，下瓣末端向腹面翻折，内缘近基部形成 1 个圆形窗口。生殖刺突外缘内凹，内缘亚基齿钝，中齿钝三角形。阳茎锥形，侧突起发达。

【发生期】 7 ~ 9 月。

【地理分布】 西藏（通麦）。

Measurements: Forewing length 14.5–15.5 mm.

Morphology: Vertex and ocellar triangle black. Rostrum yellowish brown, with two dark brown longitudinal stripes laterally. Pronotum dark brown. Mesonotum and metanotum mostly dark brown with latero-posterior corners yellowish brown. Pleura and legs yellowish brown. Wing membrane hyaline, slightly tinged with yellow. Markings yellowish brown. In forewings, apical band complete and connected to pterostigmal band anteriorly; pterostigmal band with complete basal and apical branches; basal band incomplete and only represented by a small spot on hind margin; marginal and basal spots absent. Male terga Ⅰ–Ⅴ dark brown and gradually turning into yellowish brown caudad. Notal organ on tergum Ⅲ extremely elongated and clavate, extending to middle of tergum Ⅵ. A6 darkish on basal half and yellowish brown on apical half, cylindrical. A7 and A8 yellowish brown and constricted basally. Male genital bulb oval. Epandrium long and broad, with a truncated apex. Hypandrium with a long and broad basal stalk. Hypovalves folded ventrad on distal half and forming a circular window subbasally. Gonostylus concaved on outer margin, bearing an obtuse triangular median tooth and a rounded subbasal process on inner margin. Aedeagus conical, with stout lateral processes.

Emergence: July to September.

Distribution: Tibet (Tangmai).

101. 天目山新蝎蛉 *Neopanorpa tienmushana* Cheng, 1957

【测量】 前翅长 13.0 ~ 14.0 mm。

【形态特征】 头顶黄褐色，单眼三角区黑色，喙黄褐色。前胸背板黄褐色。中、后胸背板黄褐色，中央具明显的黑色纵带；中胸背板前缘和两前侧具黑色条斑。侧板及足黄褐色。翅黄色，半透明，斑纹黄褐色。前翅端带阔，内部具大型透明窗斑，

图 3-101-1 天目山新蝎蛉 *Neopanorpa tienmushana* 雄成虫背面观（浙江临安西天目山）

Fig. 3-101-1 Male adult of *Neopanorpa tienmushana*, dorsal view (West Tianmu Mountain, Lin'an, Zhejiang)

图 3-101-2 天目山新蝎蛉 ***Neopanorpa tienmushana*** 雌成虫背面观（浙江临安西天目山，NWAFU）
Fig. 3-101-2 Female adult of ***Neopanorpa tienmushana***, dorsal view (West Tianmu Mountain, Linân, Zhejiang, NWAFU)

图 3-101-3 天目山新蝎蛉 ***Neopanorpa tienmushana*** 雄性外生殖器腹面观
Fig. 3-101-3 Male genitalia of ***Neopanorpa tienmushana***, ventral view

后角处具 1 分离的条斑；痣带完整，基支较阔，端支较窄；缘斑窄长，后部与基带相连；基斑小。雄性腹部 1 ~ 5 节背板黄褐色。第 3 节背中突略超过第 4 节背板中部。第 6 节黄褐色，筒形。第 7、8 节黄褐色，基部缢缩。雄性外生殖器卵形。第 9 节背板阔长，端部具浅 V 形缺刻；腹板具十分阔长的基柄；下瓣宽阔，内缘相互重叠，近基部形成 1 个近心形窗口。阳基侧突短，Y 形。阳茎锥形，侧突起发达。

【发生期】 6 ~ 8 月。

【地理分布】 浙江（天目山）、福建（南靖）。

Measurements: Forewing length 13.0–14.0 mm.

Morphology: Vertex yellowish brown. Ocellar triangle black. Rostrum yellowish brown.Pronotum yellowish brown. Mesonotum and metanotum yellowish brown, with distinct black mesal stripe; mesonotum black on anterior and latero-anterior margin. Pleura and legs yellowish brown.Wing membrane translucent, tinged with yellow. Markings yellowish brown. In forewings, apical band broad, with large hyaline windows innerly and a separated stripe near posterior corner; pterostigmal band complete, with a broad basal branch and a thin apical branch; marginal spot slender and connected to basal band posteriorly; basal spot small. Male terga I–V yellowish brown. Notal organ on tergum III exceeding middle of tergum IV. A6 yellowish brown, cylindrical. A7 and A8 yellowish brown and constricted basally. Male genital bulb oval. Epandrium long and broad, and emarginated in a shallow V-shaped on apex. Hypandrium with a long and broad basal stalk. Hypovalves very broad and overlapped mesally, forming a heart-shaped window subbasally. Paramere short, Y-shaped. Aedeagus conical, with stout lateral processes.

Emergence: June to August.

Distribution: Zhejiang (Tianmu Mountains) and Fujian (Nanjing).

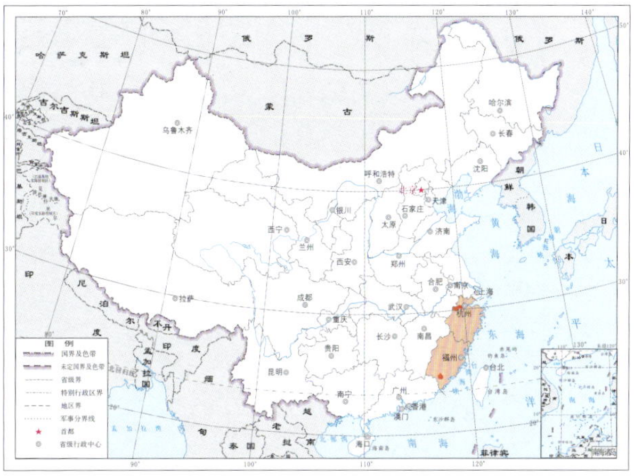

图 3-101-4 天目山新蝎蛉 *Neopanorpa tienmushana* 地理分布
Fig. 3-101-4 Distribution of *Neopanorpa tienmushana*

102. 天平山新蝎蛉 *Neopanorpa tienpingshana* Chou & Wang, 1988

图 3-102-1 天平山新蝎蛉 *Neopanorpa tienpingshana* 雄成虫侧面观（湖南桑植天平山）
Fig. 3-102-1 Male adult of *Neopanorpa tienpingshana*, lateral view (Tianping Mountain, Sangzhi, Hunan)

图 3-102-2 天平山新蝎蛉 *Neopanorpa tienpingshana* 雌成虫侧面观（湖南桑植天平山）

Fig. 3-102-2 Female adult of *Neopanorpa tienpingshana*, lateral view (Tianping Mountain, Sangzhi, Hunan)

【测量】 前翅长 11.5 ~ 12.5 mm。

【形态特征】 头顶黑色，单眼三角区黑色，喙黄褐色。胸部背板黑褐色。侧板及足基节乳黄色。足大部分浅黄褐色。翅无色透明，斑纹黑色。前翅端带阔；痣带端支缺失；基带无或仅在翅后缘有残余；无缘斑和基斑。有的个体几乎完全无斑。雄性腹部 1 ~ 5 节背板黑褐色。第 3 节上的背中突中等长度，伸达第 4 节背板后缘，盖住圆钝的后背突。第 6 节黑褐色，筒形。第 7、8 节黄褐色，基部缢缩。雄性外生殖器卵形。第 9 节背板阔长，端部稍缺刻；腹板具阔长的基柄；下瓣末端扭曲，形成 1 个近心形窗口。生殖刺突亚基齿大，碗状，着生黑色长毛；中齿钝三角形。阳基侧突末端叶片状。阳茎锥形，侧突起发达。

【发生期】 6 ~ 8 月。

【地理分布】 湖南（莽山、八大公山、壶瓶山）、湖北（巴东）、贵州（雷公山）。

图 3-102-3 天平山新蝎蛉 *Neopanorpa tienpingshana* 雄性外生殖器腹面观

Fig. 3-102-3 Male genitalia of *Neopanorpa tienpingshana*, ventral view

Measurements: Forewing length 11.5–12.5 mm.

Morphology: Vertex black. Ocellar triangle

black. Rostrum yellowish brown. Nota dark brown. Pleura and coxae creamy yellow. Legs mostly light yellowish brown. Wing membrane hyaline, with black markings. In forewings, apical band broad; pterostigmal band lacking an apical branch; basal band absent or represented only by a small spot on hind margin; marginal and basal spots absent. Male terga Ⅰ–Ⅴ dark brown. Notal organ on tergum Ⅲ moderately long, extending to hind margin of tergum Ⅳ, covering rounded postnotal organ. A6 dark brown, cylindrical. A7 and A8 yellowish brown and constricted basally. Male genital bulb oval. Epandrium long and broad, slightly emarginate at apex. Hypandrium with a long and broad basal stalk. Hypovalves twisted apically, forming a heart-shaped window subbasally. Gonostylus with an obtuse triangular median tooth, and a large bowl-shaped subbasal process with long bristles. Paramere foliate apically. Aedeagus conical and with stout lateral processes.

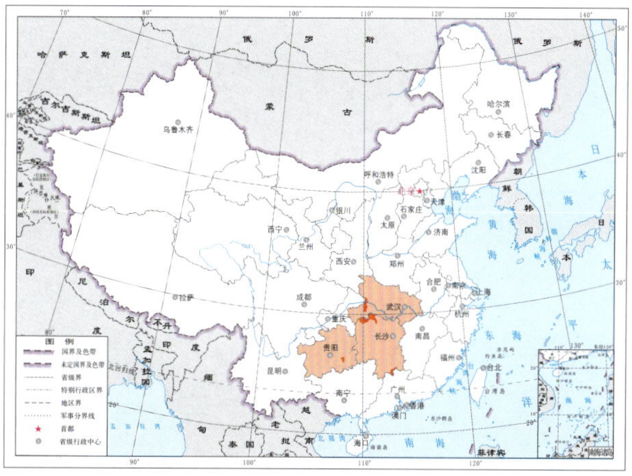

图 3-102-4 天平山新蝎蛉 *Neopanorpa tienpingshana* 地理分布

Fig. 3-102-4 Distribution of *Neopanorpa tienpingshana*

Emergence: June to August.

Distribution: Hunan (Mount Mangshan, Badagong Mountains, Mount Huping), Hubei (Badong) and Guizhou (Mount Leigong).

103. 透翅新蝎蛉 *Neopanorpa translucida* Cheng, 1957

【测量】 前翅长 14.5 ~ 15.0 mm。

【形态特征】 头顶黄褐色，单眼三角区被 1 大型黑斑包围。喙黄褐色。前胸背板黑褐色。中、后胸背板黄褐色，前缘黑色，中央具显著的黑色纵带。侧板及足黄褐色。翅黄色半透明，斑纹黄褐色较淡；翅痣红色，明显。前翅端带内缘不规则，后部具 1 透明窗斑；痣带完整，其端支基部稍收窄；缘斑后部向外与痣带基支相连，向内与基带相连；无基斑。雄性腹部 1 ~ 5 节不均匀黑褐色。第 3 节背板

图 3-103-1 透翅新蝎蛉 *Neopanorpa translucida* 雄成虫背面观（福建邵武大竹岚，NWAFU）

Fig. 3-103-1 Male adult of *Neopanorpa translucida*, dorsal view (Dazhulan, Shaowu, Fujian, NWAFU)

图 3-103-2 透翅新蝎蛉 *Neopanorpa translucida* 雄性外生殖器腹面观

Fig. 3-103-2 Male genitalia of *Neopanorpa translucida*, ventral view

后缘背中突长棒状，不超过第 4 节背板后缘。第 6 节大部分黑褐色，向后渐变为红褐色。第 7、8 节红褐色，基部缢缩。雄性外生殖器卵形。第 9 节背板阔长，端部平截；腹板具阔长的基柄；下瓣宽阔，外缘背向卷曲，端部斜截，两瓣基部分开。生殖刺突外缘凹陷，内缘具不明显的三角形中齿和大型盘状亚基齿，后者着生密集的长刚毛。阳基侧突 Y 形，细长；阳茎锥状，侧突起发达。

【发生期】 5 ~ 8 月。

【地理分布】 福建（邵武、长汀、光泽）。

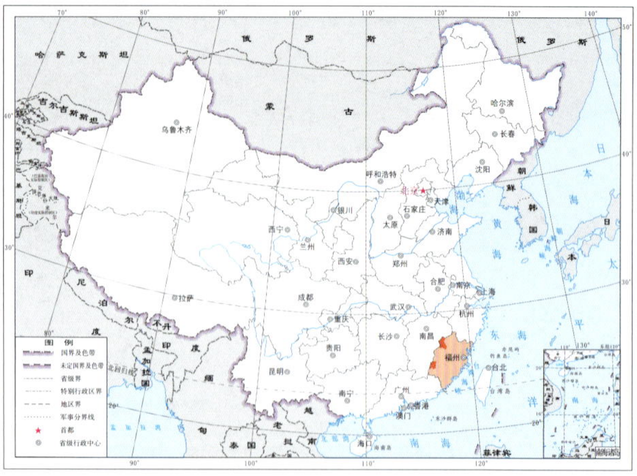

图 3-103-3 透翅新蝎蛉 *Neopanorpa translucida* 地理分布

Fig. 3-103-3 Distribution of *Neopanorpa translucida*

Measurements: Forewing length 14.5–15.0 mm.

Morphology: Vertex yellowish brown. Ocellar triangle enclosed by a large black spot. Rostrum yellowish brown. Pronotum dark brown. Mesonotum and metanotum mostly yellowish brown, with black anterior margins and a distinct black mesal stripe. Pleura and legs yellowish brown. Wing membrane translucent, tinged with yellow. Markings faint yellowish brown. Pterostigma distinct, reddish. In forewings, apical band irregular along inner margin, with a hyaline window posteriorly; pterostigmal band complete, with apical branch constricted basally; marginal spot connected to pterostigmal band apically, and connected to basal band basally; basal spot absent. Male terga Ⅰ–Ⅴ unevenly dark brown. Notal organ on tergum Ⅲ slender and stick-like, not exceeding hind border of tergum Ⅳ. A6 mostly dark brown and gradually turned into reddish brown towards apex. A7 and A8 reddish brown and constricted basally. Male genital bulb oval. Epandrium long and broad, with a truncated apex. Hypandrium with a long and broad basal stalk. Hypovalves broad, curled dorsad on lateral margin, beveled apically, and separated basally. Gonostylus concaved on outer margin, with an indistinct median tooth and a large disc-shaped subbasal process bearing dense long setae. Paramere slender and Y-shaped. Aedeagus conical, with stout lateral processes.

Emergence: May to August.

Distribution: Fujian (Shaowu, Changting, Guangze).

104. 杨氏新蝎蛉 *Neopanorpa youngi* Byers, 1994

【测量】 前翅长 13.5 ~ 15.0 mm。

【形态特征】 头顶黑褐色，单眼三角区黑色。喙黄褐色。前胸背板不均匀黄褐色。中、后胸背板黄褐色，中央具有明显的黑色纵带。侧板及足黄褐色。翅黄色，半透明，斑纹黑褐色。前翅端带完整，后缘具 1 个分离的小斑；痣带完整；缘斑向外后方延伸出 1 条窄带，与痣带的基支相连；基带完整；无基斑。雄性腹部 1 ~ 5 节背板黑褐色。第 3 节背中突长棒状，向下弯曲，超过第 4 节背板中部，并盖住第 4 节的后背突。第 6 节黄褐色，筒形。第 7、8 节黄褐色，基部缢缩。雄性外生殖器卵形。第 9 节背板末端具 U 形缺刻；腹板具阔长的基柄；下瓣近三角形，内缘近基部具 1 个耳状突起，相互重叠，形成 1 个大型窗口。阳基侧突基半部与阳茎愈合；阳茎锥形，侧突起发达。

【发生期】 4 ~ 5 月。

【地理分布】 台湾。

Measurements: Forewing length 13.5–15.0 mm.

Morphology: Vertex dark brown. Ocellar triangle black. Rostrum yellowish brown. Pronotum unevenly yellowish brown. Mesonotum and metanotum yellowish brown, with a distinct black mesal stripe. Pleura and legs yellowish brown. Wing membrane

图 3-104-1　杨氏新蝎蛉 ***Neopanorpa youngi*** 成虫背面观（台湾，林义祥摄）

1. 雄虫　2. 雌虫

Fig. 3-104-1　Adults of ***Neopanorpa youngi,*** dorsal view (Taiwan, photo by Yechan Lin)

1. Male　2. Female

图 3-104-2 杨氏新蝎蛉 ***Neopanorpa youngi*** 雄性外生殖器腹面观

Fig. 3-104-2 Male genitalia of ***Neopanorpa youngi***, ventral view

图 3-104-3 杨氏新蝎蛉 ***Neopanorpa youngi*** 地理分布

Fig. 3-104-3 Distribution of ***Neopanorpa youngi***

translucent, tinged with yellow. Markings dark brown. In forewings, apical band complete and accompanied by a separate small spot posteriorly; pterostigmal band complete; marginal spot slender and elongated, connected to basal branch of pterostigmal band; basal band complete; basal spot absent. Male terga Ⅰ–Ⅴ dark brown. Notal organ on tergum Ⅲ long stick-like and curved ventrad, exceeding middle of tergum Ⅳ, covering postnotal organ on tergum Ⅳ. A6 yellowish brown, cylindrical. A7 and A8 yellowish brown and constricted basally. Male genital bulb oval. Epandrium long and broad, emarginate terminally in a U-shape. Hypandrium with a long and broad basal stalk. Hypovalves subtriangular, with an auriculate process on inner margin subbasally, overlapped mesally and forming a large window subbasally. Paramere fused with aedeagus at basal half. Aedeagus conical and with stout lateral processes.

Emergence: April to May.

Distribution: Taiwan.

八、蝎蛉属 *Panorpa* Linnaeus, 1758

蝎蛉属 *Panorpa* 是蝎蛉科 Panorpidae 中物种最多的属，全世界共有 260 余种。中国已记载 100 余种，本图鉴收录 62 种。

Panorpa is the most specise genus in Panorpidae, with more than 260 species described worldwide. Over 100 species have been recorded from China. Sixty-two species are included in this atlas.

图 3-k 长白山蝎蛉 ***Panorpa changbaishana*** 雄虫背侧面观
Fig. 3-k Male adult of ***Panorpa changbaishana,*** dorso-lateral view

本图鉴蝎蛉属 *Panorpa* 分种检索表

（59 种，雄性。织纹蝎蛉 *Panorpa implicata*、岛生蝎蛉 *Panorpa insularis* 和山阳蝎蛉 *Panorpa shanyangensis* 雄性未知，未列入）

1. 腹部第 7 节筒形，基部不缢缩 ······ 2
   腹部第 7 节基部不同程度缢缩 ······ 17
2. 腹部第 3 节背中突强烈隆起，末端与第 4 节背板上尖锐的后背突相接触；第 6 节明显长于第 5 节 ······ 3
   腹部第 3 节背中突较扁平，盖住第 4 节背板上尖锐的后背突；第 6 节约与第 5 节等长，或稍长于后者 ······ 4
3. 生殖刺突外缘强烈内凹 ······ 107. 宝华山蝎蛉 ***Panorpa baohwashana***
   生殖刺突外缘不内凹 ······ 138. 刘氏蝎蛉 ***Panorpa liui***
4. 前翅 R_2 不分支 ······ 165. 王氏蝎蛉 ***Panorpa waongkehzengi***
   前翅 R_2 分为 2 ~ 3 支 ······ 5
5. 腹部第 6 节末端背面具深缺刻，两侧形成 1 对片状突起 ······ 6
   腹部第 6 节末端无上述缺刻及突起 ······ 7
6. 腹部第 6 节末端的缺刻 V 形，两侧片状突起三角形；侧面无尖锐的小齿突 ······ 136. 昆明蝎蛉 ***Panorpa kunmingensis***
   腹部第 6 节末端的缺刻狭窄，两侧片状突起末端稍平截；每个突起下方的侧面还具 1 个尖锐的小齿突 ······ 131. 滴蝎蛉 ***Panorpa guttata***
7. 生殖刺突长于基节，长镰刀状 ······ 141. 长蝎蛉 ***Panorpa macrostyla***
   生殖刺突短于基节 ······ 8
8. 阳基侧突分为 3 支 ······ 9
   阳基侧突不分支或分为 2 支 ······ 10
9. 阳基侧突光裸；中支在近端部又分为 2 支 ······ 137. 廖氏蝎蛉 ***Panorpa liaoi***
   阳基侧突着生长刺毛；中支不分支 ······ 139. 长支蝎蛉 ***Panorpa longiramina***
10. 阳基侧突近基部分为 2 支，“3”形，其腹支延伸至下瓣下方，背支延伸至阳茎侧面 ······ 152. 反曲蝎蛉 ***Panorpa reflexa***
    阳基侧突不分支或分为 2 支；若分为 2 支则其每支均向尾端延伸 ······ 11
11. 阳基侧突不分支 ······ 12
    阳基侧突分为 2 支 ······ 14
12. 翅斑极度退化，仅存散点状的端带和痣带；阳基侧突长棍棒状，直而平行，内侧近基部具 1 个尖齿 ······ 146. 平支蝎蛉 ***Panorpa parallela***
    翅具明显的翅斑；阳基侧突不如上述 ······ 13
13. 翅端带不明显，痣带退化为三角形斑；基带仅在后缘处形成 1 个圆形大斑；无缘斑和基斑 ······ 120. 淡色蝎蛉 ***Panorpa decolorata***
    翅端带内部具 3 ~ 4 个透明窗斑；痣带阔，端支断开，与端带融合，基支阔；缘斑斜行，与痣带的基支相连，形成 1 个显著的 V 形纹；基带前部细，后部阔；后缘基半部具一系列不规则斑块 ······ 129. 嘉理斯蝎蛉 ***Panorpa gressitti***
14. 痣带基支发达，端支退化，形成 1 条显著的斜带 ······ 15
    痣带横行或退化为翅痣处的 1 个小斑 ······ 16
15. 头顶黑色；M 脉正常；下瓣末端超出生殖肢基节端部 ······ 144. 斜带蝎蛉 ***Panorpa obliquifascia***
    头顶黄色，后头具三叉形黑纹；翅 M 在 FM 之前半透明；下瓣末端不超出生殖肢基节端部 ······ 148. 彼得蝎蛉 ***Panorpa peterseana***
16. 头顶黄色，单眼三角区黑色；翅具明显的端带和痣带，其中痣带横行，后方不分支 ······ 123. 女儿蝎蛉 ***Panorpa filina***
    头顶黑褐色，复眼缘黄褐色；翅仅在翅痣处具 1 个黑色小斑 ······ 145. 赭尾蝎蛉 ***Panorpa ochraceocauda***
17. 腹部第 7、8 节短于第 6 节，基部强烈缢缩，呈短倒锥形；生殖肢基节腹面端部各具 1 个分叉的片状突起 ······ 112. 长白山蝎蛉 ***Panorpa changbaishana***
    腹部第 7、8 节约与第 6 节等长，不呈短倒锥形；生殖肢基节无上述突起 ······ 18
18. 腹部第 6 节背面具 1 个臀角 ······ 19
    腹部第 6 节背面无臀角 ······ 20

19. 翅黄色；臀角短锥形，向上指；阳基侧突半叶形，不交叉 ………………… 125. 黄翅蝎蛉 ***Panorpa flavipennis***
翅无色；臀角指形，指向后上方；阳基侧突螺旋状，相互交叉 ……………… 135. 尤氏蝎蛉 ***Panorpa kiautai***
20. 背中突发达，棒状，超过第 4 节背板中部 ……………………………………………………………… 21
背中突扁平，不超过第 4 节背板中部 ……………………………………………………………………… 22
21. 前翅具明显的端带、痣带、缘斑、基带和基斑；下瓣细长，分歧 …… 151. 四带蝎蛉 ***Panorpa quadrifasciata***
前翅端带完整；痣带狭窄，无端支；基带仅为后缘的 1 个带状小斑；下瓣细长，在端部 1/3 处稍靠拢，端部分歧 ……………………………………………………………… 158. 痣带蝎蛉 ***Panorpa stigmalis***
22. 腹部第 7 节基部强烈缢缩呈柄状，向后突然变粗壮 ……………………………………………………… 23
腹部第 7 节基部缢缩，向端部均匀变阔，不呈柄状 ……………………………………………………… 29
23. 生殖刺突基半部着生稀疏的长毛，内侧具 1 个极大的凹陷区；阳基侧突粗长，伸入生殖刺突的凹陷区内 …… ………………………………………………………………………… 116. 支蝎蛉 ***Panorpa cladocerca***
生殖刺突基半部毛正常，内侧无上述凹陷区；阳基侧突不如上述 ……………………………………… 24
24. 阳基侧突细长，超过生殖刺突的中部 …………………………………………………………………… 25
阳基侧突较短，不伸达生殖刺突中部 ……………………………………………………………………… 26
25. 阳基侧突细线状，基部与端部粗细一致 ………………………………… 119. 斑蝎蛉 ***Panorpa deceptor***
阳基侧突基部 1/4 宽阔，端部 3/4 细长 ……………………………… 124. 黄体蝎蛉 ***Panorpa flavicorporis***
26. 痣带端支退化，与基带形成明显的 2 条横带；无其他斑纹；阳基侧突分为 2 支，Y 形 ……………… ………………………………………………………………………… 110. 双纹蝎蛉 ***Panorpa bistriata***
翅斑及阳基侧突不如上述 …………………………………………………………………………………… 27
27. 下瓣短，仅伸达生殖肢基节中部；阳基侧突与阳茎极短，仅稍超过生殖肢基节中部 ……………… ………………………………………………………………………… 111. 短突蝎蛉 ***Panorpa brevititilana***
下瓣超过生殖肢基节中部；阳基侧突与阳茎远超过生殖肢基节中部 ……………………………………… 28
28. 阳基侧突内缘具 1 列梳齿状长刺，近基部具 1 个钩状尖齿 ………………147. 栉蝎蛉 ***Panorpa pectinata***
阳基侧突内缘具 1 列梳齿状短刺，近端部具 1 个尖齿 …………………… 161. 台湾蝎蛉 ***Panorpa taiwanensis***
29. 腹部几乎全为黑褐色；生殖刺突内缘的中齿与亚基齿均十分发达，且距离较远；阳茎强烈骨化，长柱状，腹瓣与背瓣紧密愈合 ………………………………………………………………………………………… 30
腹部 1 ~ 5 节背板黄褐色或黑褐色，第 7、8 节黄褐色；生殖刺突的中齿较小，与亚基齿紧邻；阳茎不如上述 ……………………………………………………………………………………………………… 31
30. 阳基侧突细长，紧贴于阳茎侧面，基部着生 1 簇粗壮的长刺；阳茎背瓣末端具 1 对耳状突起 ……………… ………………………………………………………………………… 121. 峨眉山蝎蛉 ***Panorpa emeishana***
阳基侧突粗长，不与阳茎紧贴，端半部内缘具 1 列长刺；阳茎背瓣末端无耳状突起 ……………… ………………………………………………………………………… 153. 半带蝎蛉 ***Panorpa semifasciata***
31. 后翅后缘在 M_3 至 CuA 处显著向外扩；第 9 节腹板具长基柄，Y 形 ……… 160. 淡珀蝎蛉 ***Panorpa subambra***
后翅后缘正常；第 9 节腹板基柄较短 ……………………………………………………………………… 32
32. 阳基侧突螺旋状扭曲并相互交叉 …………………………………………………………………………… 33
阳基侧突不如上述 …………………………………………………………………………………………… 44
33. 腹部第 6 节正常 ……………………………………………………………………………………………… 34
腹部第 6 节末端稍向上隆起并被稀疏的长毛 ……………………………………………………………… 36
34. 翅斑退化为一系列散乱的小斑点 ……………………………………… 159. 斑点蝎蛉 ***Panorpa stigmosa***
翅斑消失或不明显 …………………………………………………………………………………………… 35
35. 胸部背板灰白至黄白色；生殖刺突内缘具 1 宽列稀疏的长刚毛；阳基侧突短，远离生殖肢基节端部 ………… …………………………………………………………………………113. 郑氏蝎蛉 ***Panorpa chengi***
胸部背板黑褐色，中央具黄色纵带；生殖刺突无上述刚毛；阳基侧突几乎伸达生殖肢基节端部 ……………… …………………………………………………………………………128. 淡黄蝎蛉 ***Panorpa fulvastra***
36. 生殖肢基节端部腹面内侧各具 1 个延长的突起，内缘具 1 列长刚毛 …………………………………… 37
生殖肢基节端部腹面内侧无上述突起及刚毛 ……………………………………………………………… 38
37. 前翅基带完整 ……………………………………………………………… 117. 弯曲蝎蛉 ***Panorpa curva***
前翅基带断裂成 2 个小斑 ……………………………………………………… 118. 大卫蝎蛉 ***Panorpa davidi***
38. 翅斑退化为一系列散乱的小斑点 ……………………………………………… 126. 实蝎蛉 ***Panorpa fructa***
翅斑发达，各斑、带明显 …………………………………………………………………………………… 39
39. 生殖肢基节端部腹面具 10 根以上粗刚毛 ………………………………… 164. 模蝎蛉 ***Panorpa typicoides***

生殖肢基节端部腹面具 6 根或以下粗刚毛 …… 40
40. 生殖肢基节端部腹面仅 1 ~ 2 根粗刚毛 …… 150. 秦岭蝎蛉 ***Panorpa qinlingensis***
生殖肢基节端部腹面具 4 ~ 6 根粗刚毛 …… 41
41. 翅无缘斑和基斑 …… 154. 六刺蝎蛉 ***Panorpa sexspinosa***
翅具明显的缘斑和基斑 …… 42
42. 中、后胸背板黄褐色，在前缘两侧具黑斑 …… 166. 杨氏蝎蛉 ***Panorpa yangi***
中、后胸背板黑褐色，中央具黄色纵带 …… 43
43. 阳基侧突稍超出生殖肢基节端部 …… 134. 金川蝎蛉 ***Panorpa jinchuana***
阳基侧突粗长，超出生殖肢基节端部 …… 143. 新刺蝎蛉 ***Panorpa neospinosa***
44. 生殖刺突内缘无中齿；亚基齿十分发达，长度约为生殖刺突之半；阳基侧突分为 3 支，每支上都着生有长毛刺 …… 130. 桂东蝎蛉 ***Panorpa guidongensis***
生殖刺突内缘具中齿；亚基齿不如上述；阳基侧突不分支或仅分 2 支 …… 45
45. 阳基侧突不分支 …… 46
阳基侧突分为 2 支 …… 54
46. 阳基侧突末端超过生殖刺突中部 …… 47
阳基侧突末端不超过生殖刺突中部 …… 49
47. 翅斑发达，痣带具端支，缘斑和基斑明显 …… 105. 安仁蝎蛉 ***Panorpa anrenensis***
翅端带、痣带和基带排列成明显的 3 条横带，痣带无端支，无缘斑和基斑 …… 48
48. 生殖刺突内缘的中齿钝圆；阳基侧突直，剑状，内缘着生长刺 …… 115. 周氏蝎蛉 ***Panorpa choui***
生殖刺突内缘的中齿直角形；阳基侧突在中部弯曲，内缘着生短刺 …… 163. 三带蝎蛉 ***Panorpa tritaenia***
49. 中、后胸背板大部黑色，仅小盾片黄色；下瓣短，仅伸达生殖肢基节中部 …… 122. 江崎蝎蛉 ***Panorpa esakii***
中、后胸背板不如上述；下瓣端部超过生殖肢基节中部 …… 50
50. 翅斑特殊：无端带；痣带十分淡，无端支；缘斑极小或消失；基带完整，纤细；基斑成为 1 条前方不超过 M 的长带 …… 155. 六带蝎蛉 ***Panorpa sextaenia***
翅斑不如上述 …… 51
51. 生殖肢基节腹面内缘具 1 列十分粗长的刚毛 …… 162. 三线蝎蛉 ***Panorpa trifasciata***
生殖肢基节无上述刚毛 …… 52
52. 翅端带与痣带在前缘相接 …… 127. 福建蝎蛉 ***Panorpa fukiensis***
翅端带与痣带分离 …… 53
53. 翅斑仅存端带和无端支的痣带，形成 2 条明显的横带；生殖肢基节腹面内缘正常 …… 109. 双带蝎蛉 ***Panorpa bifasciata***
翅具发达的端带、痣带、基带以及极小的缘斑和基斑；生殖肢基节腹面内缘具 1 个向后的片状突起 …… 142. 莽山蝎蛉 ***Panorpa mangshanensis***
54. 阳基侧突一支长不及另一支的 1/3；阳茎背瓣近端部分为 2 支 …… 108. 二支蝎蛉 ***Panorpa biclada***
阳基侧突一支长超过另一支的 1/3；阳茎背瓣近端部不分为 2 支 …… 55
55. 胸部背板黑褐色，中央具明显的黄色纵带；阳基侧突在近端部分为 2 支，扳手状 …… 157. 楚南蝎蛉 ***Panorpa sonani***
胸部背板黑褐色，中央无纵带；阳基侧突在近基部分为 2 支 …… 56
56. 生殖肢基节腹面内缘各具 1 个宽阔的片状突起，在端部又各具 1 个尖锐的指状突起 …… 149. 平江蝎蛉 ***Panorpa pingjiangensis***
生殖肢基节腹面内缘、端部无上述突起 …… 57
57. 阳基侧突腹支端部极其膨大，鼓槌形，具短尖 …… 114. 陈氏蝎蛉 ***Panorpa cheni***
阳基侧突腹支正常 …… 58
58. 阳基侧突分为几乎等长的 2 支 …… 106. 金身蝎蛉 ***Panorpa aurea***
阳基侧突腹支长仅为背支之半 …… 140. 黄蝎蛉 ***Panorpa lutea***

Key to species of *Panorpa* in this atlas

(59 species, males. *Panorpa implicata,Panorpa insularis* and *Panorpa shanyangensis* not included because the males are unknown)

1. A7 cylindrical, not constricted at base …… 2

A7 variously constricted at base 17

2. Notal organ on tergum Ⅲ strongly raised and in contact with acute postnotal organ on tergum Ⅳ; A6 distinctly longer than A5 3
Notal organ on tergum Ⅲ relatively flat and covering acute postnotal organ on tergum Ⅳ; A6 nearly as long as or slightly longer than A5 4
3. Gonostylus intensively concaved on outer margin .... 107. ***Panorpa baohwashana***
Gonostylus not concaved on outer margin .... 138. ***Panorpa liui***
4. R_2 in forewing unbranched .... 165. ***Panorpa waongkehzengi***
R_2 in forewing bifurcated or trifurcated 5
5. A6 intensively emarginated on apex dorsally, and forming a pair of plates laterally 6
A6 not emarginated on apex and without such plates 7
6. A6 emarginated apically in a V-shape, and forming a pair of triangular plates laterally; lacking acute tooth on lateral surface .... 136. ***Panorpa kunmingensis***
A6 shallowly emarginated on apex and forming a pair of slightly truncated plates laterally, and bearing an acute tooth under each plate on lateral surface .... 131. ***Panorpa guttata***
7. Gonostylus longer than gonocoxites and sickle-shaped .... 141. ***Panorpa macrostyla***
Gonostylus shorter than gonocoxites 8
8. Paramere trifurcated 9
Paramere unbranched or bifurcated 10
9. Paramere glabrous, with its median branch bifurcated subapically .... 137. ***Panorpa liaoi***
Paramere bearing long spines with an unbranched median branch .... 139. ***Panorpa longiramina***
10. Paramere bifurcated subbasally and 3-shaped, with its ventral branch stretched cephalad , and dorsal branch stretched caudad and aside aedeagus .... 152. ***Panorpa reflexa***
Paramere unbranched or bifurcated; both branches directed caudad if bifurcated 11
11. Paramere unbranched 12
Paramere bifurcated 14
12. Wing markings greatly reduced with scattered apical and pterostigmal bands; parameres long stick-like, straight and parallel, and each with an acute tooth on inner margin subbasally .... 146. ***Panorpa parallela***
Wing markings well-developed; paramere not shaped as above 13
13. Apical band inconspicuous; pterostigmal band reduced into a triangular spot; basal band forming a large circular spot on hind margin; marginal and basal spots absent .... 120. ***Panorpa decolorata***
Apical band with 3–4 hyaline spots innerly; pterostigmal band broad, with a separated basal branch connecting to apical band, and a broad basal branch; marginal spot obliquely connected to basal branch of pterostigmal band and forming a distinct V-shaped pattern; basal band slender anteriorly and broad posteriorly; a series of irregular spots on basal half of hind margin .... 129. ***Panorpa gressitti***
14. Pterostigmal band with a well-developed basal branch and a reduced apical branch, forming a distinct oblique band ... 15
Pterostigmal band transverse or reduced into a small spot 16
15. Vertex black; vein M normal; hypovalves exceeding apex of gonocoxites apically ...... 144. ***Panorpa obliquifascia***
Vertex yellow; a black trident-shaped pattern on occiput; vein M translucent before FM; hypovalves not exceeding apex of gonocoxites .... 148. ***Panorpa peterseana***
16. Vertex yellow with ocellar triangle black; wings with distinct apical and pterostigmal bands; pterostigmal band transverse and not furcated posteriorly .... 123. ***Panorpa filina***
Vertex dark brown but yellowish brown around compound eyes; wings with a small black spot on pterostigma .... 145. ***Panorpa ochraceocauda***
17. A7 and A8 shorter than A6, intensively constricted at base and resembling reversed short cones; gonocoxites each with an apically bifurcated plate on ventral apex .... 112. ***Panorpa changbaishana***
A7 and A8 nearly as long as A6, not resembling reversed short cones; gonocoxites lacking such plates 18
18. A6 with a single anal horn dorsally 19
A6 without anal horns 20
19. Wings yellow; anal horn short conical and directed dorsad; parameres semi-foliate and not crossed mesally .... 125. ***Panorpa flavipennis***

Wings hyaline; anal horn finger-like and directing dorso-caudad; parameres spiral and crossed mesally .... 135. ***Panorpa kiautai***

20. Notal organ well-developed, stick-shaped and extending to middle of tergum Ⅳ 21
    Notal organ flat and not extending to middle of tergum Ⅳ 22
21. Forewing with conspicuous apical, pterostigmal and basal bands and marginal and basal spots; hypovalves slender and divergent .... 151. ***Panorpa quadrifasciata***
    Forewing with a complete apical band; pterostigmal band narrow and lacking an apical branch; basal band reduced into a small spot on hind margin; hypovalves slender, convergent on distal third and divergent apically .... 158. ***Panorpa stigmalis***
22. A7 intensively constricted and stalk-like basally, and abruptly swollen towards distal portion 23
    A7 constricted basally and evenly swollen towards distal portion, not distinctly stalked 29
23. Gonostylus with sparse long setae on basal half, and a very large concaved area on inner margin; paramere long and stout, extending to concaved area of gonostylus .... 116. ***Panorpa cladocerca***
    Gonostylus with normal setae on basal half, and lacking such concaved area on inner margin; paramere not shaped as above 24
24. Paramere slender and exceeding middle of gonostylus 25
    Paramere relatively short and not exceeding middle of gonostylus... 26
25. Paramere thread-like with its basal and distal portions nearly equal in width.................... 119. ***Panorpa deceptor***
    Paramere broad at basal quarter, and slender at distal three-quarters .... 124. ***Panorpa flavicorporis***
26. Pterostigmal band with a reduced apical branch and forming two distinct transverse bands with basal band; other markings absent; paramere bifurcated and Y-shaped .... 110. ***Panorpa bistriata***
    Wing markings and parameres not shaped as above 27
27. Hypovalves short and extending to middle of gonocoxites; paramere and aedeagus short and slightly exceeding middle of gonocoxites .... 111. ***Panorpa brevititilana***
    Hypovalves exceeding middle of gonocoxites; paramere and aedeagus far exceeding middle of gonocoxites 28
28. Paramere with a row of comb-like long spines, and a hook-like acute tooth subbasally on inner margin .... 147. ***Panorpa pectinata***
    Paramere with a row of comb-like short spines, and a small acute tooth subapically on inner margin .... 161. ***Panorpa taiwanensis***
29. Abdomen almost entirely dark brown; gonostylus with a strongly developed and diverged median tooth and a subbasal process on inner margin; aedeagus strongly sclerotized and long columnar, with ventral and dorsal valves intensively fused 30
    Terga Ⅰ – Ⅴ yellowish or dark brown, A7 and A8 yellowish brown; gonostylus with a small median tooth close to subbasal process on inner margin; aedeagus not shaped as above 31
30. Paramere slender, closely clinging to aedeagus, and bearing a tuft of long stout spines basally; dorsal aedeagal valves with a pair of auriculate processes laterally .... 121. ***Panorpa emeishana***
    Paramere long and stout, not clinging to aedeagus, and bearing a row of long spines on inner margin of apical half; aedeagus lacking auriculate processes laterally .... 153. ***Panorpa semifasciata***
31. Hindwing distinctly expanded outward between M_3 and CuA on hind margin; hypandrium with a long basal stalk and Y-shape .... 160. ***Panorpa subambra***
    Hindwing normal on hind margin; hypandrium with a shortened basal stalk 32
32. Parameres spiral and crossed mesally 33
    Parameres not shaped as above 44
33. A6 normal on dorsal apex 34
    A6 projected dorsad and bearing sparse long setae on dorsal apex 36
34. Wing markings reduced into a series of scattered spots .... 159. ***Panorpa stigmosa***
    Wing markings absent or inconspicuous 35
35. Nota greyish white or yellowish white; gonostylus with a broad row of long setae on inner margin; paramere short and far away from apex of gonocoxites .... 113. ***Panorpa chengi***
    Nota dark brown with a longitudinal yellow mesal stripe; gonocoxites with normal setae oninner margin; paramere nearly extending to apex of gonocoxites .... 128. ***Panorpa fulvastra***

36. Gonocoxites with an elongated process on ventral apex innerly, and bearing a row of long setae on inner margin ... 37
Gonocoxites lacking such a process and setae ... 38
37. Basal band in forewing complete ... 117. ***Panorpa curva***
Basal band in forewing split into two small spots ... 118. ***Panorpa davidi***
38. Wing markings reduced into a series of scattered small spots ... 126. ***Panorpa fructa***
Wing markings well-developed with distinct bands and spots ... 39
39. Gonocoxites with more than ten stout setae on ventral apex ... 164. ***Panorpa typicoides***
Gonocoxites with six or fewer stout setae on ventral apex ... 40
40. Gonocoxites with 1–2 stout setae on ventral apex ... 150. ***Panorpa qinlingensis***
Gonocoxites with 4–6 stout setae on ventral apex ... 41
41. Wings lacking marginal and basal spots ... 154. ***Panorpa sexspinosa***
Wings with distinct marginal and basal spots ... 42
42. Mesonotum and metanotum yellowish brown, with black spots on anterior margin laterally ... 166. ***Panorpa yangi***
Mesonotum and metanotum dark brown, with a longitudinal yellow mesal stripe ... 43
43. Paramere slightly exceeding apex of gonocoxites ... 134. ***Panorpa jinchuana***
Paramere long and stout, far exceeding apex of gonocoxites ... 143. ***Panorpa neospinosa***
44. Gonostylus without a median tooth on inner margin; subbasal process stout and half length of gonostylus; paramere trifurcated and bearing long spines on each branch ... 130. ***Panorpa guidongensis***
Gonostylus with a median tooth on inner margin; subbasal process not as above; paramere unbranched or bifurcated ... 45
45. Paramere unbranched ... 46
Paramere bifurcated ... 54
46. Paramere exceeding middle of gonostylus apically ... 47
Paramere not exceeding middle of gonostylus apically ... 49
47. Wings with well-developed markings; pterostigmal band with an apical branch; marginal and basal spots distinct ... 105. ***Panorpa anrenensis***
Wings with apical, pterostigmal and basal bands arranged as three distinct transverse bands; pterostigmal band lacking apical branch; marginal and basal spots absent ... 48
48. Gonostylus with a rounded median tooth on inner margin; paramere straight, sword-like, and bearing long spines on inner margin ... 115. ***Panorpa choui***
Gonostylus with a right-angled median tooth on inner margin; paramere curved medially and bearing short spines on inner margin ... 163. ***Panorpa tritaenia***
49. Mesonotum and metanotum mostly black with yellow scutella; hypovalves short and only extending to middle of gonoc oxites ... 122. ***Panorpa esakii***
Mesonotum and metanotum not as above; hypovalves exceeding middle of gonocoxites ... 50
50. Wing markings peculiar: in forewings, apical band absent; pterostigmal band very faint and lacking apical branch; marginal spot very small or absent; basal band complete and slender; basal spot elongated into a long stripe, not exceeding M anteriorly ... 155. ***Panorpa sextaenia***
Wing markings not shaped as above ... 51
51. Gonocoxites each with a row of long stout setae on inner margin ventrally. ... 162. ***Panorpa trifasciata***
Gonocoxites lacking such setae ... 52
52. Apical band connected to pterostigmal band on anterior margin ... 127. ***Panorpa fukiensis***
Apical band separated with pterostigmal band on anterior margin ... 53
53. Apical band and apical-branch-lacking pterostigmal band forming two distinct transverse bands; ventro-inner margin of gonocoxites normal ... 109. ***Panorpa bifasciata***
Apical, pterostigmal and basal bands well-developed; marginal and basal spots very small; gonocoxites with a platelike process directed caudad on ventro-inner margin ... 142. ***Panorpa mangshanensis***
54. Ventral branch of paramere shorter than third as long as dorsal branch; dorsal aedeagal valves bifurcated subapically ... 108. ***Panorpa biclada***
Ventral branch of paramere longer than third as long as dorsal branch; dorsal aedeagal valves unbranched ... 55
55. Nota dark brown, with a distinct longitudinal yellow mesal stripe; paramere bifurcated subapically and wrenchshaped

.... 157. ***Panorpa sonani***
Nota dark brown, without a mesal stripe; paramere bifurcated subbasally.. 56
56. Gonocoxite with a broad plate on ventro-inner margin, and an acute finger-like process on ventral apex innerly.........
.... 149. ***Panorpa pingjiangensis***
Gonocoxite without such plate and process on ventro-inner margin, and lacking process on ventral apex innerly as above 57
57. Ventral branch of paramere greatly enlarged on apex and drumstick-shaped with an acute short tip........................
....114. ***Panorpa cheni***
Ventral branch of paramere normal.. 58
58. Two branches of paramere approximately equal in length .... 106. ***Panorpa aurea***
Ventral branch of paramere half as long as dorsal branch .... 140. ***Panorpa lutea***

105. 安仁蝎蛉 *Panorpa anrenensis* Chou & Wang, 1987

【测量】 前翅长 12.0 ~ 13.0 mm。

【形态特征】 头顶黄褐色，单眼三角区黑色。喙黄褐色。前胸背板黑褐色。中、后胸背板黑褐色，中央有 1 条灰黄色纵带。侧板及足黄褐色。翅无色透明，斑纹褐色。前翅端带后角具 1 透明斑；痣带完整，基支宽于端支；基带完整，后部较宽；缘斑、基斑明显。雄性腹部 1 ~ 5 节背板黑褐色。第 3 节背中突仅稍隆起。第 6 节黑褐色，筒形，无臀角。第 7、8 节黄褐色，基部稍缢缩。雄性外生殖器长卵圆形。第 9 节背板末端具浅 U 形缺刻；腹板基柄极短；下瓣细长，端部内缘具 1 列短刚毛。生殖刺突内缘具钝三角形中齿和短铗状亚基齿。阳基侧突细长，超过生殖刺突的中齿，内侧具密集的刺毛。阳茎腹瓣短小；背瓣末端指状延伸且相互分离。

【发生期】 9 ~ 10 月。

【地理分布】 湖南（安仁）、广西（柳州）。

Measurements: Forewing length 12.0–13.0 mm.

Morphology: Vertex yellowish brown. Ocellar triangle black. Rostrum yellowish brown. Pronotum dark brown. Mesonotum and metanotum dark brown with a greyish yellow mesal stripe. Pleura and legs yellowish brown. Wing membrane hyaline with brown markings. In forewings, apical band with a hyaline spot posteriorly; pterostigmal band complete, with a broad basal branch and a thin apical branch; basal band complete and broadened towards hind portion; marginal and basal spots conspicuous. Male terga Ⅰ–Ⅴ dark brown. Notal organ on tergum Ⅲ slightly raised. A6 dark brown, cylindrical and devoid of anal horns.

图 3-105-1 安仁蝎蛉 ***Panorpa anrenensis*** 成虫背面观（湖南安仁，NWAFU）
1. 雄虫 2. 雌虫

Fig. 3-105-1 Adults of ***Panorpa anrenensis***, dorsal view (Anren, Hunan, NWAFU)
1. Male 2. Female

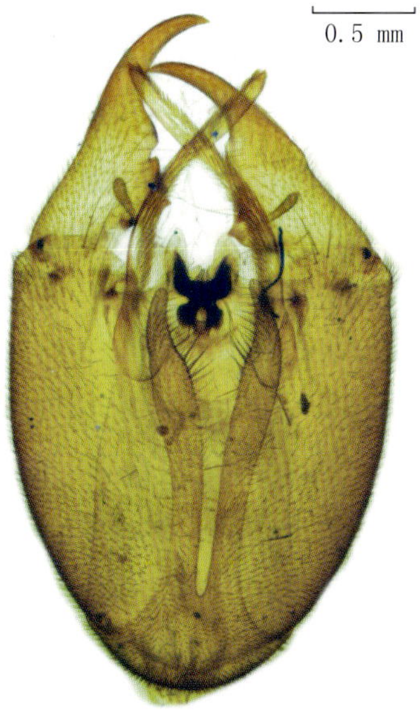

图 3-105-2 安仁蝎蛉 *Panorpa anrenensis* 雄性外生殖器腹面观

Fig. 3-105-2 Male genitalia of *Panorpa anrenensis*, ventral view

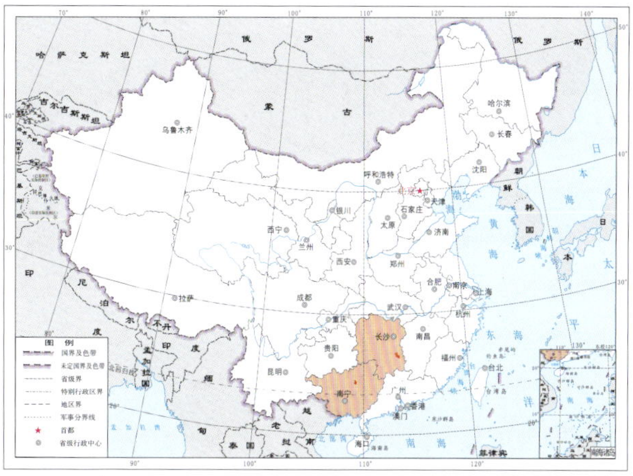

图 3-105-3 安仁蝎蛉 *Panorpa anrenensis* 地理分布

Fig. 3-105-3 Distribution of *Panorpa anrenensis*

A7 and A8 yellowish brown and slightly constricted basally. Male genital bulb long oval. Epandrium with a shallow U-shaped terminal emargination. Hypandrium with a greatly shortened basal stalk. Hypovalve slender, bearing a row of stout bristles on inner margin of apical portion. Gonostylus with an obtuse triangular median tooth and a pincer-like subbasal process on inner margin. Paramere slender, exceeding median tooth of gonostylus and bearing dense short spines on inner surface. Aedeagus with short ventral valves; dorsal valves elongated, finger-like and divergent.

Emergence: September to October.

Distribution: Hunan (Anren) and Guangxi (Liuzhou).

106. 金身蝎蛉 *Panorpa aurea* Cheng, 1957

【测量】 前翅长 15.5 ~ 16.0 mm。

【形态特征】 头顶黄褐色，单眼三角区黑色。喙黄褐色。胸部背板黑褐色，侧板及足黄色。翅略带黄色，透明，斑纹黄褐色。前翅端带阔，后缘有 1 个透明窗斑；痣带、基带完整且宽阔；缘斑缺失；基斑小。雄性腹部 1 ~ 5 节背板黑褐色。第 3 节背中突仅稍隆起，盖住第 4 节上尖锐的后背突。第 6 节黄褐色，筒形，向端部渐窄，无臀角。第 7、8 节黄褐色，基部稍缢缩。雄性外生殖器长卵形。第 9 节背板末端具深 U 形缺刻，两侧形成 1 对指形突起；下瓣细长，内缘具 1 列长鬃毛，稍超过生殖肢基节中部。生殖刺突内缘具钝三角形中齿和发达的盘状亚基齿。阳基侧突二叉，Y 形，两支内缘各具 1 列粗刚毛。阳茎腹瓣膜质不发达，背瓣指状，相互分离。

【发生期】 9 ~ 10 月。

【地理分布】 福建（邵武、武夷山）、江西（铅山）、浙江（天目山、凤阳山）。

图 3-106-1 金身蝎蛉 *Panorpa aurea* 雄成虫背面观（浙江凤阳山，NWAFU）

Fig. 3-106-1 Male adult of *Panorpa aurea*, dorsal view (Mount Fengyang, Zhejiang, NWAFU)

图 3-106-2　金身蝎蛉 *Panorpa aurea* 雄性外生殖器腹面观

Fig. 3-106-2　Male genitalia of ***Panorpa aurea***, ventral view

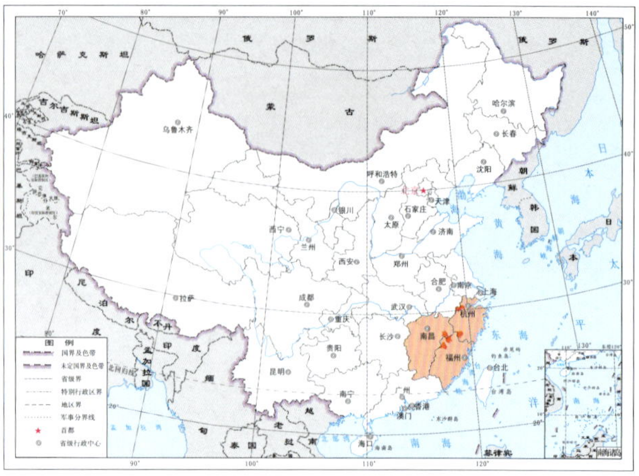

图 3-106-3　金身蝎蛉 ***Panorpa aurea*** 地理分布

Fig. 3-106-3　Distribution of ***Panorpa aurea***

Measurements: Forewing length 15.5–16.0 mm.

Morphology: Vertex yellowish brown. Ocellar triangle black. Rostrum yellowish brown. Nota dark brown. Pleura and legs yellowish. Wing membrane hyaline, slightly tinged with yellow. Markings yellowish brown. In forewings, apical band broad with a large hyaline window posteriorly; pterostigmal and basal bands complete and broad; marginal spot absent; basal spot small. Male terga Ⅰ–Ⅴ dark brown. Notal organ on tergum Ⅲ slightly raised and covering acute postnotal organ on tergum Ⅳ. A6 yellowish brown, cylindrical and tapering towards apex, devoid of anal horns. A7 and A8 yellowish brown and constricted basally. Male genital bulb long oval. Epandrium with a deep U-shaped terminal emargination, forming a pair of finger-like processes laterally. Hypovalve slender, bearing a row of long bristles along inner margin, and slightly exceeding middle of gonostylus apically. Gonostylus with an obtuse triangular median tooth and a well-developed disc-like subbasal process. Paramere bifurcated and Y-shaped, and bearing a row of stout spines on inner margin. Aedeagus with membranous poorly-developed ventral valves; dorsal valves finger-like and divergent.

Emergence: September to October.

Distribution: Fujian (Shaowu, Wuyi Mountains), Jiangxi (Yanshan) and Zhejiang (Tianmu Mountains, Mount Fengyang).

107. 宝华山蝎蛉 *Panorpa baohwashana* Cheng, 1957

【测量】 前翅长 18.0 ~ 20.0 mm。

【形态特征】 头顶黑褐色，单眼三角区黑色。喙黑褐色，两侧黄褐色。胸部背板黑褐色。侧板及足黄褐色。翅无色透明，斑纹黑褐色。前翅端带阔；痣带基支阔，端支分离为 1 个三角形小斑；缘斑小；基带断开成 2 斑；无基斑。雄性腹部黑褐色。第 3 节背中突强烈隆起，末端与第 4 节尖锐的后背突相接触。第 6 节圆筒形，长度超过第 5 节。第 7 ~ 8 节圆筒形，基部缢缩不明显。雄性外生殖器长卵形，黑色。第 9 节背板阔长，端部具深 U 形缺刻并向腹面弯折；腹板基柄极短；下瓣相当阔长，内缘波状，端部向内弯曲。生殖刺突稍长于基节，细长，基部 1/3 处强烈向内弯曲，端部 2/3 向外拱曲；内缘具钝三角形中齿和方形亚基齿。阳基侧突分 2 支，其中腹支在阳茎基部交叉。阳茎尖长，超过生殖肢基节末端；侧突起指状，发达。

【发生期】 7 月。

【地理分布】 江苏（宝华山）。

Measurements: Forewing length 18.0–20.0 mm.

Morphology: Vertex dark brown. Ocellar triangle black. Rostrum dark brown, with yellowish brown flanks. Nota dark brown. Pleura and legs yellowish. Wing membrane hyaline, with markings dark brown. In

图 3-107-1 宝华山蝎蛉 ***Panorpa baohwashana*** 雄成虫侧面观（触角、右翅和外生殖器已去除，江苏南京宝华山，NWAFU）

Fig. 3-107-1 Male adult of ***Panorpa baohwashana***, lateral view (with antennae, right wings and genitalia removed. Mount Baohua, Nanjing, Jiangsu, NWAFU)

图 3-107-2 宝华山蝎蛉 ***Panorpa baohwashana*** 雄性外生殖器腹面观（下瓣已去除）

Fig. 3-107-2 Male genitalia of ***Panorpa baohwashana***, with hypovalves removed, ventral view

forewings, apical band broad; pterostigmal band with a broad basal branch and a separated small triangular apical branch; marginal spot small; basal band split

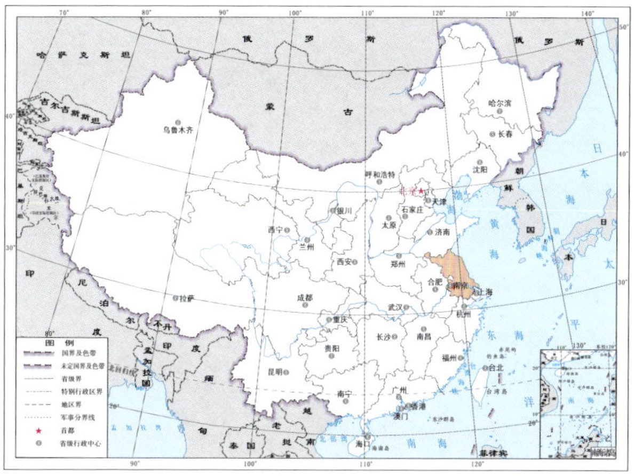

图 3-107-3 宝华山蝎蛉 ***Panorpa baohwashana*** 地理分布

Fig. 3-107-3 Distribution of ***Panorpa baohwashana***

into two spots; basal spot absent. Male abdomen dark brown. Notal organ on tergum Ⅲ greatly raised and in contact apically with acute postnotal organ on tergum Ⅳ. A6 cylindrical, longer than segment Ⅴ. A7 and A8 cylindrical and inconspicuously constricted basally. Male genital bulb black long oval. Epandrium long and broad, with a deep U-shaped terminal emargination, and bended ventrad on apical portion. Hypandrium with a greatly shortened basal stalk. Hypovalve very long and broad, wave-like on inner margin, and curved mesad on apical portion. Gonostylus slightly longer than gonocoxites, slender, intensively concaved along outer margin on basal third, arched on distal two-thirds, and bearing an obtuse triangle median tooth and a large oblong subbasal process. Parameres bifurcated, with their ventral branches crossed near base of aedeagus. Aedeagus acute and long, exceeding apex of gonocoxites; lateral processes finger-like and stout.

Emergence: July.

Distribution: Jiangsu (Mount Baohua).

108. 二支蝎蛉 *Panorpa biclada* Zhang & Hua, 2012

【测量】 前翅长 11.5 ~ 12.1 mm。

【形态特征】 头顶黑褐色。单眼三角区黑色。喙黄棕色，两侧具 2 条深色纵带。前胸背板黑色。中、后胸背板大部分黑色，具不明显的浅黄色中带。侧板及足浅黄色。翅透明无色，雄性一般无斑纹，雌性仅有且无端支的明显痣带。雄性腹部 1 ~ 5 节背板黑色。第 3 节背中突扁平，盖住第 4 节上尖锐的后背突。第 6 节黑色，长筒形，末端渐窄，无臀角。第 7、8 节黄褐色，基部缢缩。雄性外生殖器长椭圆形。第 9 节背板末端具 U 形缺刻，两侧形成 1 对指形突起；腹板具短基柄；下瓣细长，超过生殖肢基节中部，端部边缘具稀疏的长刚毛。生殖肢基节腹缘端部具整齐排列的 4 ~ 5 根长刚毛；生殖刺突长约为生殖肢基节之半，内缘具钝三角形中齿和

图 3-108-1 二支蝎蛉 ***Panorpa biclada*** 成虫背面观（陕西岚皋南宫山，NWAFU）

1. 雄虫 2. 雌虫

Fig. 3-108-1 Adults of ***Panorpa biclada***, dorsal view (Mount Nangong, Lan' gao, Shaanxi, NWAFU)

1. Male 2. Female

图 3-108-2 二支蝎蛉 ***Panorpa biclada*** 雄性外生殖器腹面观

Fig. 3-108-2 Male genitalia of ***Panorpa biclada***, ventral view

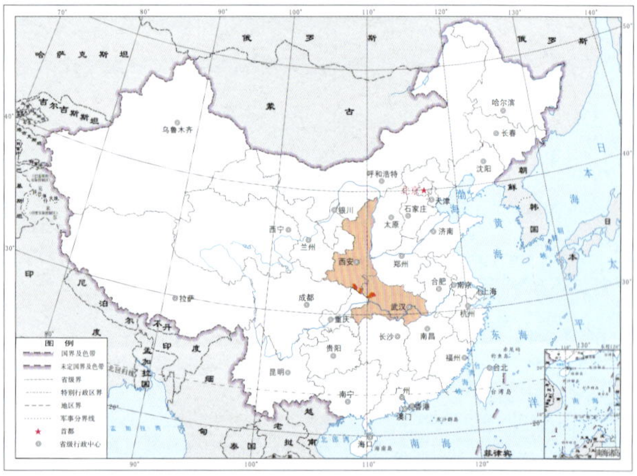

图 3-108-3 二支蝎蛉 ***Panorpa biclada*** 地理分布

Fig. 3-108-3 Distribution of ***Panorpa biclada***

盘状亚基齿。阳基侧突细长，近基部分叉，腹支短小，背支伸达生殖刺突的亚基齿。阳茎短小，侧突起发达，背瓣近末端分叉。

【发生期】 7 ~ 8 月。

【地理分布】 陕西（岚皋、化龙山）、湖北（神农架）。

Measurements: Forewing length 11.5–12.1 mm.

Morphology: Vertex dark brown. Ocellar triangle black. Rostrum yellowish brown, with two broad deep brown longitudinal stripes laterally. Pronotum black. Mesonotum and metanotum mostly black, with a indistinct light yellow median longitudinal stripe. Pleura and legs light yellow. Wing membrane hyaline. In males, markings usually absent; while a distinct pterostigmal band without an apical branch present in females. Male terga Ⅰ–Ⅴ black. Notal organ on tergum Ⅲ flat, covering acute postnotal organ on tergum Ⅳ. A6 black, long cylindrical and tapering towards apex, and devoid of anal horns. A7 and A8 yellowish brown and constricted basally. Male genital bulb long elliptical. Epandrium long and broad, with a deep U-shaped terminal emargination, and forming a pair of finger-like processes laterally. Hypandrium with a greatly shortened basal stalk. Hypovalve slender, exceeding middle of gonocoxites, and bearing sparse stout setae on apical portion marginally. Gonocoxites with 4–5 well-aligned stout setae on ventral apex. Gonostylus shorter than half length of gonocoxites, and bearing an obtuse triangular median tooth and a disc-like subbasal process. Paramere slender and bifurcated subbasally, with ventral branch short and dorsal branch extending to subbasal process of gonostylus. Aedeagus short with well-developed lateral processes; ventral valves bifurcated subapically.

Emergence: July to August.

Distribution: Shaanxi (Lan'gao, Hualong Mountains) and Hubei (Shennongjia).

109. 双带蝎蛉 *Panorpa bifasciata* Chou & Wang, 1981

【测量】 前翅长 11.0 ~ 13.0 mm。

【形态特征】 头顶黄褐色。单眼三角区黑色。1 条黑色横带穿过单眼三角区并伸达复眼内缘。喙黄褐色。胸部背板黑褐色，侧板及足黄褐色。翅无色透明，斑纹黑褐色。前翅仅具宽阔的端带和痣带，其中痣带无端支；基带仅呈 1 个小点或无，无其他斑纹。雄性腹部 1 ~ 5 节背板黑褐色。第 3 节背中突平伏，盖住第 4 节上尖锐的后背突。第 6 节黑褐色，筒形，向后渐窄，无臀角。第 7、8 节黄褐色，基部缢缩。雄性外生殖器卵圆形。第 9 节背板末端具深 U 形缺刻；腹板具极短的基柄；下瓣细长，超过生殖肢基节的中部，内缘具细长刚毛。生殖刺突内缘具钝三角形中齿和钩状亚基齿。阳基侧突细长而尖锐，基部着生细长的粗刺。阳茎粗壮，末端超过生殖肢基节端部，背瓣末端具 1 对耳状突起。

【发生期】 6 ~ 8 月。

【地理分布】 陕西（宁陕、长安、留坝、太白、周至）、河南（卢氏、嵩县）、甘肃（文县）、湖北（神农架、竹溪）。

Measurements: Forewing length 11.0–13.0 mm.

Morphology: Vertex yellowish brown. Ocellar triangle black. A black transverse stripe passing through ocellar triangle and extending to inner border of compound eyes. Rostrum yellowish brown. Nota dark brown. Pleura and legs yellowish brown. Wing membrane hyaline with dark brown markings. Forewing with broad apical and pterostigmal bands; pterostigmal band lacking apical branch; basal band reduced into a small spot or absent; no other markings. Male terga Ⅰ–Ⅴ dark brown. Notal organ on tergum Ⅲ flat and covering acute postnotal organ on tergum Ⅳ. A6 dark brown, cylindrical, tapering towards apex and devoid of anal horns. A7 and A8 yellowish brown and constricted basally. Male genital bulb oval. Epandrium emarginated terminally in a deep U-shape. Hypandrium with an extremely shortened basal stalk. Hypovalves slender, exceeding middle

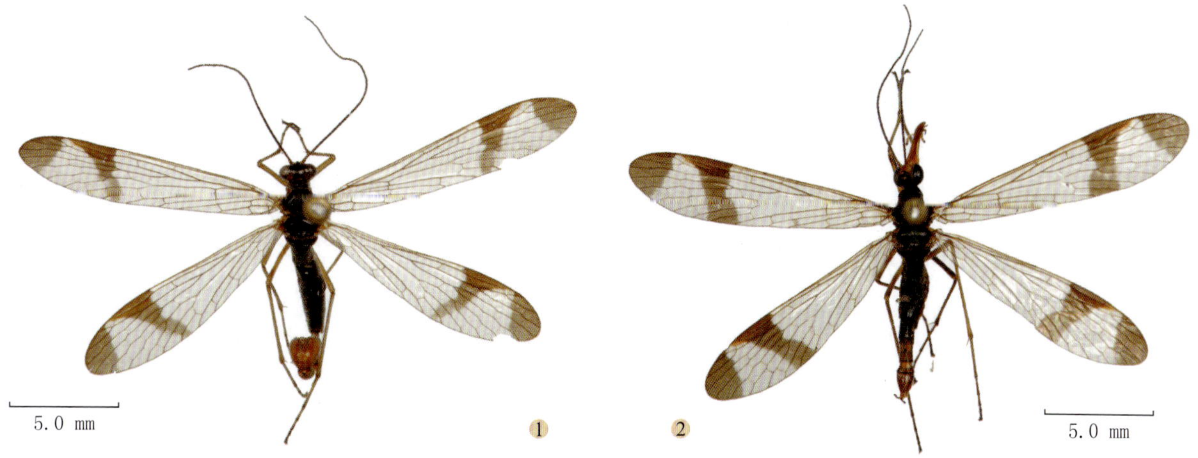

图 3-109-1　双带蝎蛉 *Panorpa bifasciata* 成虫背面观（陕西宁陕火地塘，NWAFU）
1. 雄虫　2. 雌虫

Fig. 3-109-1　Adults of *Panorpa bifasciata*, dorsal view (Huoditang, Ningshan, Shaanxi, NWAFU)
1. Male　2. Female

图 3-109-2 双带蝎蛉 ***Panorpa bifasciata*** 雄性外生殖器腹面观

Fig. 3-109-2 Male genitalia of ***Panorpa bifasciata***, ventral view

图 3-109-3 双带蝎蛉 ***Panorpa bifasciata*** 地理分布

Fig. 3-109-3 Distribution of ***Panorpa bifasciata***

of gonocoxites, and bearing slender setae on inner margins. Gonostylus with an obtuse triangular median tooth and a large hook-shaped subbasal process on inner margin. Paramere slender and acute, with several long and stout spines basally. Aedeagus stout, exceeding apex of gonocoxites; dorsal aedeagal valves forming a pair of auriculate processes apically.

Emergence: June to August.

Distribution: Shaanxi (Ningshan, Chang′an, Liuba, Taibai, Zhouzhi), Henan (Lushi, Songxian), Gansu (Wenxian) and Hubei (Shennongjia, Zhuxi).

110. 双纹蝎蛉 *Panorpa bistriata* Issiki, 1929

图 3-110-1 双纹蝎蛉 ***Panorpa bistriata*** 雌成虫背面观（台湾，林义祥摄）

Fig. 3-110-1 Female adult of ***Panorpa bistriata***, dorsal view (Taiwan, photo by Yechan Lin)

图 3-110-2 双纹蝎蛉 *Panorpa bistriata* 雄成虫背面观（台湾，林义祥摄）

Fig. 3-110-2 Male adult of ***Panorpa bistriata***, dorsal view (Taiwan, photo by Yechan Lin)

【测量】 前翅长 12.0 ~ 13.0 mm。

【形态特征】 头顶黑色，单眼三角区黑色。喙黑褐色，两侧黄褐色。胸部背板黑褐色。侧板及足黄褐色。翅无色透明，斑纹黑褐色。端带极退化；痣带明显，无端支或仅有细弱的端支；基带阔而明显；无其他斑纹。雄性腹部 1 ~ 5 节背板黑褐色。第 3 节背中突略向后突起，盖住第 4 节上尖锐的后背突。第 6 节黄褐色，向后渐窄，无臀角。第 7、8 节黄褐色，第 7 节基部强烈缢缩呈柄状。雄性外生殖器卵形。第 9 节背板末端具深 U 形缺刻；腹板几乎无基柄；下瓣细长，超出生殖肢基节端部，内缘具长鬃；生殖刺突内缘具钝三角形中齿和尖锐钩状的亚基齿；阳基侧突分为 2 叉，Y 形。阳茎短小、扁平。

【发生期】 6 ~ 7 月。

【地理分布】 台湾。

Measurements: Forewing length 12.0–13.0 mm.

Morphology: Vertex and ocellar triangle black. Rostrum dark brown with yellowish brown

图 3-110-3 双纹蝎蛉 *Panorpa bistriata* 地理分布

Fig. 3-110-3 Distribution of ***Panorpa bistriata***

flanks. Nota dark brown. Pleura and legs yellowish brown. Wing membrane hyaline, with dark brown markings. In forewings, apical band greatly reduced; pterostigmal band lacking apical branch or with a faint apical branch; basal band broad and distinct; no other markings. Male terga Ⅰ–Ⅴ dark brown. Notal organ on tergum Ⅲ slightly projecting caudad and covering acute postnotal organ on tergum Ⅳ. A6 yellowish brown, tapering towards apex and devoid of anal horns. A7 and A8 yellowish brown. A7 intensively constricted and stalk-like basally. Male genital bulb oval. Epandrium emarginate terminally in a deep U-shape. Hypandrium with an extremely shortened basal stalk. Hypovalves slender, exceeding middle of gonocoxites, and bearing long bristles on inner margins. Gonostylus with an obtuse triangular median tooth and an acute hook-shaped subbasal process on inner margin. Paramere bifurcated and Y-shaped. Aedeagus short and flat.

Emergence: June to July.

Distribution: Taiwan.

111. 短突蝎蛉 *Panorpa brevititilana* Issiki, 1929

【测量】 前翅长 11.5 ~ 12.5 mm。

【形态特征】 头顶黑褐色，单眼三角区黑色。喙黄褐色。前胸背板黑褐色。中、后胸背板黑褐色，中央具 1 条黄色中带。侧板及足浅黄褐色。翅无色透明，翅斑褐色。前翅端带阔，内缘不规则，内部具 1 ~ 2 个透明窗斑。痣带基支阔，端支细小且断开；缘斑小；基带断开成 2 小斑；基斑小。雄性腹部 1 ~ 5 节背板黑褐色。第 3 节背中突平扁，向后盖住第 4 节上尖锐的后背突。第 6 节黑褐色，筒形，背面从中部向后突然渐窄，无臀角。第 7、8 节黄褐色。第 7 节基部强烈缢缩呈柄状。雄性外生殖器长卵形。第 9 节背板末端具深 U 形缺刻，两侧形成 1 对细长的指状突起；腹板几乎无基柄；下瓣短，仅伸达生殖肢基节的中部，内缘着生 1 列长刚毛。生殖刺突细长，内缘具钝三角形中齿和大型碟状亚基齿。阳基侧突极短，末端尖。阳茎极短，扁平。

【发生期】 8 ~ 10 月。

【地理分布】 台湾（台东、台南）。

Measurements: Forewing length 11.5–12.5 mm.

Morphology: Vertex dark brown. Ocellar triangle black. Rostrum yellowish brown. Pronotum dark brown. Mesonotum and metanotum dark brown, with a distinct yellow mesal stripe. Pleura and legs light yellowish brown. Wing membrane hyaline, with brown markings.

图 3-111-1 短突蝎蛉 *Panorpa brevititilana* 雄成虫背面观（台湾台东海端，NWAFU）

Fig. 3-111-1 Male adult of *Panorpa brevititilana*, lateral view (Haituan, Taitung, Taiwan, NWAFU)

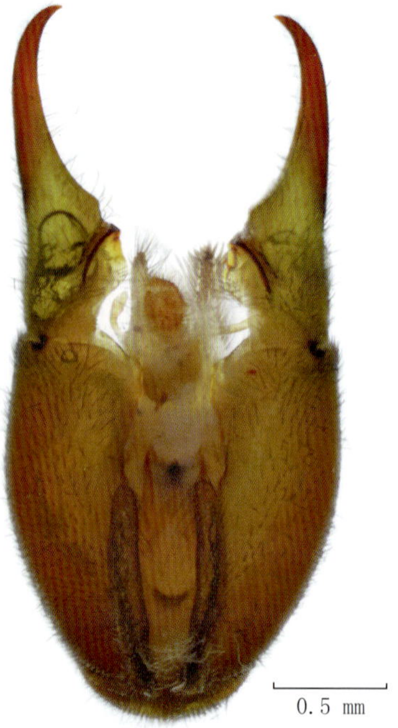

图 3-111-2 短突蝎蛉 *Panorpa brevititilana* 雄性外生殖器腹面观

Fig. 3-111-2 Male genitalia of *Panorpa brevititilana*, ventral view

In forewings, apical band broad with irregular inner margin, and with 1–2 hyaline window spots innerly; pterostigmal band broad, with a thin separated apical branch; marginal spot small; basal band split into two small spots; basal spot small. Male terga Ⅰ–Ⅴ dark brown. Notal organ on tergum Ⅲ flat and covering acute postnotal organ on tergum Ⅳ. A6 dark brown, cylindrical, tapering towards apex and devoid of anal horns. A7 and A8 yellowish brown. A7 intensively stalked basally. Male genital bulb long oval. Epandrium emarginated terminally in a deep U-shape and forming a pair of finger-like processes laterally. Hypandrium with an extremely shortened basal stalk. Hypovalves short, only extending to middle of gonocoxites, and bearing a row of long bristles on inner margins. Gonostylus with an obtuse triangular median tooth and a large disc-like subbasal process on inner margin. Paramere extremely shortened with an acute apex. Aedeagus short and flat.

图 3-111-3 短突蝎蛉 ***Panorpa brevititilana*** 地理分布

Fig. 3-111-3 Distribution of ***Panorpa brevititilana***

Emergence: August to October.

Distribution: Taiwan (Taitung, Tainan).

112. 长白山蝎蛉 *Panorpa changbaishana* Hua, 2008

图 3-112-1 长白山蝎蛉 ***Panorpa changbaishana*** 雄成虫侧面观（吉林长白山）

Fig. 3-112-1 Male adult of ***Panorpa changbaishana***, lateral view (Changbai Mountains, Jilin)

图 3-112-2 长白山蝎蛉 ***Panorpa changbaishana*** 雌成虫侧面观（吉林长白山）
Fig. 3-112-2 Female adult of ***Panorpa changbaishana***, lateral view (Changbai Mountains, Jilin)

【测量】 前翅长 12.5 ~ 13.0 mm。

【形态特征】 头顶、单眼三角区黑色。喙黑色。胸部背板黑色。侧板及足基节黑褐色。足大部分深黄褐色。翅无色透明，斑纹黑色。前翅端带阔，后缘具 1 大型窗状透明斑；痣带阔，基支阔，端支窄；基带仅存 2 个散点；缘斑小，无基斑。雄性腹部 1 ~ 5 节背板黑褐色。第 3 节背中突扁平，向后盖住第 4 节上尖锐的后背突。第 6 节黑色筒形，无臀角。第 7、8 节黑色，短小且基部缢缩，呈倒锥形。雄性外生殖器长卵形。第 9 节背板末端具深 U 形缺刻；腹板具阔长的基柄；下瓣窄长，中部稍向外扩，端半部稍内折、渐细。生殖肢基节腹面端部具 1 个端部二分叉的片状突起；生殖刺突短于基节之半，内缘基部具盘状亚基齿，中齿钝三角形。阳基侧突基部叶片状，端部渐细，末端分为尖锐的 2 叉，内缘具 1 列短刺毛。阳茎腹瓣膜质短小；背瓣指状，背缘末端具膜质膨大物。

【发生期】 6 ~ 8 月。

【地理分布】 辽宁（营口、宽甸）、吉林（长白山）、黑龙江（镜泊湖）。

Measurements: Forewing length 12.5–13.0 mm.

Morphology: Vertex, ocellar triangle and rostrum black. Nota black. Pleura and coxae dark brown. Legs mostly deep yellowish brown. Wing membrane hyaline, with black markings. In forewings, apical band broad with a large hyaline window posteriorly; pterostigmal band broad, with a broad basal branch and a narrow apical branch; basal band split into two small spots; marginal spot small; basal spot absent. Male terga Ⅰ–Ⅴ dark brown. Notal organ on tergum Ⅲ flat and covering acute postnotal organ on tergum Ⅳ. A6 dark brown, cylindrical, and devoid of anal horns. A7 and A8 black, short and constricted basally, inverted cone-shaped. Male genital bulb long oval. Epandrium emarginate terminally in a deep U-shape. Hypandrium with a long broad basal stalk. Hypovalves slender, slightly expanding laterad at middle, convergent and tapering towards apexes at distal half. Gonocoxites with a bifurcated plate on ventral apex. Gonostylus shorter than half length of gonocoxites, and bearing an obtuse triangular median tooth and a disc-like subbasal process on inner margin. Paramere foliate basally and tapering towards bifurcated acute apex, and bearing a

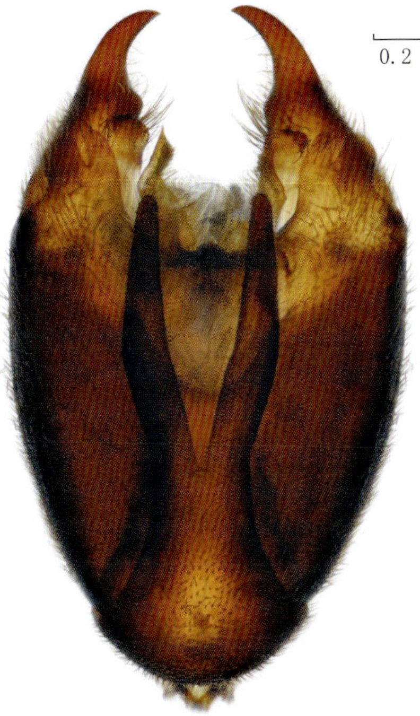

图 3-112-3 长白山蝎蛉 *Panorpa changbaishana* 雄性外生殖器腹面观

Fig. 3-112-3 Male genitalia of ***Panorpa changbaishana***, ventral view

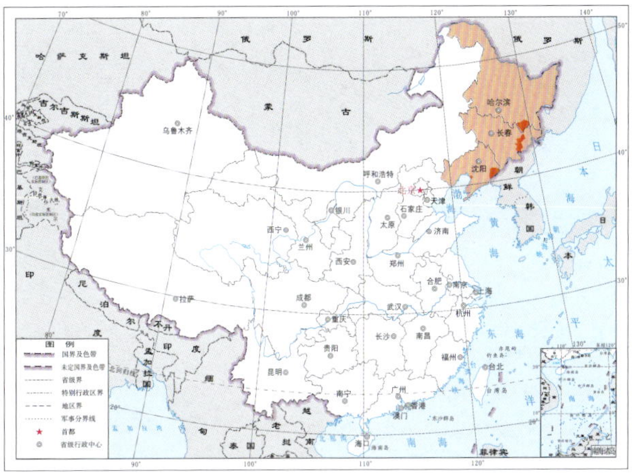

图 3-112-4 长白山蝎蛉 ***Panorpa changbaishana*** 地理分布

Fig. 3-112-4 Distribution of ***Panorpa changbaishana***

row of short spines along inner margin. Aedeagus with short membranous ventral valves; dorsal valves finger-like and bearing a membranous enlarged process on dorsal surface apically.

Emergence: June to August.

Distribution: Liaoning (Yingkou, Kuandian), Jilin (Changbai Mountains), and Heilongjiang (Jingpo Lake).

113. 郑氏蝎蛉 *Panorpa chengi* Chou, 1981

【测量】 前翅长 11.5 ~ 12.0 mm。

【形态特征】 头顶黄色，单眼三角区黑色。1 条褐色横带贯穿单眼三角区。喙红棕色。前胸背板浅黄褐色。中、后胸背板浅乳黄色，两侧加深。侧板乳黄色，足黄褐色。翅透明，略带黄色，无任何斑纹或仅端部稍加深。雄性腹部 1 ~ 5 节背板乳黄色，两侧渐深。第 3 节的背中突扁平，盖住第 4 节上尖锐的后背突。第 6 节黑褐色，筒形，向端部渐窄，无臀角。第 7、8 节黄褐色，基部缢缩。雄性外生殖器卵形。第 9 节背板末端具深 U 形缺刻，两侧形成 1 对细长指形突起；腹板基柄极短；下瓣长带状，内缘具 1 列长刚毛。生殖刺突内缘有 1 宽列稀疏的长刚毛，中齿钝三角形，亚基齿碗形。阳基侧突细而弯曲，在中部交叉。阳茎短小，背瓣叶状且分离。

【发生期】 7 ~ 8 月。

【地理分布】 陕西（太白、宁陕、鄠邑）、甘肃（宕昌）。

Measurements: Forewing length 11.5–12.0 mm.

Morphology: Vertex yellow. Ocellar triangle black. A brown transversal stripe passing through ocellar triangle. Rostrum reddish brown. Pronotum light yellowish brown. Mesonotum and metanotum light creamy yellow and darkened on two sides. Pleura creamy yellow. Legs yellowish brown. Wing membrane hyaline, slightly tinged with yellow, devoid of markings or only slightly darkened on apex. Male terga I–V creamy yellow and darkened laterally. Notal organ on tergum Ⅲ flat and covering acute postnotal organ on tergum Ⅳ. A6 dark brown, cylindrical, tapering towards apex and devoid of anal horns. A7 and A8 yellowish brown and constricted basally. Male genital bulb oval. Epandrium emarginated terminally in a deep U-shape and forming a pair of finger-like processes laterally. Hypandrium with a shortened basal stalk. Hypovalves long stripe-like, bearing a row of long bristles on inner margin. Gonostylus with a broad

图 3-113-1 郑氏蝎蛉 *Panorpa chengi* 雄成虫背面观（陕西宁陕火地塘，钟问摄）

Fig. 3-113-1 Male adult of ***Panorpa cheng,*** dorsal view (Huoditang, Ningshan, Shaanxi, photo by Wen Zhong)

row of sparse stout setae, and an obtuse triangular median tooth and a bowl-shaped subbasal process on inner margin. Parameres slender, curved and crossed mesally. Aedeagus short; dorsal aedeagal valves foliate and divergent.

Emergence: July to August.

Distribution: Shaanxi (Taibai, Ningshan, Huyi) and Gansu (Dangchang).

图 3-113-2 郑氏蝎蛉 *Panorpa chengi* 雄性外生殖器腹面观

Fig. 3-113-2 Male genitalia of ***Panorpa chengi***, ventral view

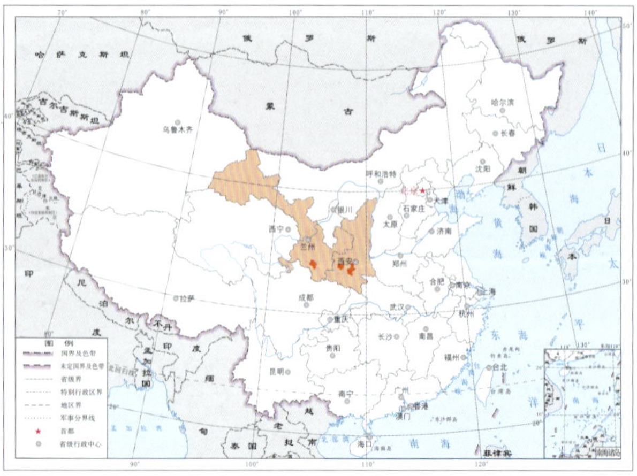

图 3-113-3 郑氏蝎蛉 *Panorpa chengi* 地理分布

Fig. 3-113-3 Distribution of ***Panorpa chengi***

114. 陈氏蝎蛉 *Panorpa cheni* Cheng, 1957

【测量】 前翅长 11.5 ~ 12.0 mm。

【形态特征】 头顶黑褐色，单眼三角区黑色。喙黄褐色。胸部背板黑褐色。侧板及足黄褐色。翅透明无色，斑纹黑褐色。前翅端带宽阔；痣带宽阔、无端支，基带狭长；缘斑小，无基斑。雄性腹部 1 ~ 5 节背板黑色。第 3 节背中突扁平，盖住第 4 节上尖锐的后背突。第 6 节黑色，筒形，无臀角。第 7、8 节黄褐色，基部缢缩。雄性外生殖器卵形。第 9 节背板末端具深 U 形缺刻，两侧形成 1 对指形突起；腹板基柄极短；下瓣细长，伸达生殖肢基节中部，内缘具 1 列长鬃。生殖刺突内缘的中齿尖锐，亚基齿宽大，具短尖。阳基侧突二分叉，其中背支短，着生一丛粗长刺毛；腹支端部呈鼓槌状膨大，具短尖。阳茎短粗，不超过生殖肢基节端部；背瓣两侧延伸出 1 对细长的指形突起。

【发生期】 8 ~ 10 月。

【地理分布】 浙江（天目山）。

Measurements: Forewing length 11.5–12.0 mm.

Morphology: Vertex dark brown. Ocellar triangle black. Rostrum yellowish brown. Nota dark brown. Pleura and legs yellowish brown. Wing membrane hyaline, with dark brown markings. In forewings, apical band broad; pterostigmal band broad, lacking apical branch; basal band narrow; marginal spot small; basal spot absent. Male terga Ⅰ–Ⅴ black. Notal organ

图 3-114-1 陈氏蝎蛉 ***Panorpa cheni*** 雄成虫背面观（浙江临安西天目山）

Fig. 3-114-1 Male adult of ***Panorpa cheni***, dorsal view (West Tianmu Mountain, Lin'an, Zhejiang)

图 3-114-2 陈氏蝎蛉 *Panorpa cheni* 雄性外生殖器腹面观

Fig. 3-114-2 Male genitalia of *Panorpa cheni*, ventral view

图 3-114-3 陈氏蝎蛉 *Panorpa cheni* 地理分布

Fig. 3-114-3 Distribution of *Panorpa cheni*

on tergum Ⅲ flat and covering acute postnotal organ on tergum Ⅳ. A6 black and cylindrical, devoid of anal horns. A7 and A8 yellowish brown and constricted basally. Male genital bulb oval. Epandrium emarginate terminally in a deep U-shape and forming a pair of finger-like processes laterally. Hypandrium with a greatly shortened basal stalk. Hypovalves slender, extending to middle of gonocoxites, bearing a row of long bristles on inner margin. Gonostylus with an acute median tooth and a broad subbasal process with a short acute tip. Paramere bifurcated: dorsal branch short and bearing a tuft of long stout spines; ventral branch strongly enlarged near apex with a short and acute tip, resembling a drumstick. Aedeagus short and stout, not exceeding apex of gonocoxites; dorsal valves with a pair of slender finger-like processes laterally.

Emergence: August to October.

Distribution: Zhejiang (Tianmu Mountains).

115. 周氏蝎蛉 *Panorpa choui* Zhou & Wu, 1993

【测量】 前翅长 14.0 mm。

【形态特征】 头顶黑色。单眼三角区黑色。喙黄褐色。胸部背板黑褐色。侧板及足黄褐色。翅无色透明，斑纹褐色。前翅端带阔；痣带阔，无端支，中部稍收狭；基带阔；无缘斑和基斑。雄性腹部 1 ~ 5 节背板黑褐色。第 3 节背板后缘背中突扁平，盖住第 4 节上的后背突。第 6 节黑褐色，圆筒形，无臀角。第 7、8 节黄褐色，基部稍缢缩。雄性外生殖器卵形。第 9 节背板末端具深 U 形缺刻，两侧形成 1 对指形突起；腹板基柄极短；下瓣窄长。生殖刺突内缘具钝三角形中齿，和发达的亚基齿；亚基齿近端部圆阔，具短尖。阳基侧突剑状，内缘具 1 列长刺，伸达生殖刺突的中齿。阳茎腹瓣叶状，端部圆阔；背瓣两侧具指状突起。

【发生期】 7 ~ 8 月。

【地理分布】 浙江（凤阳山、莫干山）。

Measurements: Forewing length 14.0 mm.

Morphology: Vertex black. Ocellar triangle black. Rostrum yellowish brown. Nota dark brown. Pleura and legs yellowish brown. Wing membrane hyaline, with brown markings. In forewings, apical band broad; pterostigmal band broad, lacking apical branch and constricted medially; basal band broad; marginal spot and spots absent. Male terga Ⅰ–Ⅴ dark brown. Notal organ on tergum Ⅲ flat and covering

图 3-115-1 周氏蝎蛉 ***Panorpa choui*** 雌成虫背面观（腹末已去除，浙江凤阳山，NWAFU）

Fig. 3-115-1 Female adult of ***Panorpa choui*** with terminalia removed, dorsal view (Mount Fengyang, Zhejiang, NWAFU)

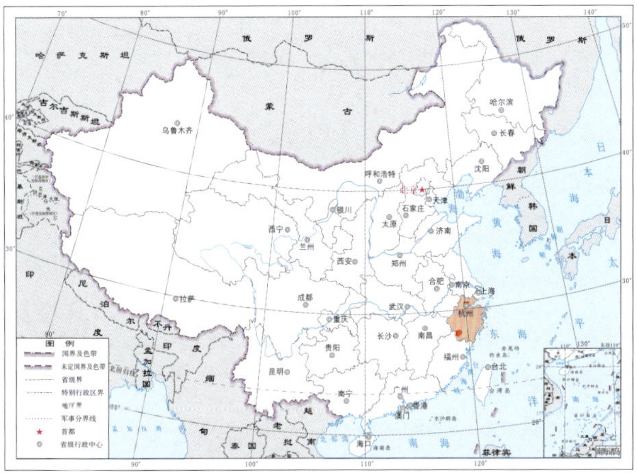

图 3-115-3 周氏蝎蛉 ***Panorpa choui*** 地理分布

Fig. 3-115-3 Distribution of ***Panorpa choui***

图 3-115-2 周氏蝎蛉 ***Panorpa choui*** 雄性外生殖器腹面观（脱水变形）

Fig. 3-115-2 Male genitalia of ***Panorpa choui***, ventral view (deformed due to dehydration)

acute postnotal organ on tergum Ⅳ. A6 dark brown and cylindrical, devoid of anal horns. A7 and A8 yellowish brown and constricted basally. Male genital bulb oval. Epandrium emarginate terminally in a deep U-shape and forming a pair of finger-like processes laterally. Hypandrium with a greatly shortened basal stalk. Hypovalves slender. Gonostylus with an obtuse triangular median tooth and a broad subbasal process on inner margin. The subbasal process rounded subapically, bearing a short acute tip. Paramere sword-like, with a row of long spines on inner margin, and extending to median tooth of gonostylus. Aedeagus with ventral valves foliate and rounded apically; dorsal valves with a finger-like process laterally.

Emergence: July to August.

Distribution: Zhejiang (Mount Fengyang, Mount Mogan).

116. 支蝎蛉 *Panorpa cladocerca* Navás, 1935

【测量】 前翅长 12.0 ~ 14.5 mm。

【形态特征】 头顶红褐色。单眼三角区黑色。喙红褐色。胸部背板黑褐色。侧板及足基节乳黄色。足大部分黄褐色。翅透明，略带黄色，斑纹黑褐色。前翅端带阔；痣带阔且缺端支，中间缢缩；有时在痣带与端带之间的后缘处具 1 个三角形小斑；基带完整；缘斑小，无基斑。雄性腹部 1 ~ 5 节黑褐色。第 3 节背中突扁平，盖住第 4 节上尖锐的后背突。第 6 节基部黑褐色，端部褐色，无臀角；第 7、8 节黄褐色。第 7 节基部强烈缢缩成柄状。雄性外生殖器卵形。第 9 节背板末端具深 U 形缺刻，两侧形成 1 对细长的指形突起；腹板基柄极短；下瓣细长，端半部稍向内弯。生殖刺突基半部内侧具 1 个极大的凹陷区，近基部内侧具 1 个指状亚基齿。阳

图 3-116-1 支蝎蛉 ***Panorpa cladocerca*** 雄成虫背面观（江西九江庐山）

Fig. 3-116-1 Male adult of ***Panorpa cladocerca***, dorsal view (Lushan Mountain, Jiujiang, Jiangxi)

图 3-116-2 支蝎蛉 ***Panorpa cladocerca*** 雄性外生殖器腹面观

Fig. 3-116-2 Male genitalia of ***Panorpa cladocerca***, ventral view

基侧突粗长，端半部向内弯曲，稍伸入生殖刺突的凹陷区内，内缘密生短毛。阳茎腹瓣短小；背瓣细长指状，分离。

【发生期】 8 ~ 10 月。

图 3-116-3 支蝎蛉 ***Panorpa cladocerca*** 地理分布

Fig. 3-116-3 Distribution of ***Panorpa cladocerca***

【地理分布】 江西（庐山）。

Measurements: Forewing length 12.0–14.5 mm.

Morphology: Vertex reddish brown. Ocellar triangle black. Rostrum reddish brown. Nota dark brown. Pleura and coxae creamy yellow. Legs mostly yellowish brown. Wing membrane hyaline, slightly tinged with yellow. Markings dark brown. In forewings, apical band broad; pterostigmal band broad, lacking apical branch and constricted medially; occasionally a small triangular spot on hind margin between apical and pterostigmal bands; basal band complete; marginal spot small; basal spot absent. Male terga Ⅰ–Ⅴ dark brown. Notal organ on tergum Ⅲ flat and covering acute postnotal organ on tergum Ⅳ. A6 dark brown basally and brown distally, devoid of anal horns. A7 and A8 yellowish brown. Base of A7 intensively constricted and stalk-like. Male genital bulb oval. Epandrium emarginate terminally in a deep U-shape and forming a pair of slender finger-like processes laterally. Hypandrium with a greatly shortened basal stalk. Hypovalves slender and slightly convergent on distal halves. Gonostylus with a large concaved area on basal half of ventral surface, and bearing a finger-like subbasal process on inner margin. Parameres long stout, curved convergently on distal halves, slightly extending into concaved areas of gonostyli, and bearing dense short spines along inner margins. Aedeagus with short ventral valves; dorsal valves slender, finger-like and separated.

Emergence: August to October.

Distribution: Jiangxi (Lushan Mountain).

117. 弯曲蝎蛉 *Panorpa curva* Carpenter, 1938

【测量】 前翅长 12.5 ~ 14.0 mm。

【形态特征】 头顶红褐色，单眼三角区黑色。喙黄褐色至红褐色。前胸背板黑褐色。中、后胸背板黑褐色，具 1 条显著的黄色中带。侧板及足浅黄褐色。翅无色透明，翅痣红褐色，斑纹褐色。前翅端带后缘具 1 透明窗斑；痣带仅存基支；基带完整；形成三条显著的横带；无其他斑纹。雄性腹部 1 ~ 5 节黑褐色。第 3 节背中突扁平，盖住第 4 节上尖锐的后背突。第 6 节基部黑褐色，端部黄褐色，末端向上隆起且被稀疏的长毛；第 7、8 节红褐色，基部缢缩。雄性外生殖器卵形。第 9 节背板末端具深 U 形缺刻，两侧形成 1 对指状突起；腹板基柄极短；下瓣长带状，伸过生殖肢基节中部，内缘具 1 列长鬃。生殖肢基节端部内侧具 1 个延长的突起，其内缘着生密集的刚毛；生殖刺突内缘具钝三角形中齿和大型盘状亚基齿。阳基侧突螺旋状弯曲，左右交叉，端部内缘具 1 列短毛。阳茎腹瓣膜质短小；背瓣骨化指状，端部背缘膨大。

【发生期】 6 ~ 8 月。

【地理分布】 陕西（宝鸡）、甘肃（文县）、四川（卧龙）。

图 3-117-1　弯曲蝎蛉 *Panorpa curva* 成虫背面观（四川卧龙，NWAFU）

1. 雄虫　2. 雌虫

Fig. 3-117-1　Adults of ***Panorpa curva***, dorsal view (Wolong, Sichuan, NWAFU)

1. Male　2. Female

图 3-117-2 弯曲蝎蛉 ***Panorpa curva*** 雄性外生殖器腹面观

Fig. 3-117-2 Male genitalia of ***Panorpa curva***, ventral view

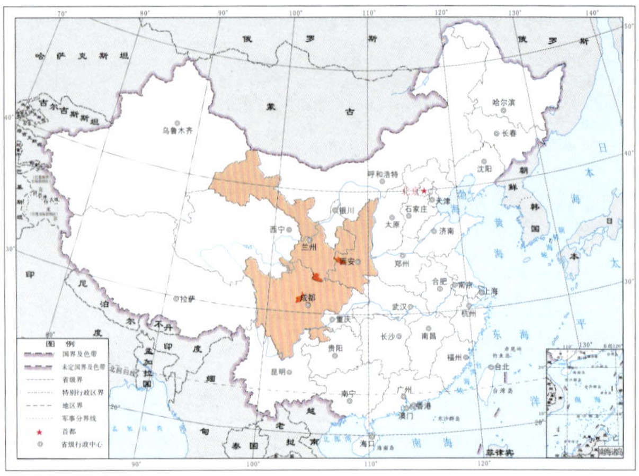

图 3-117-3 弯曲蝎蛉 ***Panorpa curva*** 地理分布

Fig. 3-117-3 Distribution of ***Panorpa curva***

Measurements: Forewing length 12.5–14.0 mm.

Morphology: Vertex reddish brown. Ocellar triangle black. Rostrum yellowish brown to reddish brown. Pronotum dark brown. Mesonotum and metanotum dark brown, with a distinct yellow mesal stripe. Pleura and legs yellowish brown. Wing membrane hyaline, with markings dark brown. Pterostigma reddish brown. In forewings, apical band with a hyaline window posteriorly; pterostigmal band lacking apical branch; basal band complete; marginal and basal spots absent. Male terga Ⅰ–Ⅴ dark brown. Notal organ on tergum Ⅲ flat and covering acute postnotal organ on tergum Ⅳ. A6 dark brown basally and yellowish brown distally, protruding dorsad and bearing long setae on apex. A7 and A8 reddish brown, and constricted basally. Male genital bulb oval. Epandrium emarginate terminally in a deep U-shape and forming a pair finger-like processes laterally. Hypandrium with a greatly shortened basal stalk. Hypovalves stripe-like, extending beyond mid-length of gonocoxite, bearing a row of long bristles on inner margin. Gonocoxites with a elongated process bearing long dense setae on inner margin of ventral apex. Gonostylus with an obtuse triangular median tooth and a large disc-like subbasal process on inner margin. Parameres spiral, crossed, and bearing a row of short spines on inner margin. Aedeagus with short membranous ventral valves; dorsal valves highly sclerotized, finger-like and enlarged on dorsal apex.

Emergence: June to August.

Distribution: Shaanxi (Baoji), Gansu (Wenxian) and Sichuan (Wolong).

118. 大卫蝎蛉 *Panorpa davidi* Navás, 1908

【测量】 前翅长 13.0 mm。

【形态特征】 头顶红褐色，单眼三角区黑色。喙黄褐色至红褐色。前胸背板黑褐色。中、后胸背板黑褐色，具 1 条显著的黄色中带。侧板及足浅黄褐色。翅无色透明，翅痣浅褐色，斑纹褐色。前翅端带后缘具 1 个透明窗斑；痣带仅存基支；基带分为 2 个斑点；无缘斑和基斑。雄性腹部 1 ~ 5 节黑褐色。第 3 节背中突扁平，盖住第 4 节上尖锐的后背突。第 6 节基部黑褐色，端部黄褐色，末端向上隆起且被稀疏的长毛；第 7、8 节红褐色，基部缢缩。雄性外生殖器卵形。第 9 节背板末端具深 U 形缺刻，两侧形成 1 对指状突起；腹板基柄极短；下瓣长带状，伸过生殖肢基节中部，内缘具 1 列长鬃。生殖肢基节端部内侧具 1 个延长的突起，内缘着生 1 列

图 3-118-1 大卫蝎蛉 *Panorpa davidi* 雄成虫背面观（四川，NWAFU）

Fig. 3-118-1 Male adult of *Panorpa davidi*, dorsal view (Sichuan, NWAFU)

图 3-118-2 大卫蝎蛉 *Panorpa davidi* 雄性外生殖器腹面观

Fig. 3-118-2 Male genitalia of *Panorpa davidi*, ventral view

密集的粗刚毛；生殖刺突内缘具钝三角形中齿和大型盘状亚基齿。阳基侧突细长，左右交叉，端部较直，内缘具 1 列短毛。阳茎腹瓣膜质短小；背瓣骨化指状，端部背缘膨大。

【发生期】 未知。

【地理分布】 四川（宝兴）。

Measurements: Forewing length 13.0 mm.

Morphology: Vertex reddish brown. Ocellar triangle black. Rostrum yellowish brown to reddish brown. Pronotum dark brown. Mesonotum and metanotum dark brown, with a distinct yellow mesal stripe. Pleura and legs yellowish brown. Wing membrane hyaline, with markings brown. Pterostigma pale brown. In forewings, apical band with a hyaline window posteriorly; pterostigmal band lacking apical branch; basal band split into two spots; marginal and basal spots absent. Male terga Ⅰ–Ⅴ dark brown. Notal organ on tergum Ⅲ flat and covering acute postnotal organ on tergum Ⅳ. A6 dark brown basally and yellowish brown distally, protruding dorsad and bearing sparse long setae on apex. A7 and A8 reddish brown and constricted basally. Male genital bulb oval. Epandrium emarginate terminally in a deep U-shape and forming a pair of finger-like processes laterally. Hypandrium with a greatly shortened basal stalk. Hypovalves long stripe-like, extending beyond middle of gonocoxite, and bearing a row of long bristles on inner margin. Gonocoxites with a protruding process bearing a row of stout setae on inner margin of ventral apex. Gonostylus with an obtuse triangular median tooth and a large disc-like subbasal process. Parameres slender, crossed, and bearing a row of short spines on inner margin. Aedeagus with short membranous ventral valves; dorsal valves highly sclerotized, finger-like and enlarged on dorsal apex.

Emergence: Unknown.

Distribution: Sichuan (Baoxing).

图 3-118-3 大卫蝎蛉 *Panorpa davidi* 地理分布

Fig. 3-118-3 Distribution of *Panorpa davidi*

119. 斑蝎蛉 *Panorpa deceptor* Esben-Petersen, 1913

图 3-119-1 斑蝎蛉 ***Panorpa deceptor*** 雄成虫（台湾，林义祥摄）
1. 侧面观 2. 背面观

Fig. 3-119-1 Male adults of ***Panorpa deceptor*** (Taiwan, photo by Yechan Lin)
1. Lateral view 2. Dorsal view

【测量】 前翅长 14.0 ~ 15.0 mm。

【形态特征】 头顶黄褐色。单眼三角区黑色。喙黄褐色。前胸背板黑褐色。中、后胸背板黑褐色，具显著的浅黄色中带。侧板及足乳黄色。翅略带黄色，透明，斑纹褐色，翅痣浅黄色明显。前翅端带完整，后缘具 1 透明窗斑；痣带完整；缘斑窄长；基带分离成 2 斑；基斑小。雄性腹部 1 ~ 5 节黑褐色。第 3 节背中突扁平，盖住第 4 节上尖锐的后背突。第 6 节黄褐色，筒形，端半部向后渐窄。第 7 节基部强烈缢缩成柄状；第 8 节基部轻微缢缩。雄性外生殖器长卵形，侧面观较扁。第 9 节背板末端具深 U 形缺刻，两侧形成 1 对指形突起；腹板基柄极短；下瓣长带状，内缘具 1 列长鬃。生殖刺突外缘稍凹陷，内缘具钝三角形中齿和大型碟状亚基齿。阳基侧突细长线状，超过生殖刺突的中部。

【发生期】 5 月。

【地理分布】 台湾。

图 3-119-2 斑蝎蛉 *Panorpa deceptor* 地理分布

Fig. 3-119-2 Distribution of *Panorpa deceptor*

Measurements: Forewing length 14.0–15.0 mm.

Morphology: Vertex yellowish brown. Ocellar triangle black. Rostrum yellowish brown. Pronotum dark brown. Mesonotum and metanotum dark brown, with a distinct pale yellow mesal stripe. Pleura and legs creamy yellow. Wing membrane hyaline, slightly tinged with yellow. Markings brown. Pterostigma light yellow, distinct. In forewings, apical band complete with a hyaline spot posteriorly; pterostigmal band complete; marginal spot slender; basal band split into two spots; basal spot small. Male terga Ⅰ–Ⅴ dark brown. Notal organ on tergum Ⅲ flat and covering acute postnotal organ on tergum Ⅳ. A6 yellowish brown, cylindrical and tapering on apical half. A7 and A8 yellowish brown. A7 strongly constricted into a stalk at base; A8 only slightly constricted basally. Male genital bulb long oval, slightly flattened in lateral view. Epandrium emarginate terminally in a deep U-shaped and forming a pair of finger-like processes laterally. Hypandrium with a greatly shortened basal stalk. Hypovalves long stripe-like, bearing a row of long bristles on inner margin. Gonostylus slightly concaved along outer margin, bearing an obtuse triangular median tooth and a large disc-like subbasal process on inner margin. Paramere slender and thread-like, exceeding middle of gonostylus.

Emergence: May.

Distribution: Taiwan.

120. 淡色蝎蛉 *Panorpa decolorata* Chou & Wang, 1981

【测量】 前翅长 14.0 ~ 15.0 mm。

【形态特征】 头顶、喙黄褐色。单眼三角区黑褐色。胸部背板浅黄褐色，两侧颜色较深。侧板及足乳黄色。翅无色透明，翅斑黑褐色；休憩时呈屋脊状叠于腹部上方。前翅端带不明显，分为 2 个模糊的条状斑；痣带前宽后窄，端支、基支十分退化；基带仅在后缘呈 1 个圆形大斑；无缘斑和基斑。雄性腹部 1 ~ 5 节背板黑褐色。第 3 节背中突扁平，盖住第 4 节上尖锐的后背突。第 6 节短筒形，无臀角。第 7、8 节短于第 6 节，圆筒形，基部不缢缩。雄性外生殖器椭圆形。第 9 节背板末端具深 U 形缺刻，两侧形成 1 对指形突起；腹板基柄阔长；下瓣短，端部钝圆。阳基侧突粗壮，指形，端部钩状弯曲且具密集的短刺。阳茎腹瓣短，膜质；背瓣叶状、分离。

【发生期】 7 ~ 8 月。

【地理分布】 陕西（宁陕、太白、南郑、周至、宝鸡）、河南（嵩县）、湖北（神农架）。

Measurements: Forewing length 14.0–15.0 mm.

Morphology: Vertex and rostrum yellowish brown. Ocellar triangle dark brown. Nota light

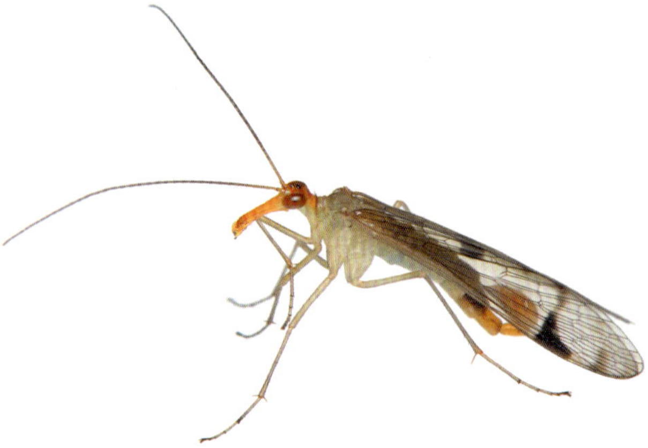

图 3-120-1 淡色蝎蛉 ***Panorpa decolorata*** 雄成虫侧面观（陕西宝鸡天台山国家森林公园）

Fig. 3-120-1 Male adult of ***Panorpa decolorata***, lateral view (Tiantaishan National Forest Park, Baoji, Shaanxi)

图 3-120-2 淡色蝎蛉 ***Panorpa decolorata*** 雄性外生殖器腹面观

Fig. 3-120-2 Male genitalia of ***Panorpa decolorata***, ventral view

yellowish brown with darkened flanks. Pleura and legs creamy yellow. Wing membrane hyaline, with markings brown. Wings held roof-like over abdomen at repose. In forewings, apical band greatly reduced into two faint stripes; pterostigmal band broad anteriorly and narrow posteriorly, with greatly reduced basal and apical branches; basal band reduced into a large circular spot near hind margin; marginal and basal spots absent. Male terga Ⅰ–Ⅴ dark brown. Notal organ on tergum Ⅲ flat and covering acute postnotal organ on tergum Ⅳ. A6 short and cylindrical, devoid of anal horns. A7 and A8 shorter than A6 and cylindrical, not constricted at base. Male genital bulb elliptical. Epandrium emarginate terminally in a deep U-shape and forming a pair of finger-like processes laterally. Hypandrium with a long broad basal stalk. Hypovalves short, with rounded apexes. Paramere stout and finger-like, with its apical portion hooked and bearing dense short spines. Aedeagus with short membranous ventral valves, and foliate separated dorsal valves.

图 3-120-3 淡色蝎蛉 ***Panorpa decolorata*** 地理分布

Fig. 3-120-3 Distribution of ***Panorpa decolorata***

Emergence: July to August.

Distribution: Shaanxi (Ningshan, Taibai, Nanzheng, Zhouzhi, Baoji), Henan (Songxian) and Hubei (Shennongjia).

121. 峨眉山蝎蛉 *Panorpa emeishana* Hua, Sun & Li, 2001

【测量】 前翅长 12.0 ~ 13.0 mm。

【形态特征】 头顶黑色，单眼三角区黑色。喙黑褐色，两侧黄褐色。胸部背板、侧板及足节黑褐色，足大部分深黄褐色。翅无色透明，翅痣棕色明显。斑纹深棕色，雄性一般退化无显著斑纹，仅端部稍加深；雌性具端带和痣带，其端带窄，内缘不规则；痣带仅为翅痣下方 1 个三角形斑。雄性腹部 1 ~ 5 节背板黑褐色。第 3 节背中突扁平，盖住第 4 节上尖锐的后背突。第 6 节长筒形，无臀角。第 7、8 节黑褐色，基部缢缩，端部平截。雄性外生殖器阔卵形。第 9 节背板末端具深 U 形缺刻，两侧形成 1 对指状突起；腹板基柄极短；下瓣细长，内缘

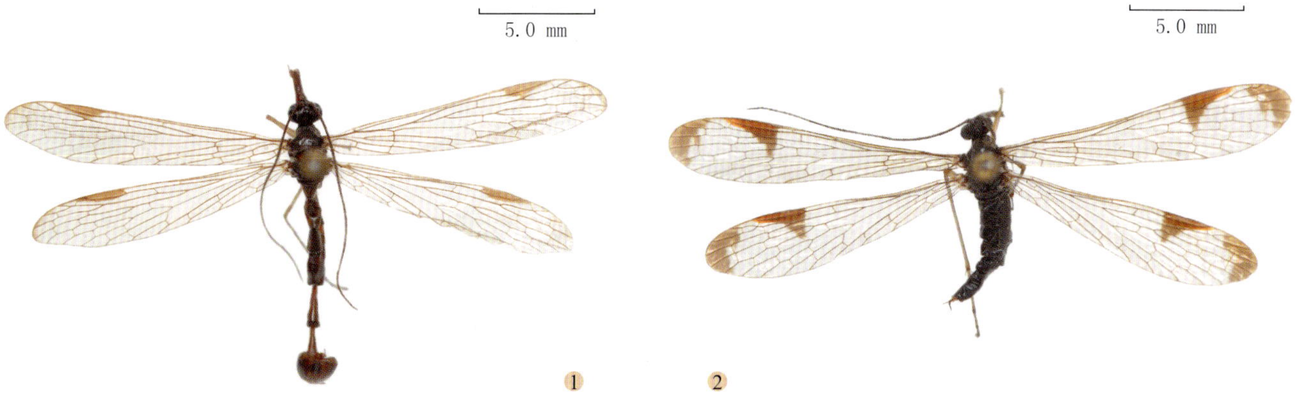

图 3-121-1 峨眉山蝎蛉 ***Panorpa emeishana*** 成虫背面观（四川峨眉山，NWAFU）
1. 雄虫 2. 雌虫

Fig. 3-121-1 Adults of ***Panorpa emeishana***, dorsal view (Mount Emei, Sichuan, NWAFU)
1. Male 2. Female

图 3-121-2 峨眉山蝎蛉 ***Panorpa emeishana*** 雄性外生殖器腹面观

Fig. 3-121-2 Male genitalia of ***Panorpa emeishana***, ventral view

图 3-121-3 峨眉山蝎蛉 ***Panorpa emeishana*** 地理分布

Fig. 3-121-3 Distribution of ***Panorpa emeishana***

具 1 列长鬃；从中部起向外弯折。生殖刺突约与基节等长，内缘的中齿粗壮，亚基齿尖细。阳基侧突紧贴阳茎，细长，基部着生 1 簇粗壮的长刺。阳茎腹瓣和背瓣紧密愈合，强烈骨化，延伸至生殖刺突中部，背瓣末端具 1 对耳状突起。

【发生期】 7 ~ 8 月。

【地理分布】 四川（峨眉山）。

Measurements: Forewing length 12.0–13.0 mm.

Morphology: Vertex and ocellar triangle black. Rostrum dark brown, with yellowish brown flanks. Nota, pleura and coxae dark brown. Legs mostly deep yellowish brown. Wing membrane hyaline, with markings deep brown. Pterostigma brown and distinct. In males, wing markings usually greatly reduced with only a slight tinge on apex; while females with an apical band (narrow and irregular on inner margin) and a greatly reduced pterostigmal band (a triangular spot below pterostigma). Male terga Ⅰ–Ⅴ dark brown. Notal organ on tergum Ⅲ flat and covering acute postnotal organ on tergum Ⅳ. A6 long cylindrical, devoid of anal horns. A7 and A8 dark brown, constricted basally and truncated apically. Male genital bulb broad oval. Epandrium emarginate terminally in a

U-shape and forming a pair of finger-like processes laterally. Hypandrium with a greatly shortened basal stalk. Hypovalves slender and bearing a row of long bristles on inner margins, with their distal halves abruptly bending laterad. Gonostylus almost as long as gonocoxites, and bearing a stout median tooth and a thin acute subbasal process on inner margin. Paramere closely adjoins aedeagus, slender, and bearing a tuft of long stout spines basally. Aedeagus strongly sclerotized with its ventral and dorsal valves intensively fused, extending to middle portion of gonostylus, and bearing a pair of auriculate processes on apex of dorsal valves.

Emergence: July to August.

Distribution: Sichuan (Mount Emei).

122. 江崎蝎蛉 *Panorpa esakii* Issiki, 1929

【测量】 前翅长 14.0 mm。

【形态特征】 头顶黄褐色；1 条宽阔的黑色横带贯穿单眼三角区；喙黑褐色，两侧黄褐色。前胸背板黑色。中、后胸背板大部分黑色，但小盾片黄色；侧板及足黄褐色。翅透明，略带黄色，无明显斑纹；翅痣棕色明显。雄性腹部 1 ~ 5 节背板黑色。第 3 节背中突平扁，向后盖住第 4 节上尖锐的后背突。第 6 节黑色，筒形，向后渐窄。第 7、8 节黄褐色，基部缢缩。雄性外生殖器长椭圆形。第 9 节背板末端具深 U 形缺刻，两侧形成 1 对细长的指形突起；腹板几乎无基柄；下瓣条状，仅伸达生殖肢基节中部，内缘着生 1 列长刚毛。生殖刺突细长，内缘具钝三角形中齿和大型碟状亚基齿。阳基侧突细长，稍超出生殖肢基节端部；阳茎背瓣细长，柳叶状。

【发生期】 8 ~ 10 月。

【地理分布】 台湾（阿里山）。

图 3-122-2 江崎蝎蛉 *Panorpa esakii* 雄性外生殖器腹面观

Fig. 3-122-2 Male genitalia of *Panorpa esakii*, ventral view

Measurements: Forewing length 14.0 mm.

Morphology: Vertex yellowish brown; a broad transverse black stripe passing through ocellar triangle;

图 3-122-1 江崎蝎蛉 *Panorpa esakii* 雄成虫背面观（台湾阿里山，NWAFU）

Fig. 3-122-1 Male adult of *Panorpa esakii*, dorsal view (Ali Mountain, Taiwan, NWAFU)

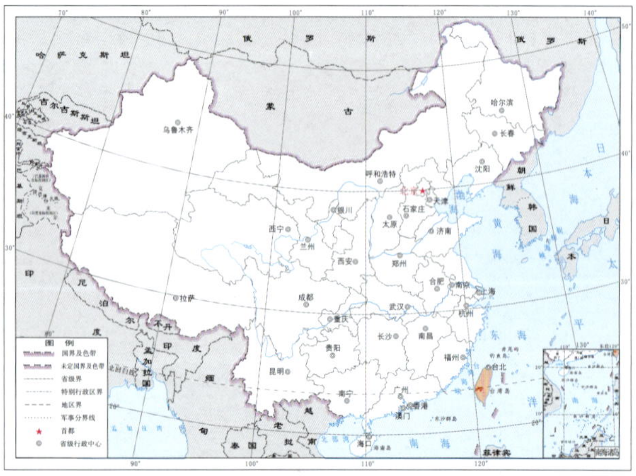

图 3-122-3 江崎蝎蛉 *Panorpa esakii* 地理分布

Fig. 3-122-3 Distribution of *Panorpa esakii*

rostrum; rostrum dark brown and yellowish brown at lateral margins. Pronotum black. Mesonotum and metanotum mostly black but yellow at scutella; pleura and legs yellowish brown. Wing membrane hyaline, slightly tinged with yellow, lacking distinct markings; pterostigma brown and distinct. Male terga Ⅰ–Ⅴ black. Notal organ on tergum Ⅲ flat and covering acute postnotal organ on tergum Ⅳ. A6 black, cylindrical and tapering towards apex. A7 and A8 yellowish brown constricted at base. Male genital bulb long elliptical. Epandrium emarginate terminally in a deep U-shape and forming a pair of finger-like processes laterally. Hypandrium with an extremely shortened basal stalk. Hypovalves short, only extending to middle of gonocoxites, and bearing a row of long bristles on inner margin. Gonostylus with an obtuse triangular median tooth and a large disc-like subbasal process on inner margin. Paramere slender, slightly exceeding apex of gonocoxites; dorsal aedeagal valves slender and foliate.

Emergence: August to October.

Distribution: Taiwan (Ali Mountain).

123. 女儿蝎蛉 *Panorpa filina* Chou & Wang, 1987

【测量】 前翅长 12.0 mm。

【形态特征】 头顶黄色。单眼三角区黑色。喙黄色。前胸背板黑褐色。中、后胸背板褐色，中央具显著的浅黄色条带。侧板及足污黄色。翅无色透明，斑纹褐色。前翅端带内部具 3 个透明窗斑；痣带前宽后窄，无分支；在 CuA 和 1A 末端各有 1 个小斑。雄性腹部 1 ~ 5 节背板深黄褐色。第 3 节背中突扁平，呈半圆形，盖住第 4 节尖锐的后背突。第 6 节短圆筒形，无臀角。第 7、8 节黄色，筒形，短于第 6 节。雄性外生殖器长卵形。第 9 节背板末端具深 U 形缺刻，两侧形成 1 对指形突起；腹板 Y 形，基柄阔长；下瓣指状，不超过生殖肢基节端

图 3-123-1 女儿蝎蛉 *Panorpa filina* 雄成虫背面观（外生殖器已去除，湖南道县，NWAFU）

Fig. 3-123-1 Male adult of *Panorpa filina* with genitalia removed, dorsal view (Daoxian, Hunan, NWAFU)

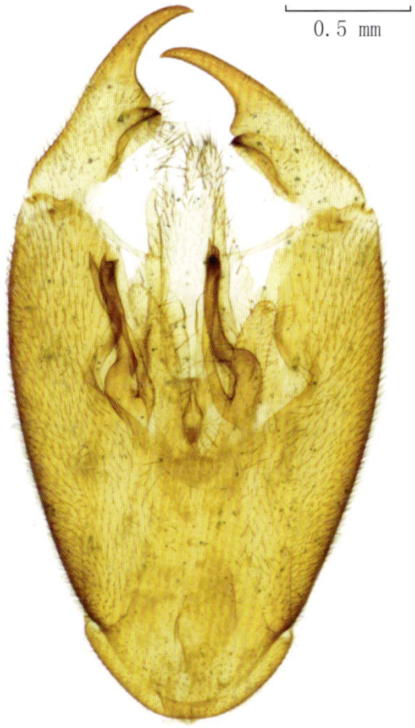

图 3-123-2 女儿蝎蛉 *Panorpa filina* 雄性外生殖器腹面观

Fig. 3-123-2 Male genitalia of *Panorpa filina*, ventral view

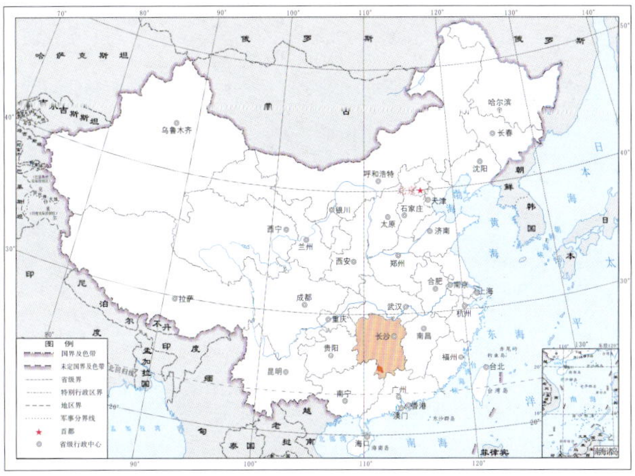

图 3-123-3 女儿蝎蛉 *Panorpa filina* 地理分布

Fig. 3-123-3 Distribution of *Panorpa filina*

部。生殖刺突长约基节的一半，内缘具大型盘状亚基齿和钝三角形中齿。阳基侧突近基部分为 2 叉，腹支短；背支粗壮，近基部强烈弯曲，端部具 2 个尖齿。阳茎背瓣指状。

【发生期】 8 月。

【地理分布】 湖南（道县）。

Measurements: Forewing length 12 mm.

Morphology: Vertex yellow. Ocellar triangle black. Rostrum yellow. Pronotum dark brown. Mesonotum and metanotum brown with a distinct pale yellow mesal stripe. Pleura and legs sordid yellow. Wing membrane hyaline, with markings brown. In forewings, apical band with three small hyaline spots; pterostigmal band unbranched, broad anteriorly and narrow posteriorly; two small spots on end of CuA and 1A, respectively. Male terga Ⅰ–Ⅴ dark yellowish brown. Notal organ on tergum Ⅲ flat, semicircular, and covering acute postnotal organ on tergum Ⅳ. A6 short cylindrical and devoid of anal horns. A7 and A8 yellow and cylindrical, shorter than A6. Male genital bulb long oval. Epandrium emarginate terminally in a U-shape and forming a pair of finger-like processes laterally. Hypandrium Y-shaped with a long broad basal stalk; hypovalves finger-like and not exceeding apex of gonocoxites. Gonostylus half length of gonocoxites, and bearing an obtuse triangular median tooth and a large disc-like subbasal process on inner margin. Paramere bifurcated subbasally, with its ventral branch short, and dorsal branch abruptly curved subbasally and bearing two acute teeth apically. Aedeagus with finger-like dorsal valves.

Emergence: August.

Distribution: Hunan (Daoxian).

124. 黄体蝎蛉 *Panorpa flavicorporis* Cheng, 1957

【测量】 前翅长 12.0 ~ 12.5 mm。

【形态特征】 头顶黄色，单眼三角区黑色。喙黄褐色。前胸背板黄褐色。中、后胸背板黄色，两侧黑褐色。侧板及足黄褐色。翅透明，略带黄色，斑纹褐色。前翅端带阔而完整；痣带基支宽，端支窄或断开；缘斑明显，细长；基带中部稍缢缩；基斑小。雄性腹部 1 ~ 5 节背板深黄褐色。第 3 节背中突扁平，盖住第 4 节上尖锐的后背突。第 6 节深黄褐色，筒形，向后渐窄，无臀角。第 7、8 节黄褐色，第 7 节基部强烈缢缩成柄状。雄性外生殖器卵形。第 9 节背板末端具深 U 形缺刻，两侧形成 1 对指形突起；腹板基柄极短；下瓣细长，稍超过生殖肢基节中部。生殖刺突内缘具钝三角形中齿和大

图 3-124-1 黄体蝎蛉 *Panorpa flavicorporis* 雄成虫背面观（福建邵武大竹岚，NWAFU）

Fig. 3-124-1 Male adult of *Panorpa flavicorporis*, dorsal view (Dazhulan, Shaowu, Fujian, NWAFU)

图 3-124-2 黄体蝎蛉 *Panorpa flavicorporis* 雄性外生殖器腹面观

Fig. 3-124-2 Male genitalia of *Panorpa flavicorporis*, ventral view

型亚基齿。阳基侧突线状、细长，末端尖细，超过生殖刺突的中齿。阳茎短，背瓣指状稍分离。

【发生期】 9 ~ 10月。

【地理分布】 福建(邵武、南平、德化、武夷山)。

Measurements: Forewing length 12.0–12.5 mm.

Morphology: Vertex yellow. Ocellar triangle black. Rostrum yellowish brown. Pronotum yellowish brown. Mesonotum and metanotum yellow, with dark brown flanks. Pleura and legs yellowish brown. Wing membrane hyaline, slightly tinged with yellow. Markings brown. In forewings, apical band broad and complete; pterostigmal band broad, with broad basal branch and thin or separated apical branch; marginal spot distinct and slender; basal band slightly constricted medially; basal spot small. Male terga Ⅰ–Ⅴ deep yellowish brown. Notal organ on tergum Ⅲ flat, covering acute postnotal organ on tergum Ⅳ. A6 yellowish brown, cylindrical, tapering towards apex, devoid of anal horns. A7 and A8 yellowish brown. A7 intensively stalked basally. Male genital bulb oval. Epandrium emarginate terminally in a U-shape and forming a pair of finger-like processes laterally. Hypandrium with a greatly shortened basal stalk; hypovalves slender, slightly exceeding middle of gonocoxites. Gonostylus with an obtuse triangular median tooth and a large subbasal process on inner margin. Paramere thread-like and slender, with an acute apex, and exceeding median tooth of gonostylus. Aedeagus short with finger-like and divergent dorsal valves.

图 3-124-3 黄体蝎蛉 *Panorpa flavicorporis* 地理分布

Fig. 3-124-3 Distribution of ***Panorpa flavicorporis***

Emergence: September to October.

Distribution: Fujian (Shaowu, Nanping, Dehua, Wuyi Mountains).

125. 黄翅蝎蛉 *Panorpa flavipennis* Carpenter, 1938

【测量】 前翅长 13.0 ~ 14.5 mm。

【形态特征】 头顶及单眼三角区黑色；喙红棕色，两侧有深色纵带。胸部背板、侧板、足基节黑褐色。足大部分深褐色。翅烟黄色半透明，翅痣红棕色明显，斑纹红棕色。前翅端带及痣带的端支水渍状；雄性无基带，但雌性有；无缘斑和基斑。雄性腹部 1 ~ 5 节背板黑色。第 3 节背中突扁平，盖住第 4 节上尖锐的后背突。第 6 节基部 2/3 圆筒形，端部 1/3 锥形；末端 1/3 处背面具 1 个锥形、向上隆起的尖角。第 7、8 节红褐色，基部缢缩，向端部逐渐变阔。雄性外生殖器卵形。第 9 节背板末端具浅 V 形缺刻；腹板基柄极短；下瓣长带状，内缘具 1 列长鬃。生殖刺突内缘的亚中齿粗壮，基齿发达而尖锐；外缘轻微内凹。阳基侧突超过生殖肢基节端部，半叶形，内缘具 1 列梳齿状长刺。阳茎 V 形，腹瓣短小；背瓣末端指状，略膨大。

【发生期】 6 ~ 10 月。

【地理分布】 四川（白鹿顶、卧龙、鞍子河）。

Measurements: Forewing length 13.0–14.5 mm.

Morphology: Vertex and ocellar triangle black. Rostrum reddish brown with two darkened longitudinal stripes laterally. Nota, pleura and coxae dark brown. Legs mostly deep brown. Wing membrane translucent, deep yellow, markings reddish brown. Pterostigma reddish brown and distinct. In forewings, apical band and apical branch of pterostigmal band water stainlike; basal band absent in males but present in females; marginal and basal spots absent. Male terga Ⅰ–Ⅴ black. Notal organ on tergum Ⅲ flat, and covering acute postnotal organ on tergum Ⅳ. A6 cylindrical at basal two thirds, and conical at distal third, and bearing a conical horn directed dorsad on distal third dorsally. A7 and A8 reddish brown and constricted basally, and enlarged towards apexes. Male genital bulb oval. Epandrium

图 3-125-1 黄翅蝎蛉 ***Panorpa flavipennis*** 成虫背面观（四川崇州鞍子河，NWAFU）
1. 雄虫 2. 雌虫

Fig. 3-125-1 Adults of ***Panorpa flavipennis***, dorsal view (Anzihe, Chongzhou, Sichuan, NWAFU)
1. Male 2. Female

图 3-125-2 黄翅蝎蛉 ***Panorpa flavipennis*** 雄虫腹部末端侧面观

Fig. 3-125-2 Male teminal abdomen of ***Panorpa flavipennis***, lateral view

图 3-125-3 黄翅蝎蛉 ***Panorpa flavipennis*** 雄性外生殖器腹面观

Fig. 3-125-3 Male genitalia of ***Panorpa flavipennis***, ventral view

图 3-125-4 黄翅蝎蛉 ***Panorpa flavipennis*** 地理分布

Fig. 3-125-4 Distribution of ***Panorpa flavipennis***

emarginate terminally in a shallow V-shape; hypandrium with a greatly shortened basal stalk. Hypovalves long stripe-like and bearing a row of long bristles on inner margin. Gonostylus with a stout median tooth and a stout and acute subbasal process on inner margin, and slightly concaved along outer margin. Paramere semi-foliate, exceeding apex of gonocoxites and bearing a row of comb-like long spines on inner margin. Aedeagus V-shaped, with short ventral valves and finger-like slightly enlarged dorsal valves.

Emergence: June to October.

Distribution: Sichuan (Bailuding, Wolong, Anzihe).

126. 实蝎蛉 *Panorpa fructa* Cheng, 1949

【测量】 前翅长 11.5 mm。

【形态特征】 头顶黑褐色。单眼三角区黑色。喙红褐色。胸部背板黑褐色。侧板及足黄褐色。翅无色透明，具散乱的褐色斑纹。前翅端带分裂为一系列模糊的小斑点；痣带不完整，基支断开，端支模糊；缘斑大；基带断开成 2 斑；基斑小。雄性腹部 1 ~ 5 节背板黑褐色。第 3 节背板后缘的背中突扁平，盖住第 4 节上尖锐的后背突。第 6 节锥形，无臀角。第 7、8 节黄褐色，基部稍缢缩。雄性外生殖器卵形。第 9 节背板末端具深 U 形缺刻，两侧形成 1 对指形突起；腹板具短阔的基柄；下瓣长带状，几乎伸达生殖肢基节端部，内缘具 1 列鬃毛。生殖肢基节端部内缘具 1 列粗壮的刚毛；生殖刺突稍短于基节，内缘具钝三角形中齿和碟形亚基齿。阳基侧突螺旋状弯曲，末端稍平直，内侧密生短刺。阳茎腹瓣短小；背瓣细长，稍分离。

图 3-126-1　实蝎蛉 *Panorpa fructa* 雄成虫背面观（四川康定贡嘎寺，NWAFU）

Fig. 3-126-1　Male adult of *Panorpa fructa*, dorsal view (Konka Temple, Kangding, Sichuan, NWAFU)

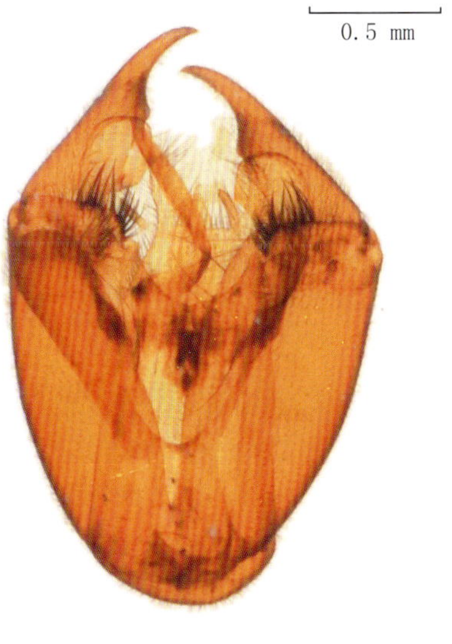

图 3-126-2　实蝎蛉 *Panorpa fructa* 雄性外生殖器腹面观

Fig. 3-126-2　Male genitalia of *Panorpa fructa*, ventral view

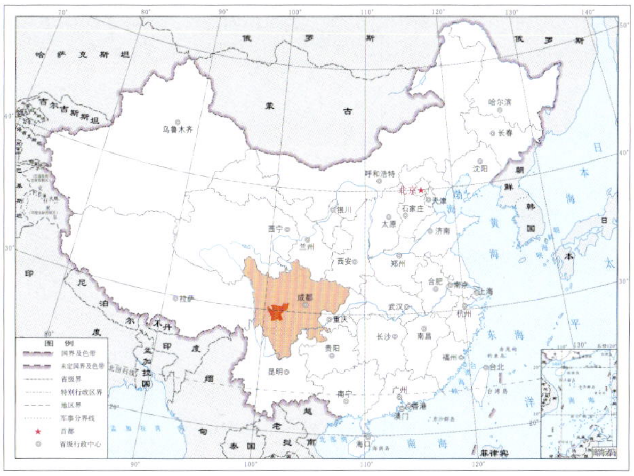

图 3-126-3　实蝎蛉 *Panorpa fructa* 地理分布

Fig. 3-126-3　Distribution of *Panorpa fructa*

【发生期】 8 ~ 9 月。

【地理分布】 四川（康定、雅江）。

Measurements: Forewing length 11.5 mm.

Morphology: Vertex dark brown. Ocellar triangle black. Rostrum reddish brown. Nota dark brown. Pleura and legs yellowish brown. Wing membrane hyaline, with scattered brown markings. In forewings, apical band split into a series of small faint spots; pterostigmal band incomplete, with a separated basal branch and a blurred apical branch; marginal spot large; basal band split into two spots; basal spot very small. Male terga Ⅰ–Ⅴ dark brown. Notal organ on tergum Ⅲ flat, covering acute postnotal organ on tergum Ⅳ. A6 conical and devoid of anal horns. A7 and A8 yellowish brown and constricted basally. Male genital bulb oval. Epandrium emarginate terminally in a deep U-shape and forming a pair of finger-like processes laterally. Hypandrium with a short broad basal stalk. Hypovalves long stripe-like, bearing a row of long bristles on inner margins, and almost extending to apex of gonocoxites. Gonocoxites with row of stout setae on ventro-apical margin. Gonostylus with an obtuse triangular median tooth and a disc-like subbasal process. Paramere spiral with a relatively straight distal portion, bearing a row of dense short spines on inner margin. Aedeagus with short ventral valves, and slender divergent dorsal valves.

Emergence: August to September.

Distribution: Sichuan (Kangding, Yajiang).

127. 福建蝎蛉 *Panorpa fukiensis* Tjeder, 1950

【测量】 前翅长 14.0 mm。

【形态特征】 头顶黑褐色，单眼三角区黑色。喙黄褐色。胸部背板黑褐色。侧板及足黄褐色。翅透明，斑纹褐色，翅痣红棕色。前翅端带阔，后缘不完整，有 1 个分离的小斑，前缘与痣带些许相接；痣带完整，端支较窄；缘斑长条形；基带宽；基斑小。雄性腹部 1 ~ 5 节背板黑褐色。第 6 节无臀角；第 7、8 节黄褐色，基部缢缩。雄性外生殖器卵形。第 9 节背板末端具深 U 形缺刻，两侧形成 1 对指形突起；腹板具长基柄；下瓣短细，仅延伸至生殖肢基节中部。生殖刺突内缘的亚基齿阔，中齿尖锐。阳基侧突具细长的柄，中部膨大，向末端渐尖，内缘着生 1 列长刺。阳茎腹瓣短，背瓣分离指状，侧突起发达。

【发生期】 9 ~ 10 月。

【地理分布】 福建（光泽）、江西（芦溪）。

Measurements: Forewing length 14.0 mm.

Morphology: Vertex dark brown. Ocellar triangle black. Rostrum yellowish brown. Nota dark brown. Pleura and legs yellowish brown. Wing membrane hyaline, with markings brown. Pterostigma reddish brown. In forewings, apical band broad with an incomplete posterior border, accompanied by a separated small spot posteriorly, and connected to pterostigmal band anteriorly; pterostigmal band complete, with a narrow apical branch; marginal spot long stripe-like; basal band broad; basal spot small. Male terga I–V blackish brown. A6 devoid of anal horns. A7 and A8 yellowish brown and constricted basally. Male genital bulb oval. Epandrium emarginate terminally in a deep U-shape and forming a pair of finger-like processes laterally. Hypandrium with a long basal stalk. Hypovalves short and narrow, extending to middle of gonocoxites. Gonostylus with an acute median tooth and a broad subbasal process on inner margin. Paramere with a slender basal stalk and enlarged medially, tapering towards apex and bearing a row of long spines on inner margin. Aedeagus with short ventral valves and finger-like divergent dorsal valves, with stout lateral processes.

图 3-127-1 福建蝎蛉 *Panorpa fukiensis* 雌成虫背面观（江西芦溪万龙山，NWAFU）

Fig. 3-127-1 Female adult of *Panorpa fukiensis*, dorsal view (Mount Wanlong, Luxi, Jiangxi, NWAFU)

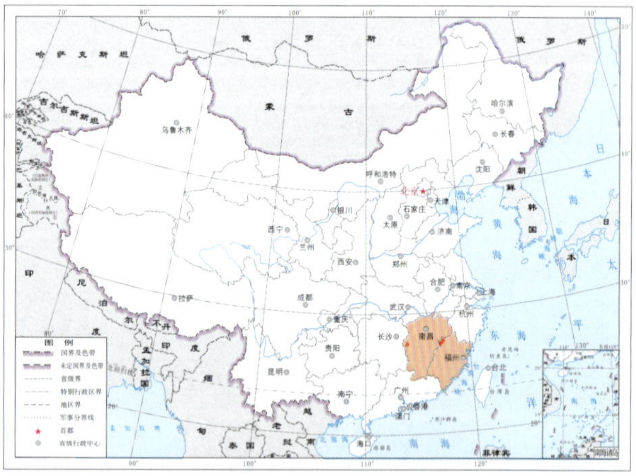

图 3-127-2 福建蝎蛉 *Panorpa fukiensis* 地理分布

Fig. 3-127-2 Distribution of *Panorpa fukiensis*

Emergence: September to October.

Distribution: Fujian (Guangze) and Jiangxi (Luxi).

128. 淡黄蝎蛉 *Panorpa fulvastra* Chou, 1981

图 3-128-1 淡黄蝎蛉 ***Panorpa fulvastra*** 成虫侧面观（陕西秦岭，姜碌摄）

1. 雄虫 2. 雌虫

Fig. 3-128-1 Adults of ***Panorpa fulvastra***, lateral view (Qinling Mountains, Shaanxi, photo by Lu Jiang)

1. Male 2. Female

【测量】 前翅长 10.5 ~ 12.0 mm。

【形态特征】 头顶、单眼三角区黑色。喙黄褐色。前胸背板黑褐色。中、后胸背板黑褐色，中央具浅黄色纵带。侧板及足乳黄色。翅透明，略带黄色，翅痣黄色，斑纹不明显，仅具端带和痣带。雄性腹部 1 ~ 5 节背板黑褐色。第 3 节背板后缘的背中突扁平，盖住第 4 节上尖锐的后背突。第 6 节黑褐色，筒形，向端部渐窄，无臀角。第 7、8 节黄褐色，基部稍缢缩。雄性外生殖器卵形。第 9 节背板末端具深 U 形缺刻，两侧形成 1 对指形突起；腹板基柄极短；下瓣带状，伸达生殖肢基节中部。生殖肢基节端部内缘有 5 ~ 6 根粗刚毛；生殖刺突具钝三角形中齿和碟形亚基齿。阳基侧突细长而螺旋形弯曲，互相交叉。阳茎腹瓣短小，背瓣指状，近端部背面膨大。

【发生期】 7 ~ 8 月。

【地理分布】 陕西（宁陕、佛坪、周至、南郑、留坝）、甘肃（宕昌）。

Measurements: Forewing length 10.5–12.0 mm.

Morphology: Vertex and ocellar triangle black. Rostrum yellowish brown. Pronotum dark brown. Mesonotum and metanotum dark brown, with a distinct pale yellow mesal stripe. Pleura and legs creamy yellow. Wing membrane hyaline, slightly tinged with yellow. Pterostigma yellow. Markings faint with only apical and pterostigmal bands present.Male terga Ⅰ–Ⅴ blackish brown. Notal organ on tergum Ⅲ flat and covering acute postnotal organ on tergum Ⅳ. A6 dark brown and cylindrical, tapering towards apex and devoid of anal horns. A7 and A8 yellowish brown and slightly constricted basally. Male genital bulb oval. Epandrium emarginate terminally in a deep U-shape and forming a pair of finger-like processes laterally. Hypandrium with an extremely shortened basal stalk. Hypovalves stripe-like, extending to middle of gonocoxites. Gonocoxites each bearing 5–6 stout setae on ventral apex innerly. Gonostylus with an obtuse triangular median tooth and a disc-like subbasal process. Parameres slender, spiral sigmoidal, and crossed mesally. Aedeagus with short ventral valves and finger-like dorsal valves; dorsal valves enlarged near dorsal apexes.

Emergence: July to August.

Distribution: Shaanxi (Ningshan, Foping, Zhouzhi, Nanzheng, Liuba) and Gansu (Dangchang).

图 3-128-2 淡黄蝎蛉 ***Panorpa fulvastra*** 雄性外生殖器腹面观

Fig. 3-128-2 Male genitalia of ***Panorpa fulvastra***, ventral view

图 3-128-3 淡黄蝎蛉 ***Panorpa fulvastra*** 地理分布

Fig. 3-128-3 Distribution of ***Panorpa fulvastra***

129. 嘉理斯蝎蛉 *Panorpa gressitti* Byers, 1970

【测量】 前翅长 11.2 mm。

【形态特征】 头顶深褐色，单眼三角区黑色。喙黄褐色。前胸背板黑褐色。中后胸背板除小盾片黄白色外，大部分黑褐色。侧板及基节乳黄色，足大部分红褐色。翅无色透明，斑纹黑褐色。前翅端带内部具3 ~ 4个透明窗斑；痣带阔，端支断开，与端带融合，基支阔；缘斑斜行，与痣带的基支相连，形成1个显著的V形纹；基带前部细，后部阔；后缘具一系列不规则斑块。停歇时，翅呈屋脊状叠于体背。雄性腹部1 ~ 5节背板黑褐色。第3节背中突扁平，盖住第4节上尖锐的后背突。第6节短筒形，无臀角；第7、8节短筒形，基部不缢缩。雄性外生殖器阔卵形。第9节背板末端具近方形缺刻，两侧形成1对细长的指形突起；腹板基柄阔长；下瓣短宽。阳基侧突细长，超过生殖刺突中部，末端爪状弯曲，端半部内缘具许多刺毛。阳茎腹瓣末端近方形；背瓣末端尖锐。

【发生期】 3 ~ 6月。

【地理分布】 广东（龙斗山、南昆山）。

图 3-129-1 嘉理斯蝎蛉 *Panorpa gressitti* 雌成虫侧面观（广东惠州南昆山，雷波摄）

Fig. 3-129-1 Female adult of *Panorpa gressitti*, lateral view (Mount Nankun, Huizhou, Guangdong, photo by Bo Lei)

Measurements: Forewing length 11.2 mm.

Morphology: Vertex dark brown, ocellar triangle black. Rostrum yellowish brown. Pronotum dark brown. Mesonotum and metanotum mostly dark brown, but yellowish white on scutella. Pleura and coxae creamy yellow. Legs mostly reddish brown. Wing membrane hyaline, with markings dark brown. In forewings, apical band with 3–4 hyaline window spots innerly; pterostigmal band broad, with a separated apical branch fused with apical band, and a broad basal branch; marginal spot obliquely elongates and connected with basal branch of pterostigmal band, forming a distinct V-shaped pattern; basal band narrow anteriorly and broad posteriorly; a series of irregular spots on hind margin. Wings held roof-like at repose.Male terga Ⅰ–Ⅴ dark brown. Notal organ on tergum Ⅲ flat and covering acute postnotal organ on tergum Ⅳ. A6 short cylindrical and devoid of anal horns. A7 and A8 short cylindrical and not constricted basally. Male genital bulb broad oval. Epandrium deeply emarginate terminally into a subrectangle and forming a pair of slender finger-like processes laterally. Hypandrium with a long broad basal stalk. Hypovalves short and broad, not extending to base of gonostylus. Paramere slender and exceeding middle of gonostylus, with an curved and claw-like apex, and bearing a row of long spines on inner margin of apical half. Ventral aedeagal valves mostly trapezoidal; dorsal valves acute

图 3-129-2　嘉理斯蝎蛉 *Panorpa gressitti* 雄性外生殖器腹面观

Fig. 3-129-2　Male genitalia of *Panorpa gressitti*, ventral view

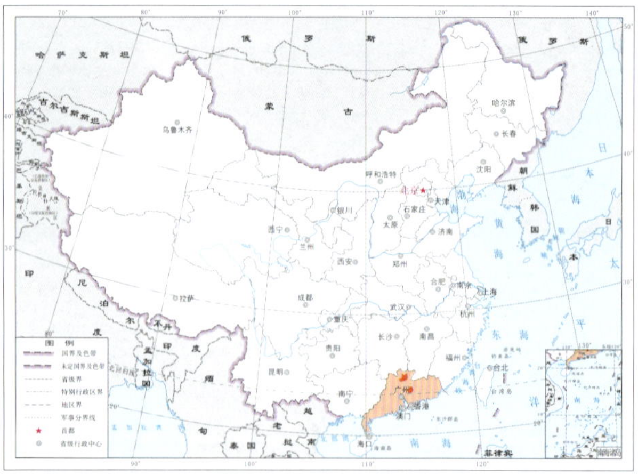

图 3-129-3　嘉理斯蝎蛉 *Panorpa gressitti* 地理分布

Fig. 3-129-3　Distribution of *Panorpa gressitti*

apically.

Emergence: March to June.

Distribution: Guangdong (Mount Longdou, Mount Nankun).

130. 桂东蝎蛉 *Panorpa guidongensis* Chou & Li, 1987

【测量】 前翅长 12.5 ~ 13.0 mm。

【形态特征】 头顶黄褐色，单眼三角区黑色。喙黄褐色。前胸背板黑褐色。中、后胸背板黑褐色，中央具宽阔的黄色纵带。侧板及足黄褐色。翅透明，略带淡黄色，斑纹黄褐色。前翅端带完整，后缘伸出 1 小角；痣带细，端支断开；基带与翅前缘断开；缘斑窄长；基斑小。雄性腹部 1 ~ 5 节背板黑褐色。第 3 节背中突扁平，盖住第 4 节上尖锐的后背突。第 6 节黑褐色，短锥形，无臀角。第 7、8 节黄褐色，基部缢缩。雄性外生殖器宽卵形。第 9 节背板末端具深 U 形缺刻，两侧形成 1 对指形突起；腹板基柄短；下瓣窄长，超过生殖肢基节的中部，内缘具 1 列长鬃。生殖刺突内缘的亚基齿十分发达，无中齿。阳基侧突分 3 支，每一支都着生长毛。阳茎强烈骨化，超过生殖肢基节的端部。

图 3-130-1　桂东蝎蛉 *Panorpa guidongensis* 雄成虫背面观（湖南桂东八面山，NWAFU）

Fig. 3-130-1　Male adult of *Panorpa guidongensis*, dorsal view (Mount Bamian, Guidong, Hunan, NWAFU)

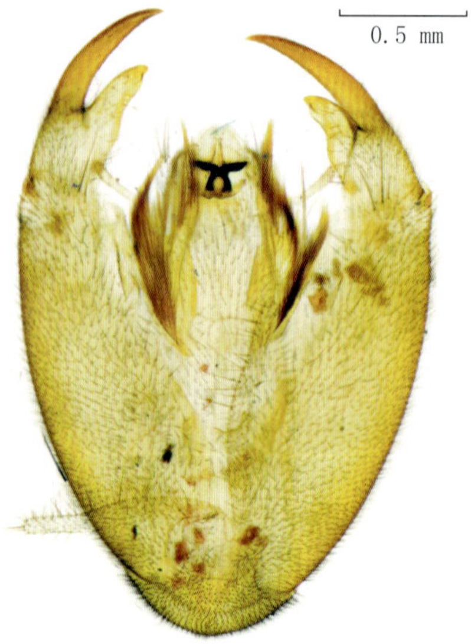

图 3-130-2　桂东蝎蛉 *Panorpa guidongensis* 雄性外生殖器腹面观

Fig. 3-130-2　Male genitalia of *Panorpa guidongensis* , ventral view

【发生期】 9 ~ 10 月。

【地理分布】 湖南（桂东）、广东（连州）。

Measurements: Forewing length 12.5–13.0 mm.

Morphology: Vertex yellowish brown. Ocellar triangle black. Rostrum yellowish brown. Pronotum dark brown. Mesonotum and metanotum dark brown with broad yellow mesal stripe. Pleura and legs yellowish brown. Wing membrane hyaline and slightly tinged with light yellow, markings yellowish brown. Forewing apical band complete and with a small elongation on hind portion; pterostigmal band slender with separated apical branch; basal band separated from anterior margin; marginal spot slender; basal spot small. Male terga Ⅰ–Ⅴ dark brown. Notal organ on tergum Ⅲ flat and covering acute postnotal organ on tergum Ⅳ. A6 dark brown, short conical and devoid of anal horns. A7 and A8 yellowish brown, and constricted basally. Male genital bulb broad oval. Epandrium deeply emarginate terminally in a U-shape and forming a pair of finger-like processes laterally. Hypandrium with a short basal stalk. Hypovalves slender, exceeding middle of gonocoxites, and bearing a row of long bristles on inner margin. Gonostylus with a extremely stout subbasal process and lacking median tooth on inner margin. Paramere trifurcated with numerous spines on each branch. Aedeagus strongly sclerotized and exceeding apex of gonocoxites.

图 3-130-3 桂东蝎蛉 *Panorpa guidongensis* 地理分布

Fig. 3-130-3 Distribution of ***Panorpa guidongensis***

Emergence: September to October.

Distribution: Hunan (Guidong) and Guangdong (Lianzhou).

131. 滴蝎蛉 *Panorpa guttata* Navás, 1908

【测量】 前翅长 11.0 ~ 11.5 mm。

【形态特征】 头顶、喙黄色，单眼三角区黑色。胸部背板黄绿色，中央具宽阔的黄色纵带。侧板及足黄色。翅略带黄色，透明，翅痣黄色显著，斑纹不明显。前翅端带仅余 1 条细长斑或消失；痣带在翅痣下方形成 1 细长带。无其他斑纹。雄性腹部 1 ~ 5 节背板黄绿色。第 3 节背板后缘的背中突扁平，后缘具密集的短毛，盖住第 4 节上尖锐的后背突。第 6 节圆筒形，向后稍变阔，背面末端具 1 狭窄的缺刻，使得两侧形成 1 对片状突起；每个突起侧下方具 1 个 U 形缺刻和 1 个尖锐的齿突。第 7 节末端向上隆起。雄性外生殖器长椭圆形。第 9 节背板末端具深 U 形缺刻，两侧形成 1 对细长的指形突起；腹板具短阔的基柄；下瓣狭长，不超过生殖肢基节端部，内缘中部稍向内凸。生殖刺突短于基节，具耳状亚基齿和钝三角形中齿，内缘的基半部长有 1 丛弯曲的长鬃毛。阳基侧突分 2 叉，其中腹支较短，背支超过生殖肢基节端部。阳茎腹瓣膜质；背瓣基半部愈合，端半部分离。

【发生期】 6 ~ 8 月。

【地理分布】 四川（宝兴、石棉、鞍子河、卧龙、西岭雪山、峨眉山）。

Measurements: Forewing length 11.0–11.5 mm.

Morphology: Vertex and rostrum yellow. Ocellar triangle black. Nota yellowish green, with a broad yellow mesal stripe. Pleura and legs yellow. Wing membrane hyaline, slightly tinged with yellow. Pterostigma yellow and distinct. Markings very faint. In forewings, apical band greatly reduced into a slender stripe or absent; pterostigmal band reduced into a slender stripe below pterostigma. Other markings absent. Male terga Ⅰ–Ⅴ yellowish green. Notal organ on tergum Ⅲ flat, with dense short setae posteriorly, and covering acute postnotal organ on tergum Ⅳ. A6 cylindrical, slightly enlarged towards

图 3-131-1 滴蝎蛉 ***Panorpa guttata*** 成虫背面观（四川石棉公益海，NWAFU）
1. 雄虫 2. 雌虫

Fig. 3-131-1 Adults of ***Panorpa guttata***, dorsal view (Gongyihai, Shimian, Sichuan, NWAFU)
1. Male 2. Female

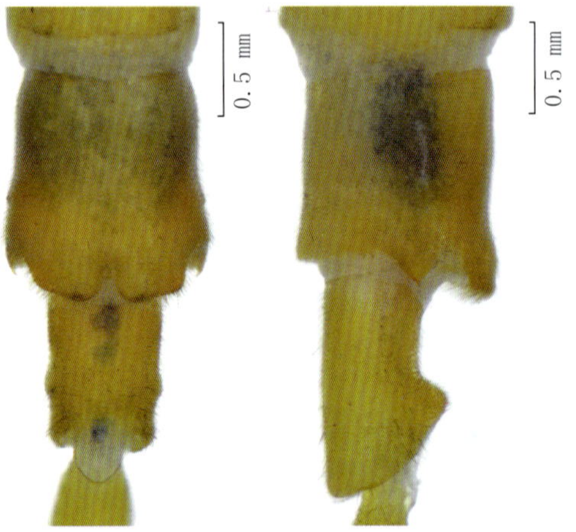

图 3-131-2 滴蝎蛉 ***Panorpa guttata*** 雄虫腹部第 6、7 节
1. 背面观 2. 侧面观

Fig. 3-131-2 Male A6 and A7 of ***Panorpa guttata***
1. Dorsal view 2. Lateral view

图 3-131-3 滴蝎蛉 ***Panorpa guttata*** 雄性外生殖器腹面观
Fig. 3-131-3 Male genitalia of ***Panorpa guttata,*** ventral view

apex, with a narrow emargination dorso-apically and forming a pair of plates laterally; below each plate are a U-shaped emargination and an acute tooth. A7 projected dorsad on dorsal apex. Male genital bulb long elliptical. Epandrium emarginate deeply in a U-shape terminally and forming a pair of finger-like processes laterally. Hypandrium with a short broad basal stalk. Hypovalves slender, not exceeding apex of gonocoxites, and slightly projected mesad on middle of inner margins. Gonostylus shorter than gonocoxites, with an obtuse triangular median tooth and an auriculate subbasal process on inner margin, and bearing a tuft of dense curved long setae on basal half of inner margin. Paramere bifurcated, with ventral

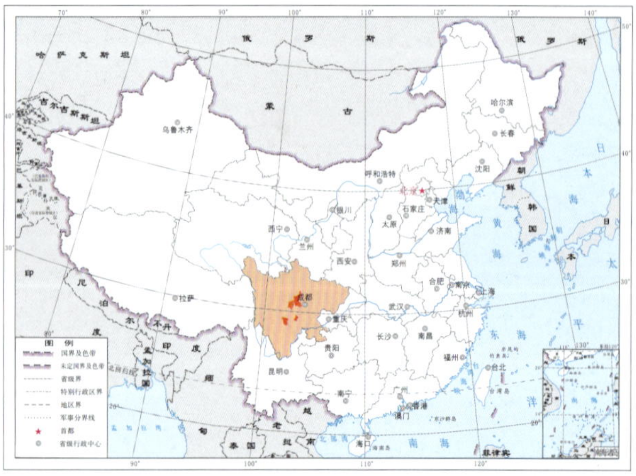

图 3-131-4 滴蝎蛉 ***Panorpa guttata*** 地理分布
Fig. 3-131-4 Distribution of ***Panorpa guttata***

branch short and dorsal branch exceeding apex of gonocoxites. Aedeagus with membranous ventral valves; dorsal valves fused basally and divergent distally.

Emergence: June to August.

Distribution: Sichuan (Baoxing, Shimian, Anzihe, Wolong, Xiling Snow Mountain, Mount Emei).

132. 织纹蝎蛉 *Panorpa implicata* Cheng, 1957

【测量】 前翅长 13.5 mm。

【形态特征】 头顶黑色，单眼三角区黑色；喙红褐色，较短粗。胸部背板黑褐色。侧板及足黄褐色。翅无色透明，斑纹褐色。前翅端带具 4 ~ 5 个大型透明窗口，与痣带断开的端支形成一系列网状纹；痣带的基支与延长的缘斑形成明显的 V 形斑；后缘具一系不规则列斑纹，在 CuP 近末端处形成 1 个透明窗口；R_1 近基部具 1 个不规则斑。雄性未知。

【发生期】 5 ~ 6 月。

【地理分布】 福建（邵武、将乐）。

Measurements: Forewing length 13.5 mm.

Morphology: Vertex and ocellar triangle black. Rostrum reddish brown and relatively stout. Nota dark brown. Pleura and legs yellowish brown. Wing membrane hyaline, with markings brown. In forewings, apical band bearing 4–5 large hyaline windows innerly; and forming a series of reticulate patterns with separated apical branch of pterostigmal band; basal branch of pterostigmal band and elongated marginal spot forming a distinct V-shaped pattern; a series of irregular markings along basal half of hind margin; a hyaline circular spot near ending of CuP; an irregular spot near base of R_1. Males unknown.

Emergence: May to June.

Distribution: Fujian (Shaowu, Jiangle).

图 3-132-1 织纹蝎蛉 *Panorpa implicata* 雌成虫背面观（福建将乐龙栖山，IZAS）

Fig. 3-132-1 Female adult of *Panorpa implicata*, dorsal view (Mount Longqi, Jiangle, Fujian, IZAS)

图 3-132-2 织纹蝎蛉 *Panorpa implicata* 地理分布

Fig. 3-132-2 Distribution of *Panorpa implicata*

133. 岛生蝎蛉 *Panorpa insularis* Hua & Chou, 1998

【测量】 前翅长 14.0 mm。

【形态特征】 头顶黄褐色；单眼三角区黑褐色，其上方有 1 个显著的 V 形黑斑；喙黄褐色，中部灰黄色。胸部背板黑褐色，中央具显著的浅黄色中带；侧板及足浅黄褐色。翅稍带黄色，透明，斑纹黑褐色，翅痣亮黄色，明显。前翅端带阔，后缘具 1 分离的小斑；痣带的基支与端支约等宽；缘斑细带状；基带完整，中部最窄，后部宽；基斑小。雄性未知。

【发生期】 5 月，11 ~ 12 月。

图 3-133-1 岛生蝎蛉 ***Panorpa insularis*** 雌成虫侧背面观（海南乐东尖峰岭，张巍巍摄）

Fig. 3-133-1 Female adult of ***Panorpa insularis***, latero-dorsal view (Jianfengling, Ledong, Hainan, photo by Weiwei Zhang)

【地理分布】 海南（尖峰岭）。

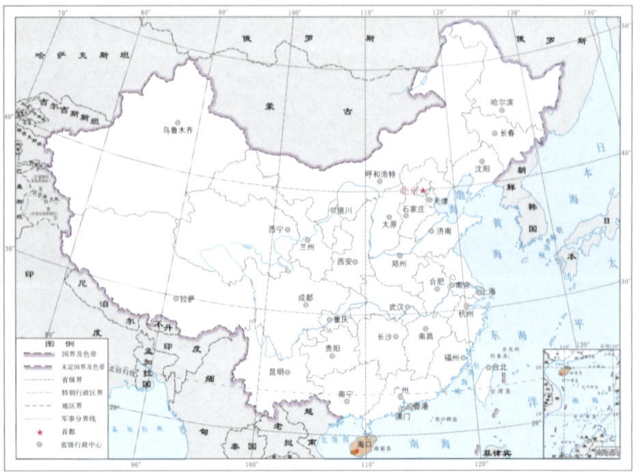

图 3-133-2 岛生蝎蛉 ***Panorpa insularis*** 地理分布

Fig. 3-133-2 Distribution of ***Panorpa insularis***

Measurements: Forewing length 14.0 mm.

Morphology: Vertex yellowish brown. Ocellar triangle dark brown, accompanied by a distinct black V-shaped pattern dorsally. Rostrum yellowish brown, but greyish yellow mesally. Nota dark brown, with a distinct pale yellow mesal stripe. Pleura and legs light yellowish brown. Wing membrane hyaline, slightly tinged with yellow. Markings dark brown. Pterostigma bright yellow and distinct. In forewings, apical band broad and accompanied by a small separated spot posteriorly; pterostigmal band with equally-wide basal and apical branches; marginal spot slender and stripe-like; basal band complete, narrow medially and broad posteriorly; basal spot small. Males unknown.

Emergence: May, November to December.

Distribution: Hainan (Jianfengling).

134. 金川蝎蛉 *Panorpa jinchuana* Hua, Sun & Li, 2001

图 3-134-1　金川蝎蛉 *Panorpa jinchuana* 成虫背面观（四川丹巴牦牛河，NWAFU）

1. 雄虫　2. 雌虫

Fig. 3-134-1　Adults of *Panorpa jinchuana*, dorsal view (Maoniu River, Danba, Sichuan, NWAFU)

1. Male　2. Female

【测量】 前翅长 12.5 ~ 13.0 mm。

【形态特征】 头顶黑褐色，单眼三角区黑色。喙黄褐色。前胸背板黑褐色。中、后胸背板黑褐色，中央具 1 条宽阔的黄色纵带。侧板及足黄褐色。翅无色透明，斑纹褐色。前翅端带阔，具有 2 个大的透明斑；痣带完整，具有阔的基支和略窄的端支；基带中断成 2 个斑点；缘斑大而明显；基斑小。雄性斑纹相对雌性些许退化，尤其痣带端支断开或缺失。雄性腹部 1 ~ 5 节背板黑褐色。第 3 节背中突扁平，盖住第 4 节尖锐的后背突。第 6 节黑褐色，筒形，无臀角。第 7、8 节黄褐色，基部缢缩。雄性外生殖器阔卵形。第 9 节背板末端具深 U 形缺刻，两侧形成 1 对指形突起；腹板基柄极短；下瓣带状，不伸达生殖刺突基部，内缘具 1 列长鬃。生殖肢基节端部内缘具 1 簇长刚毛；生殖刺突内缘具钝三角形中齿和盘状亚基齿。阳基侧突螺旋形弯曲，相互交叉，内缘具 1 列短刚毛。阳茎腹瓣膜质短小；背瓣指状，末端背缘膨大。

【发生期】 9 ~ 10 月。

【地理分布】 四川（金川、丹巴）。

图 3-134-2　金川蝎蛉 *Panorpa jinchuana* 雄性外生殖器腹面观

Fig. 3-134-2　Male genitalia of *Panorpa jinchuana*, ventral view

Measurements: Forewing length 12.5–13.0 mm.

Morphology: Vertex dark brown. Ocellar triangle black. Rostrum yellowish brown. Pronotum dark brown. Mesonotum and metanotum dark brown, with a broad yellow mesal stripe. Pleura and legs yellowish brown. Wing membrane hyaline, with markings brown. In forewings, apical band broad with two large hyaline spots innerly; pterostigmal band complete, with a broad basal branch and a thin apical branch;

basal band split into two spots; marginal spot large and conspicuous; basal spot small. Markings in males more reduced than in females, especially in separated or disappeared apical branch of pterostigmal band. Male terga Ⅰ–Ⅴ dark brown. Notal organ on tergum Ⅲ flat and covering acute postnotal organ on tergum Ⅳ. A6 dark brown, cylindrical and devoid of anal horns. A7 and A8 yellowish brown and constricted basally. Male genital bulb broad oval. Epandrium emarginate terminally in a deep U-shaped and forming a pair of finger-like processes laterally. Hypandrium with an extremely shortened basal stalk. Hypovalves stripe-like, not extending to base of gonostylus, and with a row of long bristles on inner margin. Gonocoxites with a tuft of long setae on ventral apex. Gonostylus with an obtuse triangular median tooth and a disc-like subbasal process. Parameres spiral and crossed mesally, with a row of short spines on inner margin. Aedeagus with

图 3-134-3　金川蝎蛉 ***Panorpa jinchuana*** 地理分布

Fig. 3-134-3　Distribution of ***Panorpa jinchuana***

short membranous ventral valves; dorsal valves finger-like and enlarged on distal apex.

Emergence: September to October.

Distribution: Sichuan (Jinchuan, Danba).

135. 尤氏蝎蛉 *Panorpa kiautai* Zhou & Wu, 1993

【测量】 前翅长 14.0 mm。

【形态特征】 头顶深黄褐色，单眼三角区黑色；喙黄褐色。前胸背板黄褐色；中、后胸背板黑褐色，中央具 1 黄色纵带；侧板及足黄褐色。翅无色透明，斑纹褐色。前翅端带后缘有 2 ~ 3 个透明斑；痣带斜行，缺端支；基带前宽后窄；无缘斑；基斑极小。雄性腹部 1 ~ 5 节背板黑褐色。第 3 节背中突扁平，盖住第 4 节上尖锐的后背突。第 6 节末端

图3-135-1　尤氏蝎蛉 ***Panorpa kiautai*** 雄成虫（广东乳源东山林场，NWAFU）

1. 背面观　2. 侧面观

Fig. 3-135-1　Male adult of ***Panorpa kiautai*** (Dongshan Forest Farm, Ruyuan, Guangdong, NWAFU)

1. Dorsal view　2. Llateral view

图 3-135-2　尤氏蝎蛉 ***Panorpa kiautai*** 雄性外生殖器腹面观

Fig. 3-135-2　Male genitalia of ***Panorpa kiautai***, ventral view

具 1 个显著的指形臀角。第 7、8 节黄褐色，基部稍缢缩。雄性外生殖器卵形。第 9 节背板末端具深 U 形缺刻，两侧形成 1 对指形突起；腹板基柄极短；下瓣长带状，末端显著膨大，内缘具 1 列长鬃。生殖刺突内缘具钝三角形中齿和碟形亚基齿。阳基侧突粗长，稍螺旋形弯曲，相互交叉，内侧具 1 列密集的短刺毛。阳茎腹瓣膜质短小，背瓣指状分离。

【发生期】 3 ~ 4 月。

【地理分布】 浙江（遂昌、德清）、广东（乳源）。

Measurements: Forewing length 14.0 mm.

Morphology: Vertex deep yellowish brown. Ocellar triangle black. Rostrum yellowish brown. Pronotum yellowish brown. Mesonotum and metanotum dark brown, with a yellow mesal stripe. Pleura and legs yellowish brown. Wing membrane hyaline, with markings brown. In forewings, apical band with 2–3 large hyaline spots posteriorly; pterostigmal band oblique, without apical branch; basal band broad anteriorly and thin posteriorly; marginal spot absent; basal spot very small. Male terga Ⅰ–Ⅴ dark brown. Notal organ on tergum Ⅲ flat and covering acute postnotal organ on tergum Ⅳ. A6 with a distinct finger-like anal horn on dorsal apex. A7 and A8 yellowish brown and constricted basally. Male genital bulb oval. Epandrium emarginate terminally in a deep U-shape and forming a pair of finger-like processes laterally. Hypandrium with an extremely shortened basal stalk. Hypovalves stripe-like and distinctly enlarged terminally, with a row of long bristles on inner margin. Gonostylus with an obtuse triangular median tooth and a disc-like subbasal process. Parameres long stout, spiral and crossed mesally, bearing a row of dense short spines on inner margin. Aedeagus with short and membranous ventral valves and finger-like and divergent dorsal valves.

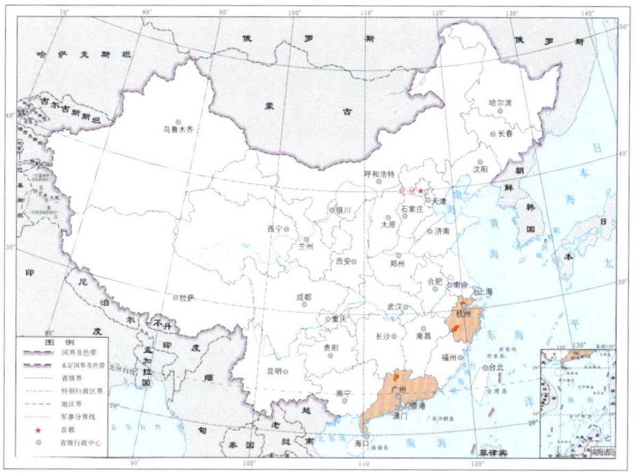

图 3-135-3 尤氏蝎蛉 *Panorpa kiautai* 地理分布

Fig. 3-135-3 Distribution of *Panorpa kiautai*

Emergence: March to April.

Distribution: Zhejiang (Suichang, Deqing) and Guangdong (Ruyuan).

136. 昆明蝎蛉 *Panorpa kunmingensis* Fu & Hua, 2009

【测量】 前翅长 10.5 ~ 12.0 mm。

【形态特征】 头顶黄色，单眼三角区黑色。喙黄色。胸部背板褐色，中央具宽阔的浅黄色纵带。侧板及足浅褐色。翅无色透明。前翅具明显的端带和散乱成许多小斑点的痣带；基带仅剩 1 个不规则小斑，位于 CuA 脉上；无缘斑和基斑。停歇时，翅呈屋脊状叠于体背。雄性腹部 1 ~ 5 节背板褐色，具黄色、向后渐细的中带。第 3 节背中突扁平，盖住第 4 节上尖锐的后背突。第 6 节背板末端中部具深 V 形缺刻，左右分裂成 2 个三角形叶片。第 7 节短小，末端向上隆起，隆起部分能嵌在第 6 节的叶片之间。雄性外生殖器长卵形。第 9 节背板末端具 U 形缺刻，两侧形成 1 对指形突起；腹板音叉状，具长基柄；下瓣较短。生殖刺突外缘强烈内凹，内缘具钝三角形中齿和大型盘状亚基齿。阳基侧突二分叉，其中腹支短细；背支叶状，向外弯曲，超过生殖肢基节端部。阳茎腹瓣膜质；背瓣骨化，较长。

【发生期】 8 ~ 10 月。

【地理分布】 云南（昆明、安宁）。

Measurements: Forewing length 10.5–12.0 mm.

Morphology: Vertex yellow. Ocellar triangle black. Rostrum yellow. Nota brown, with a broad pale yellow mesal stripe. Pleura and legs light brown. Wing membrane hyaline. In forewings, apical band

图 3-136-1　昆明蝎蛉 ***Panorpa kunmingensis*** 雄成虫侧面观（云南昆明西山）

Fig. 3-136-1　Male adult of ***Panorpa kunmingensis***, lateral view (West Hill, Kunming, Yunnan)

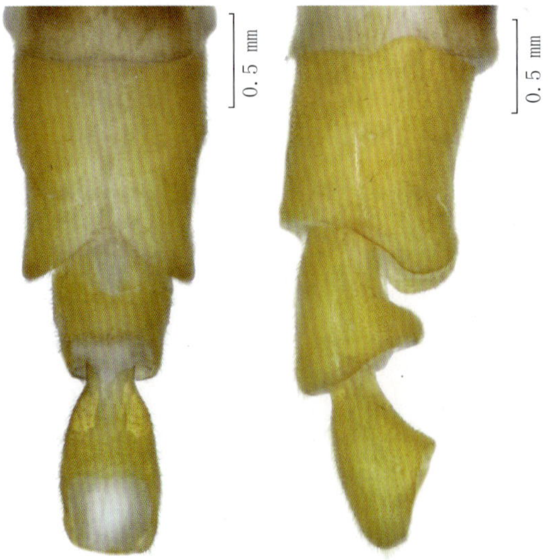

图 3-136-2　昆明蝎蛉 ***Panorpa kunmingensis*** 雄性腹部 6、7 节

1. 背面观　2. 侧面观

Fig. 3-136-2　Male A6 and A7 of ***Panorpa kunmingensis***

1. Dorsal view　2. Lateral view

图 3-136-3　昆明蝎蛉 ***Panorpa kunmingensis***

雄性外生殖器腹面观

Fig. 3-136-3　Male genitalia of ***Panorpa kunmingensis,***

ventral view

complete and distinct; pterostigmal band scattered into several small spots; basal band greatly reduced into an irregular spot on CuA; marginal and basal spots absent. Wings held roof-like at repose. Male terga I–V brown, bearing a yellow mesal stripe tapering caudad. Notal organ on tergum III flat and covering acute postnotal organ on tergum IV. A6 deeply emarginate on dorsal apex in a V-shape, and forming a pair of triangular lobes laterally. A7 very short and projected on dorsal apex, which can fit V-shaped emargination between triangular lobes on A6. Male genital bulb long oval. Epandrium emarginate terminally in a deep U-shape and forming a pair of finger-like processes laterally. Hypandrium fork-shaped with a long basal stalk. Hypovalves relatively short. Gonostylus intensively concaved along outer margin, bearing an obtuse triangular median tooth and a large disc-like subbasal process on inner margin. Paramere bifurcated with a thin and short ventral branch, and a foliate dorsal branch, which curved laterad and exceeding apex of gonocoxites. Aedeagus with membranous ventral valves and strongly sclerotized and long dorsal valves.

图 3-136-4 昆明蝎蛉 ***Panorpa kunmingensis*** 地理分布

Fig. 3-136-4 Distribution of ***Panorpa kunmingensis***

Emergence: August to October.

Distribution: Yunnan (Kunming, Anning).

137. 廖氏蝎蛉 *Panorpa liaoi* Zhou & Zhou, 2007

图 3-137-1 廖氏蝎蛉 ***Panorpa liaoi*** 雄成虫背面观（重庆南川金佛山）

Fig. 3-137-1 Male adult of ***Panorpa liaoi***, dorsal view (Mount Jinfo, Nanchuan, Chongqing)

【测量】 前翅长 10.0 ~ 12.0 mm。

【形态特征】 头、喙黄色，单眼三角区黑色。胸部背板黑褐色，具 1 条宽阔的黄色中带。侧板及足基节乳黄色，足大部分浅褐色。翅无色透明，无任何斑纹；翅痣浅棕色。停歇时屋脊状叠于体背，盖住腹部。雄性腹部 1 ~ 5 节背板黑褐色。第 3 节背中突扁平，盖住第 4 节上尖锐的后背突。第 6 节圆筒形，无臀角。第 7、8 节短筒形。雄性外生殖器长卵形。第 9 节背板末端具 U 形缺刻，两侧形成 1 对指形突起；腹板 Y 形，具较长基柄；下瓣窄长，内缘中部具 1 圆钝突起。生殖刺突外缘稍内凹，内缘近基部具粗钝中齿和大型盘状亚基齿。阳基侧突强烈骨化，分为 3 支，其中中支近端部再分 2 叉。阳茎腹瓣膜质，酒瓶状；背瓣骨化较强，侧突起钩形。

【发生期】 6 ~ 7 月。

【地理分布】 贵州（雷公山）、重庆（南川金佛山）。

图 3-137-2 廖氏蝎蛉 *Panorpa liaoi* 雄性外生殖器腹面观 [重庆（南川金佛山）]

Fig. 3-137-2 Male genitalia of *Panorpa liaoi*, ventral view (Mount Jinfo, Nanchuan, Chongqing)

Measurements: Forewing length 10.0–12.0 mm.

Morphology: Vertex and rostrum yellow. Ocellar triangle black. Nota dark brown, with a broad yellow mesal stripe. Pleura and coxae creamy yellow; legs mostly light brown. Wing membrane hyaline, devoid of markings. Pterostigma light brown. Wings held roof-like over abdomen at repose. Male terga Ⅰ–Ⅴ dark brown. Notal organ on tergum Ⅲ flat and covering acute postnotal organ on tergum Ⅳ. A6 cylindrical and devoid of anal horns. A7 and A8 short cylindrical. Male genital bulb long oval. Epandrium emarginated terminally in a deep U-shape and forming a pair of finger-like processes laterally. Hypandrium Y-shaped, with a long basal stalk. Hypovalves slender with a rounded projection medially on inner margin. Gonostylus slightly concaved along outer margin, bearing a stout median tooth and a large disc-like subbasal process on inner margin. Paramere strongly sclerotized and trifurcated, with median branch bifurcated subapically. Ventral aedeagal valves membranous and vase-shaped; dorsal valves intensively sclerotized with hook-like lateral processes.

Emergence: June to July.

Distribution: Guizhou (Mount Leigong) and Chongqing (Mount Jinfo, Nanchuan).

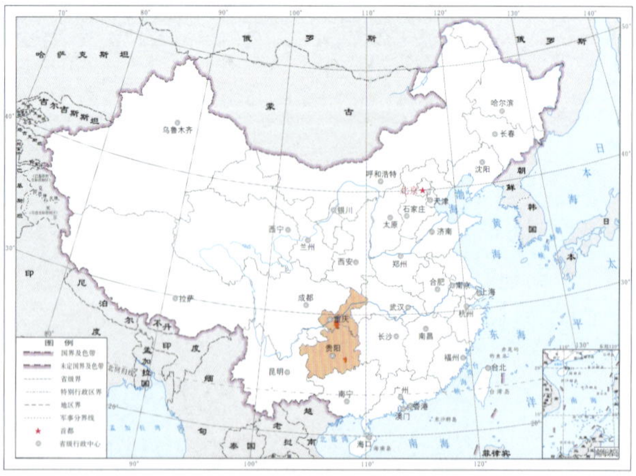

图 3-137-3 廖氏蝎蛉 *Panorpa liaoi* 地理分布

Fig. 3-137-3 Distribution of *Panorpa liaoi*

138. 刘氏蝎蛉 *Panorpa liui* Hua, 1997

【测量】 前翅长 14.0 ~ 17.5 mm。

【形态特征】 体色变异较大。大致可区分为黑型（包括春型或高海拔型）和黄型（包括夏型或低海拔型）。头顶、单眼三角区及喙黄色至黑褐色。

图 3-138-1 刘氏蝎蛉 ***Panorpa liui*** 黑色型成虫侧面观（吉林长白山）

1. 雄虫 2. 雌虫

Fig. 3-138-1 Adults of black form ***Panorpa liui***, lateral view (Changbai Mountains, Jilin)

1. Male 2. Female

图 3-138-2　刘氏蝎蛉 ***Panorpa liui*** 黄色型成虫侧背面观（黑龙江宁安镜泊湖）

1. 雄虫　2. 雌虫

Fig. 3-138-2　Adults of yellow form ***Panorpa liui***，latero-dorsal view (Jingpo Lake, Ning'an, Heilongjiang)

1. Male　2. Female

图 3-138-3 刘氏蝎蛉 ***Panorpa liui*** 黑尾型雄成虫侧面观（黑龙江宁安镜泊湖）

Fig. 3-138-3 Male adult of black form ***Panorpa liui***, lateral view (Jingpo Lake, Ning'an, Heilongjiang)

图 3-138-4 刘氏蝎蛉 ***Panorpa liui*** 黑色型雄性外生殖器腹面观

Fig. 3-138-4 Male genitalia of black form ***Panorpa liui***, ventral view

图 3-138-5 刘氏蝎蛉 ***Panorpa liui*** 黄色型雄性外生殖器腹面观

Fig. 3-138-5 Male genitalia of yellow form ***Panorpa liui***, ventral view

胸部背板黄色至黑褐色。侧板及足乳黄色至黄褐色。翅面透明，无色或淡黄色，斑纹黑褐色，变异大。一般前翅端带阔而完整，后方有1个分离的小斑或无；痣带完整，具基支和端支；缘斑明显；基带断开呈2斑；基斑呈一小点。雄性腹部第3节背中突粗壮，末端钝圆，强烈隆起，与第4节尖锐的后背突相接触。第6节长筒形，无臀角；第7、8节长筒形，基部不缢缩，向端部稍变阔。雄性外生殖器长卵形。第9节背板窄长，末端具浅V形缺刻，端部向腹面翻折；腹板Y形，基柄长；下瓣内缘近基部具1个突起，端半部具1列粗壮、排列整齐的短鬃。生殖刺突约等长于基节，内缘具阔三角形中齿和发达的亚基齿；端半部镰刀状，强烈向内侧及背侧弯曲。阳基侧突基部具长柄，中部分为2支：腹支在阳茎腹瓣下方左右交叉，末端伸达阳茎之上，膜质、膨大、多毛；背支与阳茎愈合。阳茎腹瓣短粗，向腹面隆起；背瓣短小，不明显；侧突起扁圆形。

【发生期】 5～8月。

【地理分布】 辽宁（沈阳、宽甸、千山、桓仁）、吉林（长白山）、黑龙江（镜泊湖、牡丹峰）、内蒙古（阿尔山、博克图）。

Measurements: Forewing length 14.0–17.5 mm.

Morphology: Body color highly variable. Two major forms recognized: Black form (including spring and high-elevation forms) and yellow form (including summer and low-elevation forms). Vertex, ocellar triangle and rostrum yellow to dark brown. Nota yellow to dark brown. Pleura and legs creamy yellow to yellowish brown. Wing membrane hyaline, or tinged with light yellow, with highly varied dark brown markings. In forewings, generally apical band broad and complete, and occasionally accompanied with a separated small spot posteriorly; pterostigmal band complete with a basal branch and an apical branch; marginal spot conspicuous; basal band split into two spots; basal band reduced into a small spot. Male notal organ on tergum Ⅲ stout and strongly raised with rounded apex, in contact with acute postnotal organ on tergum Ⅳ. A6 long cylindrical and devoid of anal horns. A7 and A8 long cylindrical, not constricted basally and slightly enlarged towards apex. Male genital bulb long oval. Epandrium emarginate terminally in a shallow V-shape and curved ventrad on distal portion. Hypandrium Y-shaped with a long basal stalk. Hypovalves with a projection subbasally, and a row of stout short bristles on distal half on inner margin. Gonostylus almost as long as gonocoxites, with a broad triangular median tooth and a stout subbasal process on inner margin, and sickle-shaped on distal half, which intensively curved mesad and dorsad. Paramere with a long basal stalk and bifurcated medially: ventral branch crossed below ventral aedeagal valves, with its apex extending above aedeagus, enlarged, membranous and setose; dorsal branch fused with aedeagus. Ventral aedeagal valves short and protruding ventrad; dorsal valves short and inconspicuous; lateral processes flat and rounded.

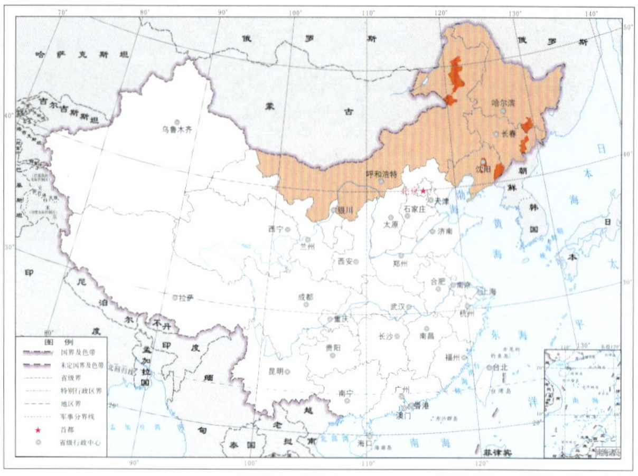

图3-138-6 刘氏蝎蛉 *Panorpa liui* 地理分布

Fig. 3-138-6 Distribution of *Panorpa liui*

Emergence: May to August.

Distribution: Liaoning (Shenyang, Kuandian, Qianshan, Huanren), Jilin (Changbai Mountains), Heilongjiang (Jingpo Lake, Mudanfeng), Inner Mongolia (Arxan, Bugt).

139. 长支蝎蛉 *Panorpa longiramina* Issiki & Cheng, 1947

【测量】 前翅长12.5 mm。

【形态特征】 头顶黑褐色。单眼三角区黑色。喙黄褐色。胸部背板黑色。侧板及足基节乳黄色，足大部分浅黄褐色。翅无色透明，翅痣褐色。在翅

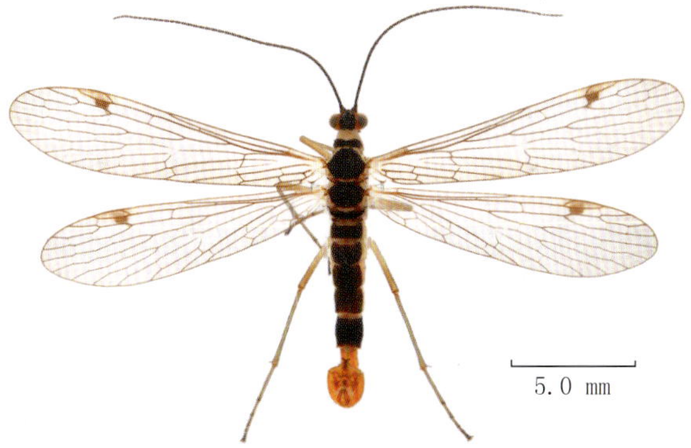

图 3-139-1　长支蝎蛉 ***Panorpa longiramina*** 雄成虫背面观（台湾台东海端，NWAFU）

Fig. 3-139-1　Male adult of ***Panorpa longiramina***, dorsal view (Haituan, Taitung, Taiwan, NWAFU)

图 3-139-2　长支蝎蛉 ***Panorpa longiramina*** 雄性外生殖器腹面观

Fig. 3-139-2　Male genitalia of ***Panorpa longiramina***, ventral view

痣后方具 1 个小斑，无其他斑纹。雄性腹部 1 ~ 5 节背板黑色。第 3 节背中突扁平，盖住第 4 节上尖锐的后背突。第 6 节黑色，筒形。第 7、8 节黄褐色，短筒形，基部不缢缩。雄性外生殖器卵形。第 9 节背板末端具深 U 形缺刻，两侧形成 1 对指形突起；腹板 Y 形，具短阔的基柄；下瓣呈钳状，短，仅伸达生殖肢基节中央。生殖刺突长约为基节之半，外缘内凹，内缘具钝三角形中齿和发达的方形亚基齿。阳基侧突分为 3 支，每支内缘具梳齿

图 3-139-3　长支蝎蛉 ***Panorpa longiramina*** 地理分布

Fig. 3-139-3　Distribution of ***Panorpa longiramina***

状长刺。阳茎腹瓣短小；背瓣阔指状，超出生殖肢基节端部。

【发生期】 5 月。

【地理分布】 台湾（南投、台东）。

Measurements: Forewing length 12.5 mm.

Morphology: Vertex dark brown. Ocellar triangle black. Rostrum yellowish brown. Nota black. Pleura and coxae creamy yellow, and legs mostly light yellowish brown. Wing membrane hyaline. Pterostigma brown. Wing markings only represented by a small spot below pterostigma. Male terga Ⅰ–Ⅴ black. Notal organ on tergum Ⅲ flat and covering acute postnotal organ on tergum Ⅳ. A6 black and cylindrical. A7 and A8 yellowish brown, short cylindrical and not constricted basally. Male genital bulb oval. Epandrium emarginate terminally in a deep U-shape and forming a pair of finger-like processes laterally. Hypandrium Y-shaped, with a short broad basal stalk. Hypovalves pincer-like, very short, only extending to middle of gonocoxite. Gonostylus almost half length of gonocoxites, concaved on outer margin, bearing an obtuse triangular median tooth and a stout subtrapezoidal subbasal process on inner margin. Paramere trifurcated, with a row of comb-like long spines on inner margin of each branch. Ventral aedeagal valves short; dorsal valves broad and finger-like, exceeding apex of gonocoxites.

Emergence: May.

Distribution: Taiwan (Nantou, Taitung).

140. 黄蝎蛉 *Panorpa lutea* Carpenter, 1945

【测量】 前翅长 17.0 ~ 17.5 mm。

【形态特征】 头顶黑红棕色，单眼三角区黑色。喙黄褐色。胸部背板黑褐色，侧板及足黄褐色。翅蜡深黄色，翅痣红棕色，斑纹黑褐色。前翅端带宽，后方有 1 个分离的三角形小斑；痣带完整，基支较宽，端支较窄；基带宽阔，呈中部缢缩的沙漏形；缘斑小，窄条形；无基斑。雄性腹部 1 ~ 5 节背板黑褐色。第 3 节背中突扁平，盖住第 4 节上尖锐的后背突。第 6 节及以后橙黄色至黄褐色。第 6 节长筒形，无臀角。第 7、8 节黄褐色，基部缢缩。雄性外生殖器卵圆形。第 9 节背板末端具深 U 形缺刻，两侧形成 1 对指形突起；腹板基柄极短；下瓣窄长，伸达生殖肢基节中部，端半部内缘具 1 列规则排列的黑鬃。生殖刺突粗壮，内缘中齿钝三角形，亚基齿三角形且尖锐。阳基侧突二分叉，腹支长约为背支之半；背支外缘有 1 列短刺毛。阳茎粗壮，不超过生殖肢基节端部。

图 3-140-1 黄蝎蛉 *Panorpa lutea* 成虫背面观（浙江临安西天目山，NWAFU）
1. 雄虫 2. 雌虫
Fig. 3-140-1 Adults of *Panorpa lutea*, dorsal view (West Tianmu Mountain, Lin'an, Zhejiang, NWAFU)
1. Male 2. Female

图 3-140-2 黄蝎蛉 *Panorpa lutea* 雄性外生殖器腹面观
Fig. 3-140-2 Male genitalia of *Panorpa lutea*, ventral view

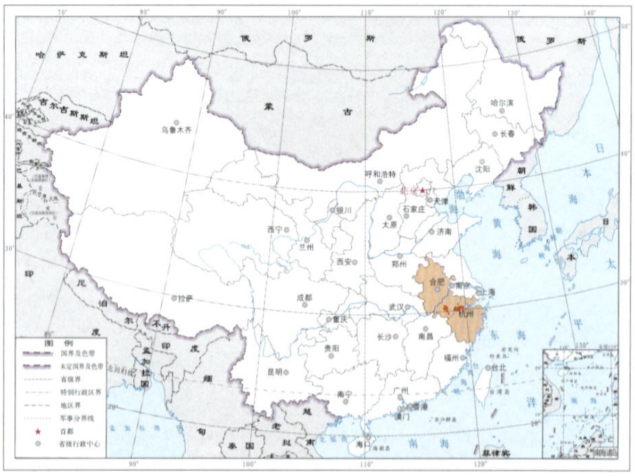

图 3-140-3 黄蝎蛉 *Panorpa lutea* 地理分布
Fig. 3-140-3 Distribution of *Panorpa lutea*

【发生期】 8 ~ 10 月。

【地理分布】 安徽（黄山）、浙江（天目山）。

Measurements: Forewing length 17.0–17.5 mm.

Morphology: Vertex dark reddish brown. Ocellar triangle black. Rostrum yellowish brown. Nota dark brown. Pleura and legs yellowish brown. Wing membrane deep wax yellow. Pterostigma reddish brown. Markings blackish brown. In forewings, apical band broad and accompanied by a separated small triangular spot posteriorly; pterostigma complete, with a broad basal branch and a thin apical branch; basal band broad and constricted medially in an hourglass-shaped; marginal spot thin and stripe-like; basal spot absent. Male terga Ⅰ–Ⅴ dark brown. Notal organ on tergum Ⅲ flat and covering acute postnotal organ on tergum Ⅳ. A6 and succeeding segments orange-yellow to yellowish brown. A6 cylindrical, devoid of anal horns. A7 and A8 yellowish brown and constricted basally. Male genital bulb oval. Epandrium emarginate terminally in a U-shape and forming a pair of finger-like processes laterally. Hypandrium with a greatly shortened basal stalk. Hypovalves slender, extending to middle of gonocoxite, and bearing a row of regularly arranged stout bristles on distal half of inner margin. Gonostylus stout, bearing an obtuse triangular median tooth and an acute and triangular subbasal process on inner margin. Paramere bifurcated with a short ventral branch, and a long dorsal branch. Dorsal branch twice as long as ventral branch, and bearing a row of short spines along outer margin. Aedeagus stout, not exceeding apex of gonocoxites.

Emergence: August to October.

Distribution: Anhui (Mount Huangshan) and Zhejiang (Tianmu Mountains).

141. 长蝎蛉 *Panorpa macrostyla* Hua, 1997

图 3-141-1 长蝎蛉 *Panorpa macrostyla* 雄成虫侧面观（吉林长白山）

Fig. 3-141-1 Male adult of *Panorpa macrostyla*, lateral view (Changbai Mountains, Jilin)

图 3-141-2 长蝎蛉 *Panorpa macrostyla* 雄性外生殖器腹面观
Fig. 3-141-2 Male genitalia of *Panorpa macrostyla*, ventral view

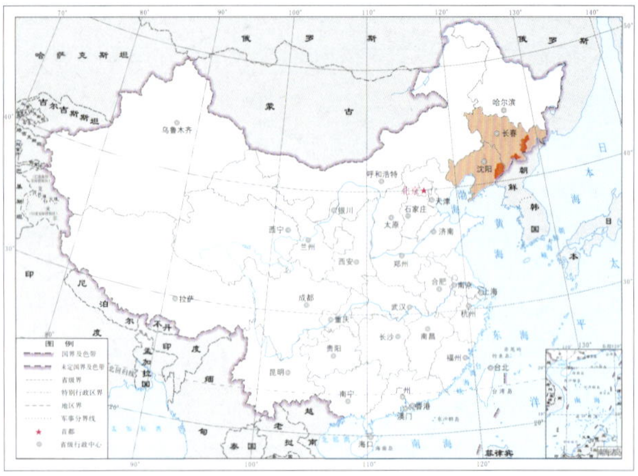

图 3-141-3 长蝎蛉 *Panorpa macrostyla* 地理分布
Fig. 3-141-3 Distribution of *Panorpa macrostyla*

【测量】 前翅长 10.0 ~ 12.0 mm。

【形态特征】 头顶、单眼三角区黑色，喙黄褐色。胸部背板褐色，具宽阔的黄色中带。侧板及足乳黄色。翅浅黄色半透明，斑纹黑褐色。前翅端带、痣带、基带形成散乱的小斑点，变异较大。停歇时，翅呈屋脊状覆于体背盖住腹部。雄性腹部 1 ~ 5 节背板黄褐色。第 3 节背中突扁平，盖住第 4 节上尖锐的后背突。第 6 节长圆筒形，无臀角。第 7、8 节长圆筒形，基部不缢缩。雄性外生殖器卵形。第 9 节背板阔长，末端圆钝；腹板 Y 形，基柄窄长；下瓣相互紧贴，末端稍分离。生殖刺突长度超过基节，长镰刀状，内缘中齿不明显，亚基齿发达。阳基侧突 2 分叉，腹支在阳茎基部左右交叉。阳茎稍超出生殖肢基节端部。

【发生期】 7 ~ 8 月。

【地理分布】 吉林（长白山、临江）、辽宁（宽甸、桓仁）。

Measurements: Forewing length 10.0–12.0 mm.

Morphology: Vertex and ocellar triangle black. Rostrum yellowish brown. Nota dark brown with a broad yellow mesal stripe. Pleura and legs creamy yellow. Wing membrane translucent, tinged with light yellow. Markings dark brown. In forewings, apical, pterostigmal and basal bands greatly reduced into a series of scattered spots with high variations. Wings held roof-like over abdomen at repose. Male terga Ⅰ–Ⅴ yellowish brown. Notal organ on tergum Ⅲ flat and covering acute postnotal organ on tergum Ⅳ. A6 long cylindrical and devoid of anal horns. A7 and A8 long cylindrical and not constricted basally. Male genital bulb oval. Epandrium long, broad and rounded at apex. Hypandrium Y-shaped and with a long narrow basal stalk. Hypovalves adjoining mesally and slightly divergent apically. Gonostylus longer than gonocoxites, sickle-shaped and bearing an inconspicuous median tooth and a stout subbasal process on inner margin. Paramere bifurcated, with their ventral branches crossed near base of aedeagus. Aedeagus slightly exceeding apex of gonocoxites apically.

Emergence: July to August.

Distribution: Jilin (Changbai Mountains, Linjiang) and Liaoning (Kuandian, Huanren).

142. 莽山蝎蛉 *Panorpa mangshanensis* Chou & Wang, 1987

【测量】 前翅长 13.5 ~ 14.0 mm。

【形态特征】 头顶、单眼三角区黑色。喙黄褐色。胸部背板黑色，小盾片黄褐色。侧板及足黄褐色。翅无色透明，斑纹褐色。前翅端带完整，后缘具 1 个分离的模糊小斑；痣带与端带在前缘相接，具宽阔的基支和狭窄的端支；缘斑和基斑极小；基

图 3-142-1 莽山蝎蛉 *Panorpa mangshanensis* 雄成虫背面观（外生殖器已去除，湖南莽山，NWAFU）

Fig. 3-142-1 Male adult of *Panorpa mangshanensis* with gentalia removed, dorsal view (Mount Mangshan, Hunan, NWAFU)

图 3-142-2 莽山蝎蛉 *Panorpa mangshanensis* 雄性外生殖器腹面观

Fig. 3-142-2 Male genitalia of *Panorpa mangshanensis*, ventral view

带完整。雄性腹部 1 ~ 5 节背板黑褐色。第 3 节背中突扁平，盖住第 4 节上尖锐的后背突。第 6 节黑褐色，筒形，无臀角。第 7、8 节黄褐色，基部稍缢缩。雄性外生殖器阔卵形。第 9 节背板末端具深 U 形缺刻，两侧形成 1 对指形突起；腹板具狭窄基柄；下瓣纤细，仅稍超过生殖肢基节中部。生殖肢基节内缘端部具 1 簇长刚毛，中部具 1 向后的指形片状突起；生殖刺突内缘具钝三角形中齿和大型盘状亚基齿。阳基侧突叶状，伸达生殖刺突中部，内缘具 1 列梳齿状短刺毛。阳茎短小；腹瓣不明显；背瓣指状，相互分离。

【发生期】 11 月。

【地理分布】 湖南（莽山、永兴）。

图 3-142-3 莽山蝎蛉 *Panorpa mangshanensis* 地理分布

Fig. 3-142-3 Distribution of *Panorpa mangshanensis*

Measurements: Forewing length 13.5–14.0 mm.

Morphology: Vertex and ocellar triangle black. Rostrum yellowish brown. Nota black with yellowish brown scutella. Pleura and legs yellowish brown. Wing membrane hyaline, with markings brown. In forewings, apical band complete and accompanied by a separated faint small spot posteriorly; pterostigmal band connected to apical band anteriorly, and with a broad basal branch and a thin apical branch; marginal and basal spots very small; basal band complete. Male terga Ⅰ–Ⅴ dark brown. Notal organ on tergum Ⅲ flat and covering acute postnotal organ on tergum Ⅳ. A6 dark brown, cylindrical and devoid of anal horns. A7 and A8 yellowish brown and constricted basally. Male genital bulb broad oval. Epandrium emarginate terminally in a deep U-shape and forming a pair of finger-like processes laterally. Hypandrium with a narrow basal stalk. Hypovalves slender and slightly exceeding middle of gonocoxites. Gonocoxites with a tuft of long setae on ventral apex, and a finger-like plate directing caudad on inner margin. Gonostylus with an obtuse triangular median tooth and a large disc-like subbasal process on inner margin. Paramere foliate, extending to middle of gonostylus, and bearing a row of comb-like spines on inner margin. Aedeagus short; ventral valves inconspicuous; dorsal valves finger-like and divergent.

Emergence: November.

Distribution: Hunan (Mount Mangshan, Yongxing).

143. 新刺蝎蛉 *Panorpa neospinosa* Chou & Wang, 1981

【测量】 前翅长 9.5 ~ 11.0 mm。

【形态特征】 头顶黄褐色。单眼三角区黑色，被 1 条黑色横带贯穿。喙黄褐色。胸部背板黑褐色，中、后胸具显著的黄色纵带。侧板及足黄褐色。翅无色透明，斑纹深褐色。前翅端带后部具 1 个透明斑；痣带完整，端支窄于基支；基带宽阔，完整或断成 2 斑；缘斑和基斑大而显著。雄性腹部 1 ~ 5 节背板黑褐色。第 3 节背中突扁平，盖住第 4 节上尖锐的后背突。第 6 节黄褐色，末端向上隆起，具稀疏的长毛。第 7、8 节黄褐色，基部稍缢缩。雄性外生殖器圆球形。第 9 节背板末端具深 U 形缺刻，两侧形成 1 对指形突起；腹板基柄极短；下瓣阔，长带状，分歧，几乎伸达生殖肢基节末端，内缘具 1 列长鬃。生殖肢基节端部内缘有 5 ~ 6 根刚毛；生殖刺突内缘具钝三角形中齿和大型碟状亚基齿。阳基侧突具长带状、螺旋形弯曲，相互交叉，内缘有密集的细毛。阳茎腹瓣膜质不发达，背瓣指状分离且近端部背面有膜质膨大区。

【发生期】 7 ~ 8 月。

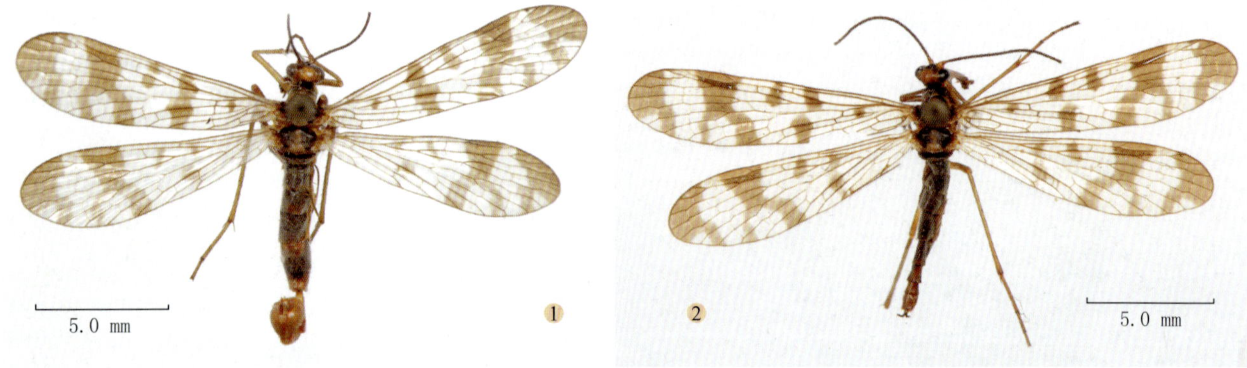

图 3-143-1 新刺蝎蛉 *Panorpa neospinosa* 成虫背面观（陕西宝鸡秦岭车站，NWAFU）

1. 雄虫 2. 雌虫

Fig. 3-143-1 Adults of *Panorpa neospinosa*, dorsal view (Qinling Raiway Station, Baoji, Shaanxi, NWAFU)

1. Male 2. Female

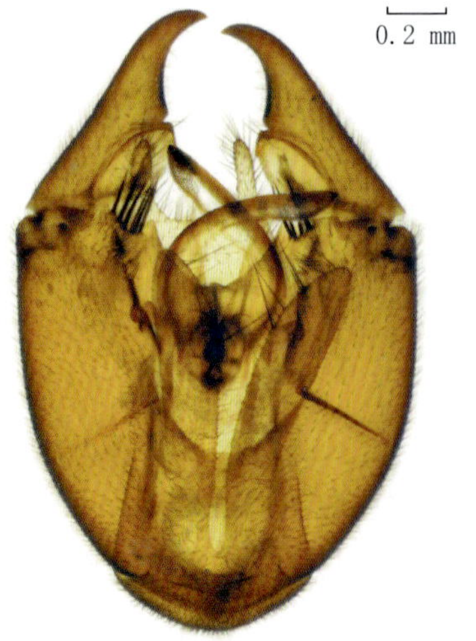

图 3-143-2 新刺蝎蛉 *Panorpa neospinosa* 雄性外生殖器腹面观

Fig. 3-143-2 Male genitalia of *Panorpa neospinosa*, ventral view

图 3-143-3 新刺蝎蛉 *Panorpa neospinosa* 地理分布

Fig. 3-143-3 Distribution of *Panorpa neospinosa*

【地理分布】 陕西（宝鸡、周至）、甘肃（文县）。

Measurements: Forewing length 9.5–11.0 mm.

Morphology: Vertex and rostrum yellowish brown. Ocellar triangle black. A broad black transverse stripe passing through ocellar triangle. Nota dark brown, with a conspicuous yellow mesal stripe on Mesonotum and metanotum. Pleura and legs yellowish brown.Wing membrane hyaline, with markings brown. In forewings, apical band complete, bearing a hyaline spot posteriorly; pterostigmal band complete, with a thin apical branch and a broad basal branch; basal band broad, complete or split into two spots; marginal and basal spots large and conspicuous. Male terga Ⅰ–Ⅴ dark brown. Notal organ on tergum Ⅲ flat and covering acute postnotal organ on tergum Ⅳ. A6 yellowish brown, projected dorsad and setose on apex. A7 and A8 yellowish brown and slightly constricted basally. Male genital bulb globular. Epandrium emarginate terminally in a deep U-shape and forming a pair of finger-like processes laterally. Hypandrium with a greatly shortened basal stalk. Hypovalves long broad stripe-like, divergen, almost extending to apex of gonocoxite, bearing a row of long bristles along inner margin. Gonocoxites with 5–6 stout setae on ventral apex. Gonostylus with an obtuse triangular median tooth and a large disc-like subbasal process on inner margin. Parameres long stripe-like and spiral, crossed mesally, and with dense short spines on inner margins. Ventral aedeagal valves membranous and poorly-developed; dorsal valves finger-like, divergent distally, and each with a membranous enlarged process on subapical portion dorsally.

Emergence: July to August.

Distribution: Shaanxi (Baoji, Zhouzhi) and Gansu (Wenxian).

144. 斜带蝎蛉 *Panorpa obliquifascia* Chou & Wang, 1987

【测量】 前翅长 12.5 ~ 13.0 mm。

【形态特征】 头顶、单眼三角区黑色，喙黄褐色。胸部背板黑褐色。侧板及足乳黄色。翅无色透明，斑纹褐色。前翅端带散乱，内缘分出 1 条窄带；痣带基支发达，明显形成 1 条斜带，端支分离、模糊；缘斑小；基带在后缘形成一系列斑纹；无基斑。雄性腹部 1 ~ 5 节背板黑褐色。第 3 节背中突扁平，盖住第 4 节上尖锐的后背突。第 6 节黑褐色，短筒形，无臀角。第 7、8 节黄褐色，短筒形，基部不缢缩。雄性外生殖器长椭圆形。第 9 节背板末端具深 U 形缺刻，两侧形成 1 对指形突起；腹板具较长基柄，Y 形；下瓣宽带状，伸过生殖肢基节末端，内缘具 1 列长鬃。生殖刺突长不及基节之半，内缘具钝三角形中齿和大型碗状亚基齿。阳基侧突分 2 叉，腹支较短，无毛；背支伸达生殖刺突中齿，内

图 3-144-1 斜带蝎蛉 *Panorpa obliquifascia* 雄成虫背面观

（广东南岭，NWAFU）

Fig. 3-144-1 Male adult of ***Panorpa obliquifascia***, dorsal view

(Nanling Mountains, Guangdong, NWAFU)

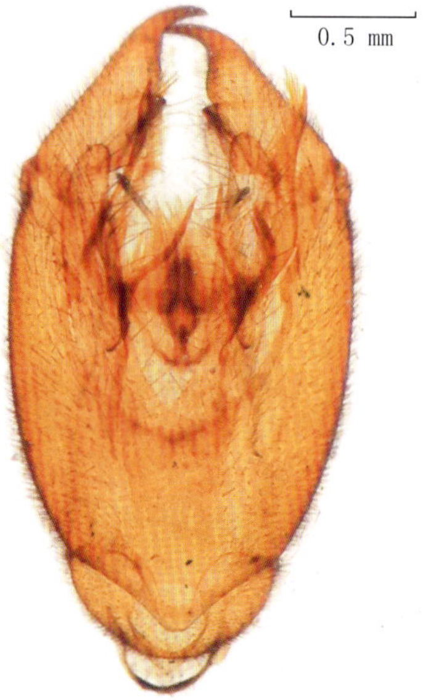

图 3-144-2 斜带蝎蛉 *Panorpa obliquifascia* 雄性外生殖器腹面观

Fig. 3-144-2 Male genitalia of ***Panorpa obliquifascia***, ventral view

缘具梳齿状长刺毛。阳茎短小，不超过生殖肢基节端部。

【发生期】 5月。

【地理分布】 湖南（莽山）、广东（南岭）。

Measurements: Forewing length 12.5–13.0 mm.

Morphology: Vertex and ocellar triangle black. Rostrum yellowish brown. Nota dark brown. Pleura and legs creamy yellow. Wing membrane hyaline, with markings brown. In forewings, apical band scattered with a separated thin stripe near inner margin; pterostigmal band with a stout basal branch and a separated faint apical branch, forming a conspicuous oblique band; marginal spot small; basal band reduced, forming a series of markings along hind margin; basal spot absent. Male terga Ⅰ–Ⅴ dark brown. Notal organ on tergum Ⅲ flat and covering acute postnotal organ on tergum Ⅳ. A6 dark brown, short cylindrical and devoid of anal horns. A7 and A8 yellowish brown, short cylindrical and not constricted basally. Male genital bulb long elliptical. Epandrium emarginate terminally in a deep U-shape and forming a pair of finger-like processes laterally. Hypandrium with a long basal stalk and Y-shape. Hypovalves long broad stripe-like, extending beyond apex of gonocoxite, and bearing a row of long bristles along inner margin. Gonostylus short, not half as long as gonocoxites, and bearing an obtuse triangular median tooth and a large bowl-shaped subbasal process on inner margin. Paramere bifurcated, with ventral branch short and glabrous; and dorsal branch long and bearing long comb-like spines on inner margin. Aedeagus short and not exceeding apex of gonocoxites.

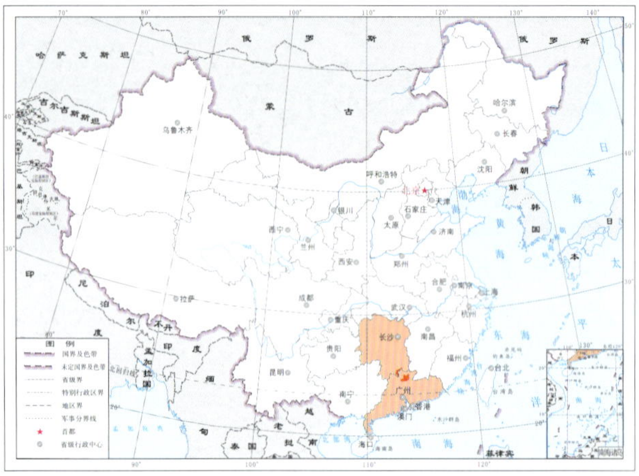

图 3-144-3 斜带蝎蛉 *Panorpa obliquifascia* 地理分布

Fig. 3-144-3 Distribution of ***Panorpa obliquifascia***

Emergence: May.

Distribution: Hunan (Mount Mangshan) and Guangdong (Nanling Mountains).

145. 赭尾蝎蛉 *Panorpa ochraceocauda* Issiki, 1927

图 3-145-1 赭尾蝎蛉 *Panorpa ochraceocauda* 雄成虫（台湾，林义祥摄）

1. 正在取食蛾类尸体，背面观 2. 头部正面观

Fig. 3-145-1 Male adult of ***Panorpa ochraceocauda*** (Taiwan, photo by Yechan Lin)

1. Feeding a dead moth, dorsal view 2. Frontal view of head

【测量】 前翅长 14.0 ~ 15.0 mm。

【形态特征】 头顶中央黑褐色，两侧接近复眼外缘处黄褐色。单眼三角区黑色。喙黄褐色，两侧稍浅。胸部背板黑褐色。侧板及足乳黄色。翅无色透明，仅翅痣处具1黑色斑点。雄性腹部1 ~ 5节黑褐色。第3节背中突扁平，盖住第4节上尖锐的后背突。第6节黑褐色，短筒形，无臀角。第7、8节黄褐色，短筒形，基部不缢缩。雄性外生殖器卵形。第9节背板末端具深U形缺刻，两侧形成1对指形突起；腹板具较长基柄，Y形；下瓣钳形。生殖刺突长仅为基节之半，内缘具钝三角形中齿和碟形亚基齿。阳基侧突短，不超出生殖肢基节端部，分为2支，每一支均着生1列刺毛。阳茎短而尖锐。

【发生期】 5 ~ 7月。

【地理分布】 台湾。

Measurements: Forewing length 14.0–15.0 mm.

Morphology: Vertex dark brown at middle, and yellowish brown near compound eyes at both sides. Ocellar triangle black. ocellar triangle black. Rostrum yellowish brown with lighter genae. Nota dark brown. Pleura and legs creamy yellow. Wing membrane hyaline. Markings only represented by a small spot near pterostigma. Male terga Ⅰ–Ⅴ dark brown. Notal organ on tergum Ⅲ flat and covering acute postnotal organ on tergum Ⅳ. A6 dark brown, short cylindrical and devoid of anal horns. A7 and A8 yellowish brown, short cylindrical and not constricted basally. Male genital bulb oval. Epandrium emarginate terminally in a U-shape and forming a pair of finger-like processes laterally. Hypandrium with a long basal stalk and Y-shape. Hypovalves pincer-like. Gonostylus half length of gonocoxites, and bearing an obtuse triangular median tooth and a disc-shaped subbasal process on inner margin. Paramere short, not exceeding apex of gonocoxites, bifurcated and bearing a row of spines on each branch. Aedeagus short and acute.

Emergence: May to July.

Distribution: Taiwan.

图 3-145-2 赭尾蝎蛉 *Panorpa ochraceocauda* 地理分布

Fig. 3-145-2 Distribution of ***Panorpa ochraceocauda***

146. 平支蝎蛉 *Panorpa parallela* Wang & Hua, 2016

【测量】 前翅长 10.0 ~ 11.3 mm。

【形态特征】 头顶、喙黄色，单眼三角区黑色。胸部背板棕色，具1条宽阔的黄色中带。侧板及足黄褐色。翅略带黄色，透明。翅痣红褐色。前翅端带退化，仅存不明显的短条带；痣带为不规则斑，后侧尖。停歇时，左右翅沿背中线紧靠，盖住腹部。雄性腹部1 ~ 5节背板污黄褐色，中央具1浅色条带。第3节背中突扁平，盖住第4节上尖锐的后背突。第6节背部黑褐色，腹面黄褐色，筒形，无臀角。第7、8节黄褐色。第7节短筒形，第8节基部缢缩。雄性外生殖器卵形。第9节背板末端具U形缺刻，两侧形成1对指形突起；腹板Y形，具较长窄基柄；下瓣较短，不伸达生殖肢基节末端。生殖刺突内缘具钝三角形中齿和发达的碟形亚基齿。阳基侧突棍棒状，直，伸达生殖刺突中部，端部1/3密被短毛，末端钩状，尖锐。阳茎腹瓣不发达，膜质；背瓣强烈骨化，端部呈熨斗状膨大。

【发生期】 7 ~ 8月。

【地理分布】 云南（永德、怒江）。

Measurements: Forewing length 10.0–11.3 mm.

Morphology: Vertex and rostrum yellow. Ocellar triangle black. Nota brown, with a broad yellow mesal stripe. Pleura and legs yellowish brown. Wing membrane hyaline, and slightly tinged with yellow. Pterostigma reddish brown. In forewings, apical band greatly reduced into an inconspicuous short stripe; pterostigmal band reduced into an irregular spot

图 3-146-1　平支蝎蛉 ***Panorpa parallela*** 雄成虫侧背面观（云南永德大雪山）

Fig. 3-146-1　Male adult of ***Panorpa parallela***, dorso-lateral view (Daxue Mountain, Yongde, Yunnan)

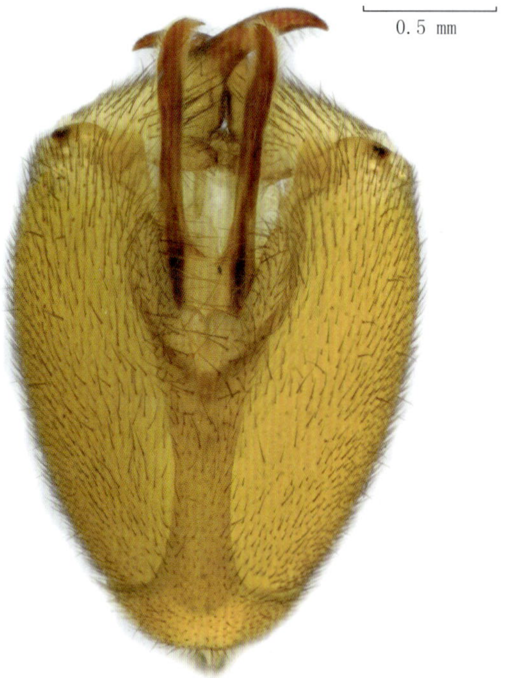

图 3-146-2　平支蝎蛉 ***Panorpa parallela*** 雄性外生殖器腹面观

Fig. 3-146-2　Male genitalia of ***Panorpa parallela***, ventral view

with acute hind apex. Wings held in close contact along mid-line over abdomen at repose. Male terga Ⅰ–Ⅴ sordidly yellowish brown, with a pale mesal stripe. Notal organ on tergum Ⅲ flat and covering postnotal organ on tergum Ⅳ. A6 dark brown dorsally and yellowish brown ventrally, cylindrical and devoid

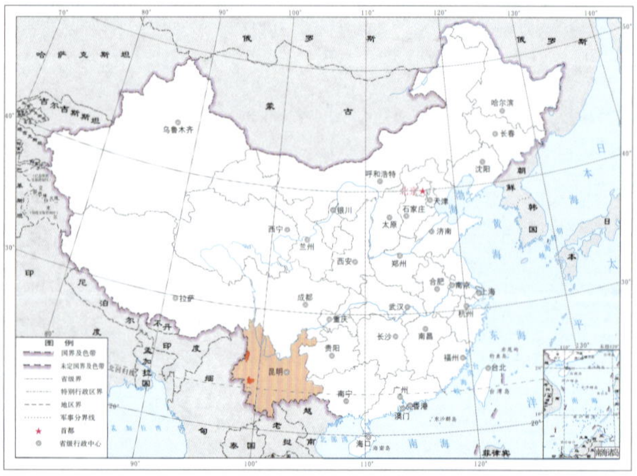

图 3-146-3　平支蝎蛉 ***Panorpa parallela*** 地理分布

Fig. 3-146-3　Distribution of ***Panorpa parallela***

of anal horns. A7 and A8 yellowish brown. A7 short cylindrical, A8 constricted basally. Male genital bulb oval. Epandrium emarginate terminally in a U-shape and forming a pair of finger-like processes laterally. Hypandrium with a long narrow basal stalk and Y-shaped. Hypovalves relatively short, not extending to apex of gonocoxite. Gonostylus with an obtuse triangular median tooth and a large disc-like subbasal process on inner margin. Paramere stick-like, extending to median tooth of gonostylus, acute and hooked on apex, and bearing numerous short spines on distal third. Ventral aedeagal valves short and membranous; dorsal valves intensively sclerotized, enlarged and flatiron-shaped distally.

Emergence: July to August.

Distribution: Yunnan (Yongde, Nujiang).

147. 栉蝎蛉 *Panorpa pectinata* Issiki, 1929

【测量】 前翅长 11.5 ~ 12.5 mm。

【形态特征】 头顶黄褐色。单眼三角区黑色。1 条宽阔的黑色横带贯穿单眼三角区并伸达复眼内缘。喙黄褐色。前胸背板黑褐色。中、后胸背板褐色，具 1 条显著的黄色中带。侧板及足浅黄褐色。翅稍带黄色，透明。斑纹褐色。前翅端带内部具 2 ~ 3 个透明小斑，后缘具 1 分离的条带；痣带纤细，具基支和端支；缘斑细长；基带分为 2 斑；无基斑。雄性腹部 1 ~ 5 节背板黑色。第 3 节背中突扁平，盖住第 4 节上尖锐的后背突。第 6 节黑褐色，锥形。第 7、8 节黄褐色。第 7 节基部强烈缢缩成柄状；第 8 节基部仅稍缢缩。雄性外生殖器长卵形。第 9 节背板末端具深 U 形缺刻，两侧形成 1 对指形突起；腹板基柄极短；下瓣细长，带状，伸达生殖肢基节中部，内缘具 1 列长鬃。生殖刺突内缘具钝三角形中齿和大型碟状亚基齿。阳基侧突短粗，近基部具 1 钩状突起，末端 2 分叉，内缘具 1 列梳齿状长刺。阳茎短小扁平。

图 3-147-1 栉蝎蛉 *Panorpa pectinata* 雄成虫背面观（台湾台东海端，NWAFU）

Fig. 3-147-1 Male adult of *Panorpa pectinata*, dorsal view (Haituan, Taitung, Taiwan, NWAFU)

【发生期】 7 ~ 8 月。

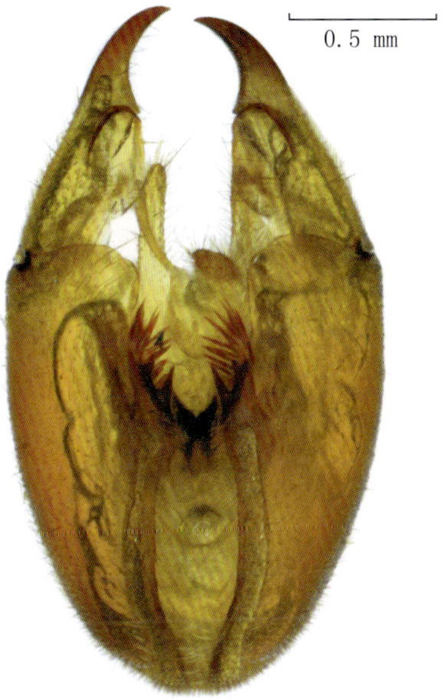

图 3-147-2 栉蝎蛉 *Panorpa pectinata* 雄性外生殖器腹面观

Fig. 3-147-2 Male genitalia of *Panorpa pectinata*,ventral view

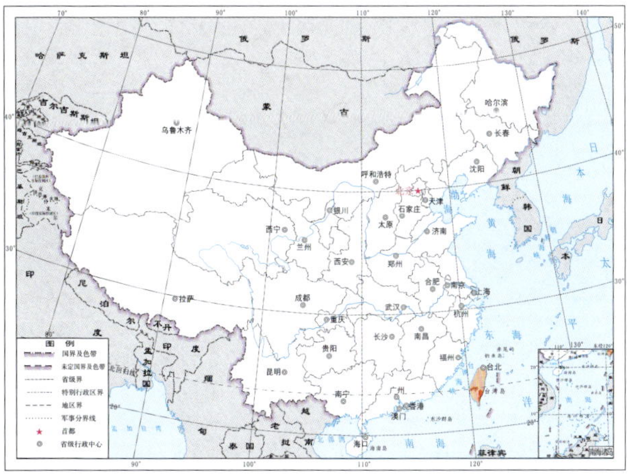

图 3-147-3 栉蝎蛉 *Panorpa pectinata* 地理分布

Fig. 3-147-3 Distribution of *Panorpa pectinata*

【地理分布】 台湾（阿里山、台东）。

Measurements: Forewing length 11.5–12.5 mm.

Morphology: Vertex yellowish brown. Ocellar triangle black. A broad black transverse stripe passing through ocellar triangle and extending to inner borders of compound eyes. Rostrum yellowish brown. Pronotum dark brown. Mesonotum and metanotum brown, with a conspicuous broad yellow mesal stripe. Pleura and legs yellowish brown. Wing membrane hyaline and slightly tinged with yellow. Markings brown. In forewings, apical band with 2–3 hyaline spots innerly, and a separated stripe posteriorly; pterostigmal band slender with a basal branch and an apical branch; marginal spot slender; basal band split into two spots; basal spot absent. Male terga Ⅰ–Ⅴ dark brown. Notal organ on tergum Ⅲ flat and covering postnotal organ on tergum Ⅳ. A6 dark brown, conical. A7 and A8 yellowish brown. A7 intensively stalked basally. A8 slightly constricted basally. Male genital bulb oval. Epandrium emarginate terminally in a deep U-shape and forming a pair of finger-like processes laterally. Hypandrium with an extremely shortened basal stalk. Hypovalves slender and stripe-like, extending to mid-length of gonocoxite, and bearing a row of long bristles on inner margin. Gonostylus with an obtuse triangular median tooth and a large disc-like subbasal process on inner margin. Paramere short and stout, with a hook-like process subbasally and bifurcated apically, and bearing a row of long comb-like spines on inner margin. Aedeagus short and flat.

Emergence: July to August.

Distribution: Taiwan (Ali Mountain, Taitung).

148. 彼得蝎蛉 *Panorpa peterseana* Issiki, 1929

【测量】 前翅长 11.5 ~ 13.0 mm。

【形态特征】 头顶黄色，单眼三角区黑色。后头具三叉形褐色斑纹。喙红棕色。胸部背板黑褐色，侧板及足浅黄褐色。翅无色透明，斑纹褐色。前翅端带形成一系列散乱的小斑点；痣带细长，基支完整，端支不明显；基带断开为 2 斑，其中后方的斑较大且与痣带的基支在后缘相连；缘斑小；无基斑；臀区后缘具一系列不规则斑。雄性腹部 1 ~ 5 节背板黑褐色。第 3 节背中突扁平，盖住第 4 节上尖锐的后背突。第 6 节黑褐色，短筒形，无臀角。第 7、8 节红褐色，短筒形，基部不缢缩。雄性外生殖器卵形。第 9 节背板末端具深 U 形缺刻，两侧形成 1 对指形突起；腹板基柄宽阔；下瓣钳状，几乎伸达生殖肢基节末端。生殖刺突内缘具钝三角

图 3-148-1 彼得蝎蛉 ***Panorpa peterseana*** 雄成虫背面观（台湾屏东大汉山，NWAFU）

Fig. 3-148-1 Male adult of ***Panorpa peterseana***, dorsal view (Mount Tahan, Pingtung, Taiwan, NWAFU)

图 3-148-2 彼得蝎蛉 ***Panorpa peterseana*** 雄性外生殖器腹面观

Fig. 3-148-2 Male genitalia of ***Panorpa peterseana***, ventral view

形中齿和方形亚基齿。阳基侧突分为 2 支，每一支内缘着生 1 列梳齿状长刺，其中背支极长，伸达生殖刺突中部。阳茎腹瓣短，背瓣指状。

【发生期】 3 ~ 5 月。

【地理分布】 台湾（南投、屏东、台东）。

Measurements: Forewing length 11.5–13.0 mm.

Morphology: Vertex yellow. Ocellar triangle black. A trident brown pattern on occiput. Rostrum reddish brown. Nota dark brown. Pleura and legs light yellowish brown. Wing membrane hyaline. Markings brown. In forewings, apical band reduced into several scattered spots; pterostigmal band slender, with a complete basal branch and an inconspicuous apical branch; basal band split into two spots, with posterior spot connected to basal branch of pterostigmal band posteriorly; marginal spot small; basal spot absent; a series of irregular markings on hind margin of anal region. Male terga Ⅰ–Ⅴ dark brown. Notal organ on tergum Ⅲ flat and covering postnotal organ on tergum Ⅳ. A6 dark brown, short cylindrical and devoid of anal horns. A7 and A8 reddish brown, short cylindrical, and not constricted basally. Male genital bulb oval. Epandrium emarginate terminally in a U-shape and forming a pair of finger-like processes

图 3-148-3 彼得蝎蛉 *Panorpa peterseana* 地理分布

Fig. 3-148-3 Distribution of ***Panorpa peterseana***

laterally. Hypandrium with an short broad basal stalk. Hypovalves pincer-like, almost extending to apex of gonocoxite. Gonostylus with an obtuse triangular median tooth and a subtrapezoidal subbasal process on inner margin. Paramere bifurcated, with a row of long comb-like spines on inner margin, dorsal branch long and extending to middle of gonostylus. Aedeagus with short ventral valves and finger-like dorsal valves.

Emergence: March to May.

Distribution: Taiwan (Nantou, Pingtung, Taitung).

149. 平江蝎蛉 *Panorpa pingjiangensis* Chou & Wang, 1987

【测量】 前翅长 13.5 ~ 15.0 mm。

【形态特征】 头顶、单眼三角区黑色，喙红棕色。胸部背板黑褐色，侧板及足黄褐色。翅无色透明，翅痣红棕色，斑纹黑棕色。前翅端带完整；痣带与端带在前缘处联合，端枝模糊且断开；基带完整；缘斑条状；无基斑。雄性腹部 1 ~ 5 节背板黑褐色。第 3 节背中突扁平，盖住第 4 节上尖锐的后背突。第 6 节黑褐色，锥状，无臀角。第 7、8 节红棕色，基部缢缩。雄性外生殖器阔卵形。第 9 节背板末端具深 U 形缺刻，两侧形成 1 对指形突起；腹板具短基柄；下瓣细长，伸达生殖肢基节中部，内缘具 1 列细长刚毛。生殖肢基节内缘中部具 1 个宽阔的片状突起，端部内缘具尖锐的长指形延伸；生殖刺突内缘中齿三角形不明显，近基部具大型盘状亚基齿。阳基侧突粗壮，分为几乎等长的 2 支，伸达生殖刺突的亚基齿，每一支内缘着生许多短刺。阳茎短小，背瓣指状、分离。

【发生期】 10 ~ 11 月。

【地理分布】 湖南（平江）、江西（庐山）。

Measurements: Forewing length 13.5–15.0 mm.

Morphology: Vertex and ocellar triangle black. Rostrum reddish brown. Nota dark brown. Pleura and legs yellowish brown. Wing membrane hyaline, with markings dark brown. Pterostigma reddish brown. In forewings, apical band complete; pterostigmal band connected to apical band anteriorly, and with a faint separated apical branch; marginal spot stripe-like; basal band complete; basal spots absent. Male terga Ⅰ–Ⅴ dark brown. Notal organ on tergum Ⅲ flat and covering postnotal organ on tergum Ⅳ. A6 dark brown, conical and devoid of anal horns. A7 and A8 reddish brown and constricted basally. Male genital bulb broad oval. Epandrium emarginate terminally in a deep U-shape and forming a pair of finger-like processes laterally. Hypandrium with a short basal stalk. Hypovalves slender, extending to middle of gonocoxite, bearing a row of long bristles on inner margin. Gonocoxites with a broad plate on middle of ventro-inner margin, and a long finger-like process on ventral apex. Gonostylus with an inconspicuous triangular median tooth and a large disc-like subbasal process on inner margin. Paramere stout, bifurcated into two equal-length branches, extending to subbasal process of gonostylus, branches bearing numerous short spines on inner margins. Aedeagus short with finger-like and divergent dorsal valves.

图 3-149-1 平江蝎蛉 ***Panorpa pingjiangensis*** 成虫背面观（湖南平江福寿山，NWAFU）

1. 雄虫 2. 雌虫

Fig. 3-149-1 Adults of ***Panorpa pingjiangensis***, dorsal view (Fushou Mountain, Pingjiang, Hunan, NWAFU)

1. Male 2. Female

图 3-149-2 平江蝎蛉 ***Panorpa pingjiangensis*** 雄性外生殖器腹面观

Fig. 3-149-2 Male genitalia of ***Panorpa pingjiangensis***, ventral view

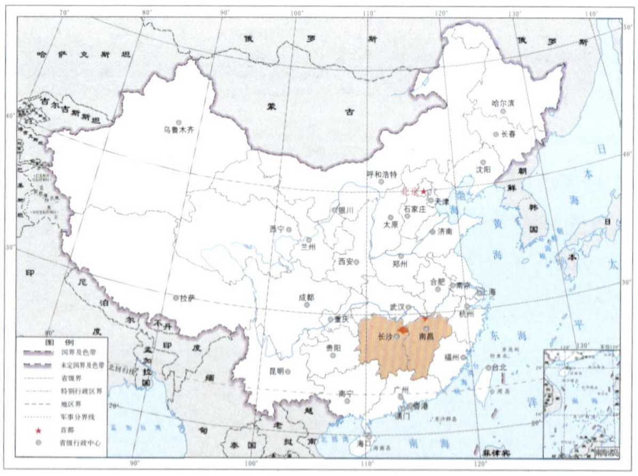

图 3-149-3 平江蝎蛉 ***Panorpa pingjiangensis*** 地理分布

Fig. 3-149-3 Distribution of ***Panorpa pingjiangensis***

Emergence: October to November.

Distribution: Hunan (Pingjiang) and Jiangxi (Lushan Mountain).

150. 秦岭蝎蛉 *Panorpa qinlingensis* Chou & Ran, 1981

【测量】 前翅长 10.5 ～ 12.0 mm。

【形态特征】 头顶及喙黄褐色，单眼三角区黑色。胸部背板黄褐色，两侧稍加深。侧板和足乳黄色。翅无色透明，斑纹黑褐色。前翅端带完整，后部具 1 个大型透明斑；痣带完整，基支阔，端支狭；缘斑小；基带断开呈 2 个小斑；无基斑。雄性腹部 1 ～ 5 节背板黄褐色。第 3 节背中突扁平，盖住第 4 节上尖锐的后背突。第 6 节黄褐色，后缘略向上突起，着生稀疏的长毛。第 7、8 节黄褐色，基部缢缩。雄性外生殖器卵形。第 9 节背板末端具深 U 形缺刻，两侧形成 1 对指形突起；腹板基柄极短；下瓣带状，伸过生殖肢基节中部。生殖肢基节内缘

图 3-150-1 秦岭蝎蛉 *Panorpa qinlingensis* 成虫背面观（陕西南郑黎坪国家森林公园）

1. 雄虫　2. 雌虫

Fig. 3-150-1 Adults of *Panorpa qinlingensis*, dorsal view (Liping National Forest Park, Nanzheng, Shaanxi)

1. Male　2. Female

图 3-150-2 秦岭蝎蛉 *Panorpa qinlingensis* 雄性外生殖器腹面观

Fig. 3-150-2 Male genitalia of *Panorpa qinlingensis*, ventral view

端部具 1 ～ 2 根长刚毛；生殖刺突内缘具钝三角形中齿和盘状的亚基齿。阳基侧突粗壮，螺旋形扭曲，

图 3-150-3 秦岭蝎蛉 *Panorpa qinlingensis* 地理分布

Fig. 3-150-3 Distribution of *Panorpa qinlingensis*

内侧密生短刺。阳茎腹瓣短小；背瓣指状，端部具膜质膨大区。

【发生期】 7 ~ 8 月。

【地理分布】 陕西（米仓山、南郑、留坝）。

Measurements: Forewing length 10.5–12.0 mm.

Morphology: Vertex and rostrum yellowish brown. Ocellar triangle black. Nota yellowish brown, with darkened flanks. Pleura and legs creamy yellow. Wing membrane hyaline, with markings dark brown. In forewings, apical band complete and bearing a large hyaline spot posteriorly; pterostigmal band complete, with a thin apical branch and a broad basal branch; marginal spot small; basal band split into two small spots; basal spot absent. Male terga Ⅰ–Ⅴ dark brown. Notal organ on tergum Ⅲ flat and covering acute postnotal organ on tergum Ⅳ. A6 yellowish brown, projected dorsad and setose on apex. A7 and A8 yellowish brown and constricted basally. Male genital bulb oval. Epandrium emarginate terminally in a U-shape and forming a pair of finger-like processes laterally. Hypandrium with a short basal stalk. Hypovalves stripe-like, extending beyond middle of gonocoxites. Gonocoxites with 1–2 long setae on ventral apex. Gonostylus with an obtuse triangular median tooth and a disc-like subbasal process on inner margin. Paramere stout and spiral, bearing dense short spines on inner margin. Ventral aedeagal valves short; dorsal valves finger-like and enlarged and membranous on dorsal apex.

Emergence: July to August.

Distribution: Shaanxi (Micang Mountains, Nanzheng, Liuba).

151. 四带蝎蛉 *Panorpa quadrifasciata* Chou & Wang, 1987

【测量】 前翅长 16.5 ~ 18.0 mm。

【形态特征】 头顶、喙红褐色，单眼三角区黑色。胸部背板红褐色，侧板及足浅红褐色。翅蜡黄色，半透明，翅痣红褐色，斑纹棕色。前翅端带窄；痣带完整，具基支和端支；在端带与痣带之间的翅后缘具 1 个三角形小斑；基带完整；缘斑和基斑小，明显。雄性腹部 1 ~ 5 节背板红褐色，第 3 节背板后缘的背中突发达，呈棒状突起，超过第 4 节中部尖锐的后背突。第 6 节长筒形，无臀角，末端背面着生密集的长刚毛。第 7、8 节延长，基部强烈缢缩成柄状。雄性外生殖器长卵形。第 9 节背板末端具深 U 形缺刻，两侧形成 1 对指形突起；腹板 Y

图 3-151-1 四带蝎蛉 *Panorpa quadrifasciata* 雄成虫背面观（湖北五峰，NWAFU）

Fig. 3-151-1 Male adult of *Panorpa quadrifasciata*, dorsal view (Wufeng, Hubei, NWAFU)

图 3-151-2 四带蝎蛉 *Panorpa quadrifasciata* 雄性外生殖器腹面观

Fig. 3-151-2 Male genitalia of *Panorpa quadrifasciata*, ventral view

形，基柄窄长；下瓣窄细而短，末端分歧，伸达生殖肢基节 2/3 处。生殖肢基节内端角具 1 向后突出的锐三角形片状突起；生殖刺突长，内缘具三角形中齿和盘状亚基齿。阳基侧突细长，结构简单，伸达生殖刺突亚基齿。阳茎腹瓣短小；背瓣柳叶形，相互分离，超出生殖肢基节端部。

【发生期】 8 ~ 10 月。

【地理分布】 湖南（道县）、广西（龙胜、柳州）、湖北（五峰）。

Measurements: Forewing length 16.5–18.0 mm.

Morphology: Vertex and rostrum reddish brown. Ocellar triangle black. Nota reddish brown. Pleura and legs light reddish brown. Wing membrane waxy yellow, translucent. Pterostigma reddish brown. Markings brown. In forewings, apical band narrow; pterostigmal band complete, with a basal branch and an apical branch; a small triangular spot on hind margin between apical and pterostigmal bands; basal band complete; marginal and basal spots small and conspicuous. Male terga Ⅰ–Ⅴ reddish brown. Notal organ on tergum Ⅲ stout clavate, exceeding acute postnotal organ on middle of tergum Ⅳ. A6 long cylindrical and devoid of anal horns, bearing numerous dense long setae on apical portion of dorsal surface. A7 and A8 elongated and intensively stalked basally. Male genital bulb long oval. Epandrium emarginate terminally in a U-shape and forming a pair of finger-like processes laterally. Hypandrium Y-shaped, with a long narrow basal stalk. Hypovalves narrow and short, extending to two-thirds of gonocoxites, divergent apically. Gonocoxites each with an acute triangular plate directed caudad on ventro-inner apex. Gonostylus long, with an triangular median tooth and a disc-like subbasal process on inner margin. Paramere slender and simple, extending to subbasal process of gonostylus. Ventral aedeagal valves short; dorsal valves willow leaf-shaped and divergent, exceeding apex of gonocoxites.

Emergence: August to October.

Distribution: Hunan (Daoxian), Guangxi (Longsheng, Liuzhou) and Hubei (Wufeng).

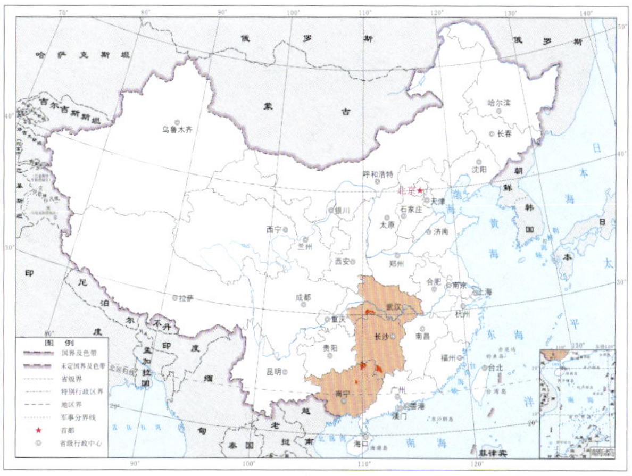

图 3-151-3 四带蝎蛉 *Panorpa quadrifasciata* 地理分布

Fig. 3-151-3 Distribution of *Panorpa quadrifasciata*

152. 反曲蝎蛉 *Panorpa reflexa* Wang & Hua, 2016

【测量】 前翅长 10.0 ~ 11.0 mm。

【形态特征】 头顶黄色。单眼三角区具 1 五角形黑色大斑，后头区亦具 1 黑色大斑。喙浅黄褐色。前胸背板黑褐色。中、后胸背板黑褐色，中央具宽阔的黄色纵带。侧板及足乳黄色。翅无色透明。前翅痣带退化为 1 个三角形小斑，无其他斑纹。停歇时，左右翅沿背中线紧靠，盖住腹部。雄性腹部 1 ~ 5 节背板黑褐色，具浅黄色、向后渐窄的中带。第 3 节背中突扁平，盖住第 4 节上尖锐的后背突。第 6 节大部黑褐色，端部黄褐色，短筒形，无臀角。第 7、8 节黄褐色，短筒形，基部不缢缩。雄性外生殖器椭圆形。第 9 节背板末端具深 U 形缺刻，两侧形成 1 对指形突起；腹板极短，下瓣仅伸达生殖肢基节基部 1/3 处，内缘具 1 列长鬃。生殖刺突短小，内缘具宽大的中齿和弯曲、端部平截的亚基齿。阳基侧突“3”形，其腹支延伸至下瓣下方，背支延伸至阳茎侧面。阳茎柱状，腹瓣膜质；背瓣强烈骨化，末端具耳状突起，侧缘波状翻卷，具 3 ~ 4 个尖齿；侧突起尖长。

【发生期】 7 ~ 8 月。

【地理分布】 云南（永德）。

Measurements: Forewing length 10.0–11.0 mm.

Morphology: Vertex yellow. Ocellar triangle

图 3-152-1　反曲蝎蛉 ***Panorpa reflexa*** 雄成虫背面观（云南永德大雪山）

Fig. 3-152-1　Male adult of ***Panorpa reflexa*,** dorsal view (Daxue Mountain, Yongde, Yunnan)

图 3-152-2　反曲蝎蛉 ***Panorpa reflexa*** 雄性外生殖器腹面观

Fig. 3-152-2　Male genitalia of ***Panorpa reflexa***, ventral view

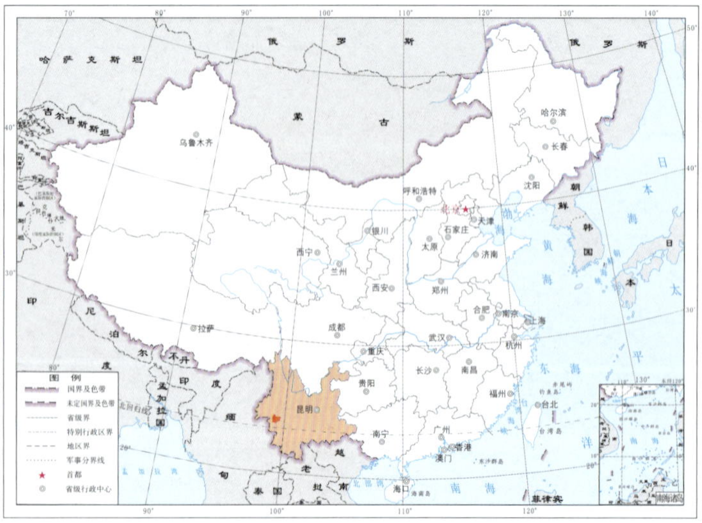

图 3-152-3　反曲蝎蛉 ***Panorpa reflexa*** 地理分布

Fig. 3-152-3　Distribution of ***Panorpa reflexa***

enclosed by a large black pentagonal spot. Another black spot on occiput. Rostrum light yellowish brown. Pronotum dark brown. Mesonotum and metanotum dark brown, with a broad yellow mesal stripe. Pleura and legs creamy yellow. Wing membrane hyaline. In forewings, pterostigmal band greatly reduced into a small triangular spot. Other markings absent. Wings held in close contact along mid-line over abdomen at repose. Male terga Ⅰ–Ⅴ dark brown, with a light yellow mesal stripe tapering caudad. Notal organ on tergum Ⅲ flat and covering acute postnotal organ on tergum Ⅳ. A6 almost dark brown, but yellowish brown on apex, short cylindrical and devoid of anal horns. A7 and A8 yellowish brown and short cylindrical, and not constricted basally.Male genital bulb elliptical. Epandrium emarginate terminally in a U-shape and forming a pair of finger-like processes laterally. Hypandrium greatly shortened, with hypovalves extending to basal third of gonocoxites and bearing a row of long bristles on inner margin. Gonostylus short, with a broad median tooth and a curved subbasal process truncated apically. Paramere 3-shaped, with a ventral branch stretched under hypovalves, and a dorsal branch aside aedeagus. Columnar aedeagus with membranous ventral valves and greatly sclerotized dorsal valves bearing a pair of auriculate processes apically. Lateral margin of dorsal valves rolled ventrad wave-like, and bearing 3–4 acute teeth. Lateral processes long and acute.

Emergence: July to August.

Distribution: Yunnan (Yongde).

153. 半带蝎蛉 *Panorpa semifasciata* Cheng, 1949

【测量】 前翅长 13.5 ~ 14.0 mm。

【形态特征】 头顶、单眼三角区黑色，喙黑褐色。胸部背板黑色，侧板及足黑褐色。翅略带浅黄褐色，透明，末端较圆；翅痣深红棕色，十分明显。雄性一般无斑纹；雌性具退化的端带和痣带；端带色浅，内部具 2 ~ 3 个透明窗斑；痣带基支断开，仅为后缘处的 1 个小斑；无端支；无其他斑纹。雄性腹部 1 ~ 5 节背板黑色。第 3 节背中突扁平，盖住第 4 节上尖锐的后背突。第 6 节黑色，长筒形，向后逐渐变细，无臀角。第 7、8 节细长，基部缢缩。雄性外生殖器阔卵形。第 9 节背板末端具深 U 形缺刻，两侧形成 1 对指形突起；腹板基柄极短；下瓣带状，延伸至生殖肢基节 2/3 处，内缘具长鬃，末端分歧。生殖刺突长于基节；内缘的中齿和亚基齿发达而尖锐。阳基侧突粗壮，延伸至生殖刺突的中齿处，内缘具长刚毛，端部向内弯曲。阳茎腹瓣与背瓣紧密愈合，强烈骨化，长柱状，延伸到生殖刺突中齿处，端部背向弯曲。

图 3-153-1 半带蝎蛉 *Panorpa semifasciata* 成虫背面观（四川石棉栗子坪，NWAFU）

1. 雄虫 2. 雌虫

Fig. 3-153-1 Adults of *Panorpa semifasciata*, dorsal view (Liziping, Shimian, Sichuan, NWAFU)

1.Male 2. Female

图 3-153-2 半带蝎蛉 *Panorpa semifasciata* 雄性外生殖器腹面观

Fig. 3-153-2 Male genitalia of *Panorpa semifasciata*, ventral view

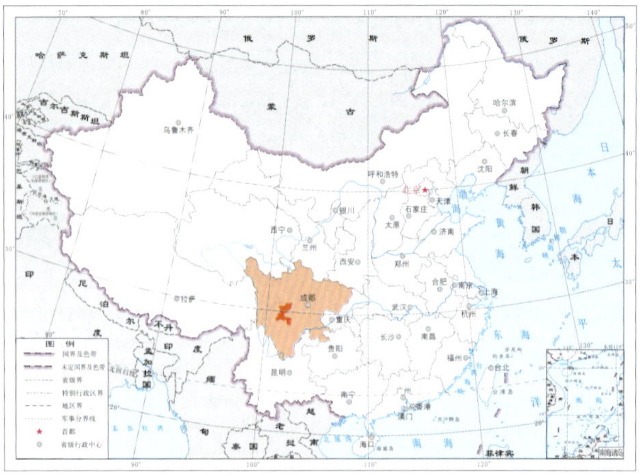

图 3-153-3 半带蝎蛉 *Panorpa semifasciata* 地理分布

Fig. 3-153-3 Distribution of *Panorpa semifasciata*

【发生期】 7 ~ 9 月。

【地理分布】 四川（康定、石棉）。

Measurements: Forewing length 13.5–14.0 mm.

Morphology: Vertex and ocellar triangle black. Rostrum dark brown. Nota black. Pleura and legs dark brown. Wing membrane hyaline, slightly tinged with yellowish brown and rounded at apex. Pterostigma reddish brown and very conspicuous. Males usually devoid of wing markings. Females with greatly reduced apical and pterostigmal bands: apical band faint with 2–3 hyaline spots innerly; pterostigmal band with a disconnected basal branch, which represented by a small spot at hind margin; apical branch absent; other markings absent. Male terga Ⅰ – Ⅴ black. Notal organ on tergum Ⅲ flat and covering acute postnotal organ on tergum Ⅳ. A6 black, long cylindrical and tapering towards apex, without anal horns. A7 and A8 slender and constricted basally. Male genital bulb broad oval. Epandrium emarginate terminally in a U-shape and forming a pair of finger-like processes laterally. Hypandrium with a greatly shortened basal stalk. Hypovalves stripe-like, extending to two-thirds of gonocoxites and bearing a row of long bristles on inner margin, divergent apically. Gonostylus longer than gonocoxites, and with stout and acute median tooth and subbasal process on inner margin. Paramere stout, extending to median tooth of gonostylus, with long spines on inner margin and curved mesad on apical portion. Aedeagus with ventral and dorsal valves intensively fused and strongly sclerotized, long columnar and extending to median tooth of gonostylus, with their apexes curved dorsad.

Emergence: July to September.

Distribution: Sichuan (Kangding, Shimian).

154. 六刺蝎蛉 *Panorpa sexspinosa* Cheng, 1949

【测量】 前翅长 12.0 ~ 15.0 mm。

【形态特征】 头顶、喙红褐色，单眼三角区黑色。前胸背板黑褐色。中、后胸背板黑褐色，中央有 1 条浅黄色纵带。侧板及足乳黄色。翅无色透明，斑纹黑褐色。翅痣棕色。前翅端带阔，后缘散乱，有时具 1 块透明斑；痣带基支宽阔，端支极窄或消失；基带阔；无缘斑和基斑。雄性腹部 1 ~ 5 节背板黑褐色。第 3 节背中突扁平，盖住第 4 节上尖锐的后背突。第 6 节基半部黑褐色，端半部黄褐色，后缘稍向上隆起，着生稀疏的长毛。雄性外生殖器卵圆形。第 9 节背板末端具深 U 形缺刻，两侧形成 1 对指形突起；腹板基柄短；下瓣带状，不伸达生殖肢基节末端，内缘具 1 列长鬃。生殖肢基节内缘端部具 4 ~ 6 根粗壮刚毛；生殖刺突内缘中齿钝

图 3-154-1 六刺蝎蛉 ***Panorpa sexspinosa*** 成虫侧面观（陕西宁陕火地塘，钟问摄）

1. 雄虫 2. 雌虫

Fig. 3-154-1 Adults of ***Panorpa sexspinosa***, lateral view (Huoditang,Ningshan,Shaanxi,photo by Wen Zhong)

1. Male 2. Female

图 3-154-2 六刺蝎蛉 ***Panorpa sexspinosa*** 雄性腹部侧面观
（红色箭头示隆起的第 6 节末端）

Fig. 3-154-2 Male abdomen of ***Panorpa sexspinosa***, lateral view (Red arrow points to the projected dorsal apex of A6)

图 3-154-3 六刺蝎蛉 ***Panorpa sexspinosa*** 雄性外生殖器腹面观

Fig. 3-154-3 Male genitalia of ***Panorpa sexspinosa***, ventral view

三角形，亚基齿大、盘状。阳基侧突细长，螺旋形扭曲，相互交叉。阳茎腹瓣短小膜质；背瓣骨化延伸，近末端背面具膜质膨大区。

【发生期】 5 ~ 8 月。

【地理分布】 陕西（宁陕、周至、太白、鄠邑、蓝田）、河南（嵩县）、甘肃（党川）。

Measurements: Forewing length 12.0–15.0 mm.

Morphology: Vertex and rostrum reddish brown. Ocellar triangle black. Pronotum dark brown. Mesonotum and metanotum dark brown, with a pale yellow mesal stripe. Pleura and legs creamy yellow.

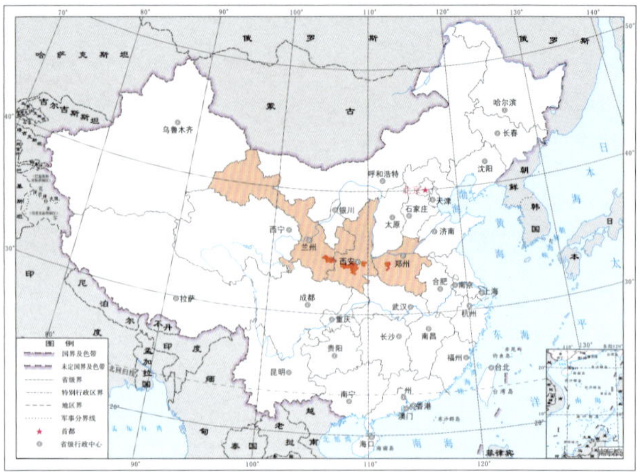

图 3-154-4 六刺蝎蛉 ***Panorpa sexspinosa*** 地理分布

Fig. 3-154-4 Distribution of ***Panorpa sexspinosa***

Wing membrane hyaline, with markings dark brown. Pterostigma brown. In forewings, apical band broad and scattered on posterior margin, and occasionally with a hyaline spot innerly; pterostigmal band with a broad basal branch and a thin or disappeared apical branch; basal band broad; marginal and basal spots absent. Male terga Ⅰ–Ⅴ dark brown. Notal organ on tergum Ⅲ flat and covering acute postnotal organ on tergum Ⅳ. A6 dark brown on basal half and yellowish brown on distal half, and projected dorsad on dorsal apex with sparse long setae. Male genital bulb oval. Epandrium emarginate terminally in a deep U-shaped and forming a pair of finger-like processes laterally. Hypandrium with a short basal stalk. Hypovalves stripe-like, not extending to apex of gonocoxite, bearing a row of long bristles on inner margin. Gonocoxites with 4–6 stout setae on ventro-apical margin innerly. Gonostylus with an obtuse triangular median tooth and a large disc-like subbasal process on inner margin. Parameres slender and spiral, and crossed mesally. Ventral aedeagal valves short and membranous; dorsal valves sclerotized and elongated, and each with a membranous enlarged process on subapical portion dorsally.

Emergence: May to August.

Distribution: Shaanxi (Ningshan, Zhouzhi, Tiaibai, Huyi, Lantian), Henan (Songxian) and Gansu (Dangchuan).

155. 六带蝎蛉 *Panorpa sextaenia* Zhou & Bao, 2002

【测量】 前翅长 16.0 ~ 19.0 mm。

【形态特征】 头顶、喙黄色，单眼三角区黑色。胸部背板、侧板及足黄色。翅蜡黄色，翅痣红褐色十分明显；斑纹黑色，十分特殊：前翅无端带；痣带十分淡，无端支；缘斑极小或消失；基带完整，纤细；基斑成为 1 条不超过 M 的长带，有时还在前缘处形成 1 个小斑。后翅仅有后方扩大的基带和水渍状的痣带基支。雄性腹部大部分黄色。第 3 节背中突扁平，盖住第 4 节上尖锐的后背突。第 6 节长筒形，向后渐细，无臀角。第 7、8 节长，稍向背面弯曲，基部缢缩。雄性外生殖器阔卵形。第 9 节背板末端具深 U 形缺刻，两侧形成 1 对指形突起；腹板基柄短；下瓣窄长，伸达生殖肢基节末端。生殖刺突与基节等长，内缘具钝圆的中齿和不明显的亚基齿。阳基侧突杆状，超过生殖刺突的亚基齿，内缘具密集的短刺。阳茎叉形，腹瓣短、膜质；背瓣骨化，细长、叶状，末端尖锐。

【发生期】 10 月。

【地理分布】 贵州（茂兰、黎平）、广西（金秀）。

Measurements: Forewing length 16.0–19.0 mm.

Morphology: Vertex and rostrum yellow. Ocellar triangle black. Nota, pleura and legs yellow. Wing membrane waxy yellow. Pterostigma reddish brown and very conspicuous. Markings black and peculiarly arranged. In forewings, apical band absent; pterostigmal band very faint and without apical branch; marginal spot very small or absent; basal band complete and slender; basal spot elongated into a slender stripe and not exceeding M, and occasionally forming a small spot on anterior margin. Hindwing only with a water stain-like basal branch of pterostigmal band, and a posteriorly enlarged basal

图 3-155-1 六带蝎蛉 *Panorpa sextaenia* 雄成虫背面观（贵州黔东南黎平，NWAFU）

Fig. 3-155-1 Male adult of *Panorpa sextaenia*, dorsal view (Liping, Qiandongnan, Guizhou, NWAFU)

图 3-155-2 六带蝎蛉 *Panorpa sextaenia* 雄性外生殖器腹面观

Fig. 3-155-2 Male genitalia of *Panorpa sextaenia*, ventral view

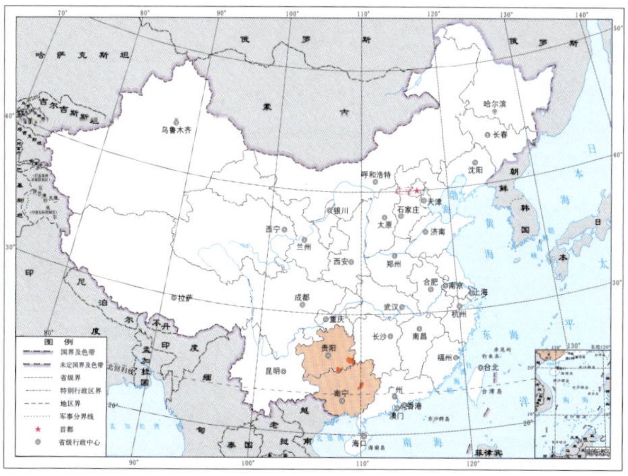

图 3-155-3 六带蝎蛉 *Panorpa sextaenia* 地理分布

Fig. 3-155-3 Distribution of *Panorpa sextaenia*

band. Male abdomen mostly yellow. Notal organ on tergum Ⅲ flat and covering acute postnotal organ on tergum Ⅳ. A6 long cylindrical, tapering towards apex and devoid of anal horns. A7 and A8 long, slightly curved dorsad and constricted basally. Male genital bulb broad oval. Epandrium emarginate terminally in a U-shape and forming a pair of finger-like processes laterally. Hypandrium with a short basal stalk. Hypovalves slender, extending to apex of gonocoxites. Gonostylus as long as gonocoxites, and bearing a rounded median tooth and a small inconspicuous subbasal process on inner margin. Paramere stick-like, exceeding subbasal process of gonostylus, and bearing dense short spines on inner margin. Aedeagus fork-shaped with short membranous ventral valves; dorsal valves sclerotized, slender and foliate with acute apexes.

Emergence: October.

Distribution: Guizhou (Maolan, Liping) and Guangxi (Jinxiu).

156. 山阳蝎蛉 *Panorpa shanyangensis* Chou & Wang, 1981

【测量】 前翅长 11.0 mm。

【形态特征】 头顶褐色，有三个黑斑；单眼三角区黑色。喙黄褐色。前胸背板黑褐色。中、后胸背板黄褐色，具宽阔的黄色中带。侧板和足黄色。翅无色透明，斑纹褐色。前翅端带不完整，内侧分出 3 个小斑；痣带完整，端支狭窄，基支宽阔；基带完整；无缘斑和基斑。雄性未知。

【发生期】 8 月。

【地理分布】 陕西（山阳）。

Measurements: Forewing length 11.0 mm.

Morphology: Vertex brown with three black spots. Ocellar triangle black. Rostrum yellowish brown. Pronotum dark brown. Mesonotum and metanotum yellowish brown and bearing a broad yellow mesal stripe. Pleura and legs yellowish brown. Wing membrane hyaline. Markings brown. In forewing, apical band incomplete with three separated small spots near inner margin; pterostigmal band complete with a narrow apical branch and a broad basal branch; basal band complete; marginal and basal spots absent. Males unknown.

Emergence: August.

Distribution: Shaanxi (Shanyang).

图 3-156-1 山阳蝎蛉 *Panorpa shanyangensis* 雌成虫背面观（腹末已去除，陕西山阳翠屏山，NWAFU）

Fig. 3-156-1 Male adult of *Panorpa shanyangensis* with terminalia removed, dorsal view (Mount Cuiping, Shanyang, Shaanxi, NWAFU)

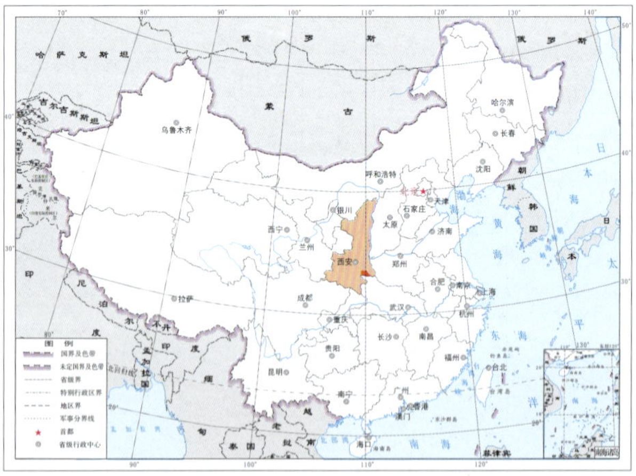

图 3-156-2 山阳蝎蛉 *Panorpa shanyangensis* 地理分布

Fig. 3-156-2 Distribution of *Panorpa shanyangensis*

157. 楚南蝎蛉 *Panorpa sonani* Issiki, 1929

【测量】 前翅长 14.5 ~ 16.0 mm。

【形态特征】 头顶黑褐色。单眼三角区黑色。喙黄褐色，两侧具深色纵纹。前胸背板黑褐色。中、后胸背板黑褐色，中央具显著的黄色中带。侧板及足黄褐色。翅无色透明，斑纹褐色。翅痣黄色明显。前翅端带完整，后缘具 1 个较大的透明斑，内部沿横脉具一系列小透明斑；痣带完整，具狭窄的基支和端支；缘斑细长；基带分裂为两个不规则斑；缘斑小而略狭长。雄性腹部 1 ~ 5 节黑褐色。第 3 节背中突扁平，盖住第 4 节上尖锐的后背突。第 6 节黑褐色，锥状，无臀角。第 7、8 节黄褐色，基部缢缩。雄性外生殖器长卵形。第 9 节背板末端具深 U 形缺刻，两侧形成 1 对指形突起；腹板基柄极短；下瓣细长，伸达生殖肢基节的 2/3 处，内缘具 1 列长鬃。生殖刺突内缘具三角形中齿和阔三角形亚基齿；近基部腹面具 1 个大型凹陷区，凹陷区外缘具 1 个向内的尖齿。阳基侧突近端部分为 2 支，呈扳手状，端半部具长刺毛。阳茎扁片状，不超出生殖肢基节端部。

【发生期】 10 ~ 11 月。

【地理分布】 台湾（屏东）。

Measurements: Forewing length 14.5–16.0 mm.

Morphology: Vertex dark brown. Ocellar triangular black. Rostrum yellowish brown, with two darkened longitudinal stripes laterally. Pronotum dark brown. Mesonotum and metanotum dark brown, with a conspicuous yellow mesal stripe. Pleura and legs yellowish brown. Wing membrane hyaline, with markings brown. Pterostigma yellow and conspicuous. In forewings, apical band complete with a large hyaline spot posteriorly, and a series of hyaline spots along crossveins; pterostigmal band complete with narrow basal and apical branches; marginal spot slender; basal band split into two irregular spots; basal spot small and slightly elongated. Male terga Ⅰ–Ⅴ dark brown. Notal organ on tergum Ⅲ flat and covering acute postnotal organ on tergum Ⅳ. A6 dark brown, conical and devoid of anal horns. A7 and A8 yellowish brown, and constricted basally. Male genital bulb long oval. Epandrium emarginate terminally in a U-shape and forming a pair of finger-like processes laterally. Hypandrium with a greatly shortened basal stalk. Hypovalves slender, extending to distal two-thirds of gonocoxites, and bearing a row of long bristles on inner margin. Gonostylus with a triangular median tooth and a broad triangular subbasal process. Gonostylus also with a large concavity on ventral surface subbasally, and an acute tooth directed mesad

图 3-157-1　楚南蝎蛉 ***Panorpa sonani*** 雄成虫背面观（台湾屏东大汉山，NWAFU）

Fig. 3-157-1　Male adult of ***Panorpa sonani***, dorsal view (Mount Tahan, Pingtung, Taiwan, NWAFU)

图 3-157-2　楚南蝎蛉 ***Panorpa sonani*** 雄性外生殖器腹面观

Fig. 3-157-2　Male genitalia of ***Panorpa sonani***, ventral view

图 3-157-3　楚南蝎蛉 ***Panorpa sonani*** 地理分布

Fig. 3-157-3　Distribution of ***Panorpa sonani***

on outer border of concavity. Paramere bifurcated subapically and wrench-like, bearing long comb-like spines on apical half. Aedeagus flat and not exceeding apex of gonocoxites.

Emergence: October to November.

Distribution: Taiwan (Pingtung).

158. 痣带蝎蛉 *Panorpa stigmalis* Navás, 1908

图 3-158-1 痣带蝎蛉 *Panorpa stigmalis* 雄成虫背面、侧面观（四川崇州鞍子河，柴义君摄）

1. 侧面观　2. 背面观

Fig. 3-158-1 Male adult of *Panorpa stigmalis* (Anzihe, Chongzhou, Sichuan, photo by Yijun Chai)

1. Lateral view　2. Dorsal view

图 3-158-2 痣带蝎蛉 ***Panorpa stigmalis*** 雄性外生殖器腹面观

Fig. 3-158-2 Male genitalia of ***Panorpa stigmalis***, ventral view

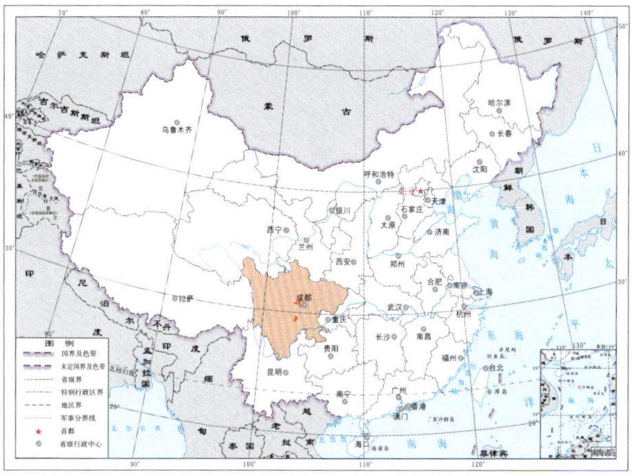

图 3-158-3 痣带蝎蛉 ***Panorpa stigmalis*** 地理分布

Fig. 3-158-3 Distribution of ***Panorpa stigmalis***

【测量】 前翅长 16.5 ~ 18.0 mm。

【形态特征】 头顶黄褐色。单眼三角区黑褐色。喙黄褐色。胸部背板黑褐色，侧板及足黄褐色。翅黄色半透明，斑纹褐色。翅痣红褐色，明显。前翅端带完整；痣带狭窄，无端支；基带仅为后缘的 1 个带状小斑。雄性腹部 1 ~ 5 节背板黄褐色。第 3 节背板后缘的背中突发达，呈短棒状，超过第 4 节背板中部。第 4 节的后背突位于背板后部 1/3 处，强烈向上隆起。第 6 节黄褐色，长筒形，末端背缘稍隆起，具 1 簇粗毛，无臀角。第 7、8 节黄褐色，第 7 节基部 1/4 强烈缢缩呈柄状。雄性外生殖器长卵形。第 9 节背板末端具深 U 形缺刻，两侧形成 1 对指形突起；腹板具窄长的基柄；下瓣细长，超过生殖肢基节的 2/3 处，端部稍分歧。生殖刺突内缘具钝三角形中齿和梯形亚基齿。阳基侧突简单，不分叉。阳茎较短，钳状，不超过生殖肢基节端部。

【发生期】 8 ~ 10 月。

【地理分布】 四川（青城山、峨眉山、鞍子河）。

Measurements: Forewing length 16.5–18.0 mm.

Morphology: Vertex yellowish brown. Ocellar triangle dark brown. Rostrum yellowish brown. Nota dark brown. Pleura and legs yellowish brown. Wing membrane translucent, tinged with yellow. Markings brown. Pterostigma reddish brown and conspicuous. In forewings, apical band complete; pterostigmal band narrow, without apical branch; basal band reduced into a small stripe-like spot on hind margin. Male terga Ⅰ–Ⅴ yellowish brown. Notal organ on tergum Ⅲ strongly developed, clavate, exceeding middle of tergum Ⅳ. Postnotal organ on distal third of tergum Ⅳ and strongly projected dorsad. A6 yellowish brown, long cylindrical, slightly projected and bearing a tuft of stout setae on dorsal apex, without anal horns. A7 and A8 yellowish brown, with A7 intensively stalked on basal quarter. Male genital bulb long oval. Epandrium emarginate terminally in a U-shape and forming a pair of finger-like processes laterally. Hypandrium with a long narrow basal stalk. Hypovalves slender, exceeding distal two-thirds of gonocoxites, slightly divergent apically. Gonostylus with an obtuse triangular median tooth and a trapezoidal subbasal process. Paramere simple. Aedeagus short and pincer-like, not exceeding apex of gonocoxites.

Emergence: August to October.

Distribution: Sichuan (Mount Qingcheng, Moung Emei, Anzihe).

159. 斑点蝎蛉 *Panorpa stigmosa* Zhou, 2006

【测量】 前翅长 11.0 ~ 12.0 mm。

【形态特征】 头顶、单眼三角区黑色。喙黑褐

图 3-159-1 斑点蝎蛉 ***Panorpa stigmosa*** 成虫背面观（云南玉龙雪山，NWAFU）

1. 雄虫 2. 雌虫

Fig. 3-159-1 Adults of ***Panorpa stigmosa***, dorsal view (Yulong Snow Mountain, Yunnan, NWAFU)

1. Male 2. Female

图 3-159-2 斑点蝎蛉 ***Panorpa stigmosa*** 雄性外生殖器腹面观

Fig. 3-159-2 Male genitalia of ***Panorpa stigmosa***, ventral view

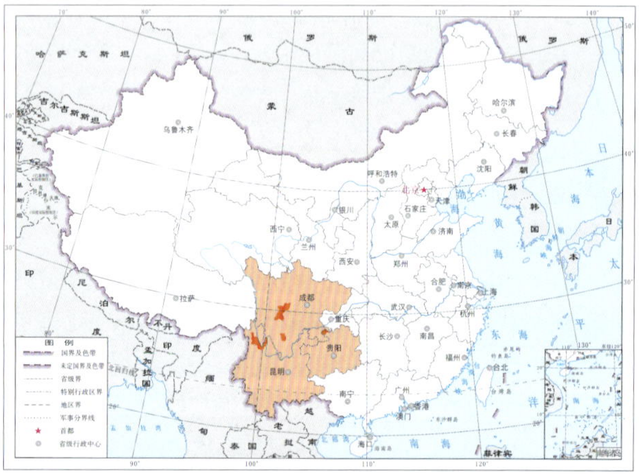

图 3-159-3 斑点蝎蛉 ***Panorpa stigmosa*** 地理分布

Fig. 3-159-3 Distribution of ***Panorpa stigmosa***

色，两颊黄褐色。胸部背板黑色。侧板及足浅黄色至黄褐色。翅无色透明，翅痣浅黄色；斑纹褐色，变异较大。前翅端带形成一系列散乱的小斑，分散于端部翅室内；痣带前部较宽且完整，后方的基支、端支均断开；基带为 CuA 和 CuP 之间的 1 个不规则斑；缘斑仅为 R_{4+5} 基部下方的 1 个小圆点；无基斑。有的雄性个体斑纹退化，仅存痣带前半部，甚至几乎无斑纹。雄性腹部 1 ~ 5 节背板黑色。第 3 节背板上的背中突扁平，盖住第 4 节上尖锐的后背突。第 6 节黑色，锥形。第 7、8 节黄褐色，基部缢缩。雄性外生殖器卵形。第 9 节背板末端具深 U 形缺刻，两侧形成 1 对指形突起；腹板基柄极短；下瓣长宽带状，伸达生殖肢基节 3/4 处，中部稍内扩，内缘端半部着生 1 列十分粗长的鬃。生殖肢基节端部内缘稍向内呈角状突起，其上方具 3 ~ 5 根长刚毛；生殖刺突内缘具钝三角形中齿和碟形亚基齿，后者近基部着生 1 列短刚毛。阳基侧突中部呈膝状向内侧弯折，端半部螺旋形扭曲，相互交叉，内缘具 1 列刺毛；阳茎腹瓣膜质，短小；背瓣在伸出生殖肢基节凹窝后强烈向背面弯曲，背面近端部具 1 个膜质膨大物。

【发生期】 6 ~ 7 月。

【地理分布】 贵州（赤水）、四川（贡嘎山、螺髻山）、云南（玉龙雪山、维西）。

Measurements: Forewing length 11.0–12.0 mm.

Morphology: Vertex and ocellar triangle black. Rostrum dark brown with yellowish brown genae. Nota black. Pleura and legs lightyellow to yellowish brown. Wing membrane hyaline. Pterostigma light yellow. Markings brown and highly variable. In forewings, apical band scattered into a series of small spots distributed in apical cells; pterostigmal band broad and complete anteriorly, with separated basal and apical branches posteriorly; basal band reduced into an irregular spot between CuA and CuP; marginal spot represented by a small rounded spot below R_{4+5}; basal spot absent. Wing markings extremely reduced, with only anterior half of apical band present, or even entirely absent in some male individuals. Male terga Ⅰ–Ⅴ black. Notal organ on tergum Ⅲ flat, and covering acute postnotal organ on tergum Ⅳ. A6 black and conical; A7 and A8 yellowish brown and constricted basally. Male genital bulb oval. Epandrium emarginate terminally in a deep U-shape and forming a pair of finger-like processes laterally; hypandrium, with an greatly shortened basal stalk. Hypovalve long stripe-like, extending to three-fourths of gonocoxite and slightly projected mesally on middle of inner margin, bearing a row of long stout bristles along apical half of inner margin. Gonocoxites slightly projected innerly on ventral apex, with 3–5 long setae. Gonostylus with an obtuse triangular median tooth and a disc-like subbasal process; subbasal process with a row of short setae subbasally. Parameres geniculate medially, with their spiral apical halves curved mesad and crossed, and each bearing a row of short spines on inner margin. Ventral aedeagal valves membranous and short; dorsal valves greatly curved dorsad and bearing a membranous enlarged process on dorsal surface subapically.

Emergence: June to July.

Distribution: Guizhou (Chishui), Sichuan (Minya Konka, Mount Luoji) and Yunnan (Yulong Snow Mountain, Weixi).

160. 淡珀蝎蛉 *Panorpa subambra* Chou & Tong, 1987

【测量】 前翅长 12.5 ~ 13.0 mm。

【形态特征】 头顶黄褐色，单眼三角区黑色。喙黄褐色。胸部背板黑褐色。侧板及足浅黄褐色。翅蜡黄色，半透明，翅痣红褐色，斑纹黄褐色，模

图 3-160-1 淡珀蝎蛉 *Panorpa subambra* 雄成虫背面观（湖南石门壶瓶山，NWAFU）

Fig. 3-160-1 Male adult of *Panorpa subambra*, dorsal view (Mount Huping, Shimen, Hunan, NWAFU)

图 3-160-2 淡珀蝎蛉 *Panorpa subambra* 雄性外生殖器腹面观

Fig. 3-160-2 Male genitalia of *Panorpa subambra*, ventral view

糊而不明显。前翅痣带无端支，其他斑纹不明显。后翅无斑纹，后缘在 M_3 至 CuA 处显著向外扩，末端较尖。雄性腹部 1 ~ 5 节背板黑褐色。第 3 节背中突扁平，盖住第 4 节上尖锐的后背突。第 6 节黑褐色，锥形，无臀角。第 7、8 节黄褐色，基部缢缩。雄性外生殖器卵形。第 9 节背板末端具深 U 形缺刻，两侧形成 1 对指形突起；腹板具较长、较窄的基柄，呈 Y 形；下瓣分歧，伸达生殖肢基节的 3/4 处。生殖肢基节内缘具半圆形突起；生殖刺突外缘稍向内凹，内缘的亚基齿喇叭状，中齿钝三角形。阳基侧突线状，细长，伸至生殖刺突的亚基齿上方。阳茎短，腹瓣膜质短小，背瓣骨化稍分离。

【发生期】 7 ~ 8 月。

【地理分布】 湖南（桑植、石门）。

Measurements: Forewing length 12.5–13.0 mm.

Morphology: Vertex yellowish brown. Ocellar triangle black. Rostrum yellowish brown. Nota dark brown. Pleura and legs light yellowish brown. Wing membrane translucent, tinged with waxy yellow. Pterostigma reddish brown. Markings faint yellowish brown. In forewings, pterostigmal band lacking apical branch; other markings inconspicuous. Hindwing devoid of markings and distinctly expanded outward from endings of M_3 to CuA, and slightly acute on apex. Male terga Ⅰ – Ⅴ dark brown. Notal organ on tergum Ⅲ flat, and covering acute postnotal organ on tergum Ⅳ. A6 dark brown, conical and without anal horns. A7 and A8 yellowish brown and constricted basally. Male genital bulb oval. Epandrium emarginate terminally in a deep U-shape and forming a pair of finger-like processes laterally; hypandrium Y-shaped, with a long narrow basal stalk, and hypovalves extending to distal three-fourths of gonocoxites. Gonocoxites with a semicicular plate on ventro-inner margin. Gonostylus concaved on outer margin, and bearing an obtuse triangular median tooth and a trumpet-shaped subbasal process on inner margin. Paramere thread-like, extending beyond subbasal process of gonocoxites. Aedeagus very short, with membranous ventral valves and sclerotized and divergent dorsal valves.

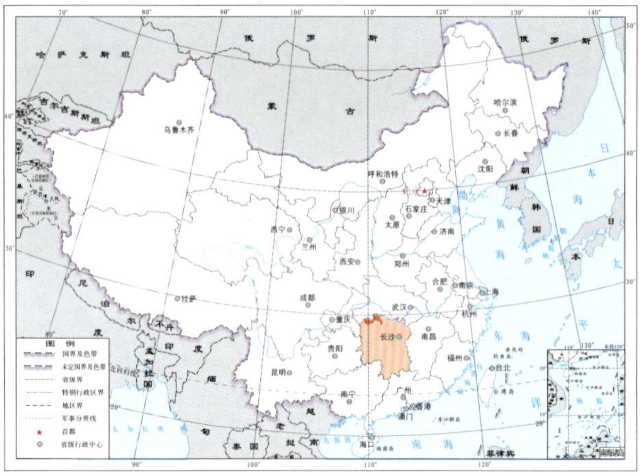

图 3-160-3 淡珀蝎蛉 ***Panorpa subambra*** 地理分布

Fig. 3-160-3 Distribution of ***Panorpa subambra***

Emergence: July to August.

Distribution: Hunan (Sangzhi, Shimen).

161. 台湾蝎蛉 *Panorpa taiwanensis* Issiki, 1929

图 3-161-1 台湾蝎蛉 ***Panorpa taiwanensis*** 雄成虫背面观（台湾花莲太鲁阁国家公园，CAU）

Fig. 3-161-1 Male adult of ***Panorpa taiwanensis***,dorsal view (Taroko National Park, Hualien,Taiwan, CAU)

【测量】 前翅长 13.0 mm。

【形态特征】 头顶浅黄色，单眼三角区黑色。1 条宽阔的黑色横带贯穿单眼三角区，并延伸至复眼内缘。喙黄褐色。前胸背板黑褐色。中、后胸背板黑褐色，中央具宽阔的浅黄色纵带。侧板及足浅黄色。翅略无色透明，具模糊的浅褐色斑纹。前翅端带成为散乱的一系列斑点；痣带在翅痣处宽大，向后变窄，无端支；基带在 CuP 处残余 1 个不规则斑；无其他斑纹。雄性腹部 1 ~ 5 节背板黑褐色。第 3 节背板上的背中突扁平，盖住第 4 节上尖锐的后背突。第 6 节大部分黑褐色，端部黄褐色，锥形。

图 3-161-2 台湾蝎蛉 ***Panorpa taiwanensis***
雄性外生殖器腹面观

Fig. 3-161-2 Male genitalia of ***Panorpa taiwanensis***,
ventral view

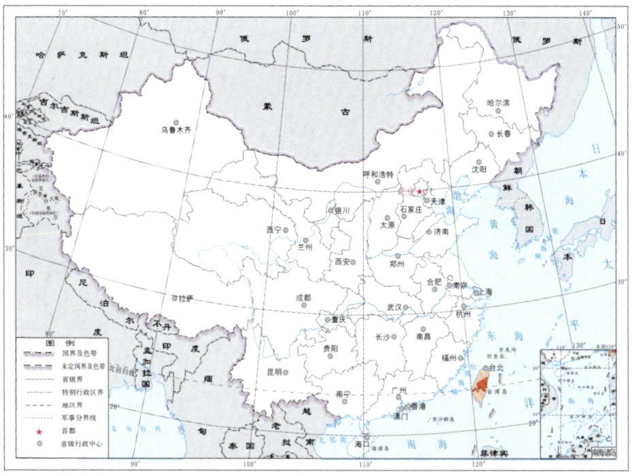

图 3-161-3 台湾蝎蛉 ***Panorpa taiwanensis*** 地理分布

Fig. 3-161-3 Distribution of ***Panorpa taiwanensis***

第 7、8 节黄褐色，基部强烈缢缩呈柄状。雄性外生殖器卵形。第 9 节背板末端具深 U 形缺刻，两侧形成 1 对指形突起；腹板基柄极短；下瓣长带状，几乎伸达生殖肢基节末端，内缘具 1 列长鬃毛。生殖刺突内缘具钝三角形中齿和大型盘状亚基齿。阳基侧突指状，近端部分为 2 支，末端尖，内缘具 1 列梳齿状长刺。阳茎腹瓣短小；背瓣宽阔，分离。

【发生期】 6 ~ 8 月。

【地理分布】 台湾（南投、阿里山、花莲）。

Measurements: Forewing length 13.0 mm.

Morphology: Vertex light yellow. Ocellar triangle black. A broad black transverse stripe passing through ocellar triangle and connected to inner borders of compound eyes. Rostrum yellowish brown. Pronotum dark brown. Mesonotum and metanotum dark brown, with a broad light yellow mesal stripe. Pleura and legs light yellow. Wing membrane hyaline. Markings faint brown. In forewings, apical band scattered into several small spots; pterostigmal band broad near pterostigma, narrowed backwards, and lacking apical branch; basal band reduced into an irregular spot on CuP; other markings absent. Male terga Ⅰ–Ⅴ dark brown. Notal organ on tergum Ⅲ flat and covering acute postnotal organ on tergum Ⅳ. A6 mostly dark brown and yellowish brown on apex, conical. A7 and A8 yellowish brown and intensively stalked basally. Male genital bulb oval. Epandrium emarginate terminally in a deep U-shape and forming a pair of finger-like processes laterally. Hypandrium with a greatly shortened basal stalk. Hypovalves long stripe-like, extending almost to apex of gonocoxite, and bearing a row of long bristles along inner margin. Gonostylus with an obtuse triangular median tooth and a large disc-like subbasal process. Paramere finger-like and bifurcated subapically with acute apexes, and bearing a row of comb-like spines along inner margin. Aedeagus with ventral valves short, dorsal valves broad and divergent.

Emergence: June to August.

Distribution: Taiwan (Nantou, Ali Mountain, Hualien).

162. 三线蝎蛉 *Panorpa trifasciata* Cheng, 1957

【测量】 前翅长 13.0 ~ 14.0 mm。

【形态特征】 头顶、单眼三角区黑色。喙黄褐色。胸部背板黑色，侧板及足浅黄褐色。翅无色透明，前翅端带、痣带和基带排列成显著的 3 条横带；痣带宽阔，中部稍缢缩，无端支；无缘斑、基斑。雄性腹部 1 ~ 5 节背板黑褐色。第 3 节背板上

图 3-162-1 三线蝎蛉 ***Panorpa trifasciata*** 雄成虫背面观（福建邵武大竹岚，NWAFU）

Fig. 3-162-1 Male adult of ***Panorpa trifasciata***,dorsal view (Dazhulan, Shaowu, Fujian, NWAFU)

图 3-162-2 三线蝎蛉 ***Panorpa trifasciata*** 雄性外生殖器腹面观

Fig. 3-162-2 Male genitalia of ***Panorpa trifasciata***, ventral view

的背中突扁平，盖住第 4 节上尖锐的后背突。第 6 节黑褐色，锥形，无臀角。第 7、8 节黄褐色，基部缢缩。雄性外生殖器圆球形。第 9 节背板末端具 U 形缺刻，两侧形成 1 对指形突起；腹板基柄极短；下瓣带状，伸达生殖肢基节中部。生殖肢基节腹面内缘端半部着生 1 列粗长的刚毛；生殖刺突具尖锐的中齿和粗壮的亚基齿。阳基侧突细长，内缘具刺毛列。阳茎腹瓣短，背瓣发达并分离。

图 3-162-3 三线蝎蛉 ***Panorpa trifasciata*** 地理分布

Fig. 3-162-3 Distribution of ***Panorpa trifasciata***

【发生期】 6 ~ 10 月。

【地理分布】 福建（邵武、武夷山）。

Measurements: Forewing length 13.0–14.0 mm.

Morphology: Vertex and ocellar triangle black. Rostrum yellowish brown. Nota black. Pleura and legs light yellowish brown. Wing membrane hyaline. In forewings, apical, pterostigmal, and basal bands distinct into three transverse bands. Pterostigmal band broad and constricted medially, lacking apical branch. Marginal and basal spots absent. Male terga Ⅰ–Ⅴ dark brown. Notal organ on tergum Ⅲ flat and covering acute postnotal organ on tergum Ⅳ. A6 dark brown, conical and devoid of anal horns. A7 and A8 yellowish brown and constricted basally. Male genital bulb broad oval. Epandrium emarginate terminally in a U-shape and forming a pair of finger-like processes laterally. Hypandrium with a greatly shortened basal stalk. Hypovalves stripe-like, extending to middle of gonocoxites. Gonocoxites with a row of long stout setae on apical half of ventro-inner margin. Gonostylus with an actue triangular median tooth and a stout subbasal process on inner margin. Paramere slender and bearing a row of spines on inner margin. Aedeagus with ventral valves short, and dorsal valves stout and divergent.

Emergence: June to October.

Distribution: Fujian (Shaowu, Wuyi Mountains).

163. 三带蝎蛉 *Panorpa tritaenia* Chou & Wang, 1987

图 3-163-1 三带蝎蛉 ***Panorpa tritaenia*** 雄成虫背面观（外生殖器已去除，湖南桑植天平山，NWAFU）

Fig. 3-163-1 Male adult of ***Panorpa tritaenia*** with genitalia removed, dorsal view (Mount Tianping, Sangzhi,Hunan, NWAFU)

图 3-163-2 三带蝎蛉 ***Panorpa tritaenia*** 雄性外生殖器腹面观

Fig. 3-163-2 Male genitalia of ***Panorpa tritaenia***, ventral view

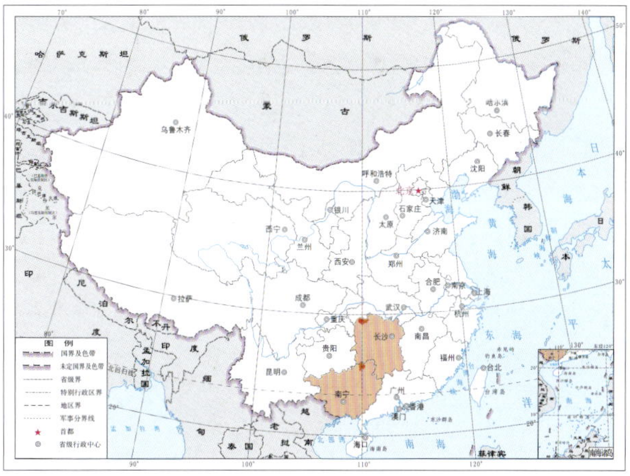

图 3-163-3 三带蝎蛉 ***Panorpa tritaenia*** 地理分布

Fig. 3-163-3 Distribution of ***Panorpa tritaenia***

【测量】 前翅长 13.0 mm。

【形态特征】 头顶、单眼三角区黑色，喙黄褐色。胸部背板黑色，侧板及足黄褐色。翅无色透明，翅痣黄色，斑纹褐色。前翅端带、痣带和基带排列成明显的 3 条横带；痣带无端支；基带从 Rs 延伸至 1A 末端；无其他斑纹。后翅与前翅相似，但缺基带。雄性腹部 1 ~ 5 节背板黑褐色。第 3 节背板上的背中突扁平，盖住第 4 节上尖锐的后背突。第 6 节黑褐色，锥形，无臀角。第 7、8 节黄褐色，基部缢缩。雄性外生殖器阔卵形。第 9 节背板末端具深 U 形缺刻，两侧形成 1 对指形突起；腹板具窄短柄；下瓣纤细，末端分歧，伸达生殖肢基节 3/4 处；生殖刺突具阔三角形中齿和发达的钩状亚基齿。阳基侧突细长，明显超出生殖刺突中齿，内缘具 1 列刺毛。阳茎粗壮；腹瓣膜质；背瓣骨化，背面的耳状突起伸达生殖刺突的中齿。

【发生期】 7 ~ 8 月。

【地理分布】 湖南（桑植）、广西（龙胜）。

Measurements: Forewing length 13.0 mm.

Morphology: Vertex and ocellar triangle black. Rostrum yellowish brown. Nota black. Pleura and legs yellowish brown. Wing membrane hyaline. Pterostigma yellow. Markings brown. In forewings, apical, pterostigmal and basal bands arranged into three distinct transverse bands. Pterostigmal band lacking apical branch; and basal band extending from Rs to end of 1A. Other markings absent. Hindwing similar to forewing but lacking basal band. Male tergal Ⅰ-Ⅴ dark brown. Notal organ on tergum Ⅲ flat and covering acute postnotal organ on tergum Ⅳ. A6 dark brown, conical and devoid of anal horns. A7 and A8 yellowish brown and constricted basally. Male genital bulb broad oval. Epandrium emarginate terminally in a U-shape and forming a pair of finger-like processes laterally. Hypandrium with a short narrow basal stalk. Hypovalves slender, divergent apically,extending to three-fourths of gonocoxites. Gonostylus with a broad triangular median tooth and a hook-like subbasal process on inner margin. Paramere slender, distinctly exceeding median tooth of gonostylus, and bearing a row of spines along inner margin. Aedeagus stout, with ventral valves membranous and dorsal valves sclerotized. dorsal valve with a pair of auriculate processes dorsally and extending to median tooth of gonostylus.

Emergence: July to August.

Distribution: Hunan (Sangzhi) and Guangxi (Longsheng).

164. 模蝎蛉 *Panorpa typicoides* Cheng, 1949

【测量】 前翅长 12.5 mm。

【形态特征】 头顶黄褐色，单眼三角区黑色。1 条宽阔的黑色横带贯穿单眼三角区。喙黄褐色。前胸背板黑褐色。中、后胸背板黑褐色，中央具 1 条黄色的中纵带。侧板及足黄褐色。翅无色透明，斑纹棕色。前翅端带阔，内部具 2 个透明斑；痣带阔，端支分离；缘斑梭形；基带断开成 2 斑；基斑极小。雄性腹部 1 ~ 5 节背板黑褐色。第 3 节背板上的背中突扁平，盖住第 4 节上尖锐的后背突。第 6 节黑褐色，无臀角，后缘稍向上隆起。第 7、8 节黄褐色，基部缢缩。雄性外生殖器宽卵形。第 9 节背板末端具深 U 形缺刻，两侧形成 1 对指形突起；腹板基柄短；下瓣长带状，伸达生殖刺突基部，内缘具 1 列长鬃；生殖肢基节端部内缘具 1 簇长鬃；生殖刺突内缘具钝三角形中齿和大型碟状亚基齿。阳基侧突端半部螺旋状扭曲，内缘具 1 列短刚毛。阳茎腹瓣膜质短小；背瓣指状，相互分离，背缘近末端具膜质膨大物。

【发生期】 7 ~ 8 月。

【地理分布】 四川（康定）。

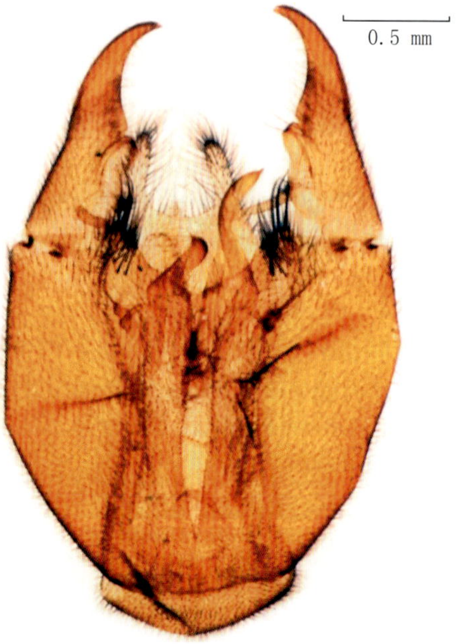

图 3-164-2 模蝎蛉 *Panorpa typicoides* 雄性外生殖器腹面观

Fig. 3-164-2 Male genitalia of *Panorpa typicoides*, ventral view

图 3-164-1 模蝎蛉 *Panorpa typicoides* 雄成虫背面观（腹部末端已去除，四川康定，NWAFU）

Fig. 3-164-1 Male adult of *Panorpa typicoides* with genitalia removed, dorsal view (Kangding, Sichuan, NWAFU)

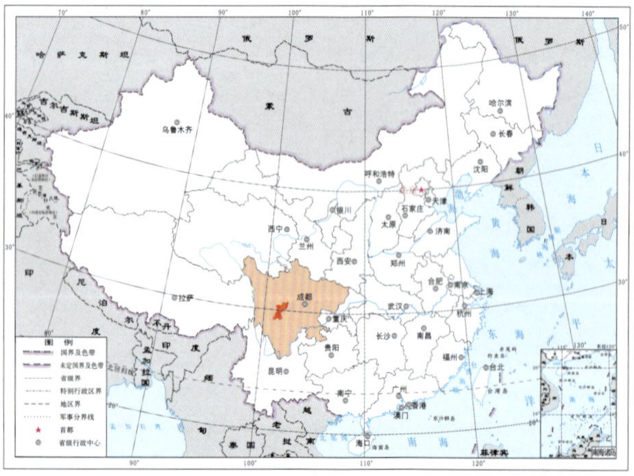

图 3-164-3 模蝎蛉 *Panorpa typicoides* 地理分布

Fig. 3-164-3 Distribution of *Panorpa typicoides*

Measurements: Forewing length 12.5 mm.

Morphology: Vertex yellowish brown. Ocellar triangle black. A broad black transverse stripe passing through ocellar triangle. Rostrum yellowish brown. Pronotum dark brown. Mesonotum and metanotum dark brown, with a yellow mesal stripe. Pleura and legs yellowish brown.Wing membrane hyaline, with markings brown. In forewings, apical band with two hyaline spots innerly; pterostigmal band broad, with a separated apical branch; marginal spot fusiform; basal band split into two spots; basal spot very small. Male terga Ⅰ–Ⅴ dark brown. Notal organ on tergum Ⅲ flat and covering acute postnotal organ on tergum Ⅳ. A6 dark brown, devoid of anal horns and slightly projected dorsad on apex. A7 and A8 yellowish brown and constricted basally. Male genital bulb broad oval. Epandrium emarginated terminally in a U-shape and forming a pair of finger-like processes laterally. Hypandrium with a short basal stalk. Hypovalves long stripe-like, extending to base of gonostylus, and bearing a row of long bristles on inner margin. Gonocoxites with a tuft of long stout setae on ventral apex. Gonostylus with an obtuse triangular median tooth and a large disc-like subbasal process on inner margin. Paramere spiral on distal half, and bearing a row of short spines on inner margin. Aedeagus with ventral valves membranous and short, and dorsal valves finger-like and divergent, bearing a membranous enlarged process on dorsal surface subapically.

Emergence: July to August.

Distribution: Sichuan (Kangding).

165. 王氏蝎蛉 *Panorpa waongkehzengi* Navás, 1935

【测量】 前翅长 8.0 ~ 9.0 mm。

【形态特征】 头顶、喙黄褐色，单眼三角区黑色。前胸背板黑褐色；中、后胸背板黑褐色，中央具宽阔的黄色纵带。侧板及足黄褐色。翅略带黄色，透明，斑纹黄褐色。翅痣黄色、明显。前翅端带分离成两条细横带；痣带完整，基支较粗，端支纤细，末端强烈向内弯曲；缘斑不规则；基带断裂成 2 斑；无基斑；Rs 共分为 4 支，其中 R_2 脉不分支（蝎蛉科 Panorpidae 大多物种 Rs 分为 5 ~ 6 支，R_2 分为 2 ~ 3 支）。雄性腹部 1 ~ 5 节背板黄褐色，中央具向后渐窄的黄色纵带。第 3 节背板上的背中突扁平，盖住第 4 节上尖锐的后背突。第 6 节黄褐色，短筒形，无臀角。第 7、8 节黄褐色，短筒形，基部不缢缩。雄性外生殖器卵形。第 9 节背板宽阔，末端渐窄，具浅缺刻；腹板具阔长的基柄；下瓣钳形，内缘具 1 钝突起。生殖刺突长约为基节之半，内缘具发达的钝三角形中齿和柱状亚基齿。阳基侧突简单，纤细，内缘具 1 列刺毛。阳茎腹瓣膜质短小；背瓣强烈延伸超过生殖肢基节端部，端部相互分离。

【发生期】 8 月。

【地理分布】 江西（庐山）、湖南（浏阳）。

图 3-165-1 王氏蝎蛉 *Panorpa waongkehzengi* 雌成虫背面观（腹末已去除，湖南浏阳，NWAFU）

Fig. 3-165-1 Female adult of *Panorpa waongkehzengi* with terminalia removed, dorsal view (Liuyang, Hunan, NWAFU)

Measurements: Forewing length 8.0–9.0 mm.

Morphology: Vertex and rostrum yellowish brown. Ocellar triangle black. Pronotum dark brown. Mesonotum and metanotum dark brown, with a broad

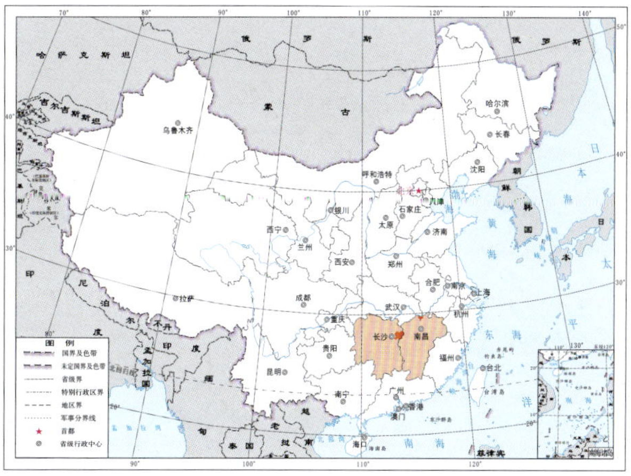

图 3-165-2 王氏蝎蛉 *Panorpa waongkehzengi* 地理分布

Fig. 3-165-2 Distribution of *Panorpa waongkehzengi*

yellow mesal stripe. Pleura and legs yellowish brown. Wing membrane hyaline, slightly tinged with yellow. Markings yellowish brown. Pterostigma yellow and conspicuous. In forewings, apical band split into two slender transverse bands; pterostigmal band complete, with a broad basal branch and a slender apical branch curving abruptly on posterior portion; marginal spot irregular; basal band split into two spots; basal spot absent. Rs quadrifurcated with an unbranched R_2 (most species in Panorpidae have Rs five-or six-branched with R_2 bifurcated or trifurcated). Male terga Ⅰ–Ⅴ yellowish brown. Notal organ on tergum Ⅲ flat and covering acute postnotal organ on tergum Ⅳ. A6 yellowish brown, short cylindrical and devoid of anal horns. A7 and A8 yellowish brown, short cylindrical, and not constricted basally. Male genital bulb oval. Epandrium broad and tapering towards slightly emarginate apex. Hypandrium with a long and broad basal stalk. Hypovalves pincer-like, bearing an obtuse projection on inner margin. Gonostylus half length of gonocoxites, and bearing an obtuse triangular median tooth and a columnar subbasal process on inner margin. Paramere simple and slender, and bearing a row of spines on inner margin. Aedeagus with ventral valves membranous and short, and dorsal valves greatly elongated and exceeding apex of gonocoxites, and divergent apically.

Emergence: August.

Distribution: Jiangxi (Lushan Mountain) and Hunan (Liuyang).

166. 杨氏蝎蛉 *Panorpa yangi* Chou, 1981

【测量】 前翅长 9.5 ~ 10.5 mm。

【形态特征】 头顶黄褐色。单眼三角区黑色，被 1 条黑色横带贯穿。喙红褐色。胸部背板黄褐色，在中、后胸背板前缘两侧具 4 个黑色斑点。侧板及足黄褐色。翅无色透明，斑纹褐色。前翅端带阔，后缘具 1 个透明斑；痣带端支窄，基支阔；基带完整；缘斑和基斑大而明显。雄性腹部 1 ~ 5 节背板黄褐色。第 3 节上的背中突扁平，盖住第 4 节上尖锐的后背突。第 6 节黄褐色，筒形，无臀角，但末端向上稍隆起，被稀疏长毛。第 7、8 节黄褐色，基部缢缩。雄性外生殖器阔卵形。第 9 节背板末端具深 U 形缺刻，两侧形成 1 对指形突起；腹板的基柄极短；下瓣阔长，伸达生殖肢基节的 3/4 处，内缘具 1 列长鬃毛。生殖肢基节腹面端部内缘具有 4 ~ 6 根长刚毛；生殖刺突内缘具钝三角形中齿，亚基齿大，盘状。阳基侧突纤细，端半部强烈弯成鱼钩状，相互交叉。阳茎腹瓣膜质短小；背瓣骨化，背向弯曲，近末端背侧具膜质膨大的突起，末端指状。

【发生期】 7 ~ 8 月。

图 3-166-1 杨氏蝎蛉 *Panorpa yangi* 成虫背面观（陕西黄陵，NWAFU）

1. 雄虫 2. 雌虫

Fig. 3-166-1 Adults of *Panorpa yangi*, dorsal view (Huangling, Shaanxi, NWAFU)

1.Male 2. Female

图 3-166-2 杨氏蝎蛉 *Panorpa yangi* 雄性外生殖器腹面观

Fig. 3-166-2 Male genitalia of *Panorpa yangi*, ventral view

【地理分布】 陕西(甘泉、黄陵、富县)、宁夏(泾源)。

Measurements: Forewing length 9.5–10.5 mm.

Morphology: Vertex yellowish brown. A black transverse stripe passing through black ocellar triangle. Rostrum reddish brown. Nota yellowish brown. Four black spots on latero-anterior corner of Mesonotum and metanotum. Pleura and legs yellowish brown. Wing membrane hyaline, with markings brown. In forewings, apical band broad, with a hyaline spot posteriorly; pterostigmal band with a narrow apical branch and a broad basal branch; basal band complete; marginal and basal spots large and conspicuous. Male terga Ⅰ–Ⅴ yellowish brown. Notal organ on tergum Ⅲ flat and covering acute postnotal organ on

图 3-166-3 杨氏蝎蛉 *Panorpa yangi* 地理分布

Fig. 3-166-3 Distribution of *Panorpa yangi*

tergum Ⅳ. A6 yellowish brown, cylindrical and devoid of anal horns, and slightly projected on dorsal apex bearing sparse long setae; A7 and A8 yellowish brown and constricte basally. Male genital bulb broad oval. Epandrium deeply emarginate terminally in a deep U-shape and forming a pair of finger-like processes laterally. Hypandrium with a greatly shortened basal stalk. Hypovalves long and broad, extending to three-fourths of gonocoxite, and bearing a row of long bristles on inner margin. Gonocoxites with 4–6 stout setae on ventral apex innerly. Gonostylus with an obtuse triangular median tooth and a large disc-like subbasal process. Parameres slender, intensively curved and hook-like on apical halves, and crossed mesally. Aedeagus with ventral valves membranous and short; dorsal valves sclerotized and curved dorsad, and each with a membranous enlarged process on subapical portion dorsally, and finger-like at apex.

Emergence: July to August.

Distribution: Shaanxi (Ganquan, Huangling, Fuxian) and Ningxia (Jingyuan).

九、华蝎蛉属 *Sinopanorpa* Cai & Hua, 2008

华蝎蛉属 *Sinopanorpa* 的主要识别特征为：翅黄色，斑纹边缘模糊；雄性腹部第 7 节基部强烈缢缩成柄状；生殖刺突长于生殖肢基节。

华蝎蛉属 *Sinopanorpawg* 全世界共有 3 种，中国已记录 3 种，本图鉴介绍 3 种。

The genus *Sinopanorpa* can be recognized mainly by the following diagnostic characters: wing membrane yellow; boundaries of wing markings blurred; A7 extremely constricted and stalk-like at the basal portion in males; and gonostylus longer than gonocoxites.

Three species have been described worldwide and endemic to China. All of them are included in this atlas.

图 3-I 染翅华蝎蛉 ***Sinopanorpa tincta*** 雄成虫侧面观（钟问摄）

Fig. 3-I Male adult of ***Sinopanorpa tincta***, lateral view (photo by Wen Zhong)

中国华蝎蛉属 *Sinopanorpa* 分种检索表

（3 种，雄性）

1. 腹部背板黑褐色；生殖刺突的亚基齿较窄，在近端部分为 2 个尖齿；阳基侧突超过生殖刺突的中齿 ………………………………………………………………………… 168. 南宫山华蝎蛉 ***Sinopanorpa nangongshana***
   腹部背板黄褐色；生殖刺突的亚基齿宽大，在近中部分为 2 个尖齿；阳基侧突不超过生殖刺突的中齿 …… 2
2. 生殖刺突亚基齿分为 2 个尖齿，其中端齿小于基齿；阳基侧突内缘具梳齿状长刺 ………………………………………………………………………… 169. 染翅华蝎蛉 ***Sinopanorpa tincta***
   生殖刺突亚基齿分为 2 个尖齿，其中端齿稍大于基齿；阳基侧突内缘具短刺 ………………………………………………………………………… 167. 指形华蝎蛉 ***Sinopanorpa digitiformis***

Key to species of *Sinopanorpa* in China

(3 species, males)

1. Terga dark brown; gonostylus with a relatively narrow subbasal process, which splits into two acute teeth subbasally; paramere exceeding median tooth of gonostylus ……………………………………… 168. ***Sinopanorpa nangongshana***
   Terga yellowish brown; gonostylus with a relatively broad subbasal process, which splits into two acute teeth nearly at middle; paramere not exceeding median tooth of gonostylys ……………………………………………………………… 2
2. Subbasal process of gonostylus split into two acute teeth, with distal tooth smaller than basal tooth; paramere with long comb-like spines on inner margin …………………………………………………………………… 169. ***Sinopanorpa tincta***
   Subbasal process of gonostylus split into two acute teeth, with distal tooth slightly larger than basal tooth; paramere with short spines on inner margin ………………………………………………………… 167. ***Sinopanorpa digitiformis***

167. 指形华蝎蛉 *Sinopanorpa digitiformis* Huang & Hua, 2008

【测量】 前翅长 17.0 ~ 18.0 mm。

【形态特征】 头顶黑褐色，单眼三角区黑色。喙黄褐色。胸部背板黑褐色，腹板及足浅黄褐色。翅蜡黄色，半透明，斑纹黄褐色。翅痣红褐色。前翅端带阔；痣带完整，具基支和端支；基带阔；无缘斑、基斑。雄性腹部 1 ~ 5 节背板黄褐色。第 3 节上的背中突扁平，盖住第 4 节上尖锐的后背突。第 6 节黄褐色，长筒形，背侧中部具 1 簇长毛；第 7、8 节黄褐色，第 7 节基部 1/3 强烈缢缩成柄状；第 8 节基部轻微缢缩。雄性外生殖器阔卵形。第 9 节

图 3-167-1 指形华蝎蛉 *Sinopanorpa digitiformis* 成虫背面观（湖北神农架，NWAFU）
1. 雄虫　2. 雌虫

Fig. 3-167-1 Adults of *Sinopanorpa digitiformis*, dorsal view (Shennongjia, Hubei, NWAFU)
1.Male　2. Female

背板阔长，末端仅稍缺刻；腹板基柄极短；下瓣长带状，端部圆阔，伸达生殖肢基节端部，内缘具1列长鬃。生殖刺突长于基节，内缘具不明显的三角形中齿和发达、分叉而尖锐的亚基齿。阳基侧突指状，中部波状弯曲，端半部直，内缘具1列短毛。阳茎扁平，不超出生殖肢基节。

【发生期】 6 ~ 8 月。

【地理分布】 湖北（神农架）。

Measurements: Forewing length 17.0–18.0 mm.

Morphology: Vertex dark brown. Ocellar triangle black. Rostrum yellowish brown. Nota dark brown. Pleura and legs light yellowish brown. Wing membrane translucent, tinged with waxy yellow. Markings yellowish brown. Pterostigma reddish brown. In forewings, apical band broad; pterostigmal band complete with a basal branch and an apical branch; basal band broad; marginal and basal spots absent. Male terga Ⅰ–Ⅴ yellowish brown. Notal organ on tergum Ⅲ flat and covering acute postnotal organ on tergum Ⅳ. A6 yellowish brown, long cylindrical and bearing a tuft of long seta on dorsal surface medially. A7 and A8 yellowish brown. A7 intensively stalked on basal third; A8 slightly constricted at base. Male genital bulb broad oval. Epandrium long, broad and slightly emarginate terminally. Hypandrium with a greatly shortened basal stalk. Hypovalves long stripe-like with rounded apexes, extending to apex of gonocoxites, and bearing long bristles along inner margin. Gonostylus longer than gonocoxites, and bearing an inconspicuous median tooth and a stout acute and bifurcated subbasal process on inner margin. Paramere finger-like, curved wave-like at middle, and straight on distal half, bearing a row of short spines on inner margin. Aedeagus flat and not exceeding apex of gonocoxites.

Emergence: June to August.

Distribution: Hubei (Shennongjia).

图 3-167-2 指形华蝎蛉 *Sinopanorpa digitiformis* 雄性外生殖器腹面观

Fig. 3-167-2 Male genitalia of *Sinopanorpa digitiformis*, ventral view

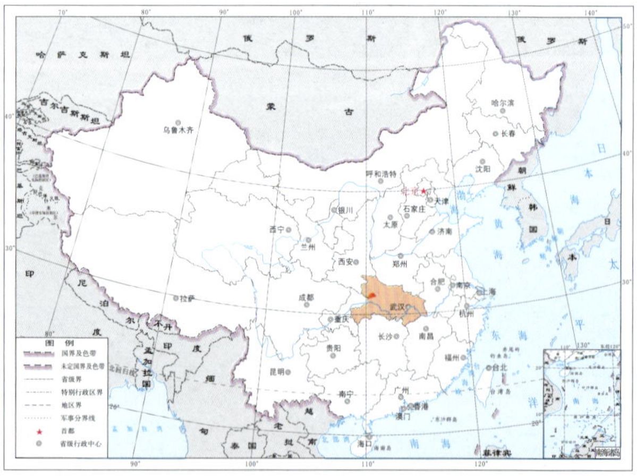

图 3-167-3 指形华蝎蛉 *Sinopanorpa digitiformis* 地理分布

Fig. 3-167-3 Distribution of *Sinopanorpa digitiformis*

168. 南宫山华蝎蛉 *Sinopanorpa nangongshana* Cai & Hua, 2008

【测量】 前翅长 16.0 ~ 17.0 mm。

【形态特征】 头顶红棕色；1条黑色横带贯穿单眼三角区。喙黄褐色。胸部背板黑褐色，侧板及足浅乳黄色。翅蜡黄色，半透明，斑纹黄褐色。翅痣红褐色。前翅端带阔；痣带完整，具基支和端支；基带断开成2斑；无缘斑、基斑。雄性腹部1 ~ 5节背板黑褐色。第3节上的背中突扁平，盖住第4节上尖锐的后背突。第6节黑褐色，长筒形，背侧中部具1簇长毛；第7、8节黑褐色，第7节基部1/3强烈缢缩成柄状；第8节基部轻微缢缩。雄性

图 3-168-1 南宫山华蝎蛉 ***Sinopanorpa nangongshana*** 成虫侧面观（陕西岚皋南宫山，姜碌摄）

1. 雄虫 2. 雌虫

Fig. 3-168-1 Adults of ***Sinopanorpa nangongshana***, lateral view (Mount Nangong Lan'gao, Shaanxi, photo by Lu Jiang)

1.Male 2. Female

外生殖器阔卵形。第9节背板阔长，末端仅稍缺刻；腹板基柄极短；下瓣长带状，端部圆阔，伸达生殖肢基节端部，内缘具1列长鬃。生殖刺突长于基节，内缘具不明显的三角形中齿和发达、分叉而尖锐的亚基齿。阳基侧突极长，超过生殖刺突的中齿，端半部具密集的短毛刺。阳茎扁平，不超出生殖肢基节。

【发生期】 7 ~ 8月。

【地理分布】 陕西（南宫山、化龙山）。

Measurements: Forewing length 16.0–17.0 mm.

Morphology: Vertex reddish brown. A black transverse stripe passing through black ocellar triangle. Rostrum yellowish brown. Nota dark brown. Pleura and legs creamy yellow. Wing membrane translucent, tinged with waxy yellow. Markings yellowish brown. Pterostigma reddish brown. In forewings, apical band broad; pterostigmal band complete, with a basal branch and an apical branch; basal band split into two spots; marginal and basal spots absent. Male terga Ⅰ – Ⅴ dark brown. Notal organ on tergum Ⅲ flat and covering acute postnotal organ on tergum Ⅳ. A6 dark brown, long cylindrical and bearing a tuft of long setae on dorsal surface medially. A7 and A8 dark brown. A7 intensively stalked on basal third; A8 slightly constricted at base. Male genital bulb broad oval. Epandrium long, broad and slightly emarginate terminally. Hypandrium with a greatly shortened basal stalk. Hypovalves long stripe-like with rounded apexes, extending to apex of gonocoxites, and bearing long bristles along inner margin. Gonostylus longer than gonocoxites, and bearing an inconspicuous median tooth and a stout acute and bifurcated subbasal process on inner margin. Paramere extremely elongated, exceeding median tooth of gonostylus, and bearing dense short spines on apical half. Aedeagus flat and not exceeding apex of gonocoxites.

Emergence: July to August.

Distribution: Shaanxi (Mount Nangong, Hualong Mountains).

图 3-168-2 南宫山华蝎蛉 ***Sinopanorpa nangongshana*** 雄性外生殖器腹面观

Fig. 3-168-2 Male genitalia of ***Sinopanorpa nangongshana***, ventral view

图 3-168-3 南宫山华蝎蛉 ***Sinopanorpa nangongshana*** 地理分布

Fig. 3-168-3 Distribution of ***Sinopanorpa nangongshana***

169. 染翅华蝎蛉 *Sinopanorpa tincta* (Navás, 1931)

【测量】 前翅长 15.5 ~ 17.0 mm。

【形态特征】 头顶黑褐色，单眼三角区黑色。喙黄褐色。胸部背板黑褐色，腹板及足乳黄色。翅蜡黄色，半透明，斑纹黄褐色。前翅端带阔；翅痣红褐色，痣带完整，具基支和端支；基带断开成2斑；缘斑小；无基斑。雄性腹部1 ~ 5节背板黄褐色。第3节上的背中突扁平，盖住第4节上尖锐的后背突。第6节黄褐色，筒形，背侧中部具1簇长毛；

图 3-169-1 染翅华蝎蛉 ***Sinopanorpa tincta*** 成虫侧面观（陕西宝鸡天台山国家森林公园，姜碌摄）

1. 雄虫 2. 雌虫

Fig. 3-169-1 Adults of ***Sinopanorpa tincta***, lateral view (Tiantaishan National Forest Park, Baoji, Shaanxi, photo by Lu Jiang)

1. Male 2. Female

第 7、8 节黄褐色；第 7 节基部 1/3 强烈缢缩成柄状；第 8 节基部轻微缢缩。雄性外生殖器卵形。第 9 节背板长，末端缺刻浅；腹板基柄短，下瓣长带状，端部钝，伸达生殖肢基节端部。生殖刺突长于基节，内缘具不明显的三角形中齿和发达、分叉的亚基齿。阳基侧突细，伸达生殖刺突的亚基齿，内缘具 1 列梳状长刚毛。阳茎扁平，不超过生殖肢基节端部。

【发生期】 7 ~ 8 月。

【地理分布】 陕西（太白、宁陕、留坝、周至、南郑、宝鸡）、甘肃（文县）、四川（平武）。

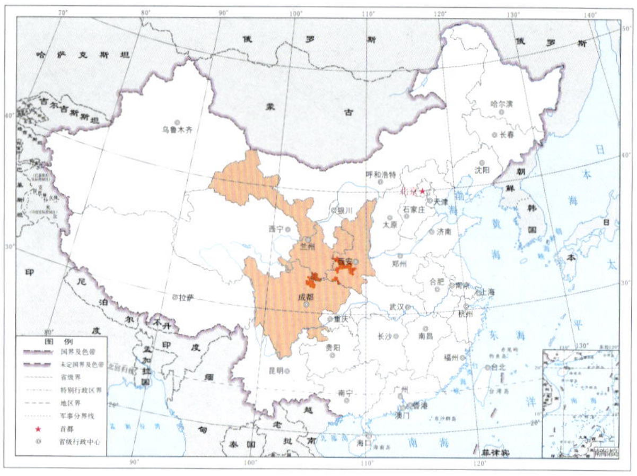

图 3-169-4 染翅华蝎蛉 *Sinopanorpa tincta* 地理分布

Fig. 3-169-4 Distribution of ***Sinopanorpa tincta***

Measurements: Forewing length 15.5–17.0 mm.

图 3-169-2 染翅华蝎蛉 ***Sinopanorpa tincta*** 雄性腹部末端侧面观

Fig. 3-169-2 Male teminal abdomen of ***Sinopanorpa tincta***, lateral view

图 3-169-3 染翅华蝎蛉 ***Sinopanorpa tincta*** 雄性外生殖器腹面观

Fig. 3-169-3 Male genitalia of ***Sinopanorpa tincta***, ventral view

Morphology: Vertex dark brown. Ocellar triangle black. Rostrum yellowish brown. Nota dark brown. Pleura and legs creamy yellow. Wing membrane translucent, tinged with waxy yellow. Markings yellowish brown. Pterostigma reddish brown. In forewings, apical band broad; pterostigmal band complete, with a basal branch and an apical branch; basal band split into two spots; marginal spot small; basal spot absent. Male terga Ⅰ–Ⅴ yellowish brown. Notal organ on tergum Ⅲ flat and covering the acute postnotal organ on tergum Ⅳ. A6 yellowish brown, cylindrical and bearing a tuft of long setae on the dorsal surface medially. A7 and A8 yellowish brown. A7 intensively stalked on the basal third; A8 slightly constricted at base. Male genital bulb oval. Epandrium long, and slightly emarginate terminally. Hypandrium with a greatly shortened basal stalk. Hypovalves long stripe-like with rounded apexes, extending to the apex of gonocoxites, and bearing long bristles along inner margin. Gonostylus longer than gonocoxites, and bearing an inconspicuous median tooth and a stout acute and bifurcated subbasal process on the inner margin. Paramere slender, extending to the subbasal process of gonostylus, and bearing a row of comb-like spines along inner margin. Aedeagus flat and not exceeding apex of gonocoxites.

Emergence: July to August.

Distribution: Shaanxi (Taibai, Ningshan, Liuba, Zhouzhi, Nanzheng, Baoji), Gansu (Wenxian) and Sichuan (Pingwu).

第四部分 拟蝎蛉科

Part Ⅳ Panorpodidae

（拟蝎蛉 short-faced scorpionflies）

拟蝎蛉科 Panorpodidae 昆虫与蝎蛉科 Panorpidae 昆虫十分相似，但喙和腹部 6 ~ 8 节均较短。全世界共有 2 属 10 余种，中国仅 1 属 3 种，本图鉴收录 1 种。

The short-faced scorpionflies are very similar to the common scorpionflies, but have relatively shorter rostrums and A6–A8. Two genera and more than 10 species have been described worldwide, and only three species in one genus are documented in China.One species is included in this atlas.

图 4-a 宽甸拟蝎蛉 ***Panorpodes kuandianensis*** 雄成虫侧面观（辽宁宽甸花脖山，姜碌摄）

Fig. 4-a Male adult of ***Panorpodes kuandianensis***, lateral view (Mount Huabo, Kuandian, Liaoning, photo by Lu Jiang)

一〇、拟蝎蛉属 *Panorpodes* MacLachlan, 1875

拟蝎蛉属 *Panorpodes* 分布于北太平洋周围地区，包括日本、朝鲜、韩国、中国东北以及北美地区，全世界共有 8 种，中国已记录 3 种，本图鉴收录 1 种。

Eight known species in the genus *Panorpodes* are distributed in the north Pacific Rim, including Japan, Korean Peninsula, northeastern China and North America. Three species are recorded in China, and only one species is included in this atlas.

图 4-b 宽甸拟蝎蛉 ***Panorpodes kuandianensis*** 雌成虫侧面观（辽宁宽甸花脖山，姜碌摄）

Fig. 4-b Female adult of ***Panorpodes kuandianensis***, lateral view (Mount Huabo, Kuandian, Liaoning, photo by Lu Jiang)

170. 宽甸拟蝎蛉 *Panorpodes kuandianensis* Zhong, Zhang & Hua, 2011

图 4-170-1 宽甸拟蝎蛉 ***Panorpodes kuandianensis*** 成虫侧面观（辽宁宽甸花脖山，姜碌摄）

1. 雄虫 2. 雌虫

Fig. 4-170-1 Adults of ***Panorpodes kuandianensis***, lateral view (Mount Huabo, Kuandian, Liaoning, photo by Lu Jiang)

1. Male 2. Female

【测量】 前翅长 12.4 ~ 15.2 mm。

【形态特征】 头顶黄色。单眼三角区黄褐色，具 4 根显著的单眼鬃。喙黄色，短粗。胸部背板黄褐色，两侧稍加深。侧板及足浅黄褐色。翅浅黄色，半透明。前翅端带不规则，内部具一系列透明大斑；痣带退化；无其他斑纹。有的个体完全无斑纹。雄性腹部第 1 ~ 5 节背板黄褐色，两侧颜色稍加深。第 3 节背板后缘的背中突极不发达，仅稍盖住第 4 节背板上尖锐的后背突。第 6 ~ 8 节与前数节相似，不像蝎蛉科 Panorpidae 一般延长。雄性外生殖器长椭圆形。第 9 节背板阔长，末端圆钝；腹板基柄阔长；下瓣伸达生殖肢基节端部以外，末端平截，内缘近基部具 1 个耳状突起，每瓣内缘具 10 ~ 16 根黑色短刺；两瓣基部之间形成新月形窗口。生殖刺突短于基节之半，内缘近基部具 1 指形突起。阳基侧突大部分与阳茎愈合。阳茎背、腹瓣紧密愈合，侧突起圆钝。

【发生期】 6 ~ 7 月。

【地理分布】 辽宁（宽甸）。

Measurements: Forewing length 12.4–15.2 mm.

Morphology: Vertex yellow. Ocellar triangle yellowish brown, bearing four conspicuous ocellar bristles. Rostrum yellow, short and stout.Nota yellowish brown, and slightly deepened towards lateral portion. Pleura and legs light yellowish brown. Wing membrane translucent, tinged with light yellow. In forewings, apical band irregular with a series of large hyaline spots innerly; pterostigmal band greatly reduced; other markings absent. Occasionally, all markings absent in some individuals. Male terga Ⅰ–Ⅴ yellowish brown, with darkened flanks. Notal organ on tergum Ⅲ poorly-developed and slightly covering the acute postnotal organ on tergum Ⅳ. A6–A8 similar to the proceeding segments and not elongated as in Panorpidae. Male genital bulb long elliptical. Epandrium long and broad, with a rounded apex. Hypandrium long and broad. Hypovalves truncated terminally, extending beyond the apex of gonocoxite, bearing an auriculate process subbasally and 10–16 black short spines on the inner margin; a crescent window formed between the two hypovalves basally. Gonostylus shorter than half the length of gonocoxites, and bearing a finger-like subbasal process on the inner margin. Paramere almost fused with aedeagus. Aedeagus with ventral and dorsal valves intensively fused with rounded lateral processes.

Emergence: June to July.

Distribution: Liaoning (Kuandian).

图 4-170-2 宽甸拟蝎蛉 ***Panorpodes kuandianensis*** 雄性外生殖器腹面观

Fig. 4-170-2 Male genitalia of ***Panorpodes kuandianensis***, ventral view

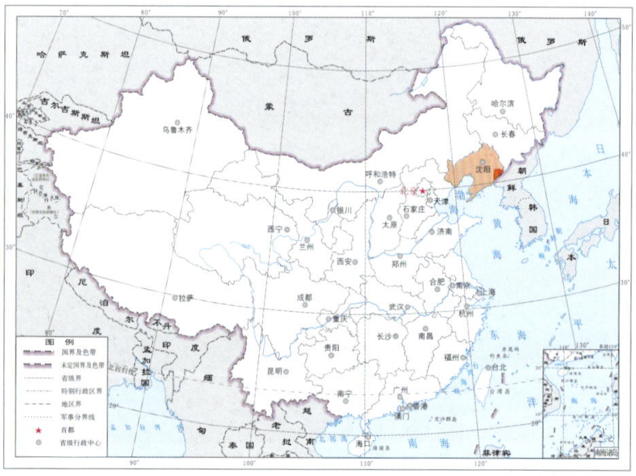

图 4-170-3 宽甸拟蝎蛉 ***Panorpodes kuandianensis*** 地理分布

Fig. 4-170-3 Distribution of ***Panorpodes kuandianensis***

参考文献
References

BENTON M J, 1995. Diversification and extinction in the history of life [J]. Science, 268(5207): 52.

BICHA W, 2018. Biodiversity of Mecoptera//Foottit RG, Adler PH. (eds) Insect Biodiversity: Science and Society, Ⅱ, PP. 705-720, John Wiley & Sons, Hoboken.

BICHA W J, 1983. A new species of scorpionfly (Mecoptera: Panorpidae) from North America [J]. Proceedings of the Entomological Society of Washington, 85(1): 152-155.

BICHA W J, 2006. New scorpionflies (Mecoptera: Panorpidae) from Jalisco, Michoacán, and Oaxaca, Mexico [J]. Proceedings of the Entomological Society of Washington, 108(1): 24-34.

BICHA W J, 2007. The first hangingfly (Mecoptera: Bittacidae) described from Vietnam [J]. Russian Entomological Journal, 16: 333-335.

BICHA W J, 2010. A review of the scorpionflies (Mecoptera) of Indochina with the description of a new species of *Neopanorpa* from northern Thailand [J]. Zootaxa, 2408(1): 61-67.

BICHA W J, 2015a. A nearly cryptic scorpionfly, *Panorpa cryptica* n. sp. (Mecoptera: Panorpidae) from North America [J]. Zootaxa, 3973(3): 591-600.

BICHA W J, 2015b. The scorpionflies (Mecoptera) of Indochina with the description of new species of *Bittacus* and *Neopanorpa* [J]. Proceedings of the Entomological Society of Washington, 117(4): 435-51.

BICHA W J, SCHIFF N, PHAM T H, et al., 2017. New species of *Neopanorpa* (Mecoptera) from Vietnam, with a key to the species of Mecoptera from Vietnam [J]. Proceedings of the Entomological Society of Washington, 119(4): 529-544.

BOSTOCK J, RILEY H T, 1855. The Natural History of Pliny, Volume Ⅲ [M]. London, H. G. Bohn, pp. 29-31.

BYERS G W, 1958. Descriptions and distributional records of American Mecoptera [J]. Journal of the Kansas Entomological Society, 31: 213-222.

BYERS G W, 1962. Descriptions and distributional records of American Mecoptera. Ⅱ [J]. Journal of the Kansas Entomological Society, 35: 299-307.

BYERS G W, 1963. The life history of *Panorpa nuptialis* (Mecoptera: Panorpidae) [J]. Annals of the Entomological Society of America, 56(2): 142-149.

BYERS G W, 1965. The Mecoptera of Indo-China [J]. Pacific Insects, 7(4): 705-748.

BYERS G W, 1966. Mecoptera from Borneo and Tioman Islands [J]. Pacific Insects, 8: 885-892.

BYERS G W, 1970. New and little known Chinese Mecoptera [J]. Journal of the Kansas Entomological Society, 43(4): 383-394.

BYERS G W, 1971. A new *Neopanorpa* from Nepal [J]. Journal of the Kansas Entomological Society, 44(4): 534-539.

BYERS G W, 1973. Descriptions and distributional records of American Mecoptera. Ⅲ [J]. Journal of the Kansas Entomological Society, 46: 362-375.

BYERS G W, 1975. Descriptions and distributional records of African Mecoptera [J]. Journal of the Entomological Society of Southern Africa, 38: 217-222.

BYERS G W, 1988. Geographical affinities of the North American Mecoptera [M]//DOWNES J A, KAVANAUGH D H (Eds.). Origins of the North American Insect Fauna [M]. Memoirs of the Entomological Society of Canada, Cambridge: Cambridge University Press, 25-30.

BYERS G W, 1989. The Nannochoristidae of South America (Mecoptera) [J]. The University of Kansas Science Bulletin, 54(2): 25-34.

BYERS G W, 1993. Autumnal Mecoptera of southeastern United States [J]. University of Kansas Science Bulletin, 55(2): 57-96.

BYERS G W, 1994. Taiwanese species of *Neopanorpa* (Insecta: Mecoptera: Panorpidae) [J]. Annals of Carnegie Museum, 63(2): 185-192.

BYERS G W, 1996. Descriptions and distributional records of American Mecoptera. Ⅳ [J]. University of Kansas Science Bulletin, 55(14): 519-547.

BYERS G W, 1997. Biology of *Brachypanorpa* (Mecoptera: Panorpodidae) [J]. Journal of the Kansas Entomological Society, 70(4): 313-322.

BYERS G W, 1999. Thirteen new Panorpidae (Mecoptera) from northern Burma [J]. Entomologica Scandinavica, 30: 197-218.

BYERS G W, 2002a. Mecoptera from Taiwan [J]. Journal of the Kansas Entomological Society, 75(4): 252-258.

BYERS G W, 2002b. Scorpionflies, hangingflies, and other Mecoptera [J]. The Kansas School Naturalist, 48(1): 1-15.

BYERS G W, 2008. New names in *Neopanorpa* (Mecoptera : Panorpidae) [J]. Journal of the Kansas Entomological Society, 81(1): 1-3.

BYERS G W, 2009, Mecoptera: scorpionflies, hangingflies [M]//RESH V H, CARDÉ R T. (Eds.), Encyclopedia of Insects. 2nd edition. Cambridge, Academic Press, 611-614.

BYERS G W, Thornhill R, 1983. Biology of the Mecoptera [J]. Annual Review of Entomology, 28(1): 203-228.

CAI L J, HUA B Z, 2009a. A new *Neopanorpa* (Mecoptera, Panorpidae) from China with notes on its biology [J]. Deutsche Entomologische Zeitschrift, 56(1): 93-99.

CAI L J, HUA B Z, 2009b. Morphology of the immature stages of *Panorpa qinlingensis* (Mecoptera: Panorpidae) with notes on its biology [J]. Entomologica Fennica, 20(3): 215-224.

CAI L J, HUANG P Y, HUA B Z, 2006. Two new Chinese *Bittacus* Latreille (Mecoptera: Bittacidae) from Michangshan Mountains [J]. Entomotaxonomia, 28(2): 127-130. [蔡立君 , 黄蓬英 , 花保祯 , 2006. 米仓山蚊蝎蛉二新种记述 [J]. 昆虫分类学报 , 28(2): 127-130.]

CAI L J, HUANG P Y, HUA B Z, 2008. *Sinopanorpa*, a new genus of Panorpidae (Mecoptera) from the Oriental China with descriptions of two new species [J]. Zootaxa, 1941: 43-54.

CARPENTER F M, 1931. Revision of the Nearctic Mecoptera [J]. Bulletin of the Museum of Comparative Zoology, 72: 205-277.

CARPENTER F M, 1935. New Nearctic Mecoptera, with notes on other Species [J]. Psyche, 42: 105-122.

CARPENTER F M, 1936. Descriptions and records of Nearctic Mecoptera [J]. Psyche, 43: 56-64.

CARPENTER F M, 1938. Mecoptera from China, with descriptions of new species [J]. Proceedings of the Entomological Society of Washington, 40(9): 267-281.

CARPENTER F M, 1945. Panorpidae from China (Mecoptera) [J]. Psyche, 52(1/2): 70-78.

CARPENTER F M, 1948. Notes on Chinese Panorpidae (Mecoptera) [J]. Psyche, 55: 28-30.

CHAU C-S, BYERS G W, 1978. The Mecoptera of Indonesia: genus *Neopanorpa* [J]. The University of Kansas Science Bulletin, 51(11): 341-405.

CHEN J, HUA B Z, 2011. A second species of *Bittacus* Latreille, 1805 (Mecoptera: Bittacidae) from Hainan Island, China [J]. Zootaxa, 3093: 64-68.

CHEN J, TAN J L, HUA B Z, 2013. Review of the Chinese *Bittacus* (Mecoptera: Bittacidae) with descriptions of three new species [J]. Journal of Natural History, 47(21/22): 1463-1480.

CHEN Q X, HUA B Z, 2016. Ultrastructure and morphology of compound eyes of the scorpionfly *Panorpa dubia* (Insecta: Mecoptera: Panorpidae) [J]. PLoS ONE, 11(6): e0156970.

CHENG F Y, 1949. New species of Mecoptera from northwest China [J]. Psyche, 56(4): 139-173.

CHENG F Y, 1952. Additions to the Mecopterous fauna of Formosa [J]. Psyche, 59(3): 89-95.

CHENG F Y, 1957a. Descriptions of new Panorpidae (Mecoptera) in the collection of the California Academy of Sciences [J]. Memoirs of the College of Agriculture, National Taiwan University, 5(1): 27-33.

CHENG F Y, 1957b. Revision of the Chinese Mecoptera [J]. Bulletin of the Museum of Comparative Zoology at Harvard College, 116(1): 1-118.

CHEUNG D K, MARSHALL S A, WEBB D W, 2006. Mecoptera of Ontario [J]. Canadian Journal of Arthropod Identification, 1: 1-13.

CHOU I, 1963. Some viewpoints about insect taxonomy [J]. Acta Entomologica Sinica, 12(5-6): 586-596. [周尧 , 1963. 有关昆虫分类学的一些观点 [J]. 昆虫学报 , 12(5/6): 586-596.]

CHOU I, RAN R B,WANG S M, 1981. Studies on the classification of Chinese Mecoptera (Ⅰ , Ⅱ) [J]. Entomotaxonomia, 3(1): 1-18.[周尧 , 冉瑞碧 , 王素梅 , 1981. 长翅目昆虫的分类研究（Ⅰ, Ⅱ）[J]. 昆虫分类学报 , 3(1): 1-18.]

CHOU I, WANG S M, LI Z D, et al., 1987. New species of Mecoptera from Hunan Province (Ⅰ)—Studies on the classification of Chinese Mecoptera (Ⅲ) [J]. Entomotaxonomia, 9(3): 201-212. [周尧 , 王素梅 , 李中定 , 等 , 1987. 湖南长翅目新种志（一）—— 长翅目昆虫的分类研究（Ⅲ）[J]. 昆虫分类学报 , 9(3): 201-212.]

CHOU I, WANG S M, LI Z D, et al., 1988. New species of Mecoptera from Hunan Province (Ⅱ)—Studies on the classification of Chinese Mecoptera (Ⅳ) [J]. Entomotaxonomia, 10(1/2): 31-42. [周尧 , 王素梅 , 李中定 , 等 , 1988. 湖南长翅目新种志（二）—— 长翅目昆虫的分类研究（Ⅳ）[J]. 昆虫分类学报 , 10(1/2): 31-42.]

DALLAI R, LUPETTI P, AFZELIUS B A, et al., 2003. Sperm structure of Mecoptera and Siphonaptera (Insecta) and the phylogenetic position of *Boreus hyemalis* [J]. Zoomorphology, 122(4): 211-220.

DUNFORD J C, SOMMA L A, 2008. Scorpionflies (Mecoptera) [M] //CAPINERA J L. (Ed.). Encyclopedia of Entomology, 2nd. Berlin:Springer Netherlands, 3304-3310.

DU W, CHEN J, HUA B Z, 2015, Discovery of *Bittacus ussuriensis* Plukento (Mecoptera: Bittacidae) from China [J]. Entomotaxonomia, 37(4): 1-5. [杜薇 , 陈静 , 花保祯 , 2015. 乌苏里蚊蝎蛉 *Bittacus ussuriensis* Plutenko 在中国的首次记录 [J]. 昆虫分类学报 , 37(4): 1-5.]

DU W, HUA B Z, 2017. Two new species of the genus *Terrobittacus* Tan and Hua, 2009 (Mecoptera: Bittacidae) from southwestern China with a key to species [J]. European Journal of Taxonomy, 294: 1-13.

DU X L, YUE C, HUA B Z, 2009. Embryonic development of the scorpionfly *Panorpa emarginata* Cheng with special reference to external morphology (Mecoptera: Panorpidae) [J]. Journal of morphology, 270: 984-995.

ENDERLEIN G, 1910. Über die Phylogenie und Klassifikation der Mecopteren unter Berücksichtigung der fossilen Formen [J]. Zoologischer Anzeiger, 35: 385-399.

ENDERLEIN G. 1912, Zur Kenntnis der Mecopteren Javas [J]. Notes from the Leyden Museum, 34: 235-239.

Erwin D H, 1990. Carboniferous-Triassic gastropod diversity patterns and the Permo-Triassic mass extinction [J]. Paleobiology, 16(2): 187-203.

ESBEN-PETERSEN P, 1912. H. Sauter's Formosa-Ausbeute: Neuroptera Planipennia [J]. Entomologische Mitteilungen, 1: 197-198.

ESBEN-PETERSEN P, 1913. H. Sauter's Formosa-Ausbeute: Planipennia Ⅱ, Megaloptera and Mecoptera [J]. Entomologische Mitteilungen, 2(7/8): 222-228.

ESBEN-PETERSEN P, 1915. A synonymic list of the order Mecoptera together with descriptions of new species [J].

Entomologiska Meddelelser, 10: 216-242.

ESBEN-PETERSEN P, 1921. Mecoptera. Monographic revision: Collections zoologiques du Baron Edm. de Selys Longchamps [J]. Catalogue Systématique et Descriptif, 5: 1-172.

ESBEN-PETERSEN P, 1927. New and little-known species of Mecoptera and Neuroptera in the Zoological Museum of Helsingfors [J]. Notulae Entomologicae, 7: 13-18.

ESBEN-PETERSEN P, 1934. Two new species of *Panorpa* Linn. (Mecoptera) [J]. Videnskabelige Meddelelser fra Dansk Naturhistorik Forening, Kjobenhaven, 99: 1-3.

FERRIS G F, REES B E, 1939. The morphology of *Panorpa nuptialis* Gerstaecker (Mecoptera: Panorpidae) [J]. Microentomology. Contributions to Entomology from the Natural History Museum of Stanford University, 4(3): 79-108.

FRIEDRICH F, POHL H, BECKMANN F, et al., 2013. The head of *Merope tuber* (Meropeidae) and the phylogeny of Mecoptera (Hexapoda) [J]. Arthropod Structure & Development, 42(1): 69-88.

FU Q, HUA B Z, 2009. A new species of *Panorpa* (Mecoptera: Panorpidae) from Yunnan, China [J]. Entomotaxonomia, 31(3): 201-205. [付强 , 花保祯 , 2009. 中国云南蝎蛉属 *Panorpa* 一新种 [J]. 昆虫分类学报 , 31(3): 201-205.]

FU Q, XIE S, HUA B Z, 2010. Mecoptera: Panorpidae, Bittacidae. In: XU H C (Ed.). Insects of Fengyangshan National Nature Reserve. Beijing, China Forestry Publishing House, 229-230. [付强 , 谢莎 , 花保祯 , 2010. 长翅目 : 蝎蛉科、蚊蝎蛉科 // 徐华潮 . 浙江凤阳山昆虫 [M]. 北京 : 中国林业出版社 , 229-230.]

GAO C, M A N, HUA B Z, 2016. *Cerapanorpa*, a new genus of Panorpidae (Insecta: Mecoptera) with descriptions of three new species [J]. Zootaxa, 4158(1): 93-104.

GAO Q H, HUA B Z, 2013. Co-evolution of the mating position and male genitalia in insects: A case study of a hangingfly [J]. PLoS ONE, 8(12): e80651.

GASSNER G, 1963. Notes on the biology and immature stages of *Panorpa nuptialis* Gerst. (Mecoptera: Panorpidae) [J]. Texas Journal of Science, 15: 142-154.

GRIMALDI D, ENGEL M S, 2005. Evolution of the Insects [M]. Cambridge: Cambridge University Press, 772.

GRIMALDI D A, ZHANG J F, FRASER N C, et al., 2005. New and revised scorpion flies of the extinct family Pseudopolycentropodidae (Mecopteroidea), including two mosquito-like species in Cretaceous amber [J]. Insect Systematics & Evolution, 36: 443-458.

HENNIG W, 1969. Die Stammesgeschichte der Insekten [M]. Frankfurt: Waldemar Kramer Verlag, 436.

HOU X Y, HUA B Z, 2008. Structures of the female reproductive systems in Panorpidae (Mecoptera) with remarks on their taxonomic significance [J]. Acta Zootaxonomica Sinica, 33(3): 427-434.

HSU T C, WU W J, CHOU L Y, 1999. A check list of Mecoptera (Insecta) of Taiwan. Journal of Agricultural Research of China [J].48(2): 149-155. [许洞庆 , 吴文哲 , 周樑镒 . 台湾长翅目（昆虫纲）名录 [J]. 中华农业研究 , 48(2): 149-155.]

HU G L, YAN G, XU H, et al., 2015. Molecular phylogeny of Panorpidae (Insecta: Mecoptera) based on mitochondrial and nuclear genes [J]. Molecular phylogenetics and evolution, 85: 22-31.

HU G L, HUA B Z, 2016. Historical biogeography of the short-faced scorpionflies (Insecta: Mecoptera: Panorpodidae) [J]. Journal of Biogeography, 43(9): 1762-1772.

HÜNEFELD F, BEUTEL R G, 2005. The sperm pumps of Strepsiptera and Antliophora (Hexapoda) [J]. Journal of zoological Systematics and evolutionary Research, 43(4): 297-306.

HUA B Z, 1997. A new species of the genus *Panorpa* (Mecoptera: Panorpidae) from northeast China [J]. Entomotaxonomia, 19(3): 213-215. [花保祯 , 1997. 蝎蛉属 *Panorpa* 东北一新种 [J]. 昆虫分类学报 , 19(3): 213-215.]

HUA B Z, 1998. *Panorpa* in Changbai Mountains with descriptions of four new species (Mecoptera: Panorpidae) [J]. Entomotaxonomia, 20(1): 37-43. [花保祯 , 1998. 吉林长白山蝎蛉记述 [J]. 昆虫分类学报 , 20(1): 37-43.]

HUA B Z, 1999. A new species of Panorpidae of Jigong Mountain in Henan (Mecoptera) [M]//SHEN X C, DENG G F (Eds.).

The fauna and taxonomy of insects in Henan. Vol. 3. Insects of the Jigong Mountain Region. Beijing: China Agricultural Scientech Press, 14-15. [花保祯 , 1999. 河南省鸡公山蝎蛉科一新种（长翅目）[M]// 申效诚 , 邓桂芬 . 河南昆虫分类区西研究第三卷 — 鸡公山区昆虫（一）. 北京 : 中国农业科技出版社 , 14-15.]

HUA B Z, 2004. Panorpodidae—A new recorded family of Mecoptera from China [J]. Entomotaxonomia, 26(4): 284-287. [花保祯 , 2004. 拟蝎蛉科 — 中国新纪录科 [J]. 昆虫分类学报 , 26(4): 284-287.]

HUA B Z, CAI L J, 2009. A new species of the genus *Panorpa* (Mecoptera: Panorpidae) from China with notes on its biology [J]. Journal of Natural History, 43(9/10): 545-552.

HUA B Z, CHOU I, 1997. The Panorpidae (Mecoptera) of Funiu Mountain in Henan Province [J]. Entomotaxonomia, 19(4): 273-278. [花保祯 , 周尧 , 1997. 河南省伏牛山蝎蛉记述 [J]. 昆虫分类学报 , 19(4): 273-278.]

HUA B Z, CHOU I, 1998a. Panorpidae (Mecoptera) in Hainan Island [J]. Entomotaxonomia, 20(2): 133-139. [花保祯 , 周尧 , 1998. 海南岛蝎蛉科昆虫记述 [J]. 昆虫分类学报 , 20(2): 133-139.]

HUA B Z, CHOU I, 1998b. The Bittacidae of Funiu Mountain in Henan (Mecoptera) [M]//SHEN X C, SHI Z Y (Eds.). The fauna and taxonomy of insects in Henan. Vol. 2. Insects of the Funiu Mountains Region. Beijing: China Agricultural Scientech Press, 64-67. [花保祯 , 周尧 , 1998. 河南伏牛山的蚊蝎蛉记述 [M]// 申效诚 , 时振亚 . 河南昆虫分类区西研究第二卷 — 伏牛山区昆虫（一）. 北京 : 中国农业科技出版社 , 64-67.]

HUA B Z, CHOU I, 1999. Panorpidae (Mecoptera) in Xizang, China, with descriptions of three new species of the genus *Neopanorpa* [J]. Entomologia Sinica, 6(1): 11-17. [花保祯 , 周尧 , 1999. 西藏新蝎蛉属昆虫三新种记述（长翅目：蝎蛉科）[J]. 中国昆虫科学 , 6(1): 11-17.]

HUA B Z, PENG T, 2005. A new synonym of *Panorpa tincta* Navas, 1931 (Mecoptera: Panorpidae) [J]. Entomotaxonomia, 27(2): 123-124. [花保祯 , 彭韬 , 2005. 染翅蝎蛉 *Panorpa tincta* Navas, 1931 一新异名 [J]. 昆虫分类学报 , 27(2): 123-124.]

HUA B Z, SUN G H, LI M L, 2001. Sichuan Panorpidae (Mecoptera) kept in the Tianjin Natural History Museum [J]. Entomotaxonomia, 23(2): 120-123. [花保祯 , 孙桂华 , 李淼琳 , 2001. 天津自然博物馆收藏的四川省蝎蛉科记述 [J]. 昆虫分类学报 , 23(2): 120-123.]

HUA B Z, TAN J L, 2007. A new species of *Bittacus* Latreille (Mecoptera, Bittacidae) from Daba Mountain in China [J]. Acta Zootaxonomica Sinica, 32(2): 455-458. [花保祯 , 谭江丽 , 2007. 中国大巴山蚊蝎蛉属一新种 [J]. 动物分类学报 , 32(2): 455-458.]

HUA B Z, TAN J L, HUANG P Y, 2008. Two new species of the genus *Bittacus* (Mecoptera: Bittacidae) from China [J]. Zootaxa, 1749(1): 62-68.

HUA L Z, 2002. List of Chinese Insects. Vol. Ⅱ [M]. Guangzhou: Zhongshan (Sun Yat-sen) University Press, 612 . [华立中 , 2002. 中国昆虫名录 . 第 2 卷 [M]. 广州 : 中山大学出版社 , 612 .]

HUANG P Y, HUA B Z, 2005. Four new species of the Chinese *Bittacus* Latreille (Mecoptera: Bittacidae) [J]. Acta Zootaxonomica Sinica, 30(2): 393-398. [黄蓬英 , 花保祯 , 2005. 中国蚊蝎蛉属四新种记述 [J]. 动物分类学报 , 30(2): 393-398.]

HUANG P Y, HUA B Z, SHEN X C, 2004. A new species of the genus *Panorpa* (Mecoptera: Panorpidae) from Wangwu Mountain [J]. Entomotaxonomia, 26(1): 29-31. [黄蓬英 , 花保祯 , 申效诚 , 2004. 蝎蛉属 *Panorpa* 一新种记述 [J]. 昆虫分类学报 , 26(1): 29-31.]

HUA Y, TAO S H, HUA B Z, 2018. An enigmatic new species of *Panorpa* Linnaeus from the Bashan Mountains (Mecoptera, Panorpidae) [J]. ZooKeys, 777: 109-118.

ISSIKI S, 1927. New and rare species of Mecoptera from Corea, Formosa and Japan [J]. Insecta Matsumurana, 2(1): 1-12.

ISSIKI S, 1929a. Descriptions of new Mecoptera from the Japanese Empire [J]. Transactions of the Natural History Society of Formosa, 19: 260-314.

ISSIKI S, 1929b. Descriptions of new species of the genus *Panorpa* from Japan and Formosa [J]. Journal of the Society of

Tropical Agriculture, 1: 182-191.

ISSIKI S, 1931. Two new species of scorpion flies (Insecta, Order Mecoptera) [J]. Annals and Magazine of Natural History, 7: 219-222.

ISSIKI S, 1933. Morphological studies on the Panorpidae of Japan and adjoining countries and comparison with American and European forms [J]. Japanese Journal of Zoology, 4: 315-416.

ISSIKI S, 1965. Notes on some Mecoptera from Formosa [J]. Special bulletin of Lepidopterological Society of Japan, 1: 245.

ISSIKI S, CHENG F Y, 1947. Formosan Mecoptera with descriptions of new species [J]. Memoirs of the College of Agriculture, national Taiwan University, 1(4): 1-17.

IWASAKI Y, 1996. Hunting and mating behavior in the Japanese hangingfly *Bittacus mastrillii* (Mecoptera: Bittacidae) [J]. Annals of the Entomological Society of America, 89: 869-874.

JIANG L, GAO Q H, HUA B Z, 2015. Larval morphology of the hanging-fly *Bittacus trapezoideus* Huang & Hua (Insecta: Mecoptera: Bittacidae) [J]. Zootaxa, 3957(3): 324-333.

JIANG L,HUA B Z, 2013. Morphology and chaetotaxy of the immature stages of the scorpionfly *Panorpa liui* Hua (Mecoptera: Panorpidae) with notes on its biology [J]. Journal of Natural History, 47(41/42): 2691-2705.

JIANG L, YUE C, HUA B Z, 2014. Larval morphology of *Panorpodes kuandianensis* (Insecta, Mecoptera, Panorpodidae) and its evolutionary implications [J]. ZooKeys, 398: 69-82.

JIANG L, HUA B Z, 2015. Morphological comparison of the larvae of *Panorpa obtusa* Cheng and Neopanorpa lui Chou & Ran (Mecoptera: Panorpidae) [J]. Zoologischer Anzeiger, 255: 62-70.

JU J S, ZHOU W B, 2003. Two new species of Mecoptera from Jiulongshan, China [J]. Jounal of Zhejiang Forestry College, 20(1): 37-40. [琚金水 , 周文豹 , 2003. 浙江九龙山长翅目 2 新种 [J]. 浙江林学院学报 , 20(1): 37-40.]

KALTENBACH A, 1978. Mecoptera (Schnabelhafte, Schnabefliegen). Handbuch der Zoologie [M], New York:Walter de Gruyter Berlin, 1-111.

KOU L X, HUA B Z, 2016. Comparative embryogenesis of Mecoptera and Lepidoptera with special reference to the abdominal prolegs [J]. Journal of Morphology, 277(5): 585-593.

KRISTENSEN N P, 1981. Phylogeny of insect orders [J]. Annual Review of Entomology, 26(1): 135-157.

KRISTENSEN N P, 1999. Phylogeny of endopterygote insects, the most successful lineage of living organisms [J]. European Journal of Entomology, 96: 237-254.

LI X, HUA B Z, CAI L J, et al., 2007. Two new species of the genus *Panorpa* (Mecoptera: Panorpidae) from Shaanxi, China with notes on their biology [J]. Zootaxa, 1542: 59-67.

LI Y L, REN D, 2009. History and development of researches on Bittacidae (Insecta: Mecoptera) [J]. Acta Geoscientica Sinica, 30(4): 554-560. [李艳丽 , 任东 , 2009. 蚊蝎蛉科（昆虫纲：长翅目）的研究历史及发展 . 地球学报 , 30(4): 554-560.]

LIEFTINK M A, 1933. De in Ned. Indie voorkomende Mecoptera (*Panorpa*) [J]. Verslagen von de vergaderingen der Afdeling nederlandsch Ost-Indie van de Nederlandsche Entomologische Vereening, 4(1): 117-120.

LIEFTINK M A, 1936. Studies in Oriental Mecoptera. The genus *Leptopanorpa* in Malaysia [J]. Treubia, Buitenzorg, 15: 271-320.

LINDGREN N K, SISSON M S, ARCHAMBEAULT A D, et al., 2015. Four forensic entomology case studies: records and behavioral observations on seldom reported cadaver fauna with notes on relevant previous occurrences and ecology [J]. Journal of Medical Entomology, 52(2): 143-150.

LINNAEUS C, 1758. Systema Naturae [M]. 10th ed. Stockholm: Laurentius Salvius, 824.

MA N, HU G L, ZHANG J X, et al., 2014. Morphological variation of the scorpionfly *Panorpa obtusa* Cheng (Mecoptera: Panorpidae) with a new synonym [J]. PLoS ONE, 9(9): e108545.

MA N, ZHONG W, GAO Q H, et al., 2012. Female genital plate diversity and phylogenetic analyses of East Asian Panorpidae

(Mecoptera) [J]. Systematics and Biodiversity, 10(2): 159-178.

MA N, HUA B Z, 2011. *Furcatopanorpa*, a new genus of Panorpidae (Mecoptera) from China [J]. Journal of Natural History, 45(35/36): 2251-2261.

MACLACHLAN R, 1875. A sketch of our present knowledge of the neuropterous fauna of Japan (excluding Odonata and Trichoptera) [J]. Transactions of the Entomological Society of London, 23(2):167-190.

MACLACHLAN R, 1887. Description de plusieurs nouvelles espèces de Panorpides provenant du Japon et de la Sibérie orientale [J]. Mitteilugen der Schweizerischen Entomologischen Gesellschaft, 7: 400-406.

MACLACHLAN R, 1894. On two small collections of Neuroptera from Tachien-lu, in the province of Szechuen, western China, on the frontier of Thibet [J]. Annals and Megazine of Natural History, 6(13): 421-436.

MIAO Y, MA N, HUA B Z, 2017. Cytotaxonomy and molecular phylogeny of the genus *Cerapanorpa* Gao, Ma & Hua, 2016 (Mecoptera: Panorpidae) [J]. Scientific Reports, 7: 4493.

MICKOLEIT G, 1976. The genital and postgenital segments of the Mecoptera females (Insecta, Holometabola) [J]. Zoomorphologie, 85(2): 133-156.

MISOF B, LIU S, MEUSEMANN K, et al., 2014. Phylogenomics resolves the timing and pattern of insect evolution[J]. Science, 346(6210): 763-767.

MIYAKÉ T, 1908, A list of Panorpidae of Japan, with descriptions of ten new species. Bulletin of the College of Agriculture [J], Imperial University of Tokyo, 8: 1-12.

MIYAKÉ T, 1913, Studies on the Mecoptera of Japan [J]. Journal of the College of Agriculture of the Imperial University of Tokyo, 4: 265-400.

MIYAKÉ T, 1912, The life-history of *Panorpa klugi* M'Lachlan [J]. Journal of the College of Agriculture, Imperial University of Tokyo, 4(2): 118-138.

NAVÁS L, 1908, Neurópteros nuevos [J]. Memorias de la real Academia de Ciencias y Artes de Barcelona, 6(25): 401-423.

NAVÁS L, 1911. Névroptères nouveaux de l'extreme Orient [J]. Russkoe Entomologicheskoe Obozreni, 11: 111-117.

NAVÁS L, 1913. Névroptères du Japon recueillis par M. Edmc Gallois [J]. Bulletin du Museue National d'Histoire Naturelle, Paris, 19: 441-451.

NAVÁS L, 1927. Commiones entomológicas 10. Mis excursiones cientificas en 1927 [J]. Revista de la Academia de Ciencias Exactas, Físicas Químicas y Naturales de Zaragoza, 11: 79-137.

NAVÁS L, 1930a. Insecta orientalia. Ⅷ series [J]. Memotie dell'Accademia pontifica dei Nuovi Lincei, Rome, 14: 409-418.

NAVÁS L, 1930b. Névroptères et insectes voisins. Chine et pays environnants [Ⅰ] serie [J]. Notes d'Entomologie Chinoise, 4: 1-12.

NAVÁS L, 1931. Decadas de insectos nuevos [J]. Revista de la Real Academia de Ciencias Exactas, Fisicasy Naturales de Madrid, 26: 60-86.

NAVÁS L, 1935. Névroptères et insectes voisins. Chine et pays environnants. Huitieme [Ⅷ] serie [J]. Notes d'Entomologie Chinoise, 2: 85-103.

NAVÁS L, 1936. Névroptères et insectes voisins. Chine et pays environnants. Neuvieme [Ⅸ] serie [J]. Notes d'Entomologie Chinoise, 3: 37-62.

PALMER C M, 2010. Diversity of feeding strategies in adult Mecoptera [J]. Terrestrial Arthropod Reviews, 3(2): 111-128.

PENNY N D, 1975. Evolution of the extant Mecoptera [J]. Journal of the Kansas Entomological Society, 48(3): 331-350.

PENNY N D, BYERS G W, 1979. A check-list of the Mecoptera of the world [J]. Acta Amazonica, 9(2): 365-388.

PLUTENKO A V, 1985. New and little-known species of Mecoptera from the Soviet Far East [J]. Entomologicheskoye Obozrenie, 64(1): 171-176.

QIAN Z X, ZHOU W B, 2001. Three new species of Mecoptera from Yunnan [J]. Journal of Zhejiang Forestry College, 18(3):

297-300.

REN D, 1993. First discovery of fossil Bittacids from China [J]. Acta Geologica Sinica, 67(4): 376. [任东 , 1993. 蚊蝎蛉科化石在我国的首次发现 [J]. 地质学报 , 67(4): 376.]

REN D, LABANDEIRA C C, SANTIAGO-BLAY J A, et al., 2009. A probable pollination mode before angiosperms: Eurasian, long-proboscid scorpionflies [J]. Science, 326(5954): 840-847.

RIEK E F, 1954. The Australian Mecoptera or scorpion-flies [J]. Australian Journal of Zoology, 2: 143-168.

RIEK E F, 1973. A revision of Australian scorpion flies of the family Choristidae (Mecoptera) [J]. Journal of the Australian Entomological Society, 12: 103-112.

RUST M K, BYERS G W, 1976. The Mecoptera of India and adjacent regions [J]. The University of Kansas Science Bulletin, 51(2): 19-90.

TAN J L, HUA B Z, 2006. Behaviors of the hangingfly *Bittacus planus* [J]. Chinese Bulletin of Entomology, 43(3): 348-351. [谭江丽 , 花保祯 , 2006. 扁蚊蝎蛉的行为学观察 [J]. 昆虫知识 , 43(3): 348-351.]

TAN J L, HUA B Z, 2008a. A new species of the genus *Bittacus* (Mecoptera, Bittacidae) from Zhejiang, China [J]. Acta Zootaxonomica Sinica, 33(3): 487-490. [谭江丽 , 花保祯 , 2008. 浙江蚊蝎蛉属一新种 [J]. 动物分类学报 , 33(3): 487-490.]

TAN J L, HUA B Z, 2008b. First discovery of Bittacidae (Mecoptera) in Hainan Island, China, with description of a new species [J]. Entomological News, 119(5): 497-500.

TAN J L, HUA B Z, 2008c. The second species of the Chinese Panorpodidae (Mecoptera), *Panorpodes brachypodus* sp. nov [J]. Zootaxa, 1751: 59-64.

TAN J L, HUA B Z, 2009a. *Bicaubittacus*, a new genus of the Oriental Bittacidae (Mecoptera) with descriptions of two new species [J]. Zootaxa, 2221: 27-40.

TAN J L, HUA B Z, 2009b. *Terrobittacus*, a new genus of the Chinese Bittacidae (Mecoptera) with descriptions of two new species [J]. Journal of Natural History, 43(47/48): 2937-2954.

THORNHILL R, 1975. Scorpionflies as kleptoparasites of web-building spiders [J]. Nature, 258(5537): 709-711.

THORNHILL R, 1976. Sexual selection and nuptial feeding behavior in *Bittacus apicalis* (Insecta: Mecoptera) [J]. The American Naturalist, 110(974): 529-548.

TILLYARD R J, 1917. Mesozoic insects of Queensland. No. 1. Planipennia, Trichoptera, and the new order Protomecoptera [J]. Proceedings of the Linnaean Society of New South Wales, 42: 175-200.

TJEDER B, 1936. Schwedish-chinensische wissenschaftliche Expedition nach den nordwestlichen Provinzen Chinas, under Leitung von Dr. Sven Hedin und Prof. Su Ping-chang. Insekten gesammelt vom schwedischen Arzt der Expedition Dr. David Hummel 1927 — 1930 [J]. Arkiv for Zoologi, 27A(33): 1-14.

TJEDER B, 1950. Mecopterens aus Fukien [J]. Bonner zoologische Beiträge, 1(2/4): 286-290.

TJEDER B, 1956. Zwei neue ost-asiatische Bittacus-Arten [J]. Beiträge zur Entomologie, 6: 45-53.

TJEDER B, 1973. Bittacidae from Burma, collected by R. Malaise (Mecoptera) [J]. Entomologisk Tidskrift, 94(3/4): 188-193.

TONG X, JIANG L, HUA B Z, 2017. A unique mating pattern of *Panorpodes kuandianensis* (Mecoptera: Panorpodidae) [J]. Contributions to Zoology, 86(3): 229-237.

VAN DER WEELE H W, 1909. Mecoptera and Planipennia of Insulinde [J]. Notes from the Leyden Museum, 31(1): 1-100.

WALKER F, 1853. Genus 5. Bittacus [M]//GRAY J E. (Ed.), Catalogue of the Specimens of Neuropterous Insects in the Collection of the British Museum, Part Ⅱ. (Sialides-Nemopterides). London: British Museum [Natural History], 465-469.

WANG J S, HUA B Z, 2016. Two new species of the genus *Panorpa* Linnaeus (Mecoptera, Panorpidae) from Yunnan, China [J]. ZooKeys, 587: 151-162.

WANG J S, HUA B Z, 2017. An annotated checklist of the Chinese Mecoptera with description of male *Panorpa guttata* Navás, 1908 [J]. Entomotaxonomia, 39(1): 24-42. [王吉申 , 花保祯 , 2017. 中国长翅目名录附滴蝎蛉 *Panorpa guttata* Navás, 1908

雄性描记 [J]. 昆虫分类学报 , 39(1): 24-42.]

WANG M, HUA B Z, 2017. Discovery of *Neopanorpa chillcotti* Byers (Mecoptera: Panorpidae) from Tibet, China, with discussion of its generic status [J]. Zootaxa, 4232(2): 241-250.

WANG M, HUA B Z, 2018. A new species of *Neopanorpa* with an extremely long notal organ from Sichuan, China (Mecoptera, Panorpidae). Zookeys,750:131-140.

WANG M, HUA B Z, 2018. High species diversity of the genus *Neopanorpa* (Mecoptera: Panorpidae) in Yunnan Province, China. Zootaxa, 4483(1): 36-66.

WANG Y, LABANDEIRA C C, Shih C, et al., 2012. Jurassic mimicry between a hangingfly and a ginkgo from China [J]. Proceedings of the National Academy of Sciences, 109(50): 20514-20519.

WHITING M F, 2002. Mecoptera is paraphyletic: multiple genes and phylogeny of Mecoptera and Siphonaptera [J]. Zoologica Scripta, 31(1): 93-104.

WHITING M F, CARPENTER J C, WHEELER Q D, et al., 1997. The Strepsiptera problem: phylogeny of the holometabolous insect orders inferred from 18S and 28S ribosomal DNA sequences and morphology [J]. Systematic Biology, 46(1): 1-68.

WILLMANN R, 1987. The phylogenetic system of the Mecoptera [J]. Systematic Entomology, 12: 519-524.

WILLMANN R, 1989. Evolution und phylogenetisches System der Mecoptera (Insecta: Holometabola) [J]. Abhandlungen der Senckenbergischen Naturforschenden Gesellschaft, 544: 1-153.

YIE S T, 1951. The biology of Formosan Panorpidae and morphology of eleven species of their immature stages [J]. Memoirs of the College of Agriculture, National Taiwan University, 2(4): 1-111.

YUE C, HUA B Z, 2010. Are abdominal prolegs serially homologous with the thoracic legs in Panorpidae (Insecta: Mecoptera): Embryological evidence [J]. Journal of Morphology, 271: 1366-1373.

ZHANG J X, HUA B Z, 2012. A new species of the genus *Panorpa* L. (Mecoptera: Panorpidae) from the Daba Mountains of central China [J]. Entomotaxonomia, 34(3): 541-547. [张俊霞 , 花保祯 , 2012. 中国大巴山蝎蛉属一新种 [J]. 昆虫分类学报 , 34(3): 541-547.]

ZHONG W, DING G, HUA B Z, 2015a. The role of male's anal horns in copulation of a scorpionfly [J]. Journal of Zoology, 295(3), 170-177.

ZHONG W, HUA B Z, 2013a. *Dicerapanorpa*, a new genus of East Asian Panorpidae (Insecta: Mecoptera: Panorpidae) with descriptions of two new species [J]. Journal of Natural History, 47(13/14): 1019-1046.

ZHONG W, HUA B Z, 2013b. Mating behaviour and copulatory mechanism in the scorpionfly *Neopanorpa longiprocessa* (Mecoptera: Panorpidae) [J]. PLoS ONE, 8(9): e74781.

ZHONG W, QI Z Y, HUA B Z, 2015b. Atypical mating in a scorpionfly without a notal organ [J]. Contributions to Zoology, 84(4): 305-315.

ZHONG W, ZHANG J X, HUA B Z, 2011. *Panorpodes kuandianensis*, a new species of short-faced scorpionflies (Mecoptera, Panorpodidae) from Liaoning, China [J]. Zootaxa, 2921: 47-55.

ZHOU W B, 2000. Six new species of Mecoptera from China [J]. Journal of Zhejiang Forestry College, 17(3): 248-254. [周文豹 , 2000. 中国长翅目 6 新种 [J]. 浙江林学院学报 , 17(3): 248-254.]

ZHOU W B, 2001. Mecoptera. In: WU H, PAN C W (Eds.). Insects of Tianmushan National Nature Reserve. Science Press, Beijing, pp. 393-396. [周文豹 , 2001. 长翅目 [M]// 吴鸿 , 潘承文 . 天目山昆虫 [M]. 北京 : 科学出版社 , 393-396.]

ZHOU W B, 2003. Four new species of Mecoptera from China [J]. Wuyi Science Journal, 19: 88-94. [周文豹 , 2003. 中国长翅目四新种记述 [J]. 武夷科学 , 19: 88-94.]

ZHOU W B, 2005. Mecoptera [M]//JIN D C, LI Z Z (Eds.). Insects from Xishui Landscape. Guiyang: Guizhou Science and Technology Publishing House,383-387. [周文豹 , 2005. 长翅目 [M]// 金道超 , 李子忠 . 习水景观昆虫 . 贵阳 : 贵州科技出版社 ,383-387.]

ZHOU W B, 2006. Mecoptera [M]//JIN D C, LI Z Z (Eds.). Insects from Chishui Suoluo Landscape[M]. Guiyang: Guizhou Science and Technology Publishing House, 73-375. [周文豹, 2006. 长翅目 [M]// 金道超, 李子忠. 赤水桫椤景观昆虫. 贵阳: 贵州科技出版社，273-375.]

ZHOU W B, BAO K H, 2002. Mecoptera [M]//LI Z Z, JIN D C (Eds.). Insects from Maolan Landscape. Guiyang: Guizhou Science and Technology Publishing House, 409-411. [周文豹, 包开洪, 2002. 长翅目 [M]// 李子忠, 金道超. 茂兰景观昆虫. 贵阳 : 贵州科技出版社, 409-411.]

ZHOU W B, FAN Z Y, 1998. Mecoptera [M]//WU H (Ed.). Insects of Longwangshan Nature Reserve. Beijing: China Forestry Publishing House, 145-147. [周文豹, 范忠勇, 1998. 长翅目 [M]// 吴鸿. 龙王山昆虫. 北京 : 中国林业出版社, 145-147.]

ZHOU W B, HU Y X, WU X P, et al., 1993a. Six new species and two new records of Mecoptera from China [J]. Journal of Zhejiang Forestry College, 10(2): 189-196. [周文豹, 胡永旭, 吴小平, 等, 1993. 中国长翅目新种新纪录种 [J]. 浙江林学院学报, 10(2): 189-196.]

ZHOU W B, HU Y X, WU X P, et al., 1993b. Two new species of *Neopanorpa* from Zhejiang (Mecoptera: Panorpidae) [J]. Journal of Zhejiang Forestry College, 10(3): 314-317. [周文豹, 胡永旭, 吴小平, 等, 1993. 浙江新蝎蛉属二新种 [J]. 浙江林学院学报, 10(3): 314-317.]

ZHOU W B, ZHOU X, 2005. Mecoptera [M]//YANG M F, JIN D C (Eds.). Insects from Dashahe Nature Reserve of Guizhou. Guiyang: Guizhou People's Publishing House, 374-379. [周文豹, 周昕, 2005. 长翅目 [M]// 杨茂发, 金道超. 大沙河昆虫. 贵阳 : 贵州科技出版社, 374-379.]

ZHOU W B, ZHOU X, 2006. Mecoptera [M]//LI Z Z, JIN D C (Eds.). Insects from Fanjingshan Landscape. Guiyang[M]: Guizhou Science and Technology Publishing House, 432-436. [周文豹, 周昕, 2006. 长翅目 [M]// 李子忠, 金道超. 梵净山景观昆虫. 贵阳 : 贵州科技出版社, 432-436.]

ZHOU X, ZHOU W B, 2007. Mecoptera[M]//LI Z Z, JIN D C (Eds.). Insects from Leigongshan Landscape. Guiyang: Guizhou Science and Technology Publishing House, 450-456. [周昕, 周文豹, 2007. 长翅目 [M]// 李子忠, 金道超. 雷公山景观昆虫. 贵阳 : 贵州科技出版社, 450-456.]

ZHOU X, ZHOU W B, 2010. Mecoptera[M]//CHEN X S, LI Z Z, JIN D C (Eds.). Insects from Mayanghe Landscape[M]. Guiyang: Guizhou Science and Technology Publishing House, 56-58. [周昕, 周文豹, 2010. 长翅目 [M]// 陈祥盛, 李子忠, 金道超. 麻阳河景观昆虫. 贵阳 : 贵州科技出版社, 56-58.]

ZHOU X, ZHOU W B, 2012. Mecoptera [M]//DAI R H, LI Z Z, JIN D C (Eds.). Insects from Kuankuoshui Landscape[M]. Guiyang:Guizhou Science and Technology Publishing House, 508-561.[周昕, 周文豹, 2012. 长翅目 [M]// 戴仁怀, 李子忠, 金道超. 宽阔水景观昆虫. 贵阳 : 贵州科技出版社, 508-561.]

ZHOU X, ZHOU W B, 2014. Mecoptera [M]//Wang Y P, TONG C L (Eds.). Insects in Qingliangfeng, Zhejiang Province.Beijing: China Forestry Publishing House, 197-199. [周昕, 周文豹, 2014. 长翅目 [M]// 王义平, 童彩亮. 浙江清凉峰昆虫. 北京 : 中国林业出版社, 197-199.]

拉丁学名索引
Latin Name Index

A

B

C

D

E

F

H

I

L

M

N

O

P

R

S

T

Z

中文名称索引
Chinese Common Name Index

N

P

Q

R

S

T

W

X

Y

Z